Taxonomy and Biogeography of Macquarie Island Seaweeds

Frontispiece: Divers packing gear to Aerial Cove. Rockhopper penguins *(Eudyptes chrysocome)* in supratidal zone and Anchor Rock on the horizon.

Taxonomy and Biogeography of Macquarie Island Seaweeds

Robert Wallace Ricker

British Museum (Natural History)

Dedicated in admiration and warmth to

Professor George F. Papenfuss,
(1903–1981)

who maintained a strong interest in the Antarctic Region and provided a starting point for modern taxonomic studies of benthic marine algae from high-latitude southern waters.

ISBN 0-565-00998-2

Typeset by Cambridge Photosetting Services
Printed by Henry Ling Ltd, The Dorset Press, Dorchester

First Published 1987
British Museum (Natural History)
Cromwell Road, London SW7 5BD

Dr R. W. Ricker
University of California
Davis, California 95616

British Library Cataloguing in Publication Data

Ricker, Robert Wallace
Taxonomy and biogeography of Macquarie Island seaweeds
1. Marine algae—Australia—Macquarie Island
I. Title II. British Museum (Natural History)
589.3994′8 QK577

ISBN 0-565-00998-2

Contents

Acknowledgements

I am deeply indebted to Dr Gerald T. Kraft for guidance and constructive criticisms that have been essential to the completion of this floristic survey. Both he and Dr Richard Wetherbee suggested the topic of study, for which I am grateful.

Description of Macquarie's benthic algal flora has required the cooperation of various specialists. Space does not permit adequate acknowledgement for all assistance received, but the following persons were especially helpful in discussions concerning specific algal groups:

Denise Bonin: Bonnemaisoniaceae
Dr Paul Gabrielson: Gigartinales
Dr Michael W. Hawkes: Bangiaceae
Dr Max H. Hommersand: Ceramiaceae, Rhodomelaceae
Dr Richard L. Moe: Desmarestiales, Corallinaceae
Dr Charles J. O'Kelly: Chaetophorales
Dr Murray J. Parsons: Dasyaceae
Dr Willem Prud'homme van Reine: Sphacelariales
Dr Paul C. Silva: Codiaceae
Dr Stephen Skinner: Elachistaceae
Dr William J. Woelkerling: Acrochaetiales, Corallinaceae
Dr Michael J. Wynne: Scytosiphonales, Delesseriaceae

I alone, however, accept full responsibility for statements made in the present account.

Certain other individuals have supplied information or material or both crucial to broader aspects of this project. I thank Mrs Iona Christianson for her generous loan of Macquarie Island algae (collected by Drs Ron Kenny and Noel Haysom in 1948–50), Dr Elise Wollaston for the use of additional Macquarie seaweeds collected during the late 1950s and early 1960s, Mrs Anna-Elise Torkelsen (O) for the use of Dr Baardseth's algal collections from Tristan da Cunha, and Drs Francoise Ardré (PC), Hannes Hertel (M) and Roland Moberg (UPS) for the loan of herbarium specimens. Discussions with Professor H. B. S. Womersley have been extremely helpful and informative in numerous taxonomic and biogeographic interpretations. Mr Peter Robins kindly supplied latin translations, and Mr Chris O'Brien assisted with some of the photography. Drs Richard L. Moe, Paul C. Silva, and Michael J. Wynne, and Messrs René Delépine and James Price supplied valuable editorial assistance as well as stylistic criticism. Mr Price additionally assisted with the location of some type specimens in the British Museum (Natural History). Dr Moe generously opened many of his resources to me—his experiences with Antarctic seaweeds and literature, and a general cleverness concerning a variety of microtechniques, were invaluable.

I thank the Antarctic Division (Australia) and the Australian Museum for supporting my field studies. For logistic support while collecting at Macquarie I., I am especially indebted to Drs James K. Lowry (Australian Museum), G. C. B. Poore (Victorian Institute of Marine Studies) and D. H. Horning; at the Vestfold Hills, to Messrs Konrad Beinssen, Paul Butler, Mark Tucker and Dr Tony Dick. I am also grateful to Dr Paul C. Silva and Professor L. Constance of the University of California Herbarium in Berkeley for gathering funds through the herbarium to offset my initial expenses to Australia.

Expenses incurred throughout this project were met largely by a grant from the University of Melbourne during the years 1979–83. For this financial support, as well as assistance with publication expenses, I thank the University. I am also grateful to the Australian Antarctic Division for additional monies used to offset publication costs.

Last in order but not in magnitude, I thank my parents and friends for their encouragements, which gave me the energy to continue working after numerous setbacks, and my wife Karin for careful and repeated readings of the manuscript.

Summary

Macquarie Island is remotely located in the Southern Ocean some 1500 km south-southeast of Tasmania. In spite of the island's continuous isolation since its volcanic emergence above sea-level less than one million years ago, it has been colonized by numerous land and sea plants. As a result of the present survey, 103 species representing 81 genera and 43 families of benthic marine algae are documented from this vicinity. Of these species, 15 belong to the Chlorophyta, 28 to the Phaeophyta and 60 to the Rhodophyta. Approximately 70 are reported from Macquarie I. for the first time, with 11 of these newly described. The new taxa include one species in the Chlorophyta *(Derbesia furcata)*, four in the Phaeophyta *(Dilophus decumbens, Haplogloia moniliformis, Lithoderma piliferum* var. *debile* and *Streblonema codiatile)* and four in the Rhodophyta *(Porphyra plocamiestris, P. ionae, Iridaea remuliformis* and *Antithamnionella alternans)*. Two additional unpublished taxa in the Chlorophyta *(Codium subantarcticum* and *Endophyton atroviridis)* are also included but not formally described. *Gononema pectinatum* and *G. ramosum* are transferred from the Ectocarpaceae to the Elachistaceae, and *Desmarestia rossii* is synonymized with *D. chordalis*. The new name *Rhodymenia subantarctica* is proposed for *R. cuneifolia* (Hooker f. & Harvey) Taylor, which is a later homonym of *R. cuneifolia* Okamura. *Delesseria lancifolia* is emended to include branched and unbranched thalli, with *D. belayevii* and *D. uschakovii* considered to be later synonyms. *Pseudophycodrys* is conceived as monotypic and restricted to Subantarctic waters. The new combination *Schizoseris condensata* is based on the synonym *Delesseria condensata* Reinsch, which predates the existing name *Schizoseris laciniata*. The concept of *Schizoseris dichotoma* is expanded to include forms previously known as *S. davisii, S. kerguelensis, S. papenfussii, S. polydactyla, S. pseudodichotoma* and *Myriogramme multinervis*. *Palmaria* (the species of which were previously placed into *Gracilaria, Leptosarca, Leptosomia,* and *Rhodymenia*) is reported for the first time from the Southern Hemisphere. Distributions and characteristic associations of seaweeds around Macquarie I., throughout the Southern Ocean and in other world regions are discussed. Algae occurring in both the Northern and Southern Hemispheres total 17% of Macquarie's flora. Roughly 4% of the species range throughout waters of the Southern Cold Temperate, Subantarctic, and Antarctic Zones; 9% occur in only the Southern Cold Temperate and Subantarctic Zones; and 15% are common to the Subantarctic and Antarctic Zones. The largest number of species (41% of the flora) are restricted to the Subantarctic but appear to be distributed throughout this circumpolar belt. Approximately 11% of the marine flora is made up of endemic taxa and <4% is too poorly understood to comment on geographic ranges.

Abbreviations

The following abbreviations are used throughout the text to indicate the location of specimens:

ADU	Adelaide, Botany, Department, University of Adelaide, South Australia
AKU	Auckland, University of Auckland
BM	British Museum (Natural History), London
C	Botanical Museum and Herbarium, Copenhagen
CHR	Christchurch, Botany Division, DSIR
E	Royal Botanic Gardens, Edinburgh
ER	Erlangen, Botanisches Institut der Universität Erlangen
FH	Farlow Herbarium of Cryptogamic Botany, Harvard University, Cambridge, Mass.
HBG	Hamburg, Institut für Allegemeine Botanik
K *in* BM	Specimens previously from Royal Botanic Gardens, Kew now on permanent loan to BM
L	Rijksherbarium, Leiden
LD	Botanical Museum, Lund
LINN	Linnean Society, London
M	München, Botanische Staatssammlung
MEL	National Herbarium of Victoria
MELU	Melbourne, University of Melbourne, Victoria
NMW	Department of Botany, National Museum of Wales, Cardiff
O	Botanisk Museum, Oslo
PC	Laboratoire de Cryptogamie, Muséum National d'Histoire Naturelle, Paris
S	Naturhistoriska Riksmuseum, Stockholm
TCD	Department of Botany, Trinity College, Dublin
TRH	Museum of the Royal Norwegian Society for Science and Letters, Trondheim
UC	Herbarium of the University of California, Berkeley, California
UPS	Institute of Systematic Botany, University of Uppsala, Uppsala
WELT	Wellington, National Museum

1: Macquarie Island

> The further an island is from a continent the smaller is its flora numerically, the more peculiar is its vegetation . . . In the case of very isolated islands moreover, the generic types are often those of very distant countries, and not of the nearest land.
>
> J. D. Hooker *Flora Antarctica*, 1845

1.1 Introduction

Unlike the algae of some regions where extensive marine floras have been published and persons generally interpret and name taxa similarly (e.g., California—Abbott & Hollenberg, 1976; Britain— Parke & Dixon, 1964, 1968, 1976), benthic marine algae from the Southern Ocean are often besieged by poor and confusing taxonomic concepts and a multiplicity of current but synonymous names. A combination of once-collected genera and species, small and isolated collections, poor anatomical documentation and a premature willingness to make sweeping biogeographic statements has led to contrasting notions of biological affinities between widely separated localities in the Southern Hemisphere. Further confusion has been fueled by a considerable volume of literature consisting largely of obscure publications, independent and unknowingly repetitious studies of identical floras, and the assumption that biogeographic relationships of one type of organism will be the same as those of others.

Most biogeographic knowledge of the Southern Ocean comes from zoological and oceanographical studies, with inputs from marine phytogeography lagging far behind. This is mainly because the correct identification of many benthic marine algae collected from this region has never been made. The southern high latitudes have been classified and fragmented into numerous small regions or provinces based primarily on animal assemblages (Powell, 1955; Knox 1960), but there is no clear agreement on the borders of these many subdivisions (Hedgpeth, 1970). Two distinct zones (Antarctic and Subantarctic) separated at the Antarctic Convergence (a circumpolar region where north-flowing cold water meets south-flowing warm water and a 1–2°C change in surface water is noticeable), however, are recognized on the basis of differences in physical and plankton data from locations throughout high-latitude southern waters (Hedgpeth, 1970). Kennett (1978), when separating the Antarctic and Subantarctic Zones, noted a predominance of siliceous planktonic microfossils south of the Antarctic Convergence and calcareous assemblages to the north. Another more northerly border, the Subtropical Convergence (with an abrupt 4–5°C change in surface water temperatures), lies between the Subantarctic and Cold Temperate Zones (Knox, 1960). The Subantarctic thereby forms a continuous circumpolar belt, with mostly uniform water temperatures, between the Antarctic and Cold Temperate Zones.

Papenfuss (1964*b*) discussed algal species that occur in both the Antarctic and Subantarctic and, because of geographical overlap of seaweed distributions, preferred to recognize the Antarctic and Subantarctic as subunits of a single biogeographic province. Skottsberg (1960), however, believed that the Subantartic is a geographical region physically and biologically distinct from the Antarctic. More recent studies by Delépine (1966) and Neushul (1968), and reviews by Knox (1979) and South (1979*a*), support Skottsberg's views. South, however, agrees with Papenfuss' (1964*b*) warning about broad biogeographic statements and points to the uselessness of phytogeographic speculations until species identities and southern floras, especially from isolated and poorly surveyed lands, are better understood.

Studies of southern seaweeds are hindered by largely unknown plant distributions and questionable algal systematics. Two of the most recent algal biogeographic studies of the Subantarctic have tried to avoid these problems either by discussing distributions of well-known species (Delépine, 1966), or through thoroughly revising the taxonomy and distributions of all species in a single genus (e.g., *Durvillaea*—Hay, 1979*b*). However, these studies combined, represent less than 2% of the taxa listed in Papenfuss' (1964*a*) catalogue.

Macquarie Island was chosen for study because: it is currently regarded as a typical Subantarctic island located close to the border between the Antarctic and Subantarctic Zones (the Antarctic Convergence); some preliminary work on the seaweeds in this vicinity has already been completed (Zinova, 1958, 1963; Kenny & Haysom, 1962); and the island is readily accessible through the annual voyages of the *Nella Dan* and *Thala Dan* conducted by Australia's Antarctic Division, Department of Science. The present account has three principal aims: (1) to produce a taxonomic and floristic report on the marine algae at Macquarie I.; (2) to elaborate on particular elements of the flora that have verified distributions elsewhere within and/or outside the Subantarctic Zone; and (3) to consider the origins of Macquarie's marine flora. The final aim is the most speculative because of its reliance at times upon unverified distribution data and questionable identifications of species by early phycologists.

1.2 Review of Subantarctic and Antarctic research

Information concerning the weather-beaten lands and surrounding tumultuous seas of the southern high latitudes has gradually accumulated from the earliest expeditions, mounted primarily to find new sources of wealth, to later voyages built around scientific or territorial/ expansionist motives (Cameron, 1974). Papenfuss (1964*b*) described the Antarctic as 'perhaps the most remote and inaccessible part of the world' and expressed surprise 'that such a large body of information about the marine algae of the region should have accumulated during the past 135 years'. This surprise would be well warranted had science been completely divorced from early explorations of Antarctic waters, but such was not the case. Even the initial voyages seeking new sources of wealth usually had naturalists on board who attended to both animals and plants (Godley, 1965). Some of the first biological observations were made on the vast numbers of penguins and seals encountered during the expeditions of Bouvet (1738–39). The giant kelp *Macrocystis pyrifera* featured in narratives of early explorers and sealers, but Gaudichaud's (1825, 1826) descriptions of seaweeds collected during the 1817–20 expedition of the French corvettes *Uranie* and *Physicienne* are the first indications of the marine plant diversity in the Antarctic and Subantarctic Zones. Phycological knowledge of these regions grew slowly until Joseph Dalton Hooker undertook the first extensive sampling and study of benthic marine algae throughout the Antarctic and Subantarctic during the Ross expeditions in the *Erebus* and *Terror* (1839–43). Hooker gathered seaweeds along the shores of Antarctica as well as from many Subantarctic localities (Fuegia, Falkland Is, Îles Kerguelen, South Shetland Is, South Georgia, and the Auckland and Campbell Is). These collections are the basis of Hooker's *Flora Antarctica* (published during the years 1844–47, with sections concerning seaweeds appearing in 1845 (Harvey & Hooker *in* Hooker) and 1845–47 (Hooker)). Subsequent exploration (e.g., de Gerlache, 1897–99; Nordenskjold, 1901–4; Scott, 1901–4; Charcot, 1908–10; Mawson, 1911–14) increased knowledge of benthic marine algae from the Antarctic and certain Subantarctic islands, the latter frequently used as points of anchorage for ships trying to reach Antarctica through the relentlessly stormy seas.

Most seaweed collections returned to the ships' points of origin and became grist for the phycological mills of the involved nations. This resulted in an impressive array of new algal taxa that appeared in species lists and taxonomic descriptions published in various societal, national, and private publications. Because essentially identical algal taxa were studied by taxonomists who were frequently in poor contact with one another, taxonomic synonyms, both generic and specific, multiplied. The abundance of names for single entities still presents many problems, especially for biogeographers.

Benthic marine algae in the Southern Hemisphere are known chiefly through: (1) floras of particular southern localities (e.g., S. Georgia—Reinsch, 1890; Falkland Is—Cotton, 1915; Patagonia—Skottsberg, 1921, 1923; Tristan da Cunha—Baardseth, 1941; Chile—Levring, 1960; Argentina—Pujals, 1963, 1977; Antarctic Peninsula—Lamb & Zimmermann, 1977); (2) less-comprehensive treatments of broad geographic areas (e.g., Harvey & Hooker, 1845; Hooker, 1846–47; Hooker & Harvey, 1845; Askenasy, 1888; Skottsberg, 1907, 1953; Gain, 1912; Kylin & Skottsberg, 1919); and (3) descriptive ecological accounts (Skottsberg, 1906, 1941*a*, 1941*b*; Zaneveld, 1964). Reports from any of the above three categories have potential for confusion in present-day taxonomic, floristic, and biogeographic analyses. Problems arise where authors have either split species into too many entities, thereby having created nomenclatural redundancy, or have applied more than one name to several different taxa. Few algal taxonomists have returned to original descriptions or adequately examined type specimens of high-latitude southern seaweeds when formulating concepts of species, which has often resulted in faulty taxonomy. Some recent works, in contrast, have lead to unambiguous concepts for particular groups of algae. For example, Moe & Silva (1977*a*, 1981) elegantly demonstrated, through culture and morphological studies of embryonic to adult thalli, that plants previously described under five different specific epithets encompassing three brown algal genera, *Himantothallus, Phaeoglossum*, and *Phyllogigas*, actually belong to one species, *Himantothallus grandifolius*. Moe's & Silva's removal of *Himantothallus* from its equivocal position in the Laminariales, to the Desmarestiales, has revealed that the Antarctic is 'uniquely devoid of kelps'.

Papenfuss (1964*a*) catalogued the voluminous taxonomic, descriptive and ecological literature concerning marine algae from the Antarctic and Subantarctic. Along with an extensive bibliography, he included a list of algae from the region annotated with helpful comments on species distributions, nomenclatural synonymies and, in some cases, queries concerning the validity of species records. This work is a starting point for any who study benthic marine algae in the Southern Ocean. Since 1964, however, many additional important works have dealt with seaweeds in the Antarctic and Subantarctic.

Prescott (1979) listed references, published prior to 1978, for freshwater and marine algae including over 300 seaweed-related articles. He incorporated all references listed by Papenfuss (1964*a*) but omitted two publications pertinent to Macquarie I. seaweeds, (Law & Burstall, 1956; Kenny & Haysom, 1962), and essential Subantarctic algal papers by Mendoza (1969*a*, 1969*b*, 1969*c*, 1970*a*, 1970*b*, 1973, 1974, 1975, 1976*a*, 1976*b*, 1976*c*, 1977). Continued interest in Southern Ocean plants since 1978 has resulted in more taxonomic and nomenclatural revisions (Hay, 1979*b*; Moe & Silva, 1980; Searles, 1980), descriptions of new taxa (Ricker & Kraft, 1979; Moe & Silva, 1979, 1983; Zaneveld & Sanford, 1980) and developmental studies (Delépine & Lambert, 1981; Moe & Henry, 1982).

Biogeographic provinces in the Southern Ocean have been circumscribed largely according to the distributions of various animals (Powell, 1951, 1955, 1960; Knox, 1960) but sometimes also by the distributions of certain plants (Gain, 1912; Skottsberg, 1960). Powell (1955, 1960) divided the Subantarctic into four provinces: the Forsterian, Antipodean, Kerguelenian, and Magellanic (Table 1). Knox (1960) followed Powell, with the notable exception of placing Heard I. (formerly in the Kerguelenian Province) into the Rossian sub-province of the Antarctic. Placement of some Subantarctic islands (e.g., Macquarie and Heard Is) into particular provinces (Kerguelenian and Antarctic respectively) is based on small samples that often completely omit algae from consideration. Discussion of these provinces has been given in several symposia on biogeography (e.g., Knox, 1960; Skottsberg, 1960; Currie, 1968; Holdgate, 1970), with the most recent symposium (Int. Symp. Mar. Biogeogr. & Evol. in the S. Hemis., Auckland, N.Z., 17–20 July 1978) including a summary of algal (South, 1979*a*) and marine animal (Knox, 1979) geography in the Antarctic and Subantarctic. South discussed data, which had been collected by scuba diving, that supported earlier ideas (Hedgpeth, 1969; Dell, 1972) about the Antarctic benthic region extending north of its generally accepted limit (the Antarctic Convergence) to include South Georgia and Bouvet I.

Division of the Antarctic and Subantarctic into provinces presupposes relationships between lands that in some cases are widely separated. Association of a select landmass with a particular biogeographic province may be based upon any one of several criteria which, taken collectively, may seem contradictory or confusing. For example, on the basis of vascular plant distributions, Cheeseman (1919) considered Macquarie I. to be part of the New Zealand Subantarctic (the Antipodean Province). Based on mollusc distributions, however, Powell (1955, 1960) put Macquarie I. in the Kerguelenian Province, thereby forming an alliance between two widely separated islands (Macquarie and Kerguelen) that have relatively few vascular plants in common.

Table 1. Biogeographic provinces. (From Powell 1955, 1960)

FORSTERIAN:	Otago, Stewart I., Snares Is
ANTIPODEAN:	Antipodes, Bounty, Auckland, and Campbell Is
KERGUELENIAN:	Kerguelen, Heard, Marion, Prince Edward, Crozet, and Macquarie Is
MAGELLANIC:	Patagonia from Chiloe I. (west coast) and Cape Blanco (east coast), Tierra del Fuego, East Patagonian continental shelf including Falkland Is. and Burwood Bank

1.3 The Subantarctic and Macquarie Island

South America and Antarctica formed one continuous landmass that blocked water circulation around Antarctica until the early Oligocene (37–35 m.y.). Northward tectonic movements of Australia and South America away from Antarctica during the mid-Oligocene (32 m.y.) opened a high-latitude pathway for circumpolar water circulation (Frakes & Kemp, 1972). Development of a circum-Antarctic current prevented warm (lower-latitude) waters from reaching high latitudes and in part caused temperatures around Antarctica to drop during the Oligocene (Frakes & Kemp, 1972). Circum-Antarctic provinciality (formation of the Subantarctic belt) probably occurred concurrently with the isolation of Antarctica (Kennett, 1978). Polar migration also contributed to the drop in temperatures around Antarctica and is probably partly responsible for the onset of the various ice ages (Black, 1931). Glaciation, which extended to and covered Macquarie I., occurred within comparatively recent times, but the magnitude and time of glaciation remains unclear (Jenkin, 1975). Following recession of the glaciers back to Antarctica, the Subantarctic rose in temperature.

Antarctica is presently encircled by two main currents. Immediately north of the continent is a band of water moving from east to west. North of this is a broader band of eastward-moving water known as the West Wind Drift. The more northerly current dominates water movements in the Subantarctic and receives input from three major oceans—the Atlantic, Pacific, and Indian. It also forms an effective barrier between the cold southern waters and warmer northern waters.

Macquarie I. (54°30′ S, 158°57′ E), approximately 1500 km south-southeast of Tasmania (Fig. 1), is exposed to harsh seas of the Southern Ocean that are driven eastwards by the West Wind Drift. The island, oriented north to south, is 34 km long by 5 km at its greatest width and is dominated by a central island plateau 250–350 m high, with the highest point 434 m above sea-level (Horning, 1979; Lowry, 1979). Macquarie I. represents a major peak along a submarine mountain range known as the Macquarie Rise (Frakes & Kemp, 1972) or Macquarie Ridge (Zaneveld & Sanford, 1980). It is composed of igneous oceanic crust, a reminder of its volcanic origin, of Miocene (26–8 m.y.) (Quilty *et al.*, 1973) or Pliocene (7–2.6 m.y.) (Varne *et al.*, 1969) age, which emerged during the Pleistocene, sometime between 1 million (Colhoun & Goede, 1973) to 100,000 years ago (Quilty, pers. comm.).

Shallow waters and a chain of several islets extend to the north and south of Macquarie I. About 13 km north lie the Judge and Clerk Rocks, and 32 km south, in the same line, are the Bishop and Clerk. Immediately east and west of Macquarie I. ocean depths exceeding 2000 fathoms occur. Soundings taken both north and south of these islets have revealed depths ranging between 2000 and 2500 fathoms (Cheeseman, 1919), thus suggesting a continuous isolation of Macquarie I. from the shallow (400 fathoms) Campbell Plateau, which joins New Zealand with the Auckland and Campbell Is.

Both winds and water currents are seemingly constant forces in the Subantarctic and strongly influence Macquarie's coastal morphology. A broad coastal flat (elevated one to several meters above sea-level; Fig. 6a) and extensive reef platforms (Fig. 6b) along the west coast of the island reflect the rugged conditions on this coastline. The east coast has little, if any, coastal flat intervening between the steep slopes rising to the central island plateau and the fringing shingle beach (Figs 2b, 4a) that is generally exposed to slight winds and small waves. Storms bringing strong winds and high seas from an easterly direction, though, are not uncommon. Winter storms predominate from the southwest and can bring waves that break over 30 m high Anchor Rock (Figs 3a, 5b) and continue on to wash across the isthmus (Fig. 2a). Storms from the east, although less frequent than ones from the west, also can be extremely forceful and cause waves to break over rocks 10–15 m high on the east coast (Figs 2b, 3b).

Cheeseman (1919) remarked on the large impact of the island's physical features, geology, and climate on the terrestrial vegetation—factors that are crucial to the distribution and occurrence of benthic marine algae at Macquarie I. Physical features, including coastal morphology, are

---

Fig. 1 Maps. **a.**Polar projection showing location of Macquarie I. (arrow) in relation to southern continental landmasses and other Subantarctic islands: ---- = convergences; inner line represents Antarctic Convergence, outer line is Subtropical Convergence. **b.** Macquarie I. with names of hut locations. [NB maps drawn by Australian Antarctic Division.]

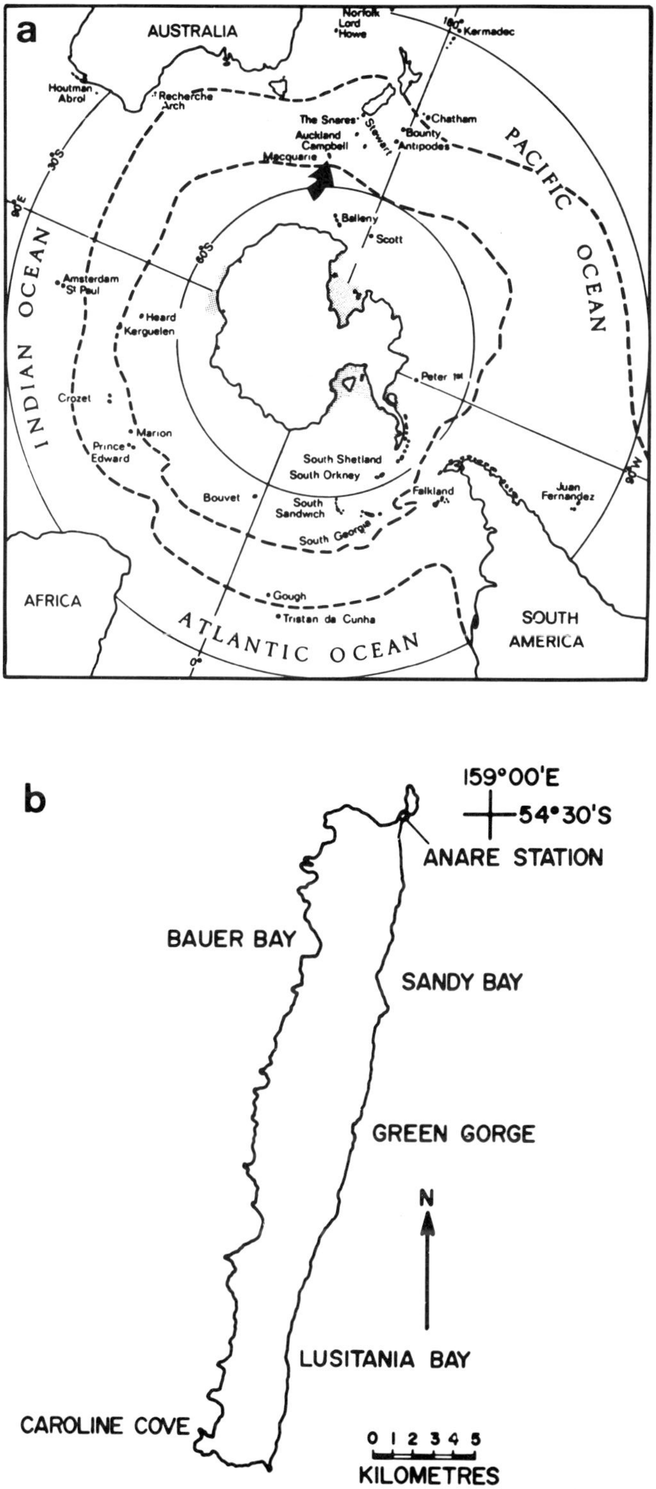
a
AUSTRALIA
Norfolk
Lord Howe
Kermadec
Houtman Abrol
Recherche Arch
The Snares
Stewart
Chatham
Auckland
Campbell
Bounty
Antipodes
Macquarie
PACIFIC OCEAN
Balleny
Scott
INDIAN OCEAN
Amsterdam
St Paul
Heard
Kerguelen
Peter 1st
Crozet
Marion
Prince Edward
South Shetland
South Orkney
Bouvet
South Sandwich
Falkland
Juan Fernandez
South Georgia
AFRICA
Gough
Tristan da Cunha
ATLANTIC OCEAN
SOUTH AMERICA
b
159°00'E
54°30'S
ANARE STATION
BAUER BAY
SANDY BAY
GREEN GORGE
N
LUSITANIA BAY
CAROLINE COVE
0 1 2 3 4 5
KILOMETRES

discussed along with plant associations and community formations in Chapter 3. Jenkin (1975) summarized and briefly discussed means and general climatic trends from a twenty-year span of Commonwealth Bureau of Meteorology records. The climate of Macquarie I. is oceanic and, as is typical of regions just north of the Antarctic Convergence, extremely uniform regarding temperature. Surface water temperatures range from 4–7°C and air temperatures (at 6 m altitude) have a similarly narrow range (3–7.3°C). Relative humidity remains high (*c.* 90%) throughout the year and an orographic cloud layer (low-lying clouds associated with the windward side of abrupt mountainous rises), common over some parts of the west coast, reduces both insolation and desiccation for intertidal plants. Surface light in general is diffuse either because of fog or clouds or both. The mean total cloud cover recorded between 1960 and 1967 was 83%. A yearly mean for sunshine duration between 1948 and 1953, and 1964 and 1969 was 814 hours.

1.4 History of Macquarie Island

Macquarie Island!—a name, for those familiar with the Subantarctic, that recalls a windswept, isolated and bleak refuge for seals, penguins and a handful of men—rather a dismal picture to most. The island, however, is considerably more attractive than this desolate description would indicate, especially to those persons accustomed to the starkness and inhospitable climes of Antarctica. Bellingshausen, during his brief visit to Macquarie I. in November 1820, wrote of his great surprise upon finding the island 'beautifully green, with the exception of its cliffs which were of a dark and dreary colour'. Even at that time, only 10 years after the discovery of the island by Captain Frederick Hasselborough in the brig *Perseverence*, Macquarie showed noticeable signs of man's presence (Debenham, 1945). Sealers hunting fur seals (*Arctocephalus forsteri*) rapidly established quarters near the major seal populations. In 18 months, nearly 180,000 fur seals were killed for their skins. Within two years, *Arctocephalus* no longer existed around Macquarie and sealers shifted their interests to new quarry. The largest seal in the world, the Southern Elephant Seal (*Mirounga leonina*), became a principal target, as did some of the penguins. Both animals died mercilessly—elephant seals for their blubber, and penguins, when seals were scarce, for their scant fat reserves. The small yields of oil from birds and depleted seal populations soon made sealing and oil collection at Macquarie I. a less attractive venture. Still, some sealers persisted with thinning seal herds on the island throughout the late 1800s (Cumpston, 1968).

Voyages to Macquarie I. in the first 100 years after its discovery focused on economic gains and made little effort toward satisfying scientific curiosity. Over the past 80 years more and more scientific expeditions have visited the island and studied many of its biological and physical phenomena. Knowledge of Macquarie's vegetation has expanded slowly through the collection of terrestrial plants. Occasionally, either incidental reports or anecdotes on marine algae would accompany accounts of terrestrial plants, although the main emphasis was directed toward understanding the land flora rather than the marine flora. Interest in marine algae at Macquarie I. has grown over the past 70 years, but the number of publications pertaining to the island's plants still reflects the early bias toward terrestrial studies.

The first recorded botanical collection, a selection of seven species of flowering plants and one fern, was sent in 1824 by C. Fraser, then superintendent of the Sydney Botanic Gardens, to Sir William J. Hooker in Glasgow. It is unknown how Fraser acquired these plants, whether he visited the island or received material from a sealer. These same eight plants were referred to by Hemsley (1884) in his report on the flora of Macquarie I. An earlier publication by Hooker (*Flora Antarctica* (1844, p.1)) alludes to the unknown flora in this locality. Hooker remarked that 'Upon M'Quarrie's Island . . . a very few plants have been collected . . . I am not aware that any account has been published of these islands . . . the botany of which is entirely unknown, but which probably in this meridian constitutes the southern extreme of terrestrial vegetation'. Scott (1883) stressed the apparent close association of Macquarie's terrestrial flora with that of New Zealand, and Kirk (1891) expanded Scott's list of 18 vascular taxa from Macquarie I. to 30. In 1894, Professor A. Hamilton made a short visit to the island to study its natural history. His account of landing on the island (1894) tells of 'passing through great beds of kelp which cover the rocks near the shore'. This most certainly refers to *Durvillaea antarctica*, which is prominent on intertidal boulders and the sides of rock walls. Hamilton's report on the terrestrial flora includes a list of 32 vascular plants.

Fig. 2 Localities around Macquarie Island. **a.** Wireless Hill and the isthmus viewed from the northern edge of the island plateau. ANARE base on the northern half of isthmus. **b.** Rock outcrop (*c.* 15 m tall) at Green Gorge.

Fig. 3 Localities around Macquarie Island. **a.** West coast showing Anchor Rock and North Head. **b.** Gorilla Head Rock (*c.* 25 m tall) on the north-east side of Wireless Hill.

Information on Macquarie's flora first came through cooperation with sealing gangs travelling to the island (e.g., Scott, 1883; Hamilton, 1894). Later reports resulted from specific scientific expeditions. Short visits, such as those by Scott in *Discovery* (November 1901) and Davis (Shackleton Expedition) in *Nimrod* (1909), contributed little, if anything, to the knowledge of plants around the island. In 1911 Douglas Mawson, leader of the Australian Antarctic Expedition (AAE), established a base and stationed a small scientific party on the north end of the island. The party's purpose, aside from maintaining radio linkage between Antarctica and Australia, was to gather data on biological and physical aspects of Macquarie I. On the basis of these biological studies, Mawson later advised the government of Tasmania to declare (in 1933) Macquarie I. a biological sanctuary where sealing would be illegal.

Other expeditions by Russians, Americans, New Zealanders, and Swedes have sampled the biota at Macquarie I., but none have conducted biological programs as extensive as those run during the AAE or the more recent Australian National Antarctic Research Expeditions (ANARE). The latter continue to visit the island each year and serve as the primary transport to and from it. Both the AAE and ANARE represent major research efforts and each has contributed to the knowledge of Macquarie's flora. H. Hamilton, AAE biologist and son of A. Hamilton, collected both terrestrial and marine plants. His vascular plant collections formed the body of Cheeseman's (1919) report on the vascular flora of Macquarie I. ANARE biologist B. W. Taylor (1955) added to Hamilton's (1894) and Cheeseman's (1919) accounts by thoroughly documenting the island's vascular flora, plant associations and communities, and soil types, whereas other ANARE expeditioners have reported on its lichens (Dodge & Rudolph, 1955), fungi (Bunt, 1965), mosses and liverworts (Clifford, 1953; Seppelt, 1977), and the ecology and physiology of various terrestrial plants (Ashton, 1965; Ashton & Gill, 1965; Jenkin, 1972, 1975).

The first list of Macquarie I. seaweeds, based on A. Hamilton's collections, was given by Laing (1896), who reported that most specimens were 'too fragmentary for adequate determination'. Among the algae, however, were six recognizable plants: *Ballia callitricha* (Agardh) Montagne, (?) *Melobesia antarctica* (Hooker f. & Harvey) Harvey, two species of *Plocamium*, *Polysiphonia* sp. and a new variety (var. *minor*) of *Delesseria lancifolia*. A later list (Laing, 1909) contains only one species, *Plocamium hookeri* Harvey, from Macquarie I. Although he discussed distributions of two taxa that he earlier (Laing, 1896) reported from Macquarie I. (*B. callitricha* and *M. antarctica*), in 1909, Laing did not attribute any seaweeds (aside from *P. hookeri*) to the island's flora. Gain (1912), seemingly unaware of Laing's 1896 paper, also recorded only *P. hookeri* from Macquarie I.

Most of H. Hamilton's AAE marine algal collections from Macquarie I. apparently have been lost, although some specimens were examined by Lucus. For example, in an account of the algae from Commonwealth Bay, Antarctica, Lucus (1919) reported *Geminocarpus geminatus* (Hooker f. & Harvey) Skottsberg from Macquarie I. and the Antarctic. The Lucus collection, which is kept at the National Herbarium, Sydney, includes two species from the 1911 Hamilton collection (i.e. *Bostrychia vaga* Hooker f. & Harvey and *Polysiphonia anisogona* Hooker f. & Harvey) that were not documented in early publications. Hamilton (1926) illustrated *Durvillaea* and described it 'writhing in the surge of the ocean swell amongst rocky reefs bordering Hasselborough Bay . . . [where] This giant marine alga is a most conspicuous feature of the . . . coastline. It grows attached to the rocks at one end and floats upwards, yards in length, towards the surface. So thickly and rankly is it matted . . . that the sea-elephants often find difficulty to force a passage through it to the land'.

Levring (1945), on the basis of collections by Dr Sten Vallin (member of the Ross Sea expedition led by C. A. Larsen, 1923–24), reported three species from Macquarie I., of which one (*Corallina officinalis* Linnaeus) has never again been recorded from this locality.

Table 2 lists the algal species previously reported from Macquarie I. Law & Burstall (1956) presented Haysom's (ANARE biologist, 1949–50) ecological studies, in which he characterized intertidal zonation using six seaweeds. Records for four of these taxa, *Phyllophora appendiculata* Skottsberg, *Porphyra columbina* Montagne, *Rhodymenia* cf. *palmatiformis* and *Enteromorpha bulbosa* (Suhr) Kützing, represent distribution extensions. Zinova's (1958, 1963) record of 24 species, collected by a Russian expedition that visited the island in 1956, increased seaweed numbers to 37. Kenny & Haysom (1962) collected numerous algae during an ecological study of Macquarie's littoral zone, of which 32 species were identified by Iona Christianson (Botany Department, Monash University). The remaining algae from Kenny's and Haysom's collections were not

Table 2. Algal records from Macquarie Island.

Laing 1896/1909 (based on Hamilton specimens, 1894)	Law & Burstall 1956 (based on Haysom specimens, 1949)	Zinova 1958/1963 (based on Russian specimens from 1956–57)	Kenny & Haysom 1962 (based on authors' collections, 1948–50)
Ballia callitricha	'lithothamnion'	*Antithamnion ptilota*	*Ballia callitricha*
Delesseria lancifolia var. *minor*	*Phyllophora appendiculata*	*Bostrychia vaga*	*Bostrychia vaga*
Melobesia antarctica	*Porphyra columbina*	*Callithamnion gracile*	*Ceramium rubrum*
Plocamium hookeri	*Rhodymenia* cf. *palmatiformis*	*Chaetangium fastigiatum*	*Chaetangium fastigiatum*
Plocamium sp.	*Durvillaea antarctica*	*Delesseria uschakovii*	*Corallina officinalis*
Polysiphonia sp.	*Enteromorpha bulbosa*	*Iridaea micans*	*Delesseria spp.*
		Jania sp.	*Hildenbrandia sp.*
		Martensia sp.	*Iridaea obovata*
		Myriogramme macquariensis	*Melobesia sp.*
		Plocamium secundatum	*Porphyra columbina*
		Polycoryne radiata	*Porphyra umbilicalis*
		Porphyra columbina	*Plocamium coccineum*
		Rhodoglossum sp.	*Polysiphonia anisogona*
		Rhodymenia palmatiformis	*Rhodymenia spp.*
		Schizoseris pseudodichotoma	*Schizoseris sp.*
		Adenocystis utricularis	*Adenocystis utricularis*
		Chordariopsis capensis	*Chordaria dictyosiphon*
		Desmarestia chordalis	*Desmarestia willii*
		Desmarestia rossii	*Durvillaea antarctica*
		Desmarestia willii	*Ectocarpus confervoides*
		Dictyota dichotoma	*Geminocarpus geminatus*
		Durvillaea caepestipes	*Lessonia variegata*
		Macrocystis pyrifera	*Macrocystis pyrifera*
		Enteromorpha gunniana	*Scytosiphon lomentaria*
		Ulva angusta	*Acrosiphonia pacifica*
		Ulva rigida	*Cladophora spp.*
			Enteromorpha bulbosa
			Enteromorpha intestinalis
			Monostroma sp.
			Prasiola sp.
			Rhizoclonium sp.
			Ulva lactuca

identified or mentioned in their 1962 publication, although these plants and vouchers of the reported seaweeds are held by Mrs Christianson and have been studied in the course of the present survey. Simpson (1976) also conducted an ecological study of the shore environment and mentioned 10 algal taxa, all of which had been previously recorded from Macquarie I. The most recent algal publications (Ricker & Kraft, 1979; Zaneveld & Sanford, 1980) describe new taxa from this vicinity. Ricker & Kraft used material from collections made during 'The Australian Museum Macquarie Island Expedition, Summer 1977–78', whereas Zaneveld & Sanford based their work on collections made by the USS *Glacier* in 1965 (New Zealand and United States Expedition to the Ross Sea, Balleny Islands and Macquarie Ridge, 1965).

It is clear that anecdotes about seaweeds on Macquarie's shores (A. Hamilton, 1984; H. Hamilton, 1926), and the brief accounts by Laing (1896, 1909), Lucus (1919) and Levring (1945), presented a very incomplete picture of the marine plants in this vicinity. A more thorough view of Macquarie's marine macrophytes has come since the establishment of ANARE in 1948, when extensive biological programmes have begun compilation of comprehensive floristic lists, detailed descriptions of new taxa, and descriptive ecological accounts.

Fig. 4 Localities around Macquarie Island. **a.** Shingle beach with a group of king penguins (*Aptenodytes patagonicus* Miller). *Durvillaea antarctica* covering intertidal rocks in the background. **b.** Rocky intertidal reef platform with a diver entering the water through a narrow band of *Durvillaea antarctica*.

Fig. 5 Localities around Macquarie Island. **a.** Southern view across Aerial Cove toward the isthmus and Handspike Point. **b.** Northern view across Aerial Cove toward Anchor Rock.

Fig. 6 Localities around Macquarie Island. **a.** View south from Mawson Point along the west coast. **b.** View north toward North Head along the west coast.

1.5 Marine habitats for benthic algae

Five major types of marine habitats can be distinguished around Macquarie I., each characterized by particular depth profiles, wave kinetics and plant assemblages. Details on specific plant associations follow each species description given in Chapter 2.

Habitat 1: Rock platforms/walls on the unprotected west coast (Figs 3a, 3b, 6b). Here extreme wave conditions affect the substratum from the supratidal down to depths of 10–15 m. This region typically is dominated by the normally lower-intertidal *Durvillaea antarctica*, which extends here from the upper intertidal to depths of >10 m in the very roughest habitats.

Habitat 2: Subtidal west coast habitats below the limits of major wave surge effects. This zone commences between 10 and 15 m depths and extends downwards to >25 m. The dominant alga here is the kelp *Macrocystis pyrifera*, which shades understory algae and moderates wave turbulence. Both *Codium subantarcticum* Silva *mscr. ined.* and *Plocamium hookeri* Harvey are common understory plants.

Habitat 3: Unprotected intertidal platforms on the east coast. Here a relatively narrow (2 m wide) band is dominated by *Durvillaea antarctica* (Simpson, 1976). This brown algal cover gives way abruptly at the upper edge of the subtidal to dense clumps of *Ballia callitricha* and large numbers of *Cenacrum subsutum* that populate the shaded, coralline-encrusted rock walls to depths of 3 m.

Habitat 4: Subtidal east coast habitats below the 3 m depths most affected by wave surge. Extending to 20+ m depths, this environment is dominated by canopies of *Macrocystis* and dense, bottom-shading overstories of *Desmarestia chordalis.* A lush growth of membranous delesseriacean red algae (e.g., *Hymenena, Myriogramme, Schizoseris*) covers the rocks in these relatively calm habitats. *Cenacrum* is the largest foliose red alga in this habitat.

Habitat 5: Calm, protected, shallow (< 5 m deep) coves and inlets on both the east and west coasts. Aerial Cove (Fig. 5) is the most studied of this type of habitat at Macquarie I. Caroline Cove (Fig. 7b) has extensive stretches of substrata in <5 m depth and with a flora similar to that typical for Habitat 5, but other portions of the cove are considerably deeper (>20 m) and experience conditions more comparable to those of Habitat 1. *Codium subantarcticum* forms large mats (>5 m across) in Habitat 5, interspersed with the large foliose red algae *Cenacrum* and (at Aerial Cove) *Pachymenia.*

Tidal movements were not measured and particular zones in the intertidal have been delineated by the positions of selected plants. The *Durvillaea* band approximates the subtidal fringe. *Porphyra columbina, Bostrychia vaga* and *Blidingia minima* define the border between the upper intertidal and supratidal. *Prasiola* generally grows still higher in the supratidal. Names for zones described in this report follow the terminology of Round (1981); the terms supratidal, intertidal, and subtidal being synonymous with supra-littoral, littoral, and infra-littoral, respectively.

1.6 Materials and methods

Collecting marine algae at Macquarie I. is not easy. No roads traverse the island and all gear used for research must be carried bodily to study sites. Most accommodation and laboratory facilities are at the ANARE base (Fig. 2a) on the north end of the island, but six small huts, positioned to the south, serve as convenient field labs and refuges (Fig. 7a). A four-man team, including the author, collected marine organisms at Macquarie I. during the austral summer from November to February, 1977–78. Algal samples were gathered from the intertidal and subtidal, the latter specimens taken with the aid of self-contained underwater breathing apparatus (SCUBA; Fig. 4b). Several plants discussed in Chapter 2 come from separate collections (e.g., dried herbarium material on loan from Iona Christianson and Dr Elise Wollaston) and are noted as such. Most collections represent summer sampling, but some material is from winter collections.

All plants were preserved shortly after collection (from immediately to four hours later) and

Fig. 7 Localities around Macquarie Island. **a.** Caroline Cove Hut with tussock grass (*Poa foliosa* Hooker f.) in the foreground **b.** Northern view of Caroline Cove from the hut. Offshore *Macrocystis* beds breaking the water surface and *Durvillaea antarctica* covering the nearshore rocks.

were fixed in a large volume of 5% formalin/seawater. Some samples were then sorted on a preliminary basis, filamentous red algae being placed in individual vials, and large, coarse, brown algae (e.g., *Durvillaea* and *Desmarestia*) transferred to 50% glycerine/seawater to maintain thallus flexibility when gradually air-dried. Most material, however, was stored in small volumes of fresh 5% formalin/seawater, packed for shipment in air-tight 30 l blue plastic barrels and stowed away from direct sunlight. This procedure effectively preserved the natural color of most plants during wet storage—a span of 3–4 years in some cases.

Most anatomical studies are based upon hand-cut sections, but microtome-cut, plastic-embedded tissues better revealed the anatomy of some members of the Corallinaceae and certain other red and brown seaweeds. Coralline algae were decalcified in 5% nitric acid prior to embedding. JB-4, a glycol methacrylate made by Polysciences, was used for embedding; most sections were cut on a Sorvall JB-4 microtome. Embedding procedures are detailed in Polyscience literature accompanying each JB-4 embedding kit. One step that may be varied according to the type of tissue being embedded concerns the amount of catalyzed solution A to be mixed with solution B (activator). A 40 : 1 (A : B) ratio was used. This allowed ample time for plastic infiltration into the tissues before polymerization. Following techniques by Henry (1977), plastic ribbons of serial sections were obtained. In the case of large tissue blocks, a ralph knife made from a microscope slide (an unpublished technique developed by Dr. Richard L. Moe of the University of California, but similiar in its results to the techniques of Semba (1979)) cut numerous good sections (ribbons up to 30 cm long) before requiring replacement.

Sections of plastic-embedded tissues were cut between 2 and 4 µm thick; thinner sections generally yielding better staining resolution between tissue and plastic. Toluidine Blue 0 (0.05% aqueous solution) and Stevenel's Blue (del Cerro *et al.*, 1980) worked equally well on methacrylate-infiltrated tissues, and a 1% aqueous solution of aniline blue was used for hand-sectioned material. Nuclear studies were done using acetocarmine (a saturated solution of carmine in 1% acetic acid) on material previously placed for 30–60 seconds in an iron alum (3% aqueous solution) mordant. Starch-pyrenoid complexes were visualized using a concentrated iodine solution. Hand-sectioned material was mounted in a 25% aqueous karo (corn syrup) solution plus phenol. Gradual infiltration of the tissues was accomplished by allowing preparations to air-dry until Karo concentration neared 100%, thereby making a permanent preparation.

Some tissues, especially in members of the Cryptonemiaceae and Gigartinaceae, have mucilaginous polysaccharide slimes that rapidly stain darkly. This gives an overall diffuse stain to the tissues and renders anatomical detail indistinct. The lack of contrast between stained cells and plastic infused with slime prevents photographic rendition of plant anatomy. In these cases, the best anatomical detail is observed and recorded using sectioned, unstained, plastic-embedded material viewed under Nomarski interference microscopy.

Micrographs were taken with Zeiss microscopes, mostly using a Zeiss Photomicroscope III. Lower magnification photographs, including macrography of filamentous red algae, were made with a Wild M4 Makroscope. A 35 mm format camera (Canon Fl or Nikkormat) was used to take habit photographs of algae ≥2 cm in length, these photographs representing either dried herbarium specimens or wet plants floating in a shallow aquarium illuminated with a combination of transmitted and reflected light.

2: The Benthic Marine Algal Flora

No single differentia (character), I repeat, either by itself or with its antecedents, can possibly express the essence of a species.

Aristotle (Ogle, 1912)

2.1 Systematic account

During the roughly one million years since Macquarie's volcanic emergence, no land bridges have connected this island to any continental landmasses or to other islands. Accordingly, ancestors of Macquarie's present flora must have arrived from elsewhere. Original and more recent imports would have had to travel considerable distances, their most likely origins being in the Cold Temperate, Subantarctic and Antarctic Zones. A study of the island's flora thus benefits from familiarity with marine algae from throughout the Southern Hemisphere.

Early expeditions into southern high latitudes tended to reveal pockets of putative endemism that were largely coincident with each new locality visited. This led to more algal names than actual species represented. Many original descriptions were based upon single collections of types which were often poorly documented. Single specimens subsequently collected and similar but not absolutely identical with the original species description were described as separate species. Morphological variability within a species generally was ignored and a single taxon, genus or species often was fragmented into many taxa.

Present attempts to name algae and describe floras in southern high latitudes are complicated by the superfluity of synonyms and the large number of still poorly understood species. One can study a select flora in the Subantarctic and illustrate its components, but because of difficulties in associating names with type or authentic specimens, taxonomy often may be imprecise. Consequently, conclusions about species distributions and the relationships between floras in the Southern Hemisphere should not be based on early reports of plant occurrences at particular localities without referring to the original specimens upon which the various records are based. Direct comparison of more recent collections from all areas under study also yields useful biogeographic data about relationships between floras. Accurate nomenclature, however, follows only from studies of type collections.

The following account of Macquarie I. seaweeds is based upon 75 algal collections composed of some 3000 specimens taken from seven main sites (Figs 8, 9). Because of its proximity to the ANARE base, the general vicinity of Wireless Hill was sampled most intensively and consequently is the most extensively documented area. Species descriptions are based on observations of the island's plants. Where taxa have been described previously comparisons are made between Macquarie specimens and those described from elsewhere. For some species, comparative illustrations are made with either type material, identical species from other localities, or closely related species. Most specimens have been identified by their accession numbers. The location of individual specimens is indicated by an abbreviation preceeding each accession number or a group of acccession numbers, with abbreviations taken from *Index Herbariorum* (Holmgren & Kenken, 1974).

The Systematic Overview (Section 2.2) lists all species that are presently documented from Macquarie I. and gives the classification system followed in this survey. Family names and their respective authorities are presented in the Systematic Catalogue (Section 2.3) and have been taken from Silva (1980*a*), with descriptions of the various algal families agreeing largely with the concepts of Silva (1982—Chlorophyta), Wynne (1982*b*—Phaeophyta) and Dixon (1982— Rhodophyta). Authorities for genera and designations of type and lectotype species agree with those given in the *Index Nominum Genericorum* (Farr *et al.*, 1979) unless otherwise specified, Species authorities generally comply with those presented in Silva's unpublished *Index Nominum Algarum* (INA).

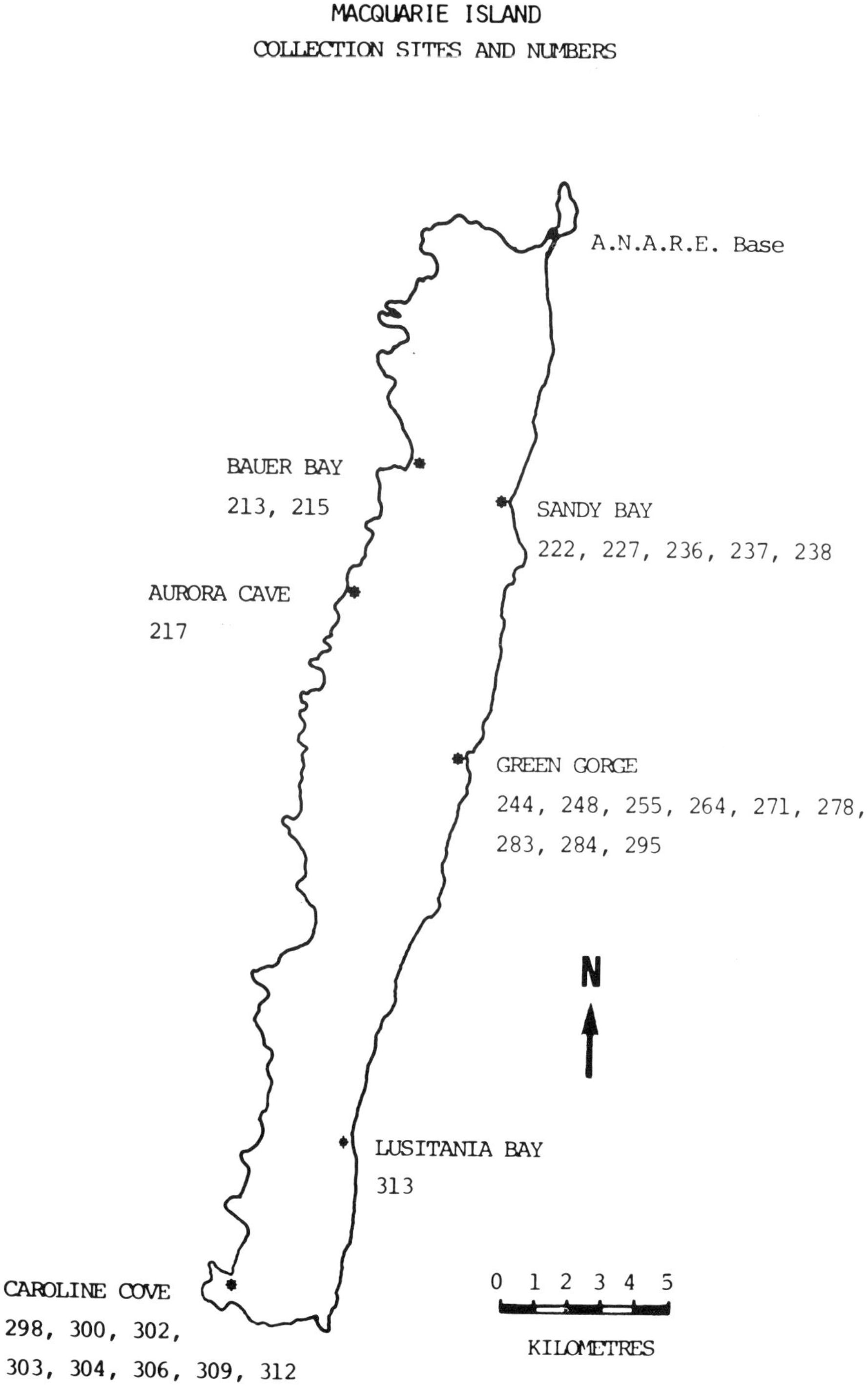

Fig. 8 Macquarie Island collection sites and numbers.

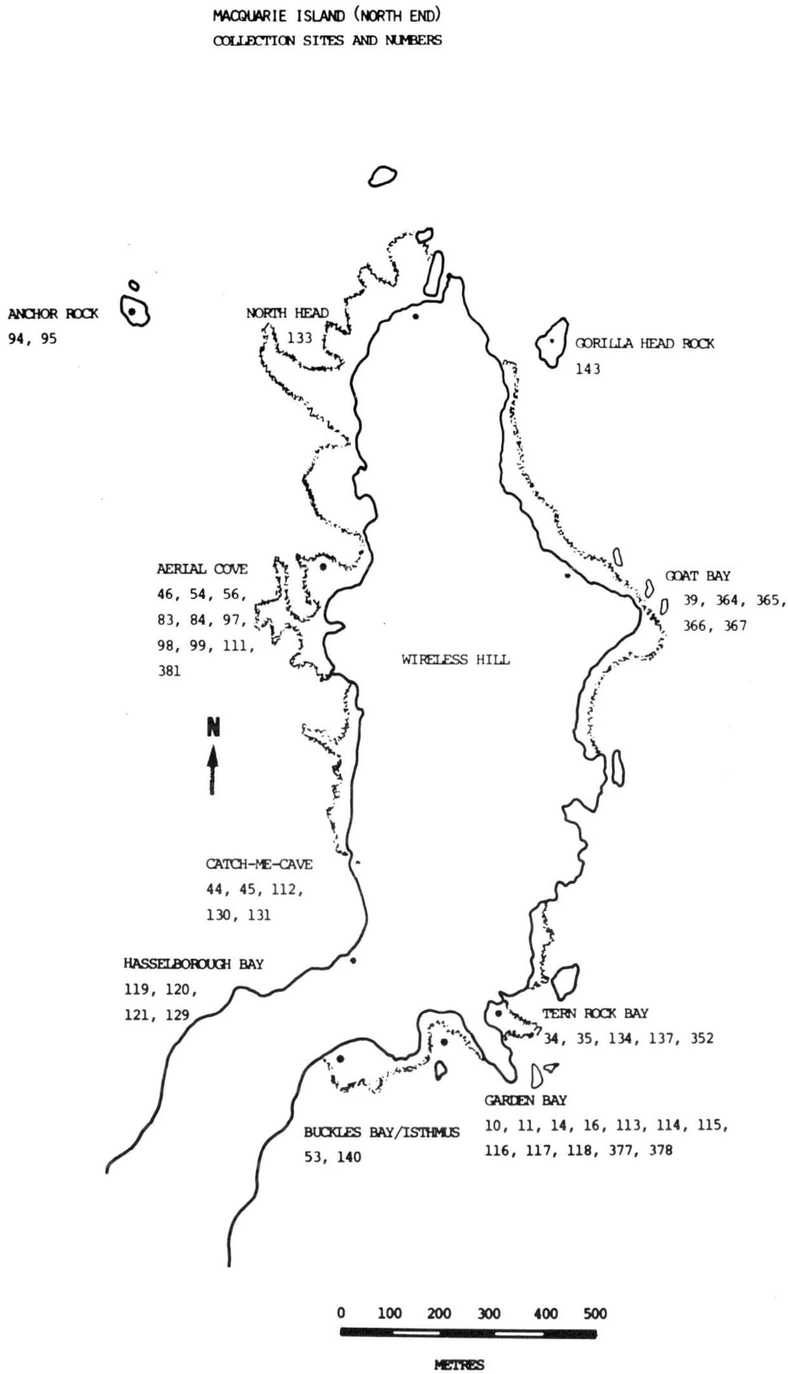

Fig. 9 Macquarie Island (north end) collections sites and numbers.

2.2 Systematic overview

On such a small and remote island it is somewhat surprising to find 103 species of marine algae representing 81 genera and 43 families—nearly three times the number of angiosperms reported from the same locality (B. W. Taylor, 1955). Most of these seaweeds were collected attached *in situ*, but in two instances drift collections have supplied the only material for study. Below is a list of Macquarie I. marine algae with notations on their drift (d), intertidal (i) or subtidal (s) origins. The list is intended to indicate collection localities, but not to reflect algal distribution limits around the island. For example, *Durvillaea antarctica*, an especially large alga that is abundant on all Macquarie shores, is listed here only from the northern portion of the island, where it was convenient to collect, preserve, and store.

Taxon	North end		Bauer Bay	Aurora Cave	Sandy Bay	Green Gorge	Lusitania Bay	Caroline Cove
	East	West						
CHLOROPHYTA								
Chaetophorales								
Chaetophoraceae								
Endophyton atroviridis		i						
Entocladia maculans		s						
Ulvales								
Monostromataceae								
Blidingia minima		i						
Monostroma sp. 'a'		i	i					
Monostroma sp. 'b'	s							
Ulvaceae								
Enteromorpha bulbosa	i	i		i				
Enteromorpha intestinalis		i		i				
Ulva angusta	s	i			s	s		
Ulva rigida		i						
Prasiolales								
Prasiolaceae								
Prasiola crispa subsp. *antarctica*		i						
Acrosiphoniales								
Acrosiphoniaceae								
Acrosiphonia pacifica	i,s	i,s	i		i	i,s		
Urospora penicilliformis		i				i		
Cladophorales								
Cladophoraceae								
Rhizoclonium ambiguum		i,s	i	i		i,s		
Bryopsidales								
Bryopsidaceae								
Derbesia furcata	s							
Codiaceae								
Codium subantarcticum	*s*	*s*	*i*			*s*		*s*
PHAEOPHYTA								
Ectocarpales								
Ectocarpaceae								
Asteronema ferruginea	s	s				i		s
Ectocarpus constanciae	s	s						
Ectocarpus siliculosus	s	s				s		s
Geminocarpus geminatus	i,s	i,s			i	i		
Pilayella littoralis	i	i						
Streblonema codiatile	s	s				s		s

Taxon	North end East	North end West	Bauer Bay	Aurora Cave	Sandy Bay	Green Gorge	Lusitania Bay	Caroline Cove
Ralfsiaceae								
Lithoderma piliferum								
var. *debile*		i			i	i,s		s
Ralfsia australis		i,s						
Chordariales								
Myrionemataceae								
Myrionema incommodum			i					
Elachistaceae								
Elachista antarctica						i		
Gononema pectinatum	i							
Gononema ramosum		i	i					
Chordariaceae								
Haplogloia moniliformis	i,s					s		
Papenfussiella tristanensis	s				s	s		
Scytothamnus fasciculatus	i							
Chordariopsidaceae								
Chordariopsis capensis	i		i		i	i		
Sphacelariales								
Sphacelariaceae								
Sphacelaria bornetii						i		
Stypocaulaceae								
Alethocladus corymbosus								s
Dictyosiphonales								
Delamareaceae								
Cladothele striarioides	i							
Punctariaceae								
Adenocystis utricularis	i	i	i		i	i,s		
Scytosiphonales								
Scytosiphonaceae								
Petalonia fascia	i	i,s	i		i			
Scytosiphon lomentaria	i	i	i		i			
Desmarestiales								
Desmarestiaceae								
Desmarestia chordalis	s	s			s	s		s
Desmarestia willii	s	s			s	s		s
Cutleriales								
Cutleriaceae								
Microzonia velutina	s							s
Dictyotales								
Dictyotaceae								
Dilophus decumbens	s	s				s		s
Laminariales								
Lessoniaceae								
Macrocystis pyrifera	s	s	s		s	s		s
Durvillaeales								
Durvillaeaceae								
Durvillaea antarctica	i	i,s						

Taxon	North end East	North end West	Bauer Bay	Aurora Cave	Sandy Bay	Green Gorge	Lusitania Bay	Caroline Cove
RHODOPHYTA								
Bangiales								
Bangiaceae								
Porphyra columbina	i	i			i	i		
Porphyra ionae		d						
Porphyra plocamiestris	s	s				s		
Porphyra woolhousiae	s					s		
Porphyra sp.	s							
Erythropeltidaceae								
Porphyropsis coccinea	s							s
Acrochaetiales								
Acrochaetiaceae								
Acrochaetium daviesii		i						
Rhodochorton concrescens	s	s				s		
Rhodochorton purpureum	i	i					i	i
Rhodochorton variabile	s							
Bonnemaisoniales								
Bonnemaisoniaceae								
Delisea pulchra	s							s
Ptilona willana		s				s		s
Nemaliales								
Chaetangiaceae								
Chaetangium fastigiatum		i	i		i	i		
Cryptonemiales								
Corallinaceae								
Melobesia membranacea	s	s				s		s
Mesophyllum patena	s					s		
Mesophyllum schmitzii		s	i					
Pseudolithophyllum consociatum		i			i	i		
Pseudolithophyllum discoideum		s			s			
Halymeniaceae								
Pachymenia sp. cf. *P. stipitata*	i	s	i					
Hildenbrandiaceae								
Hildenbrandia lecannellieri	i,s	i,s			i,s	i		
Kallymeniaceae								
Callocolax neglectus		s						
Callophyllis variegata		s						
Pugetia delicatissima		i				s		s
Gigartinales								
Gigartinaceae								
Iridaea cordata	i,s	i,s	i		s	s		s
Iridaea remuliformis		i				i		
Phyllophoraceae								
Ahnfeltia plicata	i							
Phyllophora appendiculata	i,s	i,s	i		s	s		s
Plocamiaceae								
Plocamium hookeri	s				s			s
Plocamium secundatum	s	i,s	i		s	s		s

Taxon	North end East	North end West	Bauer Bay	Aurora Cave	Sandy Bay	Green Gorge	Lusitania Bay	Caroline Cove
Cystocloniaceae								
Rhodophyllis acanthocarpa	s	s			s	s		s
Rhodymeniales								
Rhodymeniaceae								
Cenacrum subsutum	i,s	s	i		s	s		s
Rhodymenia subantarctica		s						s
Palmariales								
Palmariaceae								
Palmaria decipiens	i	i				i		
Palmaria georgica		i,s	i			i		
Ceramiales								
Ceramiaceae								
Antithamnionella alternans		s						
Antithamnionella ternifolia	s	s						
Ballia callitricha	s	i,s			s	s		s
Callithamnion gracile						s		s
Ceramium strictum		s				s		s
Dasyptilon pellucidum	s	s			s	s		s
Dasyptilon sp.	s							
Medeiothamnion flaccidum	s	s				s		s
Plumariopsis eatoni	d							
Pterothamnion plumula	s	s						s
Pterothamnion simile	s	s			s	s		
Dasyaceae								
Colacodasya inconspicua		s			s			s
Heterosiphonia berkeleyi	s	s			s	s		s
Delesseriaceae								
Cladodonta lyallii		s						s
Delesseria lancifolia	i,s				s	i,s		s
Hymenena decumbens		s						s
Hymenena laciniata	s	s				s		s
Myriogramme livida	d,s	s			s	s		s
Polycoryne radiata	s				s	s		
Pseudophycodrys phyllophora	s				s	s		s
Schizoseris condensata	i,s	i,s	i		i,s	s		
Schizoseris dichotoma	s	s			s	s		s
Rhodomelaceae								
Bostrychia vaga	i	i				i		
Lophurella hookeriana		i	i		s	s		s
Polysiphonia anisogona	i,s	s	i		i,s	i,s		s
Sporoglossum lophurellae			i			s		

2.3
Systematic catalogue

Chlorophyta

Chaetophorales

The Chaetophorales traditionally has received all branching filamentous green algae with uninucleate cells. Silva's (1982) redefinition of the order, which is based on *Chaetophora* ultrastructure, includes some unbranched genera (e.g., *Uronema*) while excluding several branched ones (*Coleochaete, Trentepohlia, Microthamnion*) and probably all marine taxa. The latter are considered by Silva to be members of the Ulvellaceae, which he has placed in a new order, the Ctenocladales. Because the marine representatives from Macquarie I. have not been studied ultrastructurally, this report maintains them as provisional members of the Chaetophorales.

Thallus organization in the Chaetophorales (as traditionally conceived) is (usually) heterotrichous with prostrate and erect branching systems, of which either may be strongly or weakly developed. Cells have parietal band-shaped chloroplasts with one or more pyrenoids. Colourless hairs have been used previously in some taxonomic diagnoses, but are now known to be variably expressed depending upon the environmental conditions where the algae grow (Yarish, 1975). Fragmentation commonly occurs in addition to production of biflagellate or quadriflagellate zoospores or aplanospores. Sexual reproduction and life histories are poorly known in most chaetophoralean algae, but all three types of gametogenesis (isogamy, anisogamy and oogamy) and both haplontic and diplontic life histories have been reported among various species (Tupa, 1974; O'Kelly & Yarish, 1980, 1981).

Some systematic schemes for chaetophoralean algae have placed them in their own order (Chaetophorales—Fritsch, 1935) or in a family in the Ulotrichales (Chaetophoraceae—Abbott & Hollenberg, 1976; G. M. Smith, 1955). The Chaetophorales, as presently construed, comprises three families, the Aphanochaetaceae, Chaetophoraceae and Schizomeridaceae, of which only the Chaetophoraceae is known from Macquarie I.

Chaetophoraceae Greville, 1824: xix, 321

Members of this family have either entirely prostrate thalli, erect thalli, or both. The marine genera in the vicinity of Macquarie I. lack erect portions and are composed of wholly endophytic or epiphytic prostrate filaments. Cellular detail and asexual reproduction are the same as reported for the order. Sexual reproduction is isogamous or anisogamous, and gametes are biflagellate or quadriflagellate.

The family is represented in the Southern Ocean by *Entocladia* (Cotton, 1915; Levring, 1944, 1945; Skottsberg, 1953) and by records of *Entoderma* (Gain, 1912) and *Entonema* (Reinsch, 1890), of which the last two comprise poorly-known filamentous epiphytes possibly identical to *Entocladia*. *Endophyton* was first reported from high-latitude southern waters by O'Kelly (1980) and is common in *Iridaea cordata* from both the Antarctic and Subantarctic Zones. Branched, uniseriate green filaments with uninucleate cells similar to *Eugomontia sacculata* Kornmann from the Northern Hemisphere have been found in mollusc shells in Eastern Antarctica (Ricker, unpubl.).

Endophyton and *Entocladia* each have one species growing at Macquarie I.

Endophyton Gardner, 1909: 371

Type species: *E. ramosum* Gardner, 1909: 372.

Thalli are entirely endophytic, mostly forming discolourations near the bases of various red and brown algal hosts, but sometimes spreading across the majority of host tissue, and are composed of irregularly branched green filaments that ramify through host fronds and produce numerous short anticlinal branches that terminate near host thallus surfaces in swollen, ovoid sporangia *c.* twice the diameter of subtending filaments. Sporangia are pyriform and mostly or entirely covered by the host cortex, except for their mucronate apices, which generally protrude through the host cuticle. Each cell contains one band-shaped, parietal plastid with a single pyrenoid. Both zoospores (larger) and gametes (smaller) are biflagellate.

Endophyton resembles *Acrochaete* in the production of specialized sporangia, but those of the latter develop long delicate hairs. *Acrochaete* also has multiple pyrenoids per cell and a partially epiphytic thallus (Gardner, 1909).

At present, the genus includes only the type species (described from the northeast Pacific: Gardner, 1909; Abbott & Hollenberg, 1976). An undescribed species occurs at Macquarie I. and other localities in the Southern Hemisphere (O'Kelly, 1980). Although included here, it will be given a formal diagnosis elsewhere by O'Kelly (pers. comm.).

Endophyton atroviridis O'Kelly sp. ined.

Fig. 10

DISTRIBUTION: Heard I.; Macquarie I.; Vestfold Hills, Antarctica. Common in *Iridaea cordata* and probably circumpolar in distribution, like the host, throughout the Antarctic and Subantarctic Zones.

MATERIAL EXAMINED:

1. Heard I.: endophyte in *I. cordata*; (J. M. B. Smith, February 1983); MELU, R 3651.
2. Macquarie I.: collection MA-111; Herbarium specimen: MELU, R 3345.
3. Vestfold Hills, Antarctica: Tidegauge Pt., lower littoral to upper sub-littoral boulders; endophyte in *I. cordata* (Ricker, 14 December 1981); Herbarium specimen: MELU, R 2727.

DESCRIPTION: Thalli forming dull green to red, irregularly-shaped patches to several mm across in *I. cordata* (Fig. 10a), often surrounding host cystocarps. Filaments 3–4 μm diameter (Fig. 10d), branched, ramifying through host medulla, terminating in swollen structures (gametangia/sporangia?) with mucronate apices sometimes protruding through host frond surface (Fig. 10c). Reproductive structures pyriform (Fig. 10c–e), 7–11 μm diameter, 17–21 μm long, with single parietal, band-shaped chloroplast containing one pyrenoid. Gametes and spores unknown.

Endophyton atroviridis infects large numbers of its host, which predominates in the lower intertidal to shallow subtidal. In preserved material, pigments are readily lost from the irregularly branched filaments, thereby rendering them difficult to distinguish from *Iridaea* medullary filaments (Fig. 10b). The endophyte can be distinguished easily in the host cortex, however, where 'sporangia' intercalate with, but are morphologically distinct from files of host cortical cells.

NOTES: *Endophyton atroviridis* is tentatively separated from the only other species in the genus, *E. ramosum*, by its occurrence in a different host species and its geographical separation. O'Kelly (pers. comm.) also uses differences in sporangial diameter to separate the two species, but size ranges reported by both Abbott & Hollenberg (1976: 6–12 μm diameter) and Gardner (1909: 10–12 μm diameter) overlap with those for *E. atroviridis*.

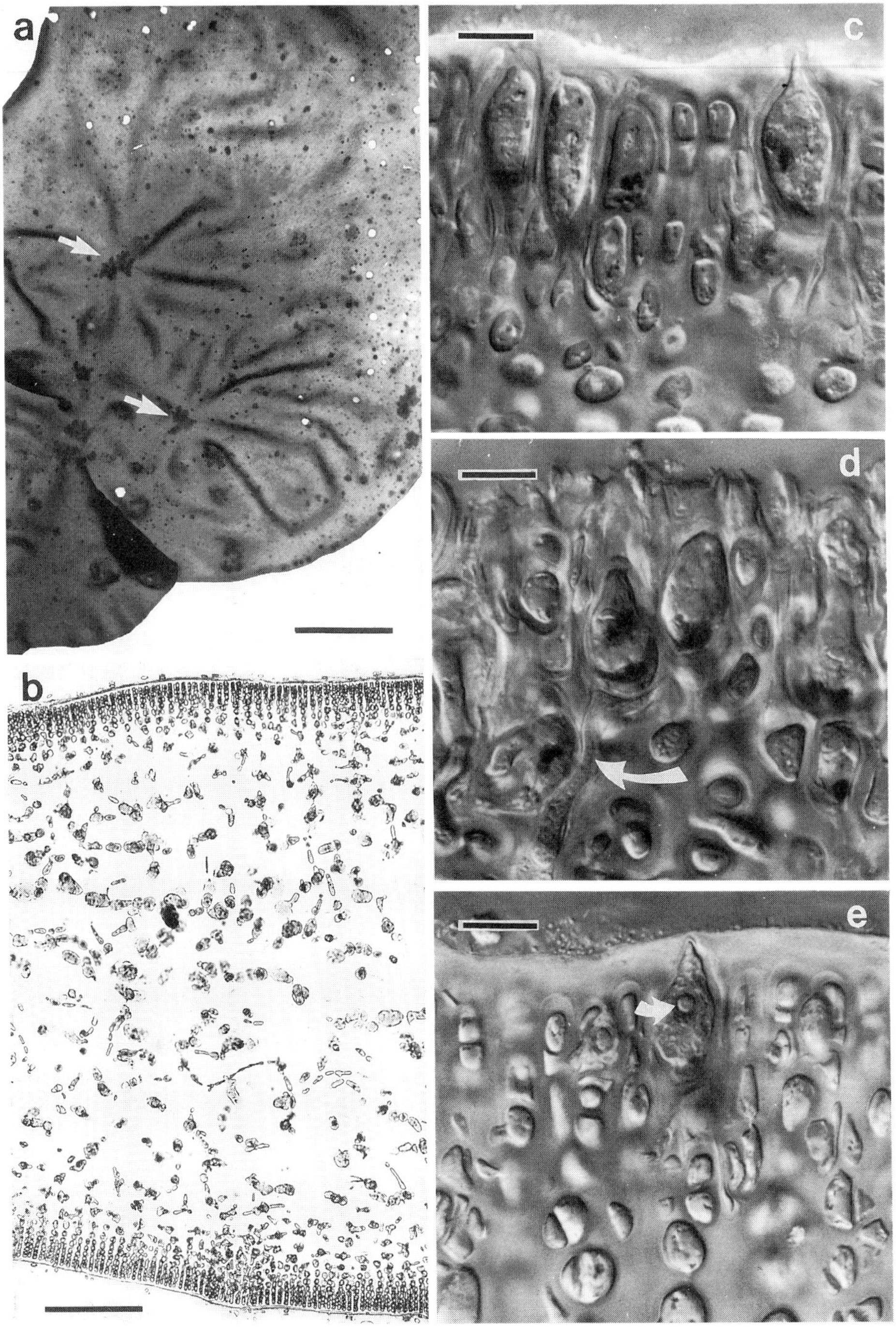

Fig. 10 *Endophyton atroviridis* O'Kelly sp. ined.: MELU, R 2727, R 3345. **a.** Endophytic patches (arrows) in *Iridaea cordata.* **b.** Section showing relationship of host cortex to medulla. **c–d.** Cortical sections of infected Antarctic *Iridaea* (R 2727); **d.** basal rhizoids (arrow) penetrating host medulla. **e.** Cortex of infected *Iridaea* from Macquarie I. (R 3345) with sporangium (?) of endophyte having single chloroplast with one pyrenoid (arrow). Scales: a = 2 cm; b = 100 µm; c–e = 10 µm.

Entocladia Reinke, 1879: 476

Type species: *E. viridis* Reinke, 1879: 476.

Thalli may be epiphytic, endophytic, or endozoic, and are entirely prostrate with uniseriate, irregularly branched filaments that may remain separate or become compacted into a pseudoparenchyma. Some thalli have both filamentous and pseudoparenchymatous portions. Each cell is uninucleate and has a single parietal, band-shaped chloroplast containing one or more pyrenoids. An alternation of isomorphic generations is characteristic, but asexual reproduction commonly occurs through quadriflagellate mitospores. Sexual reproduction is through biflagellate aniso- or isogametes produced in gametangia, which usually develop in pseudoparenchymatous portions of the thallus.

Abbott & Hollenberg (1976) refer to *Entocladia* as 'nonpiliferous' and use presence or absence of hairs to separate chaetophoraceous genera. Nielsen (1979) also considers most species that have been previously placed in *Entocladia* to be hairless, but she observed material of the type species, *E. viridis*, from the type locality to produce *Acrochaete*-like hairs. Consequently, she moved this species into *Acrochaete*, making *Entocladia* a later synonym of *Acrochaete*, and proposed *Epicladia* Reinke as the first available name for the remaining (hairless) species previously placed in *Entocladia*. O'Kelly & Yarish (1981), however, have shown hair-like (enucleate) structures (setae) in culture-grown plants to be variably produced depending upon environmental conditions and consider hair production and morphology to be taxonomically less important than the overall thallus organization. O'Kelly & Yarish also consider vegetative cell size and morphology, number of pyrenoids per cell, habitat and host selection, and the dimensions and morphology of male gametes to be too variable for separating even species. They view sporangial development, however, as taxonomically important. Nielsen (1983) disagrees with some of O'Kelly's and Yarish's conclusions and continues to separate genera and species using various combinations of uniformity in cell sizes, the presence or absence of hairs, hair morphology, and pyrenoid numbers.

In the present account, overall thallus organization is considered to be important for distinguishing chaetophoraceous genera, and *E. viridis* is maintained in *Entocladia* instead of being placed in *Acrochaete*, which traditionally has contained species with both erect and prostrate filaments (O'Kelly & Yarish, 1981).

Two species of *Entocladia*, *E. maculans* and *E. viridis*, are listed from Antarctic and Subantarctic waters (Papenfuss, 1964*a*), but only the former grows at Macquarie I.

Entocladia maculans (Cotton) Papenfuss, 1964*a*: 1

Fig. 11

Basionym: *Endoderma maculans* Cotton, 1915: 161.

Synonyms: see Papenfuss, 1964*a*.

Type Locality: Falkland Is.

Lectotype: K *in* BM, microscope slide BM 2532, coll. Vallentin (specimen annotated by J. H. Price (1 March 1985) as provisional lectotype (Price, pers. comm.)).

Distribution: Falkland Is (Cotton, 1915); Macquarie I.; South Shetland Is (Gain, 1912 (as *Entoderma sp.*)).

Material examined: Macquarie I.: collection MA-98; Herbarium specimen: MELU, R 2143 (on *Hymenena laciniata*).

Description: Thalli epiphytic; filaments irregularly branched, forming pseudoparenchymatous tissues with nearly isodiametric (9 μm diameter) to elongate (4–9 × 9–16 μm) cells (Figs 11a–c), or

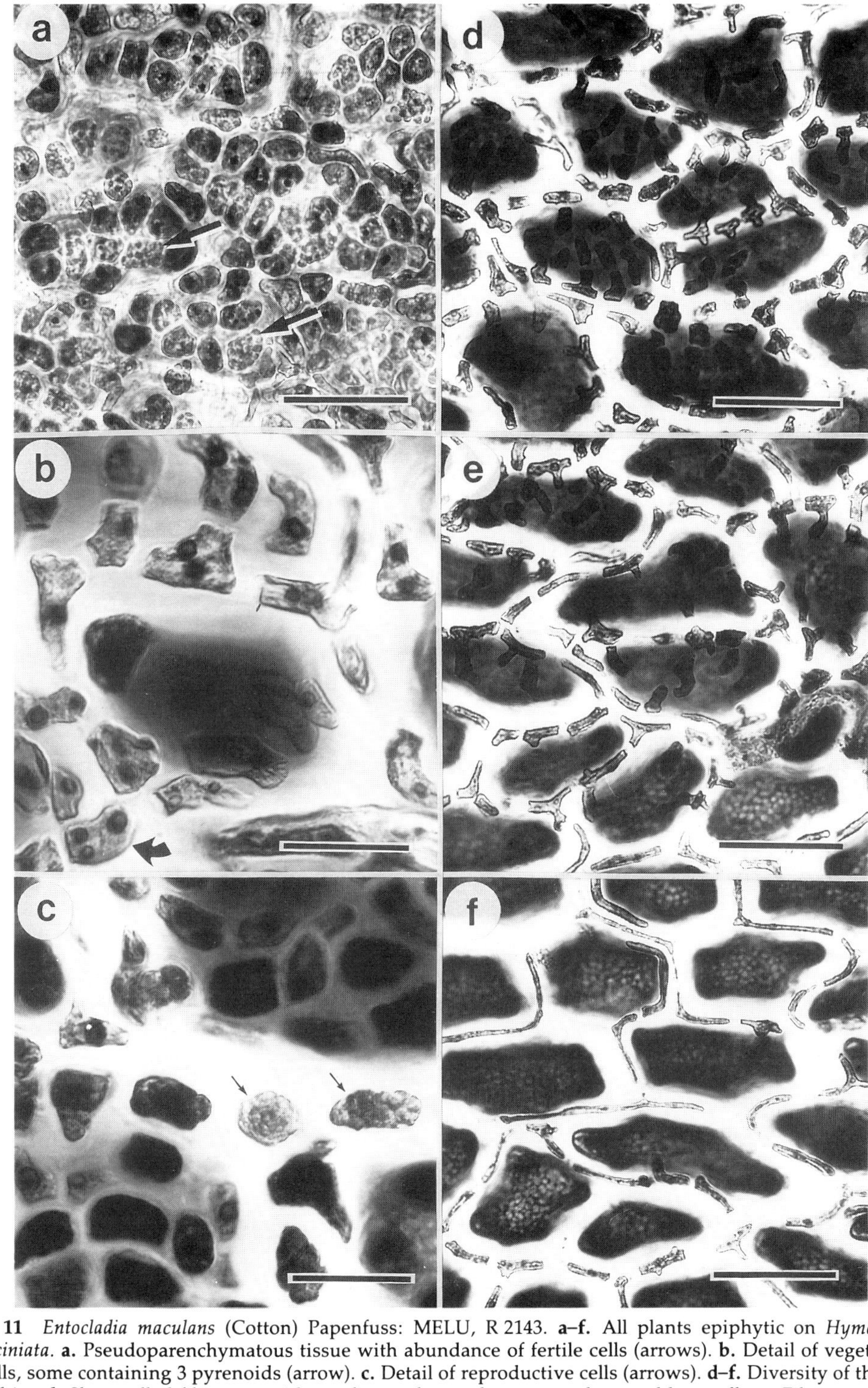

Fig. 11 *Entocladia maculans* (Cotton) Papenfuss: MELU, R 2143. **a–f.** All plants epiphytic on *Hymenena laciniata*. **a.** Pseudoparenchymatous tissue with abundance of fertile cells (arrows). **b.** Detail of vegetative cells, some containing 3 pyrenoids (arrow). **c.** Detail of reproductive cells (arrows). **d–f.** Diversity of thallus habits. **d.** Short-celled filaments with nearly equal growth across and around host cells. **e.** Filaments with slightly elongated cells mostly around host cells. **f.** Long narrow filaments strictly around host cells. Scales: a, d–f = 50 μm; b, c = 20 μm.

comprising loose network of filaments (2–3 × up to 50 μm) between or across host cells (Figs 11d–f). Cells with single, parietal, band-shaped chloroplasts containing 1–3 pyrenoids (Fig. 11b). Reproductive cells isodiametric (9 μm diameter) to elongate (4–6 × 10–15 μm), occurring only in pseudoparenchyma (Fig. 11a, c).

Entocladia maculans grows as greenish patches on membranous blades of members of the Delesseriaceae and often is associated with both necrotic and healthy portions of host fronds. It was seen only in *Hymenena* and *Myriogramme*, which grow in the subtidal, and never in such intertidal species as *Schizoseris condensata* (Reinsch) Ricker.

NOTES: Cotton (1915) separated *Entocladia (Endoderma) maculans* from *E. viridis* by 'the decidedly larger cells of the filaments, which also, for the most part, radiate across the cells of the host instead of following the outline of the cell-walls. A much larger amount of pseudoparenchyma is, moreover, developed in the present species [*E. maculans*], and in this sporangia arise in abundance'. Although some Macquarie specimens comprise large amounts of pseudoparenchyma composed of nearly isodiametric cells that cover cells of the host (Fig. 11a), others consist of slightly elongate cells that run across and around host cells (Fig. 11b, c, d), or long filamentous cells that are restricted to anticlinal walls of host cells (Fig. 11e, f). It seems, therefore, that both names might actually refer to one species, in which case *E. viridis* would have nomenclatural priority. Clearly, thallus morphology in this complex is variable and not suitable for distinguishing two entities, and cytological details such as number of pyrenoids per cell either are not documented (Baardseth, 1941; Cotton, 1915) or their variability is not discussed (Skottsberg, 1953). Southern plants are maintained here under the name *E. maculans* until critical comparisons are made with *E. viridis* from the Northern Hemisphere. Distribution records for *E. viridis* from the Antarctic Peninsula (Skottsberg, 1953), Îles Crozet and Kerguelen (Levring, 1944), Campbell I. (Levring, 1945), and Tristan da Cunha (Baardseth, 1941) should be viewed as possible records for *E. maculans*.

Ulvales

Green algae that have parenchymatous rather than filamentous thalli were first united in one order (the Ulvales) by Blackman & Tansley (1902). Subsequently, some authors (Papenfuss, 1960; V. J. Chapman & D. J. Chapman, 1973) considered the similarities between parenchymatous plants and filamentous members of the Ulotrichales sufficient to warrant unification of both groups into one order (Ulotrichales). Recent ultrastructural evidence, however, has supported recognition of the Ulvales as distinct from the Ulotrichales (Stewart & Mattox, 1975).

The Ulvales includes biseriate filaments; monostromatic blades, sacs, or tubes; and distromatic blades. All thalli are composed of uninucleate cells with parietal, cup-shaped or laminate chloroplasts usually containing one or more pyrenoids.

Sexual reproduction is isogamous, anisogamous, or oogamous. Motile gametes are biflagellate. Asexual reproduction is through quadriflagellate, rarely biflagellate, zoospores. Life histories are either haplobiontic or diplobiontic with isomorphic generations.

Of the five families in the Ulvales (Capsosiphonaceae, Gomontiaceae, Monostromataceae, Percursariaceae, Ulvaceae) recognized by Silva (1982), only two occur at Macquarie I. Members of both Monostromataceae and Ulvaceae are well represented by intertidal and subtidal plants at Macquarie I.

Monostromataceae Kunieda, 1934: 106

Members of this family have monostromatic thalli that may develop into either open (*Kornmannia*) or closed (*Blidingia*) tubes, closed sacs (*Monostroma*), or foliose fronds (*Kornmannia, Monostroma*). Thallus development, beginning with zoospores germinating into prostrate discs that later give rise to one of the above-mentioned upright forms, is the major feature distinguishing monostromatacean algae from others in the order. The three genera in the family differ in their life histories—*Kornmannia* having a strictly asexual phase, *Monostroma* being haplobiontic, and *Blidingia* exhibiting several possibilities that include isomorphic and heteromorphic phases as well as an asexual monophasic cycle (Tatewaki & Iima, 1984).

Contrary to the present interpretation of *Monostroma*'s life history as haplobiontic, some authors consider the genus to have a diplobiontic life history with heteromorphic generations (viewing the diploid zygote as a unicellular sporophyte) and place it either in its own order, the Monostromatales (Kornmann & Sahling, 1977), in the Ulotrichales (Parke & Dixon, 1976; Tanner, 1981), or in the Ulvales, thereby broadening the boundaries of this last order to include plants with heteromorphic life histories (Bliding, 1968).

Only *Blidingia* and *Monostroma* are known from high-latitude southern waters, and both grow at Macquarie I. Because the life histories of family members from the Southern Ocean are unknown, however, some reports of *Monostroma* may actually represent the fronds of another morphologically similar, monostromatic foliose genus.

Blidingia Kylin, 1947*b*: 8

Type species: *B. minima* (Nägeli *ex* Kützing) Kylin, 1947*b*: 8.

Plants consist of monostromatic, branched or unbranched, tubular fronds that develop from prostrate discs. Thalli may remain basally attached to discs or become detached and continue growth either pelagically or as epiphytes. Thallus surfaces comprise small, irregularly organized cells with single parietal chloroplasts (described sometimes as stellate), each containing one centrally located pyrenoid. Life history patterns include sexual alternations of isomorphic or heteromorphic phases, asexual alternations of heteromorphic phases and an asexual monophasic cycle (Tatewaki & Iima, 1984). Asexual zoospores are tetraflagellate.

Blidingia is a common component of most intertidal floras in the temperate to colder (ice-free) waters of the Northern Hemisphere (Kjellman, 1883; Kylin, 1949; Abbott & Hollenberg, 1976; Kornmann & Sahling, 1977) and is also known from temperate shores in the Southern Hemisphere (Womersley, 1984). This report represents the first record of its occurrence in high-latitude southern waters.

Blidingia minima (Nägeli *ex* Kützing) Kylin, 1947*b*: 8

Fig. 12

BASIONYM: *Enteromorpha minima* Nägeli *ex* Kützing, 1849: 482.

TYPE LOCALITY: Helgoland.

TYPE: presumably in L (Koster, 1969), but specimens were not seen.

SOUTHERN OCEAN DISTRIBUTION: Macquarie I.

MATERIAL EXAMINED: collection MA-119; Herbarium specimen: MELU, R 2390.

DESCRIPTION: Thalli to 8 mm tall, with fronds tubular (Fig. 12a) or compressed (Fig. 12b), yellowish-green in colour, lacking distinct points of attachment (i.e., without obvious basal discs), irregularly branched with numerous hook-like, spicate branches (Fig. 12c). Surface view of fronds with widely-spaced, rounded to elongate cells 4.5–7.5 μm diameter (Fig. 12d); thallus transections with cells mostly isodiametric to 8.5 μm diameter. Chloroplasts each with prominent centrally located pyrenoid (Fig. 12e, f). Reproductive structures unknown.

Blidingia minima grows in the upper intertidal to supratidal, where it is associated sometimes with *Acrosiphonia pacifica*, *Enteromorpha bulbosa*, *Prasiola crispa* or *Porphyra columbina*. More often, however, it grows interlocked with branches of *Bostrychia vaga* and is concealed mostly by this red alga's mat-forming thallus.

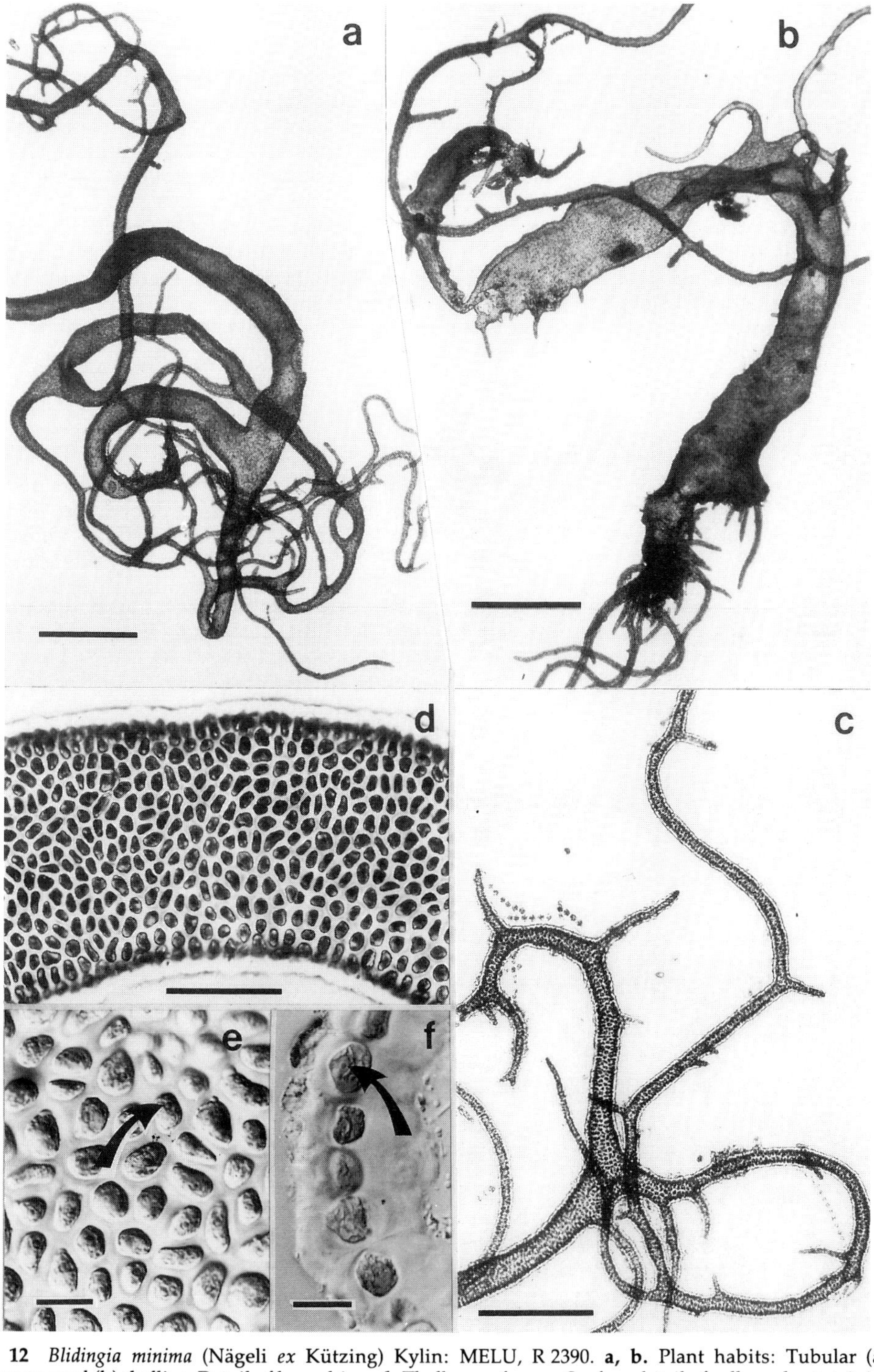

Fig. 12 *Blidingia minima* (Nägeli *ex* Kützing) Kylin: MELU, R 2390. **a, b.** Plant habits: Tubular (a) and compressed (b) thalli. **c.** Detail of branching. **d.** Thallus surface. **e.** Surface detail of cells each containing one pyrenoid (arrow). **f.** T.S. thallus showing single pyrenoid (arrow) per cell. Scales: a, b = 400 µm; c = 200 µm; d = 50 µm; e, f = 10 µm.

NOTES: Although the diagnostic feature that distinguishes *Blidingia* from *Enteromorpha* (development from a basal disc in the former vs. basal rhizoids in the latter) has not been seen in Macquarie plants, other morpholgical and anatomical characteristics (including thallus habit and thickness; cell shapes, sizes, and arrangements over thallus surfaces; and one centrally located pyrenoid per chloroplast) agree completely with previous descriptions of *B. minima* given by Kylin (1949), Bliding (1963), and Kornmann & Sahling (1977).

Blidingia minima is morphologically extremely variable depending upon the conditions where it grows. Generally, plants from the high interdial to supratidal are small (<1 cm tall), and although not seen at Macquarie I., those from sheltered localities may be largely unbranched tubes to nearly 24 cm long (Bliding, 1963). Growth without obvious attachment structures has been previously documented for Californian specimens of *B. minima* (Abbott & Hollenberg, 1976—for *B. minima* var. *subsalsa*). These plants became dislodged from their basal discs and developed without any apparent attachment structures, either intertwining among algae, much as at Macquarie I., or adrift in bays.

Monostroma Thuret, 1854: 29

Lectotype: *M. bullosum* (Roth) Thuret, 1854: 29 (see Pfeiffer, 1874).

Traditionally, *Monostroma* has included all monostromatic, foliose members of the Ulvales. At present, however, only those species with monostromatic, saccate or foliose fronds that develop from discoid bases are placed in this genus (Tanner, 1981). Blade surfaces are composed of rounded to angular cells above and elongated cells below. *Monostroma* commonly grows in temperate to colder waters and is well documented in the Northern Hemisphere through the studies of Wittrock (1866), Kornmann & Sahling (1962), Bliding (1968) and Tatewaki (1972). The genus is apparently less abundant in the Southern Hemisphere (Womersley, 1971*a*), although V. J. Chapman (1956) recorded nine species plus several varieties from New Zealand. Papenfuss (1964*a*) listed five species from southern high latitudes [*M. applanatum* Gain 1911*b*, *M. grevillei* (Thuret) Wittrock 1866, *M. hariotii* Gain 1911*a*, *M. splendens* (Ruprecht) Wittrock 1866, *M. undulatum* Wittrock 1866], although one of these (*M. applanatum*) is now considered to be a member of the Chrysophyceae (Delépine *et al.*, 1970) and another (*M. splendens*) a member of *Ulvaria* in the Ulvaceae (Tanner, 1981). Life histories are largely unknown for *Monostroma* species from southern high latitudes. Consequently, taxonomic placement of southern species in this genus (vs. other genera with monostromatic foliose fronds such as *Gayralia*, *Kornmannia*, *Protomonostroma*, *Ulvaria*) should be considered as tentative.

Monostroma species are separated using either strictly morphological criteria, such as thallus habit, blade thickness and color, cell shapes, sizes, and patterns of organization, and choloroplast shapes (Wittrock, 1866; V. J. Chapman, 1956), or morphological features plus reproductive (number of flagella per zoospore) and life history characteristics (Kornmann & Sahling, 1962).

Two species of *Monostroma* were collected from Macquarie I. Neither was obtained in sufficient quantity to permit its identification at the specific level.

Monostroma sp. 'a'

Fig. 13

SYNONYM: possibly *Monostroma latissimum* Wittrock *sensu* V. J. Chapman, 1956: 379.

SOUTHERN OCEAN DISTRIBUTION: Macquarie I.

MATERIAL EXAMINED: collections MA-11, 111, 213; Herbarium specimens: MELU, R 1467, 2362, 3268.

DESCRIPTION: Thalli to 6 cm tall, soft, pale green, with rounded or lobed margins, often splitting into several fronds, 24–27 μm thick in mid-thallus (Fig. 13a). Cells of mid-thallus surfaces often paired, angular and oblong (5–6 × 11–12 μm) or isodiametric (7–8 μm diameter), mostly

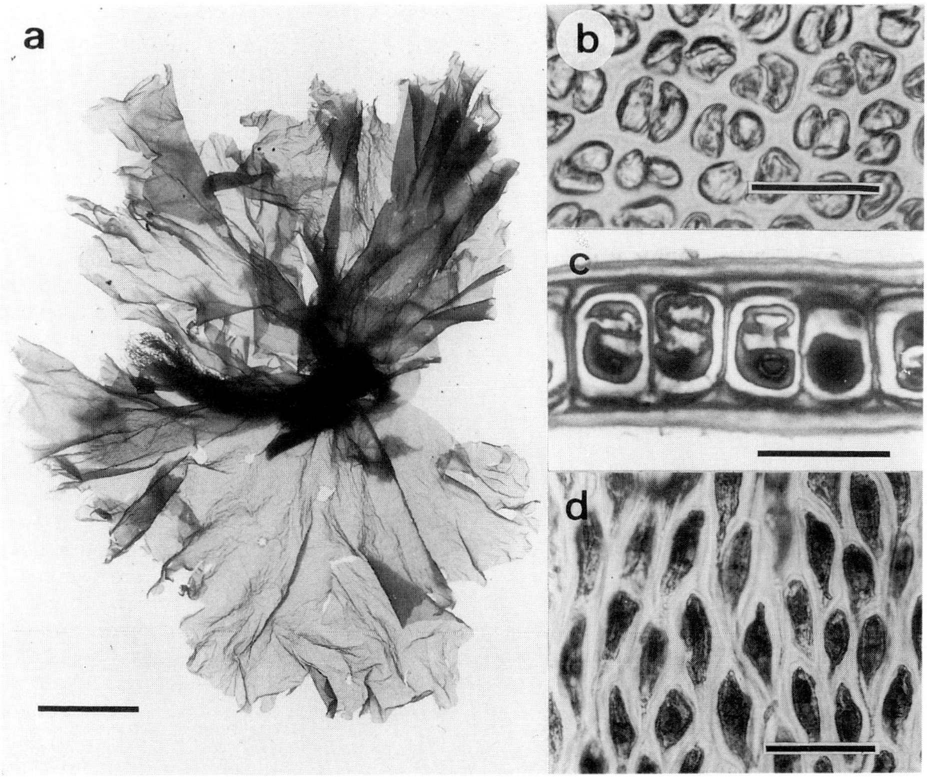

Fig. 13 *Monostroma sp.* 'a': MELU, R 1467, 3268. **a.** Thallus habit (R 3268). **b.** Surface view of cells in mid-thallus. **c.** Sectional view of mid-thallus. **d.** Surface view of cells near thallus base. Scales: a = 2 cm; b = 25 μm; c = 20 μm; d = 50 μm.

disordered but sometimes weakly organized in rows (Fig. 13b); sectional view of mid-thallus with cells 14–15 μm tall, rectangular, with single, parietal chloroplast cap-like facing both surfaces of blade (Fig. 13c); one pyrenoid per chloroplast. Cells of lower thallus with spindle-shaped cells to 50 μm long, often tapering basipetally into long filamentous portions (Fig. 13d). Reproductive structures unknown.

Although uncommon, plants were present in the mid-intertidal, where they were growing on rocky surfaces, fleshy red algal crusts (*Hildenbrandia*?) and limpets.

NOTES: Macquarie plants apparently change quickly from the saccate stage to one of flat and spreading fronds, thereby differing from the mostly saccate *Monostroma grevillei* and *M. hariotii*. Fronds of *M. undulatum* are also believed to expand in early development (Kornmann & Sahling, 1977), but these become much larger and thicker than those of Macquarie's species 'a'. Plants identified by Hylmö (1919, 1938) as *M. splendens*, a species now considered to be the type of *Ulvaria*, have fronds similar in shape and length to those of species 'a', but thicker and comprised of taller cells (as seen in thallus transections). V. J. Chapman's (1956) description of *M. latissimum* Wittrock (1866) from New Zealand encompasses attributes of species 'a', suggesting a common identity for plants from both areas. Neither entity, however, can be known by this specific epithet, which according to Bliding (1968) is predated by the earlier synonym *Ulva oxysperma* Kützing (1843)—a plant once placed in *Monostroma* (Doty, 1947), but now considered to belong in *Ulvaria* (whose members have rhizoidal bases). Thalli of *Monostroma* arise from discoid bases but also produce, especially in older individuals, elongated cells in the lower portions of fronds that resemble the rhizoids produced by species of *Ulvaria*. Consequently, distinguishing species of

Monostroma from those of *Ulvaria* is often difficult for older plants or those that have been detached from the substratum and are missing their basal discs. Developmental studies are required to confidently place species 'a' in the appropriate genus, let alone the appropriate species.

Monostroma sp. 'b'

Fig. 14

SOUTHERN OCEAN DISTRIBUTION: Macquarie I.

MATERIAL EXAMINED: collection MA-34; Herbarium specimen: MELU, R 1284 (microscope slide R/M 262).

DESCRIPTION: Thalli small (<5 mm tall), with delicate, entire to laciniate fronds having slightly sculpted margins (Fig. 14a, b). Mid-thallus blade surfaces with cells angular, elongate to trapezoidal or rectangular, 4–5 × 12–14 µm, aligned in rows or disordered (Fig. 14c). Lower thallus surfaces with cells acropetally and basipetally elongated, to 100 µm long (Fig. 14d). All cells containing single parietal chloroplast with one prominent pyrenoid. Reproductive structures unknown.

Although uncommon, plants were found as epiphytes on *Desmarestia chordalis* (Fig. 14a) in 4–5 m depth in a generally calm bay on the east coast of the island.

NOTES: Because only sterile plants were collected, it is impossible to know if mature thalli become much larger. On the basis of present measurements, however, species 'b' might be closely allied with *M. parvum* V. J. Chapman (1956), but there is insufficient information regarding Macquarie individuals to confidently place them in this or any other species.

Ulvaceae Lamouroux *ex* Dumortier, 1822: 72, 102

Plants have tubular or leafy and membranous thalli that begin development with swarmers germinating into rhizoidally attached uniseriate filaments. Members of other ulvalean families initially develop either basal discs (Monostromataceae, Percursariaceae), remain unicellular (Gomontiaceae), or form uniseriate filaments that attach by a ring of rounded, non-rhizoidal cells (Capsosiphonaceae). Six genera are recognized in the Ulvaceae on the basis of differing ontogenies that lead to either monostromatic fronds (*Protomonostroma, Gayralia, Ulvaria*), monostromatic tubes (*Enteromorpha*), or distromatic blades (*Chloropelta, Ulva*). Of these six genera, only *Enteromorpha* and *Ulva* occur in the vicinity of Macquarie I.

Enteromorpha Link *in* Nees, 1820: 5 (***nom. cons.***)

Type species: *E. intestinalis* (Linnaeus) Nees, 1820: index [2].

Thalli are monostromatic, tubular, simple or branched, rhizoidally attached, and usually hollow throughout, although lower portions sometimes are solid and terete. Fronds either remain as closed tubes, becoming compressed and partially fused with distromatic portions and cavities produced along thallus margins, or split into monostromatic blade-like portions at apices. Cells are uninucleate and contain single, plate-like, cup-shaped plastids with one or more pyrenoids. Sexual reproduction is through biflagellate isogametes or anisogametes, and asexual reproduction is through quadriflagellate zoospores.

Some species in this genus can tolerate a broad salinity range, growing in freshwater to full-strength seawater and usually exhibiting extensive morphological plasticity. Taxonomy within the genus is imprecise because species limits are based on the seemingly variable features of thallus shape and degree of branching, cell arrangements on frond surfaces, cell dimensions (including ratios of length to breadth), and pyrenoid numbers per chloroplast. At Macquarie I.,

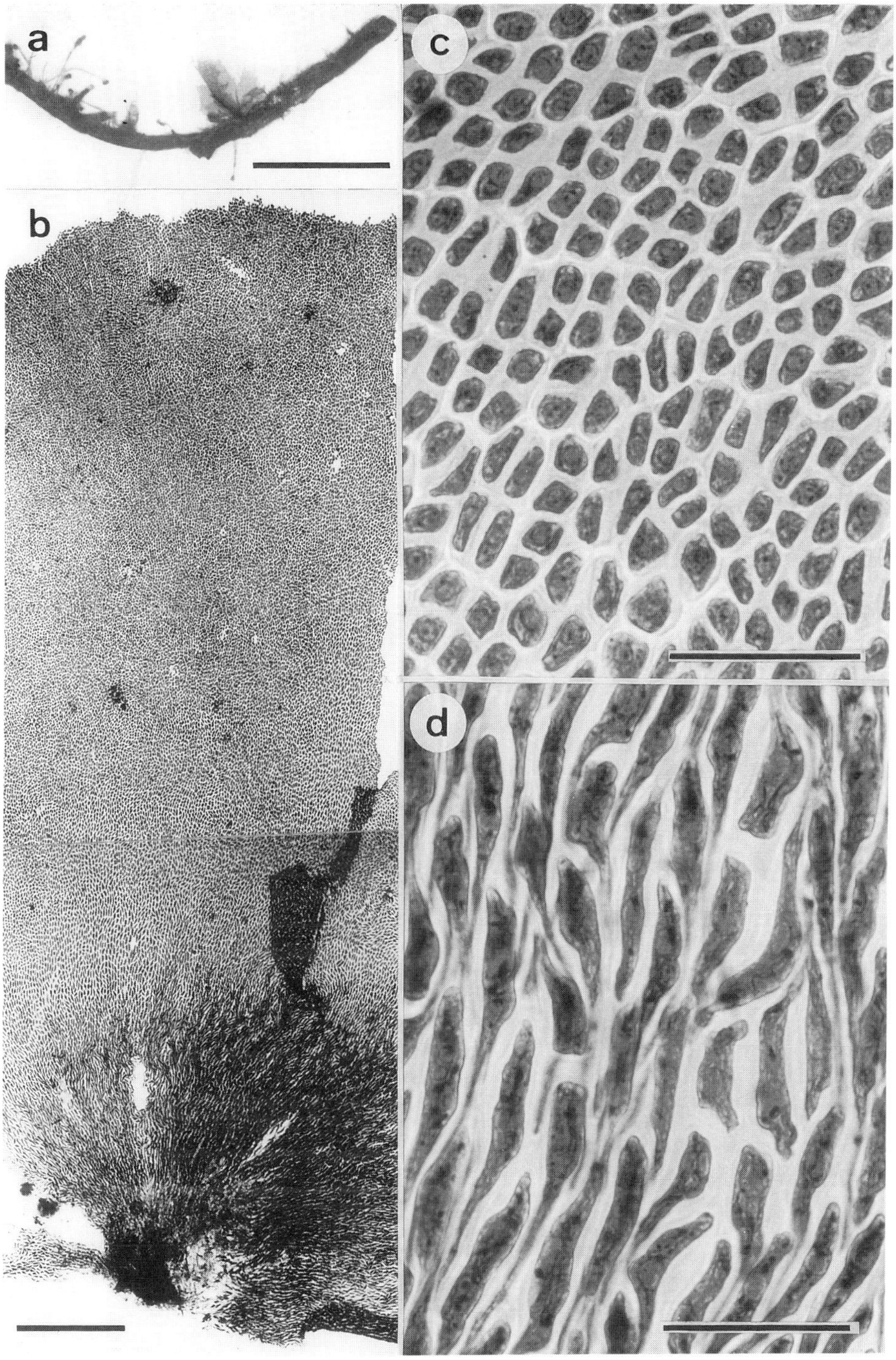

Fig. 14 *Monostroma sp.* 'b': MELU, R 1284. **a.** Epiphytic habit on *Desmarestia chordalis*. **b.** Frond surface. **c.** Surface view of mid-thallus cells. **d.** Surface view of cells near thallus base. Scales: a = 1 cm; b = 500 μm; c, d = 50 μm.

three species have been reported, *E. gunniana* (Zinova, 1958) and *E. bulbosa* and *E. intestinalis* (Kenny & Haysom, 1962), and three morphologically distinct forms are currently recognized. The Tasmanian entity *E. gunniana* J. Agardh, however, is actually *Blidingia minima* (Womersley, 1956). Two of the forms from Macquarie I. are similar to two morphological variations of *E. bulbosa* illustrated by Zaneveld (1969).

Enteromorpha bulbosa (Suhr) Montagne, 1846: 3

Figs 15, 16

BASIONYM: *Solenia bulbosa* Suhr, 1839: 72.

SYNONYMS: see Papenfuss, 1964*a*.

TYPE LOCALITY: Peru.

TYPE: L (L. 1391, sheet no. 40).

SOUTHERN OCEAN DISTRIBUTION: Antarctic Peninsula (Gain, 1912; Skottsberg, 1953; Lamb & Zimmermann, 1977); Argentina (Boraso, 1975); Auckland, Campbell and Chatham Is (V. J. Chapman, 1956); Îles Bouvet and Kerguelen (Reinbold, 1907); Îles Crozet (Levring, 1944); Falkland Is (Cotton, 1915); Fuegia (Hariot, 1907); Macquarie I. (Law & Burstall, 1956; Kenny & Haysom, 1962); South Georgia (Hylmö, 1919).

MATERIAL EXAMINED: Macquarie I.: collections MA-134, 217; Herbarium specimens: MELU, R 3265, 3273.

DESCRIPTION: Plants to 3 cm tall, with fronds basally branched, remaining narrow and tubular throughout, or dilating, rupturing above, and becoming cornucopia-shaped (Fig. 15a). Surface view of upper thallus with cells 11–17 μm diameter either disordered or arranged in rows perpendicular to one another (Fig. 15b); lower thallus cells either isodiametric (to 18 μm diameter) or broader (18 μm) than long (10 μm), arranged in vertical rows parallel to thallus axis (Fig. 15c). Upper thallus 45–50 μm thick, lower thallus 60–65 μm thick. Cells from upper and lower thallus 3–4 times tall as wide (Fig. 15e, f). Stipe to 450 μm diameter, filled with rhizoids (Fig. 15d). Reproductive structures unknown.

Enteromorpha bulbosa grows abundantly in upper intertidal to supratidal tide pools and sometimes is associated with the distinctly longer, unbranched *E. intestinalis*.

NOTES: Macquarie plants are shorter than those (10–12.5 cm tall) originally described by Suhr (1839), but otherwise the two agree in most respects, including the lack of adherence to paper upon drying. Hylmö (1938) stated that cells of stipes and narrow branches are usually aligned in longitudinal rows, whereas ones from broader thallus portions are more often disordered, and he distinguished this species from most others in the genus because the aligned cells are broader than long when seen in surface view. This feature is only partly distinctive for plants from Macquarie I., as cell breadth to length ratios are highly variable. Boraso (1975) distinguished *E. bulbosa* from *E. intestinalis* because the latter is unbranched. V. J. Chapman (1956), however, reports some branching in *E. intestinalis* and thinks that it may be better to consider *E. bulbosa* as a distinct variety of *E. intestinalis*. In the present study, *E. intestinalis* is circumscribed as unbranched. *Enteromorpha bulbosa* is maintained as a distinct species that may vary morphologically from tubular monostromatic fronds, through cornucopia-like saccate blades (Fig. 15a), to compressed thalli with distromatic portions (Fig. 16a–g). Plants with distromatic portions may belong to *E. compressa* (Linnaeus) Greville, which also has compressed fronds of similar thickness to Macquarie plants (48 μm for *E. compressa* var. *compressa* (Bliding, 1963)), but tentatively they are placed in *E. bulbosa* because of similarities in basal branching, cell height to width ratios, and rhizoid abundance inside stipes.

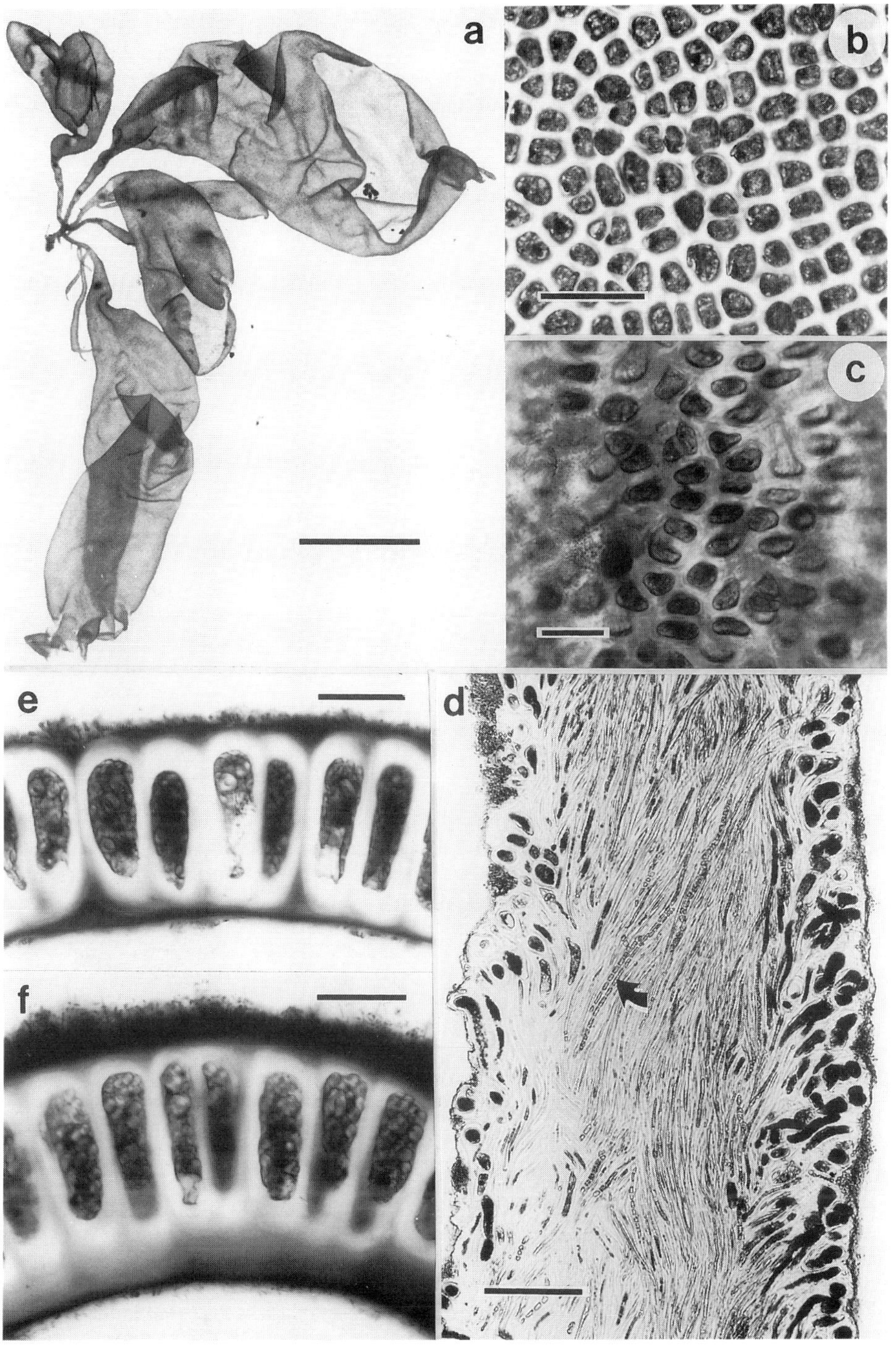

Fig. 15 *Enteromorpha bulbosa* (Suhr) Montagne: MELU, R 3273. **a.** Saccate habit. **b.** Surface view of upper thallus. **c.** Surface view of stipe. **d.** L.S. stipe filled with rhizoids and unidentified endophytic filaments (arrow). **e.** T.S. upper thallus. **f.** T.S. stipe. Scales: a = 5 mm; b, c = 40 μm; d = 100 μm; e, f = 20 μm.

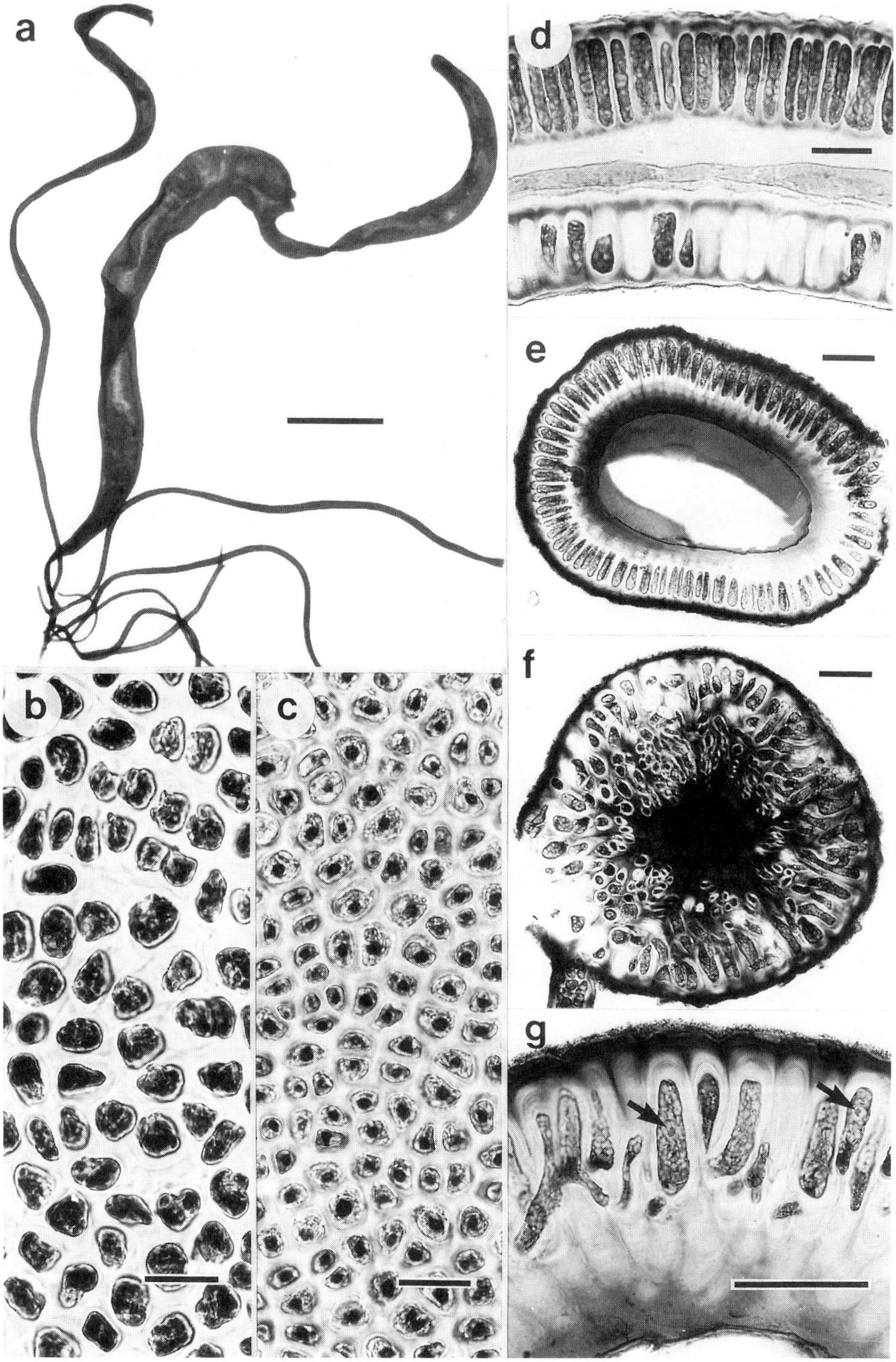

Fig. 16 *Enteromorpha sp.* cf. *E. bulbosa:* MELU, R 3265. **a.** Thallus habit with compressed fronds. **b.** Surface view of upper thallus. **c.** Surface view of lower thallus. **d.** T.S. upper thallus. **e.** T.S. lower thallus. **f, g.** T.S. stipe. **g.** Detail of cells showing one pyrenoid per cell (arrows). Scales: a = 2 mm; b–d = 20 μm; e–g = 50 μm.

Enteromorpha intestinalis (Linnaeus) Nees, 1820: index [2]

Fig. 17

BASIONYM: *Ulva intestinalis* Linnaeus, 1753: 1163.

SYNONYMS: see Papenfuss, 1964*a*.

TYPE LOCALITY: probably northern Europe.

TYPE: presumably in LINN (Koster, 1969), but specimens were not seen.

DISTRIBUTION: This species is known from most temperate to colder waters. In the Southern Ocean, it is reported from: Argentina (Boraso, 1975); Auckland Is (see Papenfuss, 1964*a*); Falkland Is (Cotton, 1915); Fuegia & Îles Kerguelen (Hooker, 1847); Macquarie I. (Kenny & Haysom; 1962); Tristan da Cunha (Baardseth, 1941).

MATERIAL EXAMINED: Macquarie I.: collections MA-133, 217; Herbarium: MELU, R 3272, 3274.

DESCRIPTION: Thalli to 15 cm tall, unbranched, composed of compressed or open tubes twisted into irregular spirals, rhizoidally attached (Fig. 17a). Frond surfaces with cells of mid-thallus rounded to angular, 4.5–6.5 × 7–8 μm, often paired, aligned in short, irregularly arranged rows (Fig. 17b); cells of lower thallus sometimes in vertical files, with cell breadth (= file breadth: 8–9 μm) > length (4 5 μm) (Fig. 17c). Blade transections 35–40 μm thick in mid-thallus, with cells 2–3 times as tall as wide (Fig. 17d); 55–60 μm thick in lower thallus, with cells of nearly equal height to width (Fig. 17e). Rhizoids produced internally from cells of lower thallus. Reproductive structures not observed.

Plants commonly become entangled to form large clumps or mats on rocky surfaces that line shallow pools in the high intertidal, where salinities have been measured between 17 and 22 ppt.

NOTES: Link (1820) described *Enteromorpha (Ulva) intestinalis* as having wrinkled thalli that can be hairy or spinose, compressed, sac-like, or fine and thread-like. According to more recent descriptions, the thallus may be branched (V. J. Chapman, 1956) or strictly simple (Abbott & Hollenberg, 1976; Womersley, 1984). Bliding (1963) noted the presence of small proliferations from the lower thallus but still considered the fronds as simple.

Hylmö (1938) identified nearly all Subantarctic *Enteromorpha* plants found during the Swedish expedition of 1907–9 as *E. bulbosa*, which he characterized as having stipes and narrow branches with vertical rows of cells whose breadths exceed their lengths. Using Hylmö's criteria, the plants described here would be considered as a form of *E. bulbosa* rather than *E. intestinalis*. Because of consistent differences in habit, however, unbranched thalli are identified as *E. intestinalis* and distinguished in the present account from the branched thalli of *E. bulbosa*.

Ulva Linnaeus, 1753: 1163 (*nom. cons.*)

Type species: *Ulva lactuca* Linnaeus, 1753: 1163.

Thalli are foliose, generally broadly expanded, distromatic throughout (excepting some monostromatic margins in which the cell layers have not fused) and attached by multinucleate basal rhizoids. All non-rhizoidal cells are uninucleate and contain a single, cup-shaped or sheet-like plastid with one to several pyrenoids. Life histories have isomorphic phases, with sporophytes producing tetraflagellate zoospores and gametophytes bearing biflagellate anisogametes. Gametes may develop parthenogenetically into gametophytes of their own sex (Bliding, 1968). *Ulva* develops from a monostromatic tube that becomes compressed into a distromatic blade and thus is distinguished from the only other distromatic foliose member of the Ulvales (*Chloropelta*) that arises from a distromatic tube splitting open to produce a two-cell thick leafy blade (Tanner, 1980).

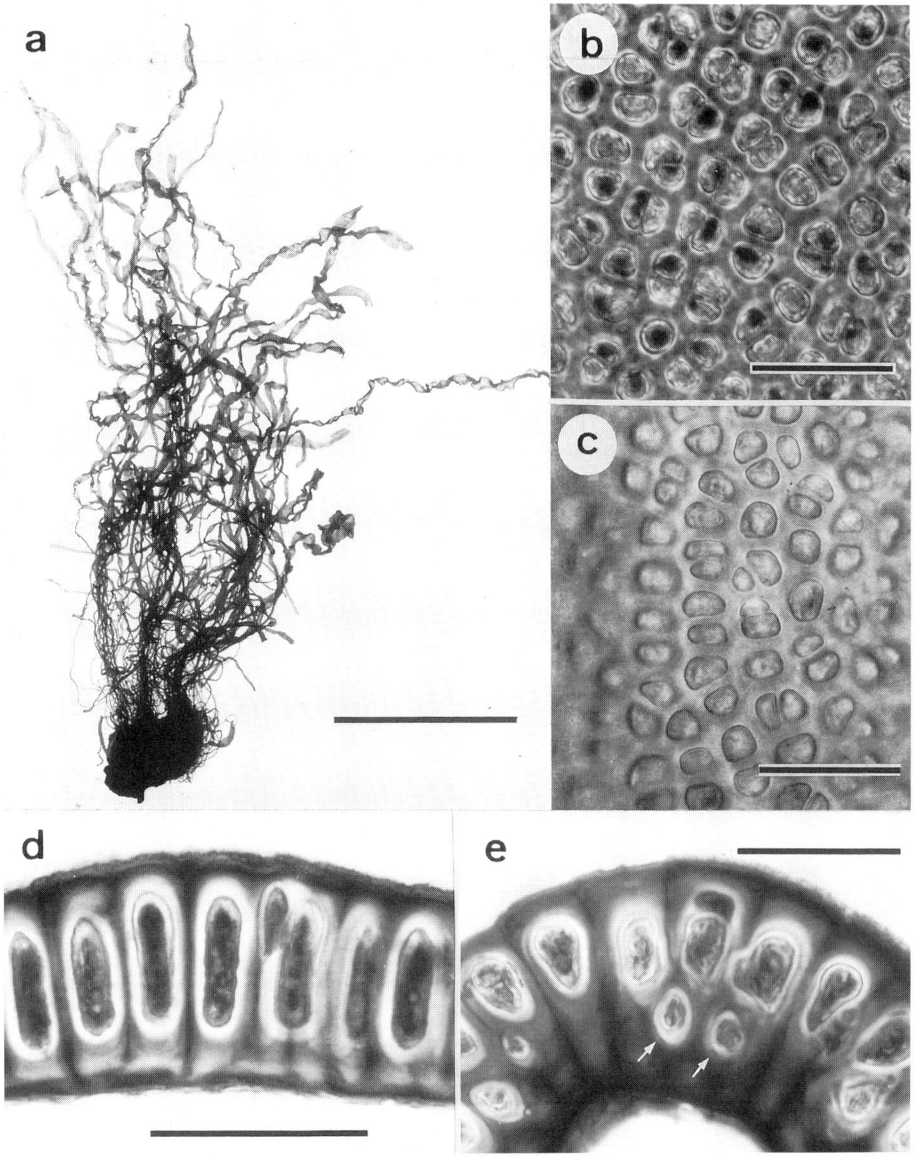

Fig. 17 *Enteromorpha intestinalis* (Linnaeus) Nees: MELU, R 3272, 3274. **a.** Habit. **b.** Surface view of cells from mid-thallus. **c.** Surface view of cells from lower thallus. **d.** T.S. mid-thallus. **e.** T.S. stipe showing rhizoids (arrows). Scales: a = 3 cm; b, c = 30 μm; d, e = 40 μm.

Species in *Ulva* have been distinguished by Koeman & Hoek (1981) using blade morphology (including colour and texture), cell patterns on blade surfaces, structure of blade bases, pyrenoid numbers in vegetative cells, superficial views of chloroplasts, cell sizes, cell height to width ratios as seen in thallus transections, thallus thickness, young germling morphologies, and modes of reproduction. Bliding (1968) did not separate species using basal structures of thalli, but he considered interfertility as a taxonomically useful character along with the others mentioned above. In the present treatment, criteria dealing with reproduction, interfertility, and development have not been considered.

Papenfuss (1964*a*) listed seven species from the Southern Ocean, with one represented by two varieties. Of these species, only two occur at Macquarie I.

Ulva angusta Setchell & Gardner, 1920: 283

Figs 18, 20a–c

SYNONYM: *Enteromorpha angusta* (Setchell & Gardner) Doty, 1947: 20.

TYPE LOCALITY: San Francisco, California.

HOLOTYPE: UC, no. 205623 (= Gardner no. 4430). (N.B. type specimens illustrated by Setchell & Gardner (1920, pl. 27)).

SOUTHERN OCEAN DISTRIBUTION: Macquarie I. (Zinova, 1958).

MATERIAL EXAMINED: Macquarie I.: collections MA-133, 140, 227, 271, 381; Herbarium specimens: MELU, R 0893, 0895, 3263, 3266, 3267.

DESCRIPTION: Thalli linear, strap-like, usually unbranched, to 55 cm tall by 3 cm wide, arising from cuneate bases, having straight or ruffled margins above (Fig. 18a, b) and composed largely of uninucleate cells with single cup-shaped chloroplasts, each containing one pyrenoid (Fig. 18c, d). Upper thallus 60–65 µm thick, with cell outlines on blade surface mostly rounded, 7– 12 µm diameter (Fig. 20a). Mid-thallus similarly thick, with cell outlines on blade surface more angular, 9–17 µm diameter (Fig. 20b). Lower thallus 85–95 µm thick, with rhizoids between surface layers (Fig. 18e). Cells of lower thallus with angular outlines, and either small or large (18– 55µm) diameter (Fig. 20c). Sectional views of upper and mid-thallus with box-shaped cells 18– 30 µm tall. Sectional views of lower thallus with more rectangular cells (20–25 µm × 35–40 µm) producing rhizoids from innermost surfaces (Fig. 18e). Reproductive structures not observed.

Plants grow in tidepools *c.* 1 m above the *Durvillaea* zone, but also commonly occur on rocky outcrops 5–16 m deep. The subtidal plants often cover isolated boulders located in sand plains or grow on submerged rocky reefs in association with *Delesseria* and *Polysiphonia* under canopies of *Desmarestia* and *Macrocystis.*

NOTES: Macquarie plants grow more than three times the size of the strictly intertidal plants that Setchell & Gardner (1920) originally described from California. Other measurements of Macquarie thalli generally are greater than those presented in the original description of the species, but Abbott & Hollenberg (1976) have since reported on Californian specimens nearly twice the length of those from Macquarie I. These authors also speculated on the alga's probable occurrence in the California subtidal. The presence of *U. angusta* in the Southern Ocean implies a perplexing disjunction in distribution, but on the basis of thallus habit, cross-sectional views of thalli and pyrenoid numbers per chloroplast, there is no reason to taxonomically separate the two geographic groups of plants.

Ulva rigida C. Agardh, 1823: 410

Figs 19, 20d–f

SYNONYMS: see Bliding (1968); Papenfuss (1964*a*) (as *U. rigida* (C. Agardh) Thuret).

TYPE LOCALITY: Cape of Good Hope, S. Africa.

TYPE: LD. (Specimen no. 14449 has been tentatively selected for lectotypification by R. B. Searles (10 October 1975) in unpublished Lund Museum typification records).

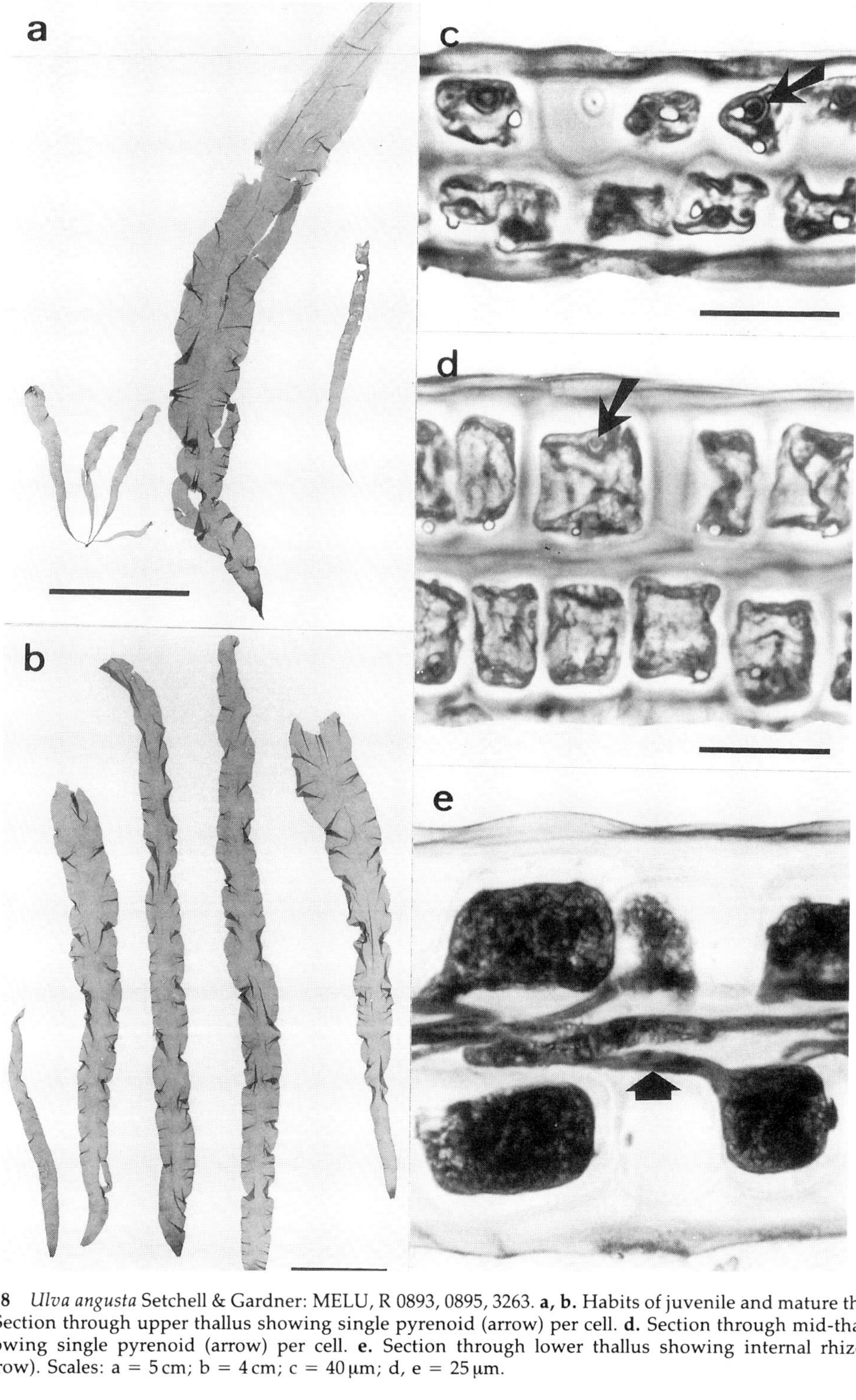

Fig. 18 *Ulva angusta* Setchell & Gardner: MELU, R 0893, 0895, 3263. **a, b.** Habits of juvenile and mature thalli. **c.** Section through upper thallus showing single pyrenoid (arrow) per cell. **d.** Section through mid-thallus showing single pyrenoid (arrow) per cell. **e.** Section through lower thallus showing internal rhizoids (arrow). Scales: a = 5 cm; b = 4 cm; c = 40 μm; d, e = 25 μm.

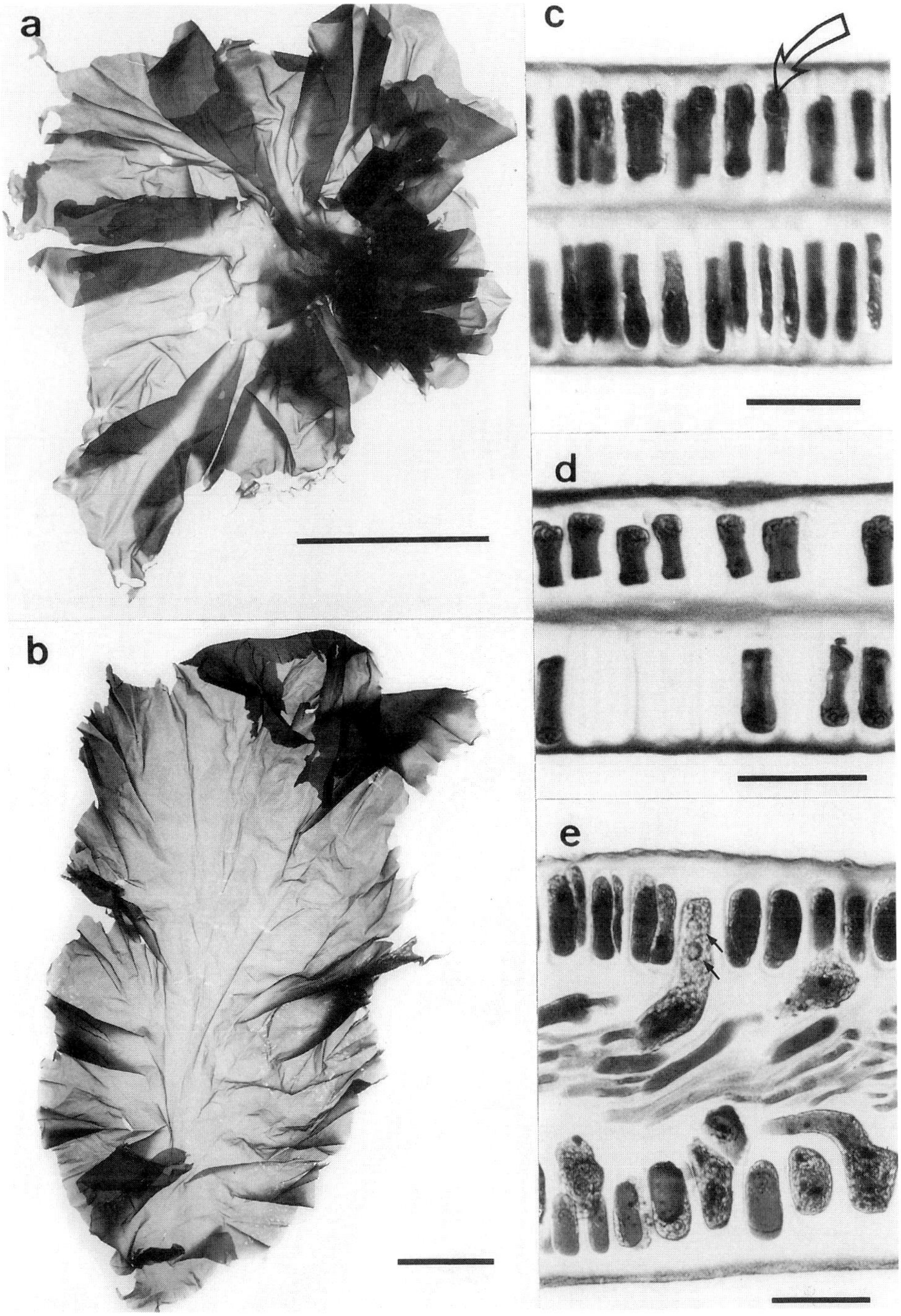

Fig. 19 *Ulva rigida* C. Agardh: MELU, R 1699, 3262. **a, b.** Frond habits. **c.** Section through upper thallus near blade margin showing at least one pyrenoid (arrow) per cell. **d.** Section through mid-thallus. **e.** Section through lower thallus showing internal rhizoids and two pyrenoids (arrows) in one chloroplast. Scales: a, b = 5 cm; c = 30 μm; d, e = 50 μm.

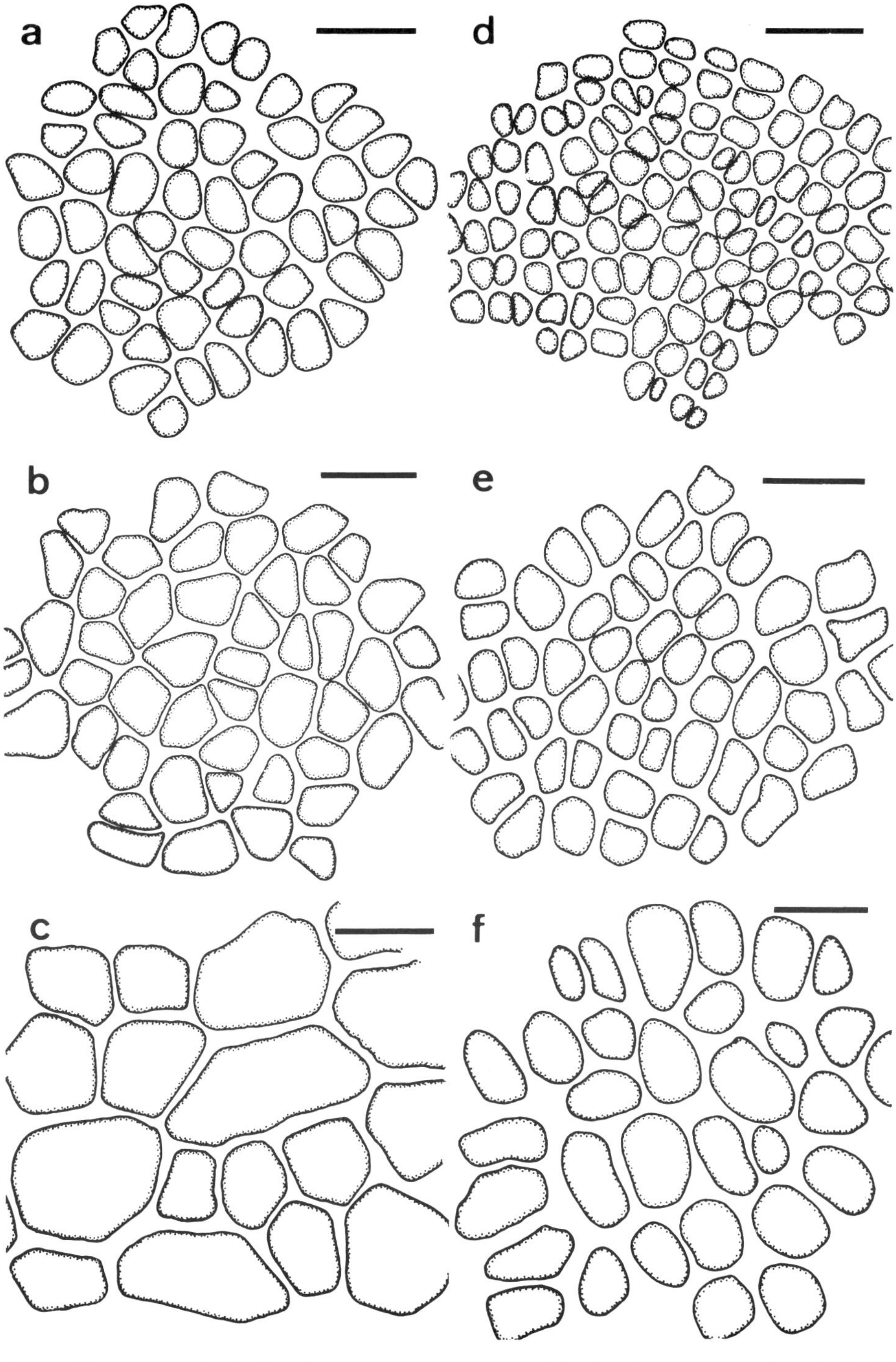

Fig. 20a–c *Ulva angusta* Setchell & Gardner: MELU, R 3263. **d–f** *Ulva rigida* C. Agardh: MELU, R 3262. **a–f.** Surface views of cells from upper (a, d), middle (b, e), and lower (c, f) portions of thalli. Scales: a–f = 25 μm.

SOUTHERN OCEAN DISTRIBUTION: Auckland Is (Laing, 1909); Campbell I. (Levring, 1945); Fuegia (Hylmö, 1919); Falkland Is (Cotton, 1915); Îles Crozet and Kerguelen (Levring, 1944); Macquarie I. (Zinova, 1958); Tristan da Cunha (Baardseth, 1941).

MATERIAL EXAMINED: Macquarie I.: collections MA-46, 83, 84, 111, 364, 381; Herbarium specimens: MELU, R 1699, 3262, 3264.

DESCRIPTION: Thalli to 40 cm tall, broad and foliose, usually irregularly lobed, expanding immediately above and obscuring points of attachment (Fig. 19a, b), composed of uninucleate cells with single chloroplasts each containing 1–2 pyrenoids (Fig. 19c, e). Upper thallus 75–80 μm thick, with surface view of mostly rounded to square-shaped cells (7–9 μm diameter) arranged in rows perpendicular to one another (Fig. 20d). Mid-thallus 95–105 μm thick, with surface view of angular cells (12–16 μm diameter) in non-perpendicular rows (Fig. 20e). Lower thallus to 200 μm thick, filled with rhizoids between surface layers. View of lower thallus surfaces with rounded (11–27 μm diameter) disordered cells (Fig. 20f). Height to width ratios of cells from sections of upper thallus 3.5–5 (Fig. 19c), mid-thallus 2.5–3.5 (Fig. 19d), lower thallus 2–3 (Fig. 19e). Lower thallus cells generally producing rhizoidal elongations from inner surfaces. Reproductive structures not observed.

Plants have crisp, bright green fronds that are often the dominant foliage of pools in the mid- to upper intertidal. *Ulva rigida* also may grow attached to rocky surfaces in the lower intertidal to shallow subtidal but is less common subtidally than *U. angusta*.

NOTES: *Ulva rigida* is a component of many temperate to cold temperate marine floras and has been described from the Atlantic (Bliding, 1968; Koeman & Hoek, 1981), Pacific (Abbott & Hollenberg, 1976) and Southern (Boraso, 1977; Hylmö, 1919, 1938) Oceans. Plants from Macquarie I. are similar to ones described by Hylmö (1919) from Fuegia and the Falkland Is., but differ from most Northern Hemisphere specimens which generally have fronds with macroscopic ribs at their bases, small teeth along their margins and less elongate cells (as seen in sectional view of the thallus) containing a greater number of pyrenoids per chloroplast (Koeman & Hoek, 1981). The Southern Ocean plants presently ascribed to *U. rigida* might best be separated from this chiefly Northern Hemisphere taxon and described as a new species, but such a move awaits more thorough analyses of Agardh's type specimen and a better understanding of morphological plasticity within the species.

Prasiolales

Multicellular plants comprising uninucleate cells with axile stellate chloroplasts, each containing one centrally positioned pyrenoid, are unique among the Chlorophyta and are placed in a single family (Prasiolaceae) and order (Prasiolales; Kornmann & Sahling, 1977). Bold & Wynne (1978), however, consider these cytological features insufficient to justify separate ordinal status and therefore retain the Prasiolaceae in the Ulvales.

Plants consist of either filaments, packet-like clusters of cells, or foliose fronds, the last of which superficially resemble those of monostromatic members of the Ulvales. Constituents of the Prasiolales differ, however, in having blade surfaces with distinctive, mosaic-like patterns of cell blocks with the smallest clusters of cells being diads or tetrads. Reproduction is often through vegetative fragmentation, akinetes, or aplanosporangia. Sexual reproduction is mostly unknown, although oogamy has been reported for *Rosenvingiella* (Kornmann & Sahling, 1974) and *Prasiola* (Friedmann & Manton, 1960).

Prasiolaceae Blackman & Tansley, 1902: 138

Members of this family live in either terrestrial, freshwater or marine environments. In many instances, plants are distinctive of areas with high nitrogen levels such as near bird colonies. Species show considerable morphological plasticity which, combined with poor understanding of life histories, has led to some taxonomic confusion at the generic level. Depending upon one's interpretation of life histories, three (*Prasiola, Prasiococcus, Prasiolopsis*) or five genera (additionally including *Rosenvingiella* & *Schizogonium*) may be recognized for the family, totalling approximately 40 species.

Prasiola is the only genus in the family listed by Papenfuss (1964*a*) from the Antarctic and Subantarctic Zones, although *Rosenvingiella* has recently been seen growing in the Antarctic intertidal near the Vestfold Hills (pers. obs.).

Prasiola (C. Agardh) Meneghini, 1838: 360

Lectotype: *P. crispa* (Lightfoot) Kützing, 1843: 295 (see Silva, 1980*b*).

Thalli of *Prasiola* range from simple or branched uniseriate filaments (hormidium stage) through pluriseriate filaments (schizogonium stage) to monostromatic blades (prasiola stage) (Knebel, 1935). Foliose thalli, however, usually are most prominent. Typically, species are separated on: habitat preference (freshwater, marine, or terrestrial); thallus shape and dimensions (pertaining mostly to foliose portions); and means of attachment (either by rhizoids, a discoid base subtending stipes, or along thallus margins). These three criteria, however, are seemingly too variable for accurately delimiting species and differences in life histories apparently better separate taxa (Kornmann & Sahling, 1974).

Prasiola is a common component of supratidal communities in the Antarctic and Subantarctic (Delépine & Hureau, 1966; Delépine *et al.*, 1966), where at least 14 different entities (either species or varieties) have been reported. Most of these algae grow in freshwater habitats and have been listed by Prescott (1979), although Hylmö's (1938) record of a new freshwater species (*P. skottsbergii*) has been omitted. Papenfuss (1964*a*) listed four species and one variety from southern marine waters, some of which also are included in Prescott's list as freshwater species. *Prasiola crispa* subsp. *antarctica* is the only member of the genus known from Macquarie I. It could conceivably be classified as either a freshwater, marine or terrestrial species.

Prasiola crispa (Lightfoot) Kützing subsp. *antarctica* (Kützing) Knebel, 1935: 20

Fig. 21

Basionym: *Prasiola antarctica* Kützing 1849: 473.

Synonyms: see Papenfuss, 1964*a*.

Type locality: Antarctic Peninsula.

Type: presumably in L, but specimens were not seen.

Distribution: Antarctic Peninsula (Hooker, 1846–47); Commonwealth Bay (Lucas, 1919); Falkland Is (Hooker, 1846–47; Cotton, 1915); Fuegia (Hariot, 1889); Îles Kerguelen (Reinbold, 1908); Macquarie I. and Vestfold Hills, Antarctica (pers. obs.); South Georgia (?): listed in Reinsch's (1890) index to the algal flora from South Georgia, but not discussed in the text; South Orkney Is (Holmes, 1912—as *Prasiola crispa*); Victoria Land (Hylmö, 1919).

Material examined: Macquarie I.: Herbarium specimen: MELU, R 1475 (including filamentous and foliose stages, non-infected and fungal-infected thalli).

Description: Plants either filamentous or more prominent and foliose, with blades to 1.5 cm tall × 3 cm broad, orbicular to kidney-shaped (Fig. 21e), pale to bright green, 26–28 μm thick (Fig. 21f). Frond surfaces comprising cells in tetrads clustered in yet larger groups forming mosaic pattern of squares (Fig. 21g). Sectional view of blades with cells rectangular, 5–11 μm broad × 19–20 μm tall, appearing similar to cells in filamentous (hormidium) stage (Fig. 21 h); cells from filaments

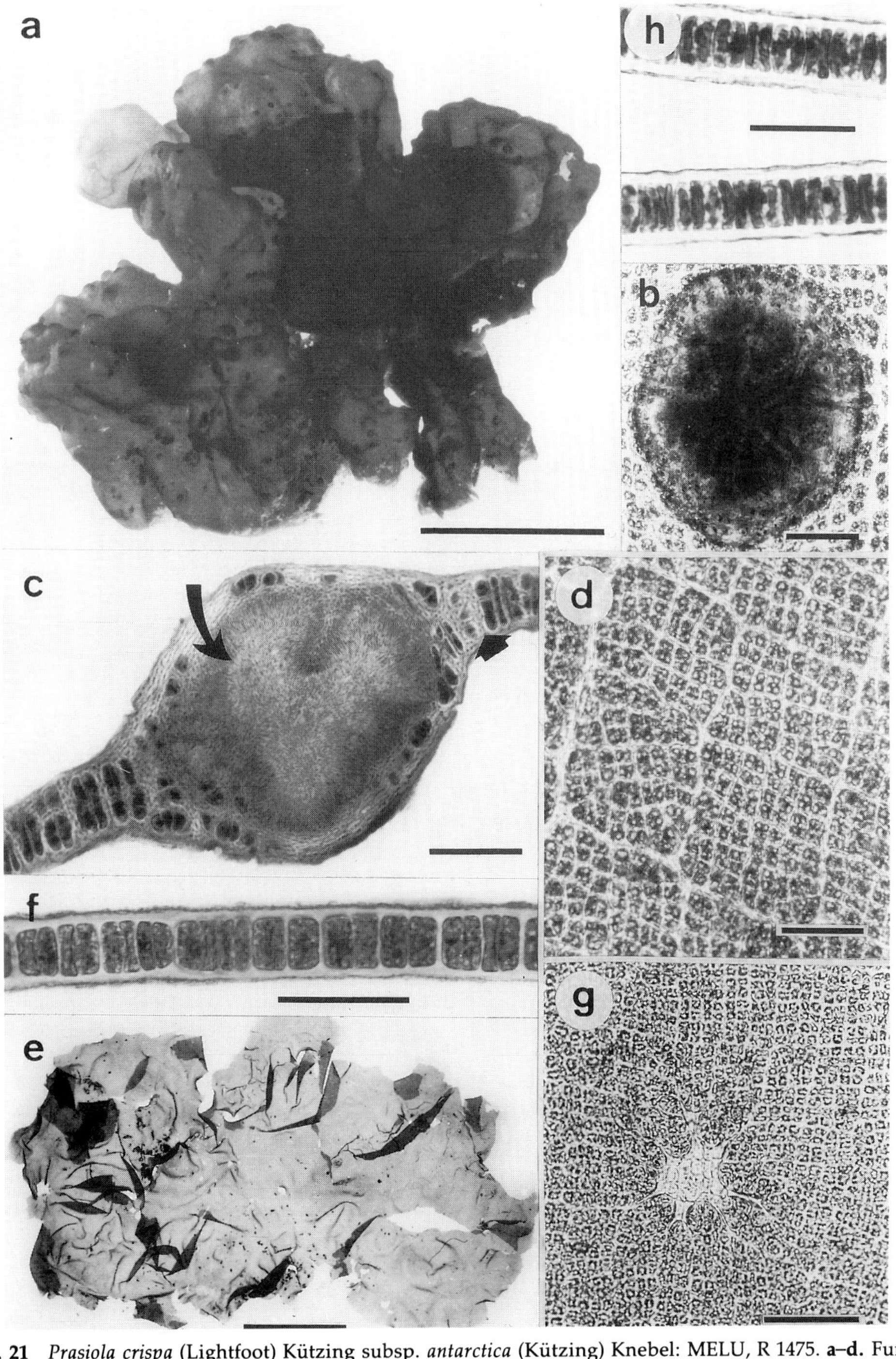

Fig. 21 *Prasiola crispa* (Lightfoot) Kützing subsp. *antarctica* (Kützing) Knebel: MELU, R 1475. **a–d.** Fungus-infected thallus. **a.** Globose habit. N.B. Black spots on thallus surface are ascocarps. **b.** Close-up surface view of ascocarp. **c.** Section through frond showing ascocarp (long arrow) and fungal hyphae (short arrow) between host cells. **d.** Frond surface. **e–g.** Non-infected thallus. **e.** Habit. **f.** Section through frond. **g.** Frond surface. **h.** Filaments of hormidium stage. Scales: a, e = 5 mm; b–d, f, g = 50 μm; h = 25 μm.

sometimes biconcave in side view. Reproductive structures unknown.

Plants grow in the supratidal, usually near penguin or cormorant colonies where guano is abundant, and form thick mats atop granitic surfaces. Although *Prasiola* at Macquarie I. would rarely be immersed in saltwater and regularly is exposed to drying winds and periodic deluges of freshwater, it undoubtedly is a euryhaline alga because of its constant exposure to sea-spray. In the Vestfold Hills (Antarctica), however, this species often grows 30–50 m above sea-level (pers. obs.) in guano-rich melt-water streams below bird colonies, where sea-spray probably would have little effect.

NOTES: Thalli of *Prasiola* generally become infected with the ascomycete *Kohlmeyera complicatula* (Nylander) Schatz = *Turgidosculum complicatulum* (Nylander) J. Kohlmeyer et E. Kohlmeyer (1979); (J. Kohlmeyer, pers. comm.) and as a result develop globose fronds to 18 mm across, with thickened and convoluted surfaces (Fig. 21a). Thallus colour changes from bright to dull green, almost olive. Black spots (ascocarps) 160–225 μm across (Fig. 21b) occur on thallus surfaces, and when sectioned, fungal hyphae can be seen filling the fruiting structures and interwoven among host cells (Fig. 21c). Thallus surfaces (of fungus-infected plants) away from ascocarps show prominent mosaic patterns of cell groupings to various orders, the basic unit being a tetrad (Fig. 21d). Infected thalli have been widely known as the lichen genus *Mastodia* (Hooker & Harvey, 1845; Hariot 1887*b*), but Kohlmeyer & Kohlmeyer (1979) do not consider this a true lichen symbiosis, but rather a mycophycobiosis. The alga in this relationship usually is considered to be *Prasiola tessellata* (Hooker f. & Harvey) Kützing. Infected plants from Macquarie I. closely resemble the anatomy of infected *P. tessellata* from Cape Horn (Hariot, 1889), but they differ in having blade surfaces with mostly rounded cells (vs. angular ones for *P. tessellata*), a feature considered diagnostic for separating the two species (Knebel, 1935).

Acrosiphoniales

Members of this order attach rhizoidally, have branched or unbranched, uniseriate gametophytic thalli with either localized or diffuse growth. Cells are uninucleate or multinucleate and have a single, perforate, parietal chloroplast with many pyrenoids. Plastids strongly resemble the fused masses of discoid chloroplasts seen in morphologically similar species of the Cladophorales, but taxa from both orders are separated readily on differences in pyrenoid-starch organization, cell wall components, types of cell division, and life histories. Acrosiphonalean pyrenoids are surrounded by many polygonal-shaped starch grains that collectively form tessellate-surfaced structures differing from those bilobed, starch-pyrenoid complexes typical for cladophoralean algae (van den Hoek, 1981). Cell walls of acrosiphonalean plants, with the main components of mannan, xylan, glucan and rhamnan, are distinct from the principally crystalline cellulose I walls of cladophoralean plants (van den Hoek, 1981). Hudson & Waaland (1974) described a special pattern of nuclear divisions in multinucleate thalli, in which a ring of nuclei forms adjacent to a developing septum and karyokinesis occurs simultaneously with cytokinesis. Life histories typically have heteromorphic generations with unicellular sporophytic (*Codiolum* or chlorochytrium-like) stages.

A *'Codiolum'* stage occurs in several orders (Acrosiphoniales, Codiolales, Ulotrichales, and Monostromatales) and has prompted Kornmann (1973) to create the class Codiolophyceae to contain them. The first three of these orders include filamentous green algae. The Ulotrichales, broadly interpreted by Abbott & Hollenberg (1976) as comprising branched and unbranched filamentous plants and blade-like thalli of the Monostromataceae, is narrowly construed here, as in Bold & Wynne (1978), to include only filamentous unbranched thalli with uninucleate cells. Additionally, ulotrichalean plants mostly have modified basal cells instead of producing rhizoids, as typically occurs in acrosiphonalean algae.

As circumscribed here and by Silva (1982), members of Kornmann's Codiolales constitute the family Codiolaceae, which together with the Acrosiphoniaceae form the Acrosiphoniales. Both families occur at Macquarie I.

Acrosiphoniaceae Jónsson, 1959: 1567

Members of this family have highly branched thalli, whereas members of the Codiolaceae remain unbranched. Two genera, *Acrosiphonia* and *Spongomorpha*, constitute the Acrosiphoniaceae and both have been reported from the Southern Ocean. *Acrosiphonia* and *Spongomorpha* have some morphologically similar species but are distinguished by their respectively multinucleate and uninucleate cells. Because species of both genera can appear morphologically alike, records of either from other Southern Ocean localities need to be confirmed through nuclear studies. It seems likely that most records will represent one species of *Acrosiphonia* because reports of *Spongomorpha* either are dubious and believed to represent *A. pacifica* (Hylmö, 1938), do not pertain to a green alga but rather to a member of the Acrochaetiaceae (*Camontagnea*—Pujals, 1981), or have not been documented through nuclear studies (W. R. Taylor, 1939). *Acrosiphonia pacifica* has also been confirmed at widely separated localities, where it is dominant in the intertidal.

Acrosiphonia J. Agardh, 1846: 104

Type species: Not designated.

Thalli usually have hook-like determinate branches that, along with rhizoids, interlock main axes to form matted skeins. Branching is alternate to pectinate to irregular and sometimes verticillate along main axes. Growth is localized and synchronous in upper branches (Kornmann & Sahling, 1977). Cells are multinucleate and contain a single parietal perforate plastid with numerous pyrenoids, each surrounded by many polygonal, plate-like starch grains.

The total number of species in *Acrosiphonia* remains to be determined, such a count awaiting monographic study of the genus. A thorough survey of *Acrosiphonia* should better distinguish this genus from *Spongomorpha*. At present, some authors still consider members of the former as constituting a sub-section of the latter (*Index Nonimum Genericorum*). Possibly the most comprehensive account yet produced for *Acrosiphonia* is that by Kornmann (1962*b*), in which he reduced eight taxa into three morphologically distinct species. More recent studies of this genus have focused on thallus development (Kornmann, 1965) or life histories (Kornmann, 1970*a*, 1970*b*) of only a few Northern Hemisphere species.

The genus is represented at Macquarie I. by *A. pacifica*.

Acrosiphonia pacifica (Montagne) J. Agardh ***in*** Hohenacker 1857, no. 256

Fig. 22

Basionym: *Conferva pacifica* Montagne, 1842*a*: 16.

Synonyms: see Papenfuss, 1964*a* (as *Spongomorpha pacifica* (Montagne) Kützing).

Type locality: Auckland Is.

Type: Herb. Kützing, specimen no. 199, L (illustrated by Zaneveld, 1969, pl. 84).

Distribution: Auckland Is (Montagne, 1842*a*); Campbell I. (Laing, 1909; Levring, 1945); Chile (north to laguna San Rafael; Aisen Province: Levring, 1960); Falkland Is (Cotton, 1915); Fuegia (Hylmö, 1938); Heard I. (Chittleborough, 1956); Îles Crozet (Levring, 1944); Îles Kerguelen (Reinbold, 1907); Macquarie I.; Melchoir I. (Delépine, 1966); South Georgia (Reinsch, 1890—as *Cladophora arcta*); South Shetland Is (Moe, pers. comm.); Stewart I. (Adams *et al.*, 1974).

Material examined: Macquarie I.: collections MA-10, 11, 83, 84, 111, 129, 131, 134, 137, 213, 238, 244, 264, 381; Herbarium specimens: MELU, R 0506, 0815, 1367, 3261, 3270.

a
b
c
d
f
e

DESCRIPTION: Thalli composed of rope-like tangles of uniseriate filaments, forming skeins to 10 cm long (Fig. 22b) or short-cropped mats of dense hemispherical tufts 2–3 cm tall. Numerous hook-like determinate lateral branchlets (Fig. 22c) and interweaving descending rhizoids interlock filaments, maintaining spongy frayed-rope appearance of long skeins. Main axes with opposite or verticillate branching; laterals with pectinate branching. Branch apices rounded when immature (Fig. 22e), acute when mature (Fig. 22f). Cells of filaments multinucleate, 2–4 times as long as broad in mid-filament, 80–100 μm broad on average, with single reticulate parietal chloroplast containing many spherical pyrenoids (Fig. 22d). Pyrenoids surrounded by numerous starch grains resulting in tessellated structures (Fig. 22d, inset) approximately 8–10 times size of nuclei. Reproductive structures unknown.

This species is extremely common around Macquarie I. and usually grows on rocks in the mid- to upper-intertidal (Fig. 22a). It also occurs on limpet and chiton shells, becomes entangled with *Polysiphonia anisogona*, and grows as an epiphyte on coralline algae, *Delesseria lancifolia*, and *Schizoseris condensata*.

NOTES: There has been such confusion regarding the generic limits of *Acrosiphonia*, *Cladophora* and *Spongomorpha* that *A. pacifica* has been placed in each genus at one time or another (Hohenacker (1857), Kützing (1849), Kützing (1854), respectively). The diagnostic features of multinucleate cells, plastid type, and pyrenoid-starch organization have not been documented clearly in earlier accounts (Montagne, 1842*a*; Kützing, 1854), which tended to emphasize the less diagnostic hook-like lateral branchlets that occur in several genera.

Hylmö (1938) believed that material from the Southern Ocean previously referred to *Cladophora (Spongomorpha) arcta*, a hook-forming, filamentous green alga from the Northern Hemisphere, actually represented *A. pacifica*, and he suggested that this species might be distributed throughout the entire Subantarctic Zone. The confirmed presence of this species at the widely separated Macquarie and South Shetland Is (Moe, pers. comm.) supports Hylmö's belief that this alga is a circumpolar species.

Codiolaceae den Hartog, 1959: 111

Algae in this family attach rhizoidally and have unbranched filamentous thalli with multinucleate cells. Life histories generally are considered diplontic with a unicellular '*Codiolum*' sporophyte. The filamentous gametophyte is haploid and produces anisogametes; the diploid sporophyte produces quadriflagellete zoospores.

Urospora (including *Hormiscia*) alone represents the family and is known from intertidal environments in cold marine and brackish waters in both Northern and Southern Hemispheres.

Urospora Areschoug 1866: 15 (***nom. cons.***)

Type Species: *U. mirabilis* Areschoug, 1866: 16 [= *U. penicilliformis* (Roth) Areschoug, 1874: 4].

Urospora is common throughout the Antarctic and Subantarctic in a variety of forms mostly referable to *U. penicilliformis*. Lokhorst & Trask (1981), however, believe that one Antarctic record (Gain, 1912: as *U. penicilliformis*) represents the North Atlantic species *U. wormskioldii* (Mertens *in* Hornemann) Rosenvinge.

The broad range of thallus morphologies within this genus has allowed some taxonomic confusion with other green algae, such as *Ulothrix* and *Chaetomorpha*, which also have unbranched, uniseriate, filamentous thalli composed of cells each containing a single parietal reticulate chloroplast (Lokhorst & Vroman, 1974*a*, 1974*b*; Lokhurst, 1978). *Urospora*, however, has

Fig. 22 *Acrosiphonia pacifica* (Montagne) J. Agardh: MELU, R 0506, 0815, 3261, 3270. **a.** Habit of matted plants covering intertidal boulders. **b.** Thallus habit with tangled skeins of filaments. **c.** Branches with hook-like determinate lateral branches. **d.** Cells with abundant pyrenoids (arrow) in parietal reticulate chloroplasts. Inset: Detail of tessellated surface of pyrenoid surrounded by starch grains (arrowhead). **e.** Apices of immature branches. **f.** Apices of mature branches. Scales: b = 2 cm; c = 1 mm; d = 100 μm, inset = 5 μm; e, f = 500 μm.

a rhizoidal holdfast developed from the lower cells of erect filaments, whereas *Ulothrix* and *Chaetomorpha* both attach using single modified basal cells. The last two genera usually have life histories with isomorphic phases.

The genus is represented at Macquarie I. by *U. penicilliformis.*

Urospora penicilliformis (Roth) Areschoug, 1874: 4

Fig. 23

BASIONYM: *Conferva penicilliformis* Roth, 1806: 271.

SYNONYMS: see Lokhorst & Trask, 1981.

TYPE LOCALITY: Eckwarden, Germany.

TYPE: Herb. Agardh, LD, specimen no. 7417.

DISTRIBUTION: A cosmopolitan species found in both Northern and Southern Hemispheres, this alga probably occurs along most or all colonizable shores in the Antarctic and Subantarctic, but at present is reported only from: Antarctic Peninsula, Wilkes Land (Gain, 1912); Falkland Is, Fuegia (Hylmö, 1919); Îles Crozet (Skottsberg, 1953); Îles Kerguelen (Reinbold, 1908); Macquarie I.; Mawson Coast, Vestfold Hills (Ricker, unpubl. obs.); South Shetland Is (Moe, pers. comm.); Tasmania (Australia) (Womersley, 1984).

MATERIAL EXAMINED:

1. Macquarie I.: collections MA-111, 131, 278; Herbarium specimens: MELU, R 1764, 1891. Hasselborough Bay, upper littoral (Taylor, 8 January 1950): MELU, R 3314–3316 (= Taylor collection no. 540). Nugget Point, intertidal rocks (Taylor, 6 February 1951): MELU, R 3317– 3319 (= Taylor collection no. 536).
2. Heard I.: Atlas Cove, fouling buoy (Chittleborough, 14 February 1950): MELU, R 3313 (= Chittleborough specimen no. 816).
3. Antarctica: Vestfold Hills, growing on intertidal boulder (Ricker, 8 December 1981): MELU, R 2551. Mawson, on vertical rock wall 30 cm above sea-level (Ricker, 26 November 1981): MELU, R 2566, 2567.

DESCRIPTION: Plants filamentous, with unbranched, uniseriate axes to 5–6 cm long, solitary or clustered (Fig. 23a), often intertwining (Fig. 23b), forming epiphytic or epilithic tufts or lawns. Attachment by basipetal rhizoids from lower cells of filaments (Fig. 23c). Rhizoids produced inside or outside thickened, somewhat lamellated, cell walls of lower filament. Filament cells morphologically variable, ranging from ovoid to barrel-shaped to elongate and linear; degree of constriction between cells related to cell length, with elongate cells having little to no nodal constriction. Plastids of young cells parietal, band-shaped, smooth or lobed along periclinal edges of bands (Fig. 22d); plastids of mature cells more densely pigmented, parietal or axile and irregularly shaped, seemingly lobed, overlapping, reticulate. Pyrenoids present in young and old plastids, more numerous in latter but obscured by increased pigment concentration and abundance of free starch grains. All cells, excluding ones with rhizoids, potentially becoming zoosporingia (Fig. 23f); emptied sporangia occurring among mature and immature sporangia (Fig. 23g, h). Zoospores numerous, tetraflagellate, posteriorly pointed, laterally released through

---

Fig. 23 *Urospora penicilliformis* (Roth) Areschoug: MELU, R 1764, 1891. **a.** Habit of plants growing epiphytically on *Chaetangium fastigiatum*. **b.** Habit of unbranched filaments winding together to form rope-like strands. **c.** Thallus base with downward-growing rhizoids (arrow). **d.** Filaments with a single band-shaped chloroplast per cell. **e.** Zoospores being released from individual cells in the filament. **f.** Zoosporangia crowded with spores. **g, h.** Filaments with full and empty zoosporangia. Scales: a = 100 μm; b = 500 μm; c–h = 50 μm.

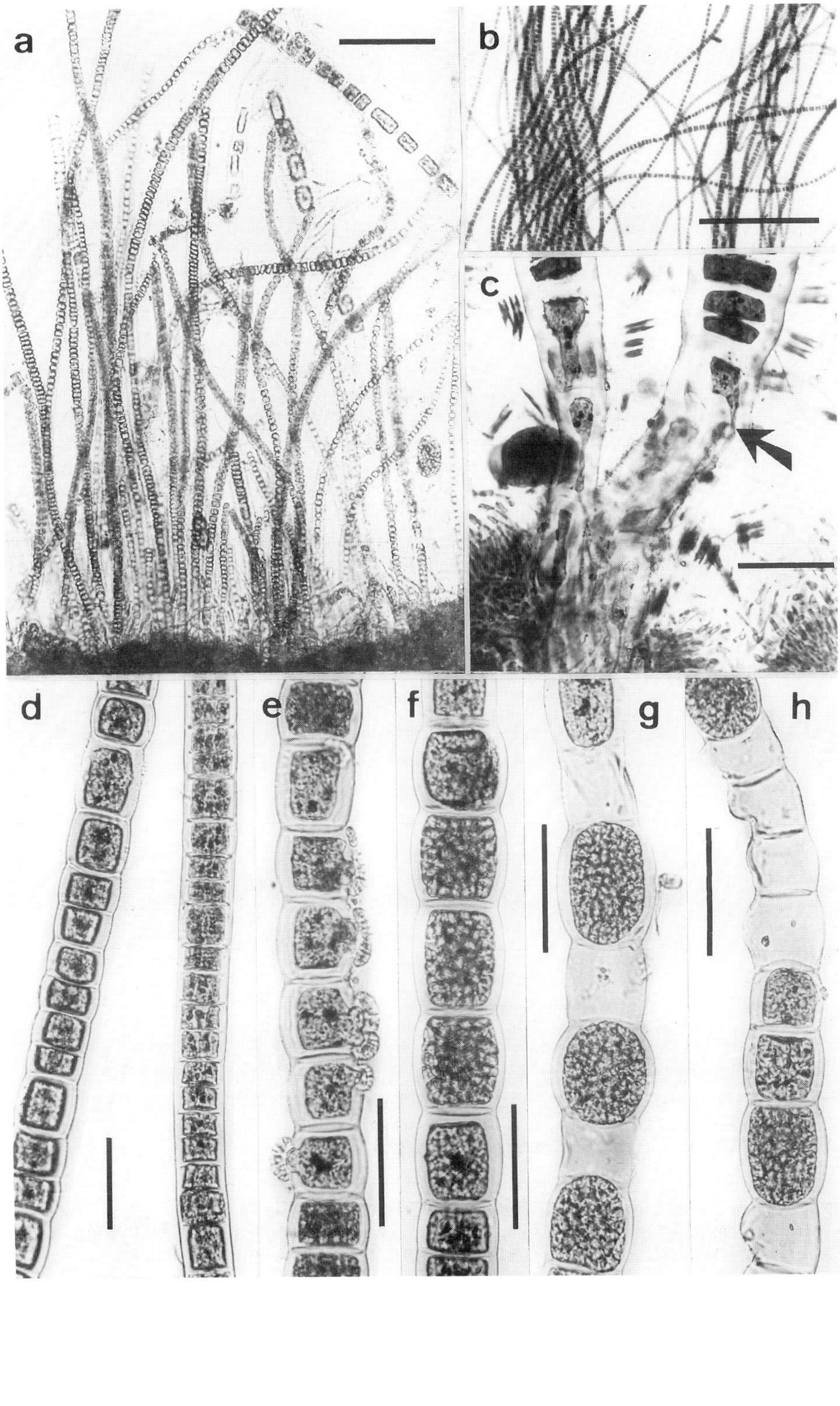
a
b
c
d
e
f
g
h

1–2 pores (appearing as cytoplasmic bulges) in sporangial walls (Fig. 23e). Sexual reproduction not observed in Macquarie plants.

I observed production of gametangia in plants that I collected and grew in the Antarctic. Male gametangia were yellow-olive green, containing many ovoid to spindle-shaped biflagellate gametes per cell, whereas female gametangia were darker green, containing numerous ovoid-elliptical to clavate-shaped biflagellate gametes with pointed or blunt posteriors. The filamentous phase alternated with unicellular plants that agree with *Codiolum gregarium* Braun. The *'Codiolum'* phase was highly variable in its cell shape and rhizoid length (up to 175 μm).

NOTES: *Urospora penicilliformis* is less abundant on Macquarie's shores than along the Vestfold Hills coastline of Antarctica (pers. obs., 1982). In one instance at Macquarie I. virtually a pure stand of this alga covered a boulder. In the Antarctic plants commonly formed nearly unialgal patches and often dominated the intertidal community with a continuous band along rock faces. These Antarctic plants are comparable in size and detail with Lokhorst's & Trask's (1981) findings on European specimens of *U. penicilliformis* and were observed in culture to alternate with a *Codiolum* phase. The sporophytic phase was also seen in field-grown *Cruoriopsis*-like crusts. Macquarie I. specimens were not studied in culture and no *Codiolum* phase was observed in collections of the most likely encrusting host, *Hildenbrandia lecannellieri* Hariot.

Both Macquarie and Antarctic plants have highly variable cell shapes and filament lengths. Some Antarctic plants were morphologically similar to *Chaetomorpha mawsonii* Lucas (1919), which was described from Eastern Antarctica. Although type material of this species has not been studied, Lucas' illustrations match the morphology of some of the larger culture-grown plants of *U. penicilliformis* from the Vestfold Hills. The possible common identity of these two taxa might explain the unexpected occurrence of *Chaetomorpha*, a genus with primarily cold temperate to warmer water affinities, in the Antarctic.

Cladophorales

Members of the Cladophorales have filamentous thalli with rhizoidal bases. The filaments are uniseriate, simple or branched, and composed of multinucleate cells. Growth generally is diffuse but is sometimes localized distally in branched thalli. Karyokinesis is independent of cytokinesis. Individual cells have numerous discoid or polygonal chloroplasts usually united in a parietal reticulate layer. Each chloroplast generally contains a bilenticular pyrenoid with both hemispheres separated by a thylakoid (Hoek, 1981). Most species that have been studied exhibit an alternation of isomorphic sporophytic and gametophytic generations, but an alternation of heteromorphic generations (Archer & Burrows, 1960) and zygotic meiosis (Siddique & Faridi, 1977) have been reported for two species of *Cladophora*. Vegetative reproduction by fragmentation is most common, but where sexual reproduction occurs gametes are isogamous and not produced in specialized sexual organs.

Superficial resemblances between cladophoralean and acrosiphonialean algae, as well as diagnostic features like cell walls composed of cellulose I, bilobed starch-pyrenoid complexes and isomorphic generations for members of the Cladophorales, have already been discussed (cf. Acrosiphoniales). In spite of these differences, some authors (Abbott & Hollenberg, 1976) continue to interpret the Cladophorales broadly as including genera from the Acrosiphoniales (*Spongomorpha*—Kornmann & Sahling, 1977), Codiolaceae (*Codiolum, Urospora*—Kornmann & Sahling, 1977) and Ulotrichales (*Gomontia*—Kornmann, 1963, 1965).

The Cladophorales comprises the Cladophoraceae and Anadyomenaceae (Silva, 1982), the latter including strictly tropical to subtropical marine species. The Cladophoraceae, in contrast, includes marine and freshwater species and is represented in equatorial to high-latitude waters.

Cladophoraceae Wille *in* Warming, 1884: 30

Papenfuss (1964*a*) listed eight genera representing this family in the Antarctic and Subantarctic. Two of these, *Codiolum* and *Gomontia*, are alternate stages in the life histories of other algae, and another two, *Spongomorpha* (Kornmann & Sahling, 1977) and *Urospora* (Tanner, 1981), belong to the Acrosiphoniales. Of the four remaining genera (*Chaetomorpha, Cladophora, Rama,* and *Rhizoclonium*), *Cladophora* has the greatest number of species attributed to it in high-latitude southern waters. It is probably the least understood genus, however, because morphological characteristics used to separate its species from one another, or from those of the similar-appearing genera *Acrosiphonia* and *Spongomorpha,* are variable or not sufficiently diagnostic. The only record of *Cladophora* from Macquarie I. (Kenny & Haysom, 1962) has not been confirmed through the present study, in which all *Cladophora*-like plants are assigned to *Acrosiphonia pacifica.*

The Cladophoraceae is represented at Macquarie I. by *Rhizoclonium ambiguum* (Hooker f. & Harvey) Kützing.

Rhizoclonium Kützing, 1843: 261

Lectotype: *R. jürgensii* Kützing, 1843: 261 [= *R. riparium* (Roth) Harvey, 1849: pl. 238. (see Koster, 1955: 341)].

Thalli comprise uniseriate filaments that are largely unbranched above and attached by rhizoids below. Cells contain 1–4 nuclei and have single, parietal, reticulate plastids with multiple pyrenoids. Cell shapes are either broader than long, nearly isodiametric, or longer than broad. Reproduction is either vegetative through fragmentation, asexual through biflagellate or quadri-flagellate zoospores, or sexual through biflagellate isogametes that are produced in gametangia, which have been transformed from vegatative cells. Life histories have isomorphic phases.

Filament and cell dimensions, both extremely variable within individuals and between species, have been used to distinguish the nearly 100 freshwater and marine species in the genus. Of the three species (*R. ambiguum* (Hooker f. & Harvey) Kützing, *R. riparium* (Roth) Harvey, *R. tortuosum* (Dillwyn) Kützing) listed by Papenfuss (1964*a*) from the Subantarctic, all are largely defined by the dimensions of cells and filaments and therefore require critical re-evaluation. Only *R. ambiguum* is discussed in the present survey.

Rhizoclonium ambiguum (Hooker f. & Harvey) Kützing, 1849: 387

Fig. 24

BASIONYM: *Conferva ambigua* Hooker f. & Harvey, 1845: 295.

SYNONYMS: see Papenfuss, 1964*a*.

TYPE LOCALITY: Îles Kerguelen.

TYPE: BM, sheet no. 638 (designated by Y. M. Chamberlain as isotype material).

SOUTHERN OCEAN DISTRIBUTION: Campbell and Snares Is (Laing, 1909); Fuegia (Hariott, 1889); Heard I. (Ricker, unpubl.); Îles Kerguelen, (Hooker, f. & Harvey, 1845); Macquarie I. (N.B. Levring (1945) indicated distribution at the Auckland Is, but the original published record for this location has not been located for the present report.)

MATERIAL EXAMINED:

1. Macquarie I.: collections MA-45, 98, 111, 120, 130, 133, 213, 217, 244, 264, 278, 302; Herbarium specimens: MELU, R 1287, 2041, 2110.

2. Heard I.: Cape Gazert, upper littoral rock pools (R. G. Chittleborough, 25 April 1949): MELU, R 3325 (= Chittleborough specimen no. 97).

DESCRIPTION: Plants to 23 cm long, filamentous, tufted, with uniseriate axes mostly simple, lax, erect (Fig. 24a), arising from basal turf of coarse, wiry, highly branched filaments (Fig. 24c, d). Erect axes 56–84 µm diameter (Fig. 24g, h), tapering slightly toward rounded apices (Fig. 24g), occasionally laterally bearing short rhizoids. Compact basal mat composed of filaments 140–

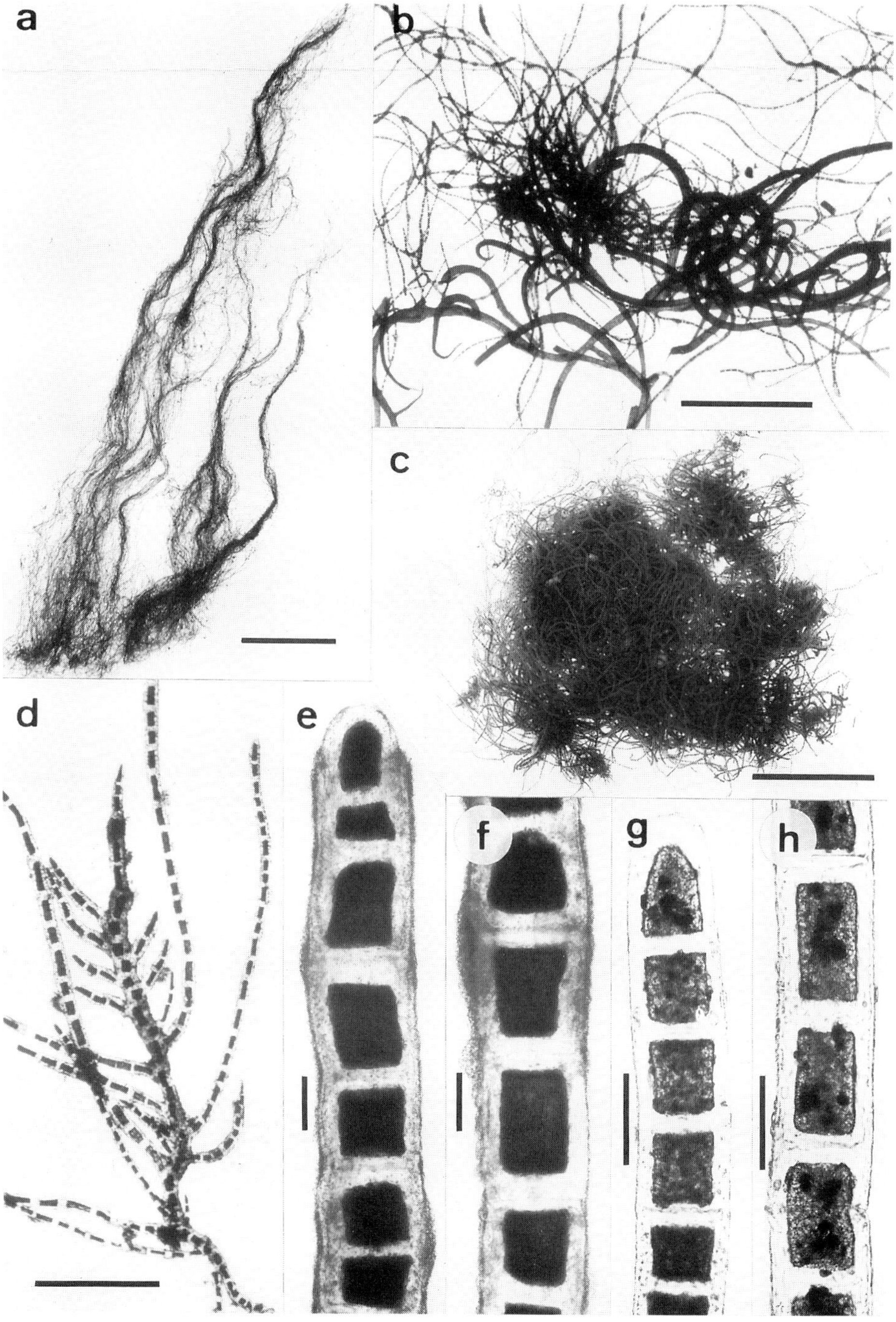

Fig. 24 *Rhizoclonium ambiguum* (Hooker f. & Harvey) Kützing: MELU, R 1287, 2041, 2110. **a.** Habit of 'skein' form. **b.** Habit of basal portion with long filaments growing intermingled with *Bostrychia*. **c.** Habit of 'wiry turf' form. **d.** Detail of branched basal filaments. **e, f.** Apical and mid-filament portions of 'wiry turf' thalli. **g, h.** Apical and mid-filament portions of 'skein' thalli. Scales: a = 2 cm; b = 4 mm; c = 1 mm; d = 500 μm; e–h = 100 μm.

230 µm diameter (Fig. 24e, f). Both erect and basal filaments comprising multinucleate cells of variable length, generally 1–2 times as long as broad. Each cell with single parietal chloroplast containing multiple pyrenoids. Vegetative reproduction apparently largely by fragmentation; sexual reproduction unknown.

Plants usually grow in shallow crevices in boulders in the upper intertidal. In this habitat *R. ambiguum* sometimes forms a basal turf alongside or mixed with *Bostrychia* (Fig. 24b). *Rhizoclonium* typically does not form skeins of erect filaments in the upper intertidal. Lower in the intertidal, where *Acrosiphonia*, *Enteromorpha* and *Porphya* are abundant, crack-dwelling turfs produce loose skeins that apparently detach from their bases and grow entangled with other algae.

NOTES: Some authors (Abbott & Hollenberg, 1976) describe the cells of *Rhizoclonium* as multinucleate. Others (e.g., Womersley, 1984) indicate a maximum of four nuclei per cell, thereby distinguishing *Rhizoclonium* cells from those of other filamentous green algae, such as members of *Cladophora*, containing many nuclei. Based on the number of nuclei per cell, *R. ambiguum* is inappropriately placed in *Rhizoclonium* and would be better assigned to *Cladophora*. It is presently maintained in *Rhizoclonium*, however, until this genus is monographed and variation in the number of nuclei per cell is known for its constituent species.

Plants from Macquarie I. agree with isotype material, which has filaments *c.* 75 µm diameter intertwined in skeins and attached below with rhizoids. Another collection of this species also stored in the British Museum of Natural History (Gough I.—Wace, no. 1557), is coarse, branched and similar to the 'wiry turf' form known from rock crevices at Macquarie I. Examination of either basal turfs or erect skeins separately suggests two distinct taxa because of the gross morphological differences between the forms. As observed at Macquarie I., however, a single plant can comprise both types of filaments. Such variation in filament morphology was noted earlier by Laing (1909), but most subsequent descriptive accounts have ignored the basal turf in favor of the more conspicuous 'skeins'.

Bryopsidales

Also known as the Caulerpales, Codiales, or Siphonales, this order includes plants composed of simple or branched, mostly non-septate tubes (siphons). These siphonous tubes may grow singly (as uniaxial thalli) or in aggregates (as multiaxial thalli) and give rise to a broad morphological range of marine and (less commonly) freshwater forms. Plants are considered to be unicellular and to have multinucleate cells filled with numerous discoid to fusiform plastids with or without pyrenoids. Two types of plastids (amylogenic leucoplasts and chloroplasts) occur in members of one family, the Udoteaceae. Two xanthophyll pigments characteristic for the order are siphonein and siphonoxanthin. Cell walls of some genera (e.g., *Bryopsis*, *Derbesia*) are non-cellulosic in sporophytes and composed primarily of mannan, and cellulosic mixed with xylan in gametophyes (Huizing & Rietema, 1975).

Vegetative reproduction most commonly occurs through rhizome-like propagation, but sometimes also by fragmentation or propagule formation. Sexual reproduction is mostly isogamous or anisogamous, although rarely oogamous. Life histories are documented poorly throughout the order, but both haplobiontic and diplobiontic genera are known.

Although the Bryopsidales comprises six families including approximately 24 genera, only three (Bryopsidaceae, Caulerpaceae, Codiaceae), with four genera (*Bryopsis*, *Derbesia*, *Caulerpa*, *Codium*), occur in the Subantarctic Zone. Of these families, the Caulerpaceae, which extends marginally along the northern edge of the Subantarctic Zone, generally is more characteristic of tropical and some temperate floras and does not occur as far south as Macquarie I.

Bryopsidaceae Bory de Saint-Vincent, 1829: 203

Plants are uniaxial, often delicate, and constructed of siphonous tubes that may be constricted to various degrees throughout their lengths. Siphons are simple or profusely branched, with branches arising irregularly, dichotomously, alternately, oppositely, or radially. Many species have chloroplasts with pyrenoids. Life histories, where known, are diplobiontic, usually with either the gametophyte or sporophyte dominant and filamentous, whereas the alternate stage is smaller and vesicular. Cell walls of gametophytes are composed chiefly of cellulose and xylan, whereas those of sporophytes are of mannan. Reproduction, as studied in *Bryopsis* and *Derbesia*, is uniform (thus supporting the merger of the traditionally separate Derbesiaceae with the Bryopsidaceae), sporophytes producing zoospores with a subapical crown of many flagella, and gametophytes giving rise to biflagellate anisogametes.

Of the six genera in this family, only *Bryopsis*, *Derbesia*, and *Lambia* have distribution ranges that extend south into high latitudes of the Southern Ocean, although the generally warm-water *Pedobesia* does occur in Tasmania (Womersley, 1984). *Bryopsis* is the most widely distributed of these algae and is represented by the greatest number of species. Only *Derbesia*, however, is present at Macquarie I.

Derbesia Solier, 1846: 452

Lectotype: *D. marina* (Lyngbye) Solier, 1846: 453.

Sporophytes are constructed of sparsely branched siphonous filaments, whereas gametophytes develop into spherical vesicles that have been named the *Halicystis* stage. Cells contain numerous discoid plastids that, depending upon the species, may or may not have pyrenoids.

Of the more than 20 species in the genus, most are known from the Northern Hemisphere, and none have been previously reported from the Subantarctic and Antarctic Zones.

Derbesia furcata R. W. Ricker sp. nov.

Fig. 25

Holotype: (Fig. 25a): MELU, R 1509; with isotypes in ADU, BM, Herb. Ricker, and UC.

Type locality: Small unnamed bay south of Tern Rock, Macquarie I. (54°29′56″ S. lat., 158°57′35″ E. long.); growing on sponge in 10 m depth (Ricker, 29 November 1977).

Distribution: Known only from the type collection.

Material examined: collection MA-34; Herbarium specimen: MELU, R 1509.

Diagnosis: Plantae ad 2.5 mm altae, compositae filis atrovirentibus siphoniis erectis et prostratis ex basi crassa (60–65 μm diametris) orientibus. Axes erecti plerumque semel dichotomi, 30–45 μm diametris, versus apices rotundatos leviter angustati, plastidia multa discoidea vel angularia continentes, 2.0–3.5 μm diametris, cum pyrenoidibus 1–2, ab iodo fuscatis. Axis prostrati irregulariter ramosi, a rhizoideis thallum suffigentes, erectis minus pigmentosi, prope basin crassam obturamentum cytoplasmica plerumque habentes. Granula cytoplasmica magna amorpha abundantia. Reproductio ignota.

The species name refers to the alga's characteristic dichotomies in its upright axes.

Description: Plants to 2.5 mm tall, composed of dark green, erect and prostrate, siphonous filaments arising from stout bases (60–65 μm diameter). Erect axes usually once dichotomous (Fig. 25a, b), 30–45 μm diameter, tapering slightly toward rounded apices, with numerous discoid to angular plastids, each 2.0–3.5 μm diameter with 1–2 pyrenoids that stain dark brown with iodine. Prostrate axes irregularly branched, anchoring thallus rhizoidally, less pigmented than uprights, commonly separated from erect filaments by cytoplasmic plug near stout base (Fig. 25c). Large amorphous cytoplasmic granules abundant. Reproduction unknown.

Derbesia furcata is presumed to be very rare at Macquarie I. and is known only from a single clump growing on and in a white sponge in the subtidal.

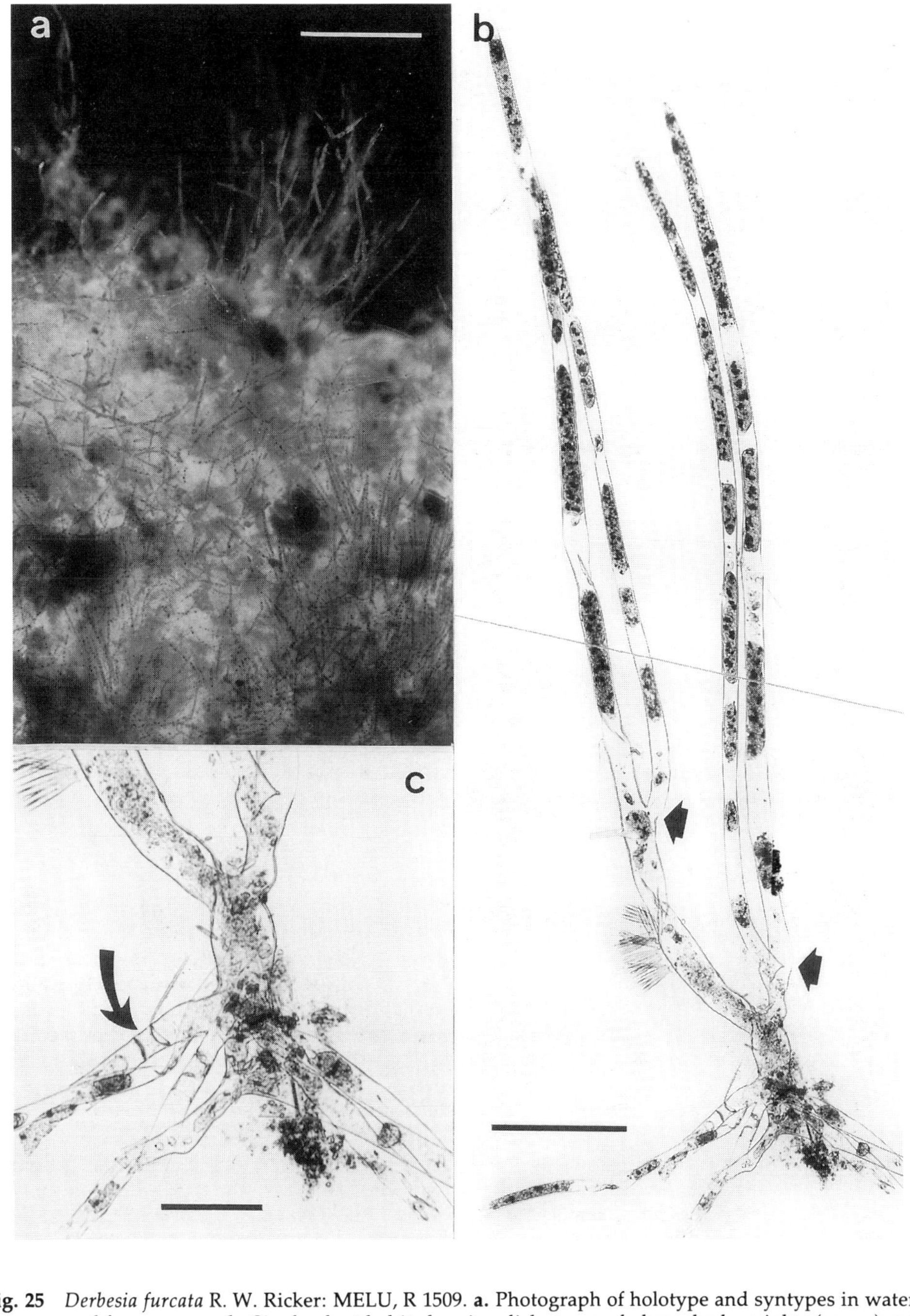

Fig. 25 *Derbesia furcata* R. W. Ricker: MELU, R 1509. **a.** Photograph of holotype and syntypes in water prior to removal from sponge. **b.** Single plant habit showing dichotomously branched uprights (arrow). **c.** Detail of basal portion showing cytoplasmic plug (arrow) near juncture of a prostrate axis with the stout base. Scales: a = 1 mm; b = 300 µm; c = 100 µm.

NOTES: Because reproductive structures were not observed, the taxonomic placement of this alga remains somewhat uncertain. The morphology of the new species is unlike that of other closely related taxa previously reported from southern high latitudes and generic placement in *Derbesia* seems best in light of the siphonous construction and sparse branching pattern of the new species. *D. furcata* is much smaller than previously reported species in the genus, but because it lacks reproductive structures and might be immature, size alone probably is not a reliably distinctive character. The alga's branching pattern and isolated location, however, justify its provisional description as a new species.

Lambia antarctica (Skottsberg) Delépine (1967), formerly considered to be a species of *Derbesia*, is distinct from the present alga in having siphons up to 20 cm tall by 2–3 mm diameter and plastids lacking pyrenoids (Lamb & Zimmermann, 1977). It is known from the Antarctic Peninsula (Moe & DeLaca, 1976) and Eastern Antarctica (Ricker, pers. obs.). Another uniaxial siphonous alga with possible affinities to *D. furcata* is *Bryopsis*, but species in this genus are radially or pinnately branched along percurrent upright axes.

Codiaceae Kützing, 1843: 302, 308

The diagnostic features for this family, as exemplified by its only genus, *Codium*, include: free-living multiaxial thalli composed of siphonous filaments terminating peripherally in swollen structures called utricles; cytoplasmic plugs at or near the bases of utricles and gametangia; primarily mannan cell walls; and an absence of pyrenoids and leucoplasts. Reproduction haplobiontic; anisogametes in simple gametangia lateral on utricles.

Members of other families in the Bryopsidales share some of these features, but only those in the Udoteaceae also have multiaxial thalli anatomically resembling individuals of *Codium*. No species in the Udoteaceae grow in the Subantarctic, however, thus leaving *Codium* as the one multiaxial siphonous representative in this zone.

Codium Stackhouse, 1797: xvi, xxiv

Type species: *C. tomentosum* (Hudson) Stackhouse, 1797: xxiv.

Although a large genus with more than 100 species constituting a broad range of habits and morphologies, *Codium* is easily identified anatomically. Thalli are anchored by numerous rhizoids. Above the base is a multiaxial core of siphons peripherally surrounded by sympodially derived utricles. Cytoplasmic plugs (septa) usually occur at or near the bases of utricles and gametangia. Gross thallus morphology and anatomical details including the sizes and shapes of utricles and gametangia are the principal features used to separate species in the genus. Modifications to the apices of utricles, the presence or absence and positioning of hairs or hair scars and septa, and the patterns of utricle development are also valuable taxonomic criteria.

Codium is unknown in the Antarctic Zone, but eight Cold Temperate species are listed by Papenfuss (1964*a*) from Subantarctic waters. He believed that names for six of these species are misapplied, however, and inferred that the southern plants probably are distinct from previously described taxa with which they have been commonly allied. Setchell (1937), when discussing the likelihood of a Japanese species (*C. contractum* Kjellman) growing along the shores of West Patagonia, suggested that the South American plants be re-studied in view of such a dubious species range.

A revision of all *Codium* species growing in the southern high latitudes is necessary but has not been undertaken for the present account. Instead, a new entity, *C. subantarcticum*, is compared with other mat-forming *Codium* species that have been reported from the Southern Hemisphere. In addition to the three prostrate species listed by Papenfuss (1964*a*: *C. adhaerens* C. Agardh, *C. difforme* Kützing, *C. spongiosum* Harvey), five others (*C. cerebriforme* Setchell, *C. dimorphum* Svedelius, *C. lucasii* Setchell, *C. mamillosum* Harvey and *C. pomoides* J. Agardh) are discussed. The name '*C. antarcticum* Silva', based on a specimen of *C. subantarcticum* from the Falkland Is, appeared as a *nomen nudum* in Zaneveld (1969) and, therefore, is illegitimate (Articles 6 & 39, International Code of Botanical Nomenclature). Although *C. subantarcticum* is included here, it will be given a formal diagnosis elsewhere by Silva (pers. comm.).

Codium subantarcticum Silva sp. ined.

Fig. 26

SYNONYM: *Codium antarcticum* Silva *in* Zaneveld, 1969: pl. cvii.

DISTRIBUTION: Macquarie I.; Falkland Is and the tip of South America (Silva, pers. comm.).

MATERIAL EXAMINED: Macquarie I.: collections MA–39, 46, 54, 83, 98, 111, 215, 238, 244, 271, 283, 284, 295, 300, 302, 303, 309, 364, 377, 378, 387; Herbarium specimens: MELU, R 0042, 1325, 1447, 3494–3503.

DESCRIPTION: Thalli to several centimetres thick, flattened, cushion-like, overlapping to form extensive, dark forest-green mats of irregular shape up to a meter across; often cerebriform and loosely adherent to rock substrata subtidally (Fig. 26a), more securely attached and applanate intertidally. Primary (parent) utricles often directly bearing multiple secondary (daughter) utricles along lower halves, thus forming clusters (Fig. 26c, d), but sometimes remaining solitary without lateral connections (Fig. 26e). Septa separating cytoplasm of connected utricles. Numerous small discoid chloroplasts and larger amorphous granules present. Utricles cylindrical, occasionally constricted subapically, clavate, 1.4–3.0 mm long, 120–350 µm diameter throughout, tapering gradually to abruptly at bases, without hairs or hair scars, apices flattened and angular to symmetrically rounded, without thickened or modified end walls. Gametangia fusiform, 320–520 µm long, 64–100 µm diameter, borne singly or in pairs laterally on the upper half to one third of a utricle (Fig. 26d, e). Medullary filaments mostly 40–85 µm diameter (Fig. 26e).

This species is extremely common around Macquarie I., occurring both intertidally in tidepools and subtidally to depths of at least 21 m. It is abundant on the floors of many protected shallow coves. Subtidal plants usually are epiphytized by *Ceramium strictum* and often have various red algae (e.g., *Cenacrum subsutum, Plocamium secundatum* and several members of the Delesseriaceae) embedded in or partially covered by the mats. In some individuals, *Streblonema codiatile* grows endophytically near the host thallus surface, thus forming a lattice interlocking adjacent utricles. When growing on the floors of kelp forests, *C. subantarcticum* mats partly cover or encircle holdfasts of *Macrocystis* and *Desmarestia*.

NOTES: *Codium subantarcticum*, like *C. cerebriforme* (Setchell, 1937) and *C. spongiosum* (Silva & Womersley, 1956), produces thickened, firm but spongy mats that often have superficial ridges. Another similarity between *C. spongiosum* and *C. subantarcticum*, but apparently not *C. cerebriforme*, is the change in thallus morphology following liquid preservation in 5% formalin/seawater. Both species are reasonably firm when fresh but become gelatinous in preservative (Fig. 26b) and later lustrous when dried and pressed. According to Setchell (1937), *C. cerebriforme* remains in excellent condition, presumably suggesting that liquid fixation does not alter the morphology of the plants.

Thallus consistency in *Codium* species probably is determined largely by utricle dimensions and the organization of utricles, i.e., solitary or laterally joined with other utricles to form clusters. Very dense and solid thalli, as typical of *C. pomoides*, have extremely fine utricles, whereas the spongy cushions of *C. mamillosum* have larger swollen ones. The dense mats of *C. lucasii* are composed of slender utricles and the spongy mats of *C. spongiosum* are composed of extremely long (to 6 mm), broad utricles (Silva & Womersley, 1956). *Codium subantarcticum* has shorter and somewhat narrower utricles than *C. spongiosum* and produces a firmer, thinner mat that is relatively less spongy.

The utricles of *C. subantarcticum* differ, in their lack of hairs or hair scars, from those of *C. cerebriforme* and *C. spongiosum*. Also, utricles of *C. subantarcticum* are mostly simple (very rarely branched), where those of *C. cerebriforme* apparently branch more frequently and form complex utricle clusters. Other species, such as *C. difforme* and *C. dimorphum*, have utricles with helmet-shaped thickenings at their tips (Setchell, 1937), which differ from the unmodified utricle apices of *C. subantarcticum*. *Codium adhaerens*, also believed to have utricle tips with helmet-shaped thickenings (Setchell, 1937), was first described by Bailey and Harvey (1874) based on plants from

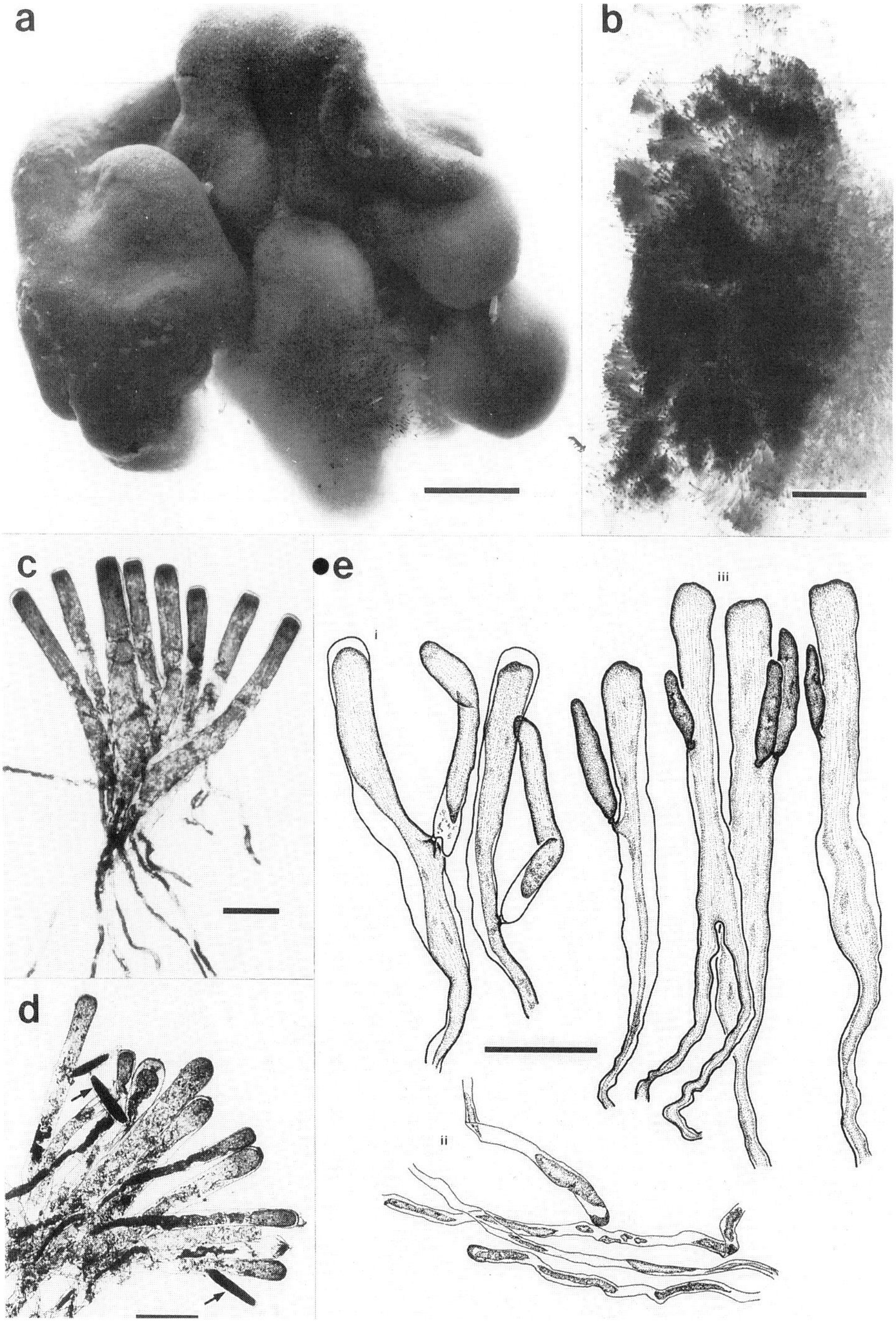

Fig. 26 *Codium subantarcticum* Silva sp. ined.: MELU, R 3494. **a.** Habit of mat-forming plants in water prior to drying as herbarium specimen. **b.** Lubricous mat with dissociated utricles resulting from liquid preservation . Some utricle clusters are held in positions relative to one another by an endophytic lattice of *Streblonema codiatile*. **c.** Sterile utricle cluster. **d.** Fertile utricle cluster with laterally attached gametangia (arrows). **e.** Camera lucida drawings: (i) Primary utricles bearing secondary utricles; (ii) Medullary filaments; (iii) Solitary utricles bearing gametangia. Scales: a = 1 cm; b = 5 mm; c–e = 500 μm.

Valparaiso, South America. This feature, however, presumably was not clarified in the original description of *C. adhaerens*.

C. subantarcticum can be distinguished from other *Codium* species by its applanate to cerebriform thallus morphology, utricle dimensions, unmodified utricle apices and absence of hairs. Superficially, this species most closely resembles the spongy mats of the tropical to subtropical species *C. spongiosum*.

Phaeophyta

Ectocarpales

All filamentous brown algae that produce unilocular and plurilocular sporangia, including uniseriate and pluriseriate forms, were first grouped together in the Ectocarpaceae by Oltmanns (1904). The family was raised later to ordinal status independently by Oltmanns (1922) and Setchell & Gardner (1922), although the latter authors (1925) soon narrowed the concept of the group by removing some of the morphologically more complex members to the Chordariales and Dictyosiphonales. Kylin (1933) decreased species numbers further by limiting the Ectocarpales to (mostly) uniseriate filamentous taxa with isomorphic phases in their life histories. These plants generally grow diffusely, but a few species have localized meristems that in some cases give rise to colourless hairs. Pseudoparenchymatous forms were moved into either the Chordariales, Dictyosiphonales, or Punctariales.

Some authors (Clayton, 1981; Fritsch, 1945; Parke & Dixon, 1968; Russell & Fletcher, 1975) believe that a broad interpretation of the order is appropriate, but life history studies (Clayton, 1974—Ectocarpaceae; Nakamura, 1972—Ralfsiaceae; Sauvageau, 1929—Punctariaceae) have shown the Ectocarpales *sensu lato* to be a heterogenous group whose members display several different developmental and reproductive strategies. The narrow concept of the order strictly includes those plants with an alternation of isomorphic generations (Papenfuss, 1951), in which sporophytes bear either plurilocular sporangia of unilocular and plurilocular sporangia, and gametophytes bear only plurilocular gametangia.

Some recent floristic accounts (Abbott & Hollenberg, 1976; Silva, 1979) have considered the Ectocarpales as comprising one family of primarily uniseriate filamentous taxa, the Ectocarpaceae, whereas others (Kornmann & Sahling, 1977) have interpreted the order broadly to include numerous families. The present scheme, which is similar to that of Wynne (1982*b*), includes the Ectocarpaceae, Ralfsiaceae and Sorocarpaceae. Wynne's concept of the Ralfsiaceae includes *Analipus*, which some (Nelson, 1982) would place in the monotypic family Heterochordariaceae. Only the Ectocarpaceae and Ralfsiaceae are represented at Macquarie I. and throughout southern high latitudes.

Ectocarpaceae C. Agardh, 1828: 9

Members of this family are constructed simply, have microscopic or macroscopic thalli and exist in most habitats around Macquarie I. They often grow epiphytically, forming small filamentous balls on larger algae, but also occur as tufts on rocks and shells and as endophytes. A loose aggregation of free filaments probably is the most diagnostic characteristic for members of this family. Intraspecific cytological and morphological variation, which typically is extensive, led Ravanko (1970), using field-collected material and culture studies, to lump many genera in a single *Ectocarpus* complex, the different forms of which reflected individual growing conditions of each plant. Classically, however, genera are separated using a combination of plant habit (epiphytic or epilithic vs. endophytic), branching pattern (opposite, alternate, irregular, or with specialized branchlets), types of growth (diffuse or localized), presence or absence of true hairs, plastid shape (band- vs. disc-shaped), and location of unangia and plurangia.

In the classical treatment employed here, five ectocarpacean genera, totalling six species, are recognized using only field collected material. Some similarities existing between these plants and those of two closely related families, the Elachistaceae and Myrionemataceae, have led other workers to intercharge species within these three families.

Kuckuck (*in* Skottsberg, 1921), when classifying *Gononema*, stated that its habit is closest to that of *Elachista* in the Elachistaceae, and, although both genera have similar 'ectocarpoid' characteristics, he kept them separate from the Ectocarpaceae. Skottsberg (1921), however, considered *Gononema* more like *Ectocarpus* and consequently placed it in the Ectocarpaceae. This placement thereby lessened the importance of thallus habit as a family characteristic. *Gonomena*, with its 'elachistoid' pseudoparenchymatous basal cushion, is considered here as a member of the Elachistaceae, and thallus construction with loosely aggregated uniseriate filaments is maintained as the principal criterion for inclusion in the Ectocarpaceae.

Distinctions between the Ectocarpaceae and Myrionemataceae have been clouded similarly by placing *Streblonema*, with its 'myrionematoid' endophytic habit, apical or subapical growth, and sometimes uniseriate plurangia, in the Ectocarpaceae. *Streblonema*, unlike *Gononema*, is maintained in the Ectocarpaceae because of its greater resemblance to other members of this family. Many ectocarpacean genera have endophytic parts. *Asteronema*, presently placed in the Ectocarpaceae, also has subapical to apical growth. Although uniseriate plurangia do occur in *Streblonema*, multiseriate structures similar to those borne by all members of the Ectocarpaceae are seen most frequently.

Of the eight ectocarpacean genera listed by Papenfuss (1964*a*) as occurring in the Subantarctic, two (*Actinema* and *Entonema*) are insufficiently known for taxonomic placement, one (*Gononema*) is transferred in the present account to the Elachistaceae, and another (*Spongonema*) is a dubious record, probably representing a species of *Ectocarpus* (Skottsberg, 1907). All four of the remaining genera (*Ectocarpus, Geminocarpus, Pilayella, Streblonema*) grow around Macquarie I. Additionally, *Asteronema*, which was previously reported by Delépine & Asensi (1975) from Kerguelen, also grows along Macquarie's shores.

Asteronema Delépine & Asensi, 1975: 296

Type species: *A. australe* Delépine & Asensi, 1975: 296.

Thalli form epilithic, epiphytic, or epizoic mats that consist of dense basal layers of prostrate filaments bearing uniseriate, sparsely branched erect filaments. Branching is alternate to unilateral. Growth is apical to subapical and diffuse. Cells have multiple ribbon-like plastids, each terminated by an internal pyrenoid. The plastids are generally arranged in 1–2 star-like clusters with the pyrenoid ends centrally positioned (Delépine *et al.*, 1976). The generic name refers to the stellate disposition of plastids within the cells. Unangia and plurangia are sessile or stalked, lateral or terminal, and may occur on the same thallus.

The known distribution for this genus is the Subantarctic (Îles Kerguelen, Macquarie I., Patagonia) and the south coast of Australia. Two species have been described, but the Australian alga, *A. ferruginea* (Harvey) Delépine & Asensi (Clayton, 1974; Womersley, 1967), is similar if not identical to the type species, *A. australe* (Delépine & Asensi, 1975). Because *A. australe* has not been studied during the present survey, the two species are maintained as distinct, and Macquarie plants are allied with the Australian entity on the strength of similarities to Harvey's type material, (wet) formalin-preserved plants from Victoria, Australia, and Womersley's (1967) illustrations of *A. (Sphacella) ferruginea.*

Asteronema ferruginea (Harvey) Delépine & Asensi, 1975: 296

Fig. 27

Basionym: *Cladophora ferruginea* Harvey, 1860: 340.

Synonym: *Sphacella ferruginea* (Harvey) Womersley, 1967: 194.

Type locality: Safety Cove, Port Arthur, Tasmania.

Holotype: Herb. Harvey, TCD (unnumbered sheet with type written in pencil). (N.B. TCD has three large sheets with *c.* eight smaller sheets attached that presumably are isotypes.)

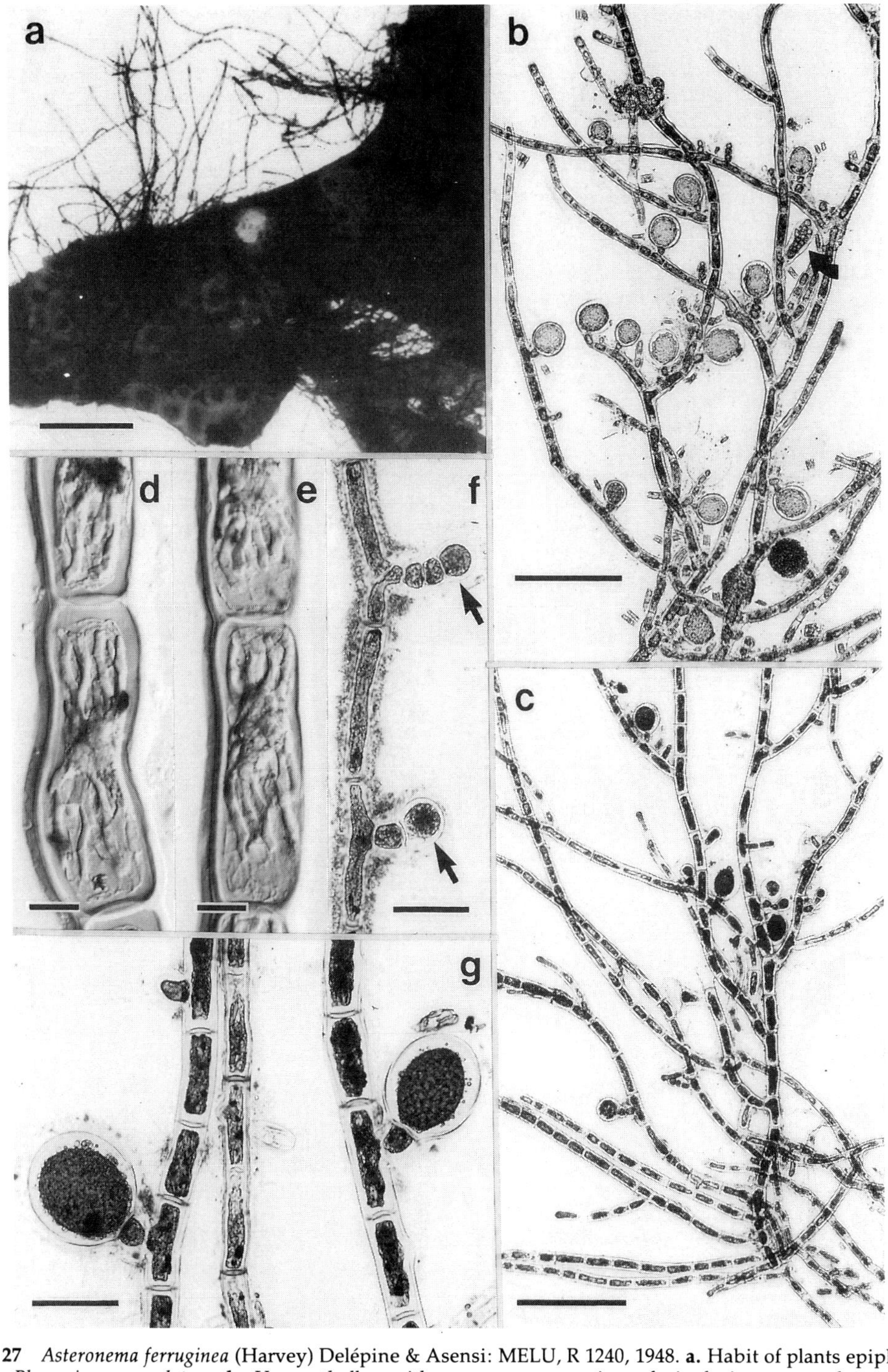

Fig. 27 *Asteronema ferruginea* (Harvey) Delépine & Asensi: MELU, R 1240, 1948. **a.** Habit of plants epiphytic on *Plocamium secundatum*. **b.** Upper thallus with numerous unangia and single immature plurangium (arrow). **c.** Lower thallus. **d, e.** Detail of cells with 'stellate' plastids. **f.** Developing stipitate unangia (arrows). **g.** Mature unangia. Scales: a = 1 mm; b, c = 200 μm; d, e = 10 μm; f = 40 μm; g = 50 μm.

DISTRIBUTION: Macquarie I.; southern Australia (Victoria and Tasmania: Womersley, 1967).

MATERIAL EXAMINED:

1. Macquarie I.: collections MA-11, 46, 264, 302; Herbarium specimens: MELU, R 1240, 1948.
2. Tasmania, Australia: Holotype and isotype specimens. Port Arthur, VDL (= Tasmania); on *Hormosira*: TCD, unnumbered specimens.
3. Victoria, Australia: Cape Otway, Crawfish Bay; mid-littoral pool, epiphytic on *Hormosira banksii* (Kraft, 5 April 1975): MELU, K 5181.

DESCRIPTION: Sparsely branched, pale-brown, filamentous thalli epiphytic on *Plocamium* and *Melobesia* (another *Plocamium* epiphyte) (Fig. 27a). Plants composed of loose aggregation of uncorticated upright filaments, 1–3 mm long, produced from few prostrate axes. Branching alternate or secund (Fig. 27b, c); main axes and lateral branches nearly equal diameter (17–26 µm); cells 2.5–4(–4.5) times as long as broad, filled with multiple ribbon-like plastids generally arranged in 1–2 stellate groups (Fig. 27d, e); plastids sometimes appearing parietal band-shaped or as compact mass of discs. Unangia and plurangia sometimes on same thallus. Plurangia rare, usually absent; unangia spherical to subspherical, 54–66 µm × 60–70 µm (Fig. 27b, g), terminal on modified branchlets (Fig. 27b) or lateral on 1–2-celled stalks (Fig. 27f). Young plurangia clavate, to 80–85 µm long, sessile, laterally borne (Fig. 27b); mature plurangia unknown.

Collection MA-264 contained a dense turf of olive-coloured plants epiphytic on the base of *Palmaria*. These thalli are cytologically and morphologically similar to the epiphytes on *Plocamium* but, in contrast, produce a thick mat of prostrate filaments that give rise to erect filaments which are slightly more branched than those on *Plocamium*.

NOTES: *Asteronema ferruginea* is separated from other ectocarpalean taxa at Macquarie I. by its delicate filaments with elongate, narrow cells, formation of a prostrate mat, and abundance of spherical unangia.

Delépine & Asensi (1975) separated *A. australe* from *A. ferruginea* on differences of plurangia shape, *A. australe*'s epilithic or epiphytic growth on encrusting coralline algae vs. *A. ferruginea*'s growth on *Hormosira* or *Pterocladia*, and on differences in the compactness of the basal mat. Macquarie plants, which have characteristics of both species, suggest the union of *A. australe* and *A. ferruginea* in a single, broadly circumscribed species. Plurangia on Macquarie thalli, presumed here to be immature, are similar to those illustrated by Womersley (1967) for *A. ferruginea*. Substrata occupied are similar to those supporting *A. australe* and plant habit embraces forms reported for both species. However, extensive collections from both type localities, assembled to guage intraspecific morphological and cytological variation, plus critical studies of type specimens of each entity are necessary before the two species can be confidently considered identical or distinct.

Ectocarpus Lyngbye, 1819: 130

Lectotype: *E. siliculosus* (Dillwyn) Lyngbye, 1819: 131.

Plants form tufts or mats of branched uniseriate filaments arising from prostrate axes, which attach by rhizoids to a variety of substrata. Branching of erect axes sometimes is irregular, often alternate or pectinate, and rarely opposite. Growth is mostly intercalary without localized meristems. Cells contain one or more ribbon-shaped plastids with one to several pyrenoids. Sporophytes sometimes produce unilocular sporangia alone but more commonly in addition to multiseriate plurilocular sporangia. Gametophytes bear plurilocular gametangia (Abbott & Hollenberg, 1976).

Eight species from this genus are listed in Papenfuss' (1964*a*) catalogue of Antarctic and Subantarctic seaweeds, but it seems likely that several of these are growth forms of the same species. For example, *E. confervoides* (Roth) Le Jolis and the type species appear to be identical. The opinion here, similar to that held by Rosenvinge & Lund (1941), is that *E. confervoides* is a growth form of *E. siliculosus*.

Two species of *Ectocarpus*, *E. siliculosus* and *E. constanciae* Hariot, are recognized from Macquarie I. Both express features characteristic for other taxa reported from the southern high latitudes, thus suggesting further reduction in the number of Antarctic and Subantarctic species. A comprehensive study of members of this genus is necessary before the actual number of *Ectocarpus* species in the Southern Ocean can be ascertained. This is beyond the scope of the present survey and, consequently, species plasticity is only discussed, no synonymies are formally proposed, and a figure for the total number of *Ectocarpus* species occurring in the southern high latitudes is not given.

Ectocarpus constanciae Hariot, 1887*a*: 56

Figs 28, 29

Type locality: Orange Bay, Grevy I., Fuegia.

Type: Herb. Hariot (PC), but specimens were not seen.

Distribution: Falkland Is, Fuegia, South Georgia (Skottsberg, 1907); Îles Kerguelen (Hariot, 1887*a*); Macquarie I.

Material examined: Macquarie I.: collections MA-34, 46, 95, 215, 302, 365; Herbarium specimens: MELU, R 1240, 1280, 1516, 3328, 3388.

Description: Plants to 1(–1.5) cm tall, filamentous, epiphytic, forming shaggy turfs or loose spherical tufts. Main filaments with irregular to subdichotomous branching below (Fig. 28c), alternate to pectinate above (Fig. 28b), minimally tapering toward apices, usually beset with numerous short, hook-like lateral branchlets (Fig. 28g). Cells in upper axes 20–30 μm diameter, 1–2(–3) times as long as broad, longer (–4.5 times) below, usually strongly constricted at nodes between cells of equal length and breadth, weakly so between more elongate cells. Cells usually with several ribbon-shaped plastids, each with one or more pyrenoids, lying parallel to filament axis (Fig. 28f, i), sometimes with one parietal, band-shaped plastid (Fig. 28d–f). Cells of lower branches frequently corticated by narrow, basipetally directed rhizoids (Fig. 29h, i), sometimes laterally bearing plurangia (Fig. 29i). Plurangia mostly stipitate on 1–2-celled stalks, adaxial or terminal on short lateral branchlets, produced singly from lateral branch cells (Figs 28h, i; 29d, g). Plurangia occasionally sessile, either adaxial or abaxial (Fig. 29c), or growing in fascicles with multiple plurangia arising from individual cells of lateral branches (Fig. 29e). Unangia unknown.

Plants grow in the subtidal between 5–10 m depth and most commonly occur on *Desmarestia chordalis* (Fig. 28a), *Plocamium secundatum* and members of the Delesseriaceae. Clusters of filaments form hairy balls similar to those produced by *E. siliculosus*.

Notes: Plants from Macquarie I. mostly resemble those first described by Hariot from Fuegia. Variation in thallus morphology, however, is greater than originally depicted for the species and, consequently, forms that otherwise might be considered as belonging to *E. fasciculatus* Harvey and *Spongonema tomentosum* (Hudson) Kützing have been incorporated in *E. constanciae*. *Ectocarpus fasciculatus* is the only species reported from the Subantarctic with plurangia borne in fascicles and, therefore, would be a likely candidate for the plant illustrated in Fig. 29. A highly twisted branching pattern, where lateral branches hook around main axes (Fig. 29 a, b), is characteristic for *Spongonema*, a genus reported from the Subantarctic but whose record is believed to represent *E. siliculosus* (Skottsberg, 1907). The morphology of a single plant displaying features of all three above taxa, plus short hook-like lateral branchlets characteristic for *E. constanciae*, inclines me to place all of these forms in one species. Future studies may show a continuum in form between *E. siliculosus* and *E. constanciae*, two obviously closely related species, whereby both would be united under *E. siliculosus*. At present, however, the two are recognized as distinct on the basis of larger cell diameters, apically tapering branches, helically coiled ribbon-like plastids, and more elongate, solitary plurangia in *E. siliculosus* vs. minimally tapering

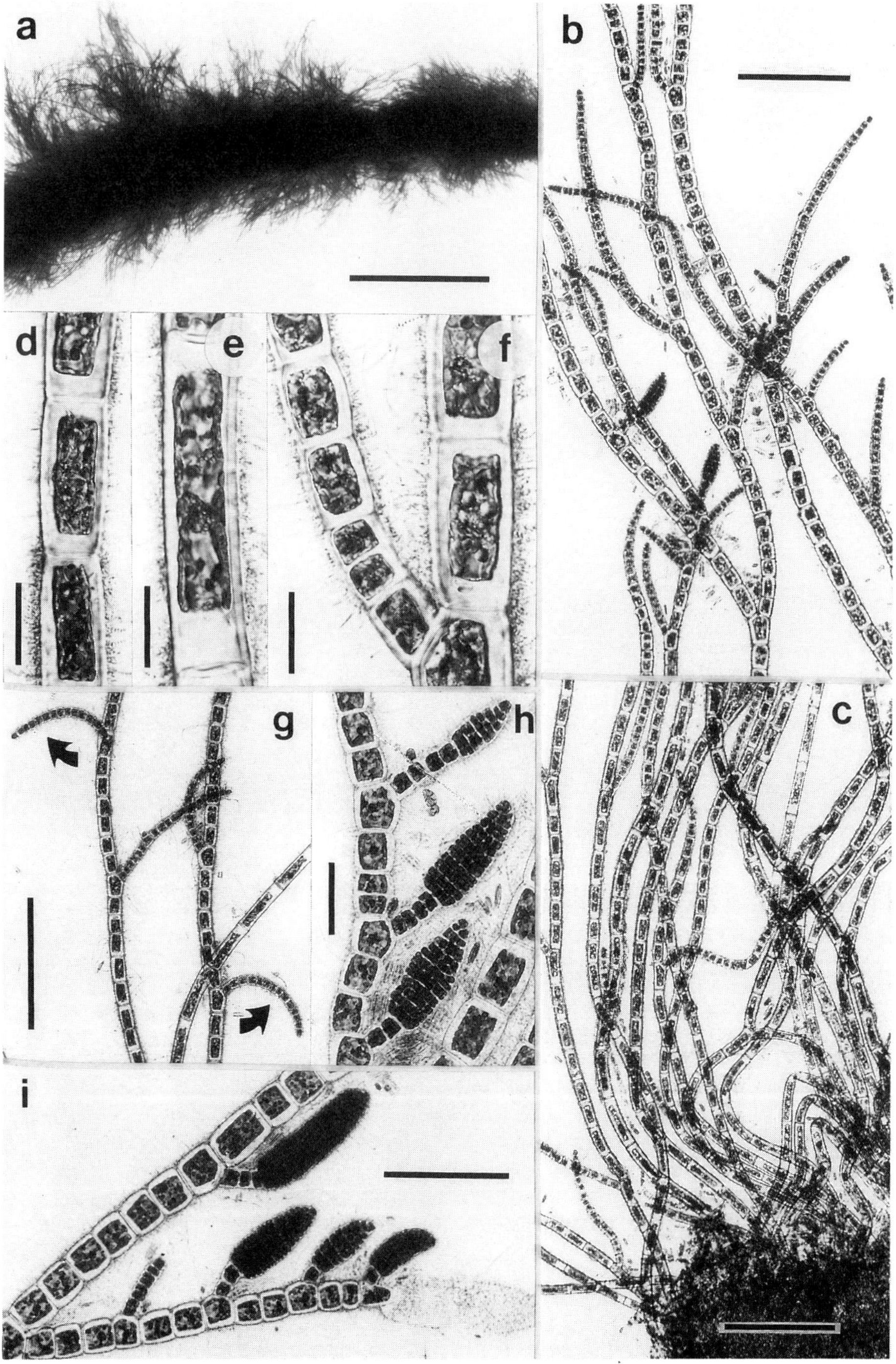

Fig. 28 *Ectocarpus constanciae* Hariot: MELU, R 1240, 1280, 1516. **a.** Habit of plants epiphytic on *Desmarestia chordalis*. **b.** Upper thallus. **c.** Lower thallus. **d–f.** Detail of cells. **g.** Determinate hook-like branchlets (arrows). **h, i.** Stipitate and terminal plurangia. Scales: a = 2 mm; b, c, g = 200 μm; d–f = 20 μm; h = 50 μm; i = 100 μm.

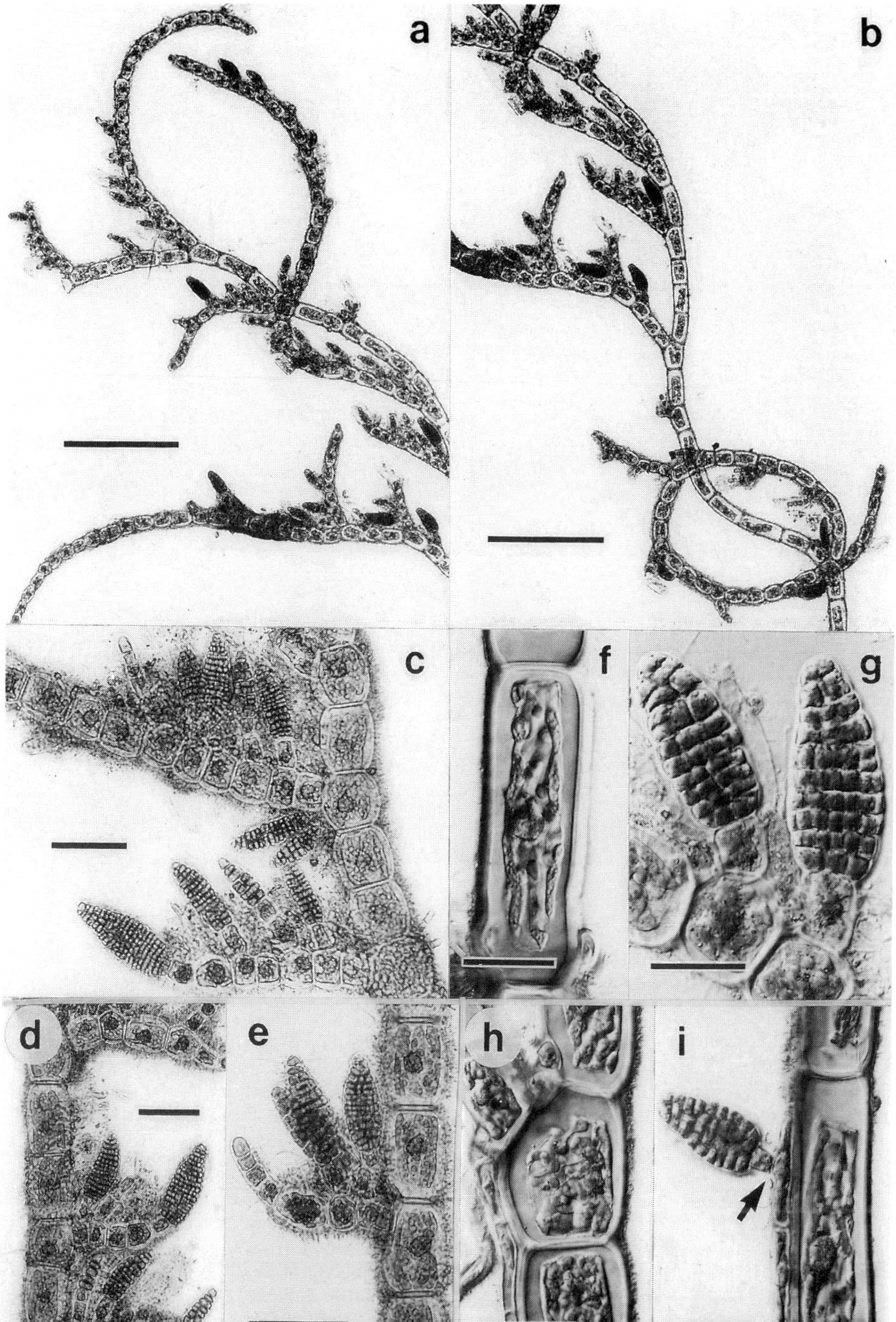

Fig. 29 *Ectocarpus constanciae* Hariot: MELU, R 3388. **a.** Habit of upper thallus. **b.** Habit of mid-thallus. **c.** Adaxial and abaxial stipitate plurangia. **d, e.** Fasciculate clusters of plurangia. **f.** Cell detail. **g.** Detail of plurangia. **h.** Corticating rhizoids. **i.** Plurangium borne on rhizoid (arrow). Scales: a, b = 200 µm; c–e = 50 µm; f–i = 20 µm.

branches, the presence of lateral hook-like branchlets, linearly arranged ribbon-like plastids, an abundance of corticating rhizoids covering lower axes, and fasciculate or non-fasciculate, relatively squat, lanceolate plurangia in *E. constanciae*.

Ectocarpus siliculosus (Dillwyn) Lyngbye, 1819: 131

Fig. 30

BASIONYM: *Conferva siliculosa* Dillwyn, 1809: 69.

SYNONYM: *Ectocarpus confervoides* (Roth) Le Jolis 1863: 75.

TYPE LOCALITY: England.

TYPE: not located but possibly in NMW (see Dixon, 1966).

SOUTHERN OCEAN DISTRIBUTION: Falkland Is, Fuegia, Juan Fernandez I. (Levring, 1941: as *E. confervoides*); Macquarie I. (Kenny & Haysom, 1962); Patagonia (Asensi, 1966*a*); South Georgia (Skottsberg, 1921).

MATERIAL EXAMINED: Macquarie I.: collections MA-140, 248, 271; Herbarium specimens: MELU, R 0788, 0816, 0832, 0833, 3384. Garden Cove, drift specimens (Taylor, 21 February 1951): MELU, R 3331, 3332, 3333, 3334 (= Taylor collection no. 543). Buckle's Bay, drift specimens (Haysom, 30 January 1950): MELU, R 3330 (= Haysom collection no. MI/49/x34).

DESCRIPTION: Highly variable thallus habit, branching pattern, cell dimensions; plants often forming dense filamentous balls, 1–2 cm diameter, epiphytic on *Desmarestia chordalis*. Filaments usually loosely clustered into brush-like tufts above, tightly twisted into ropy strands below. Filament branching irregular to alternate below, frequently secund above. Axes uncorticated, tapered toward apices, composed of barrel-shaped cells to 50 μm diameter, 1–2 times as long as broad or longer toward the thallus base. Each cell with several (generally helically arranged) band-shaped plastids containing one or more pyrenoids (Fig. 30e). Plurangia abundant, pyriform to lanceolate-elongate (65–215 μm × 22–30 μm), sometimes sessile, usually stipitate on 1–2-celled stalks, unilateral on adaxial side of lateral branches or frequently terminal on short branchlets (Fig. 30b–d, f). Intercalary and aberrant plurangia (similar to those described by G. M. Smith (1942) for *E. variabilis*) rare. Unangia not seen.

Larger, loosely organized plants to 10 cm tall (Fig. 30a) were collected occasionally from rocky surfaces 15 m deep in the subtidal. Because these thalli displayed similar branching and had comparable cell sizes and similarly shaped plurangia as seen in the smaller epiphytic balls, they are considered to be part of the same species complex.

NOTES: Although this alga occurs in many forms (similar to *f. confervoides*, *f. typica*, *f. variabilis* in Rosenvinge & Lund (1941)) around Macquarie I. and varies greatly in size and habit, in most cases it is readily separated from the complex known as *E. constanciae* by the absence of hook-like lateral branchlets. Hooks in *E. constanciae*, however, are sometimes weakly developed, and occasionally this leads to confusion between the two taxa. In these instances, differences in cell dimensions are useful for species separations.

Features such as branching angles and the presence of secondary transverse cell divisions were used by Skottsberg (1907) to distinguish some *Ectocarpus* species, but these have proved too variable in the present studies to satisfactorily separate species. Both wide (Fig. 30b) and narrow (Fig. 30c) branching angles occur within the same species. Some plants of *E. siliculosus* also produced secondary transverse cell walls in upper branches, a feature diagnostic for *E. falklandicus* Skottsberg (1907), and consequently appeared very much like the latter entity.

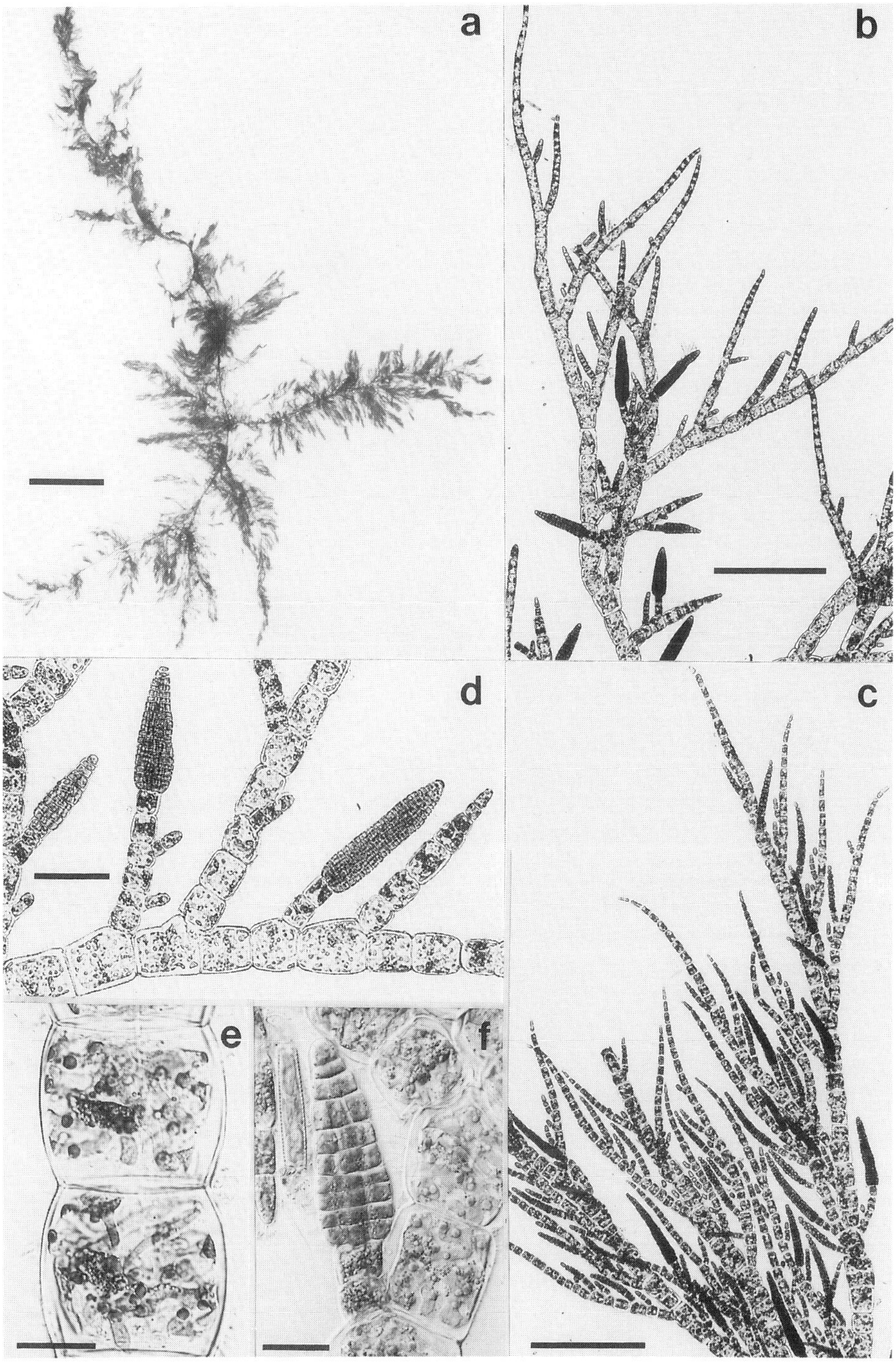

Fig. 30 *Ectocarpus siliculosus* (Dillwyn) Lyngbye: MELU, R 0788, 0816, 0832. **a.** Plant habit. **b, c.** Upper axes showing branching angles. **d.** Plurangia. **e.** Cell and plastid detail. **f.** Detail of immature sporangium. Scales: a = 1 cm; b, c = 200 μm; d = 50 μm; e, f = 20 μm.

Geminocarpus Skottsberg, 1907: 12

Type species: *G. geminatus* (Hooker f. & Harvey) Skottsberg, 1907: 13.

Plants form tufts of erect filaments to 10 cm long that attach by basal rhizoids. Filaments are initially monosiphonous but become partly polysiphonous behind branch apices. Branching is distichous-opposite, with branches arising only from the polysiphonous portions. Cells contain numerous discoid plastids. Reproductive structures are produced in unangia and plurangia, which are transformed from lateral branches, that are sessile or stipitate and may occur on the same thallus. Unangia generally are spherical, whereas plurangia are conical to lanceolate and oriented slightly toward the apices of bearing branches.

Skottsberg (1907) proposed the name *Geminocarpus* for *Ectocarpus geminatus* Hooker f. & Harvey and the new species *G. austrogeorgiae* because twin reproductive structures develop opposite one another along main axes from individuals of both taxa. The two species occur in southern high latitudes, with the former typical of Subantarctic waters and reaching north to New Zealand (Lindauer *et al.*, 1961), and the latter typical of Antarctic waters.

Geminocarpus is similar to *Pilayella* in producing true parenchyma as a result of relatively few longitudinal divisions. For this reason, Pedersen (1984) has transferred both genera, along with five others, to a new family, the Pilayellaceae, which he aligns with the Tilopteridales. One of the other five genera, *Phaearthron* (based on *Xanthosiphonia austrogeorgica* Skottsberg), also occurs in the Southern Ocean but not at Macquarie I. For the present survey, both *Geminocarpus* and *Pilayella* are maintained in the Ectocarpaceae because of their largely uniseriate filamentous thalli.

Geminocarpus geminatus (Hooker f. & Harvey) Skottsberg, 1907: 13

Fig. 31a, b, f, g

BASIONYM: *Ectocarpus geminatus* Hooker f. & Harvey, 1845: 251.

TYPE LOCALITY: San Martin's Cove, Cape Horn.

LECTOTYPE: Herb. Hookerianum (BM): unnumbered specimen. (N.B. lectotype illustrated by Lamb & Zimmermann (1977, p. 148, fig. 14).)

DISTRIBUTION: Îles Crozet and Kerguelen (Levring, 1944); Falkland Is and Fuegia (Hooker f. & Harvey, 1845); Juan Fernandez (Levring, 1941); Macquarie I. (Lucas, 1919); New Zealand (Lindauer *et al.*, 1961); South Georgia (Skottsberg, 1907, 1921).

MATERIAL EXAMINED:

1. East Falkland Is: Berkeley Sound (D. Lyall, August 1842): BM, unnumbered specimen.
2. Fuegia: St. Martin's Cove, Cape Horn (J. D. Hooker): BM, unnumbered specimen.
3. Macquarie I.: collections MA-11, 39, 83, 84, 111, 134, 137, 238, 264, 381, 397; Herbarium specimens: MELU, R 0507, 3364, 3365.

ADDITIONAL SPECIES EXAMINED:

G. austrogeorgiae Skottsberg, 1907: 14.

Vestfold Hills, Antarctica: unnamed island south of Kazak I., plants epiphytic on *Himantothallus* in 5 m depth (Ricker & Dick, 7 December 1981): MELU, R 2601, 2846.

DESCRIPTION: Plants to 2 cm tall, filamentous, with oppositely branched thalli (Fig. 31a) forming yellowish-brown epilithic, epiphytic, or epizoic tufts, attached rhizoidally from lower cells of main axes. Plants initially uniseriate (Fig. 31b), soon becoming partly multiseriate. Branches arising from multiseriate portions (Fig. 31g), comprising cells with numerous discoid plastids (Fig. 31f). Sessile multiseriate plurangia common, lateral or terminal, pyriform to 91 μm long. Unangia rare, infrequently borne with plurangia on same plant.

Plants grow in the intertidal and shallow subtidal, and most commonly occur as epiphytes on larger brown algae but also cover small rocks and chiton shells.

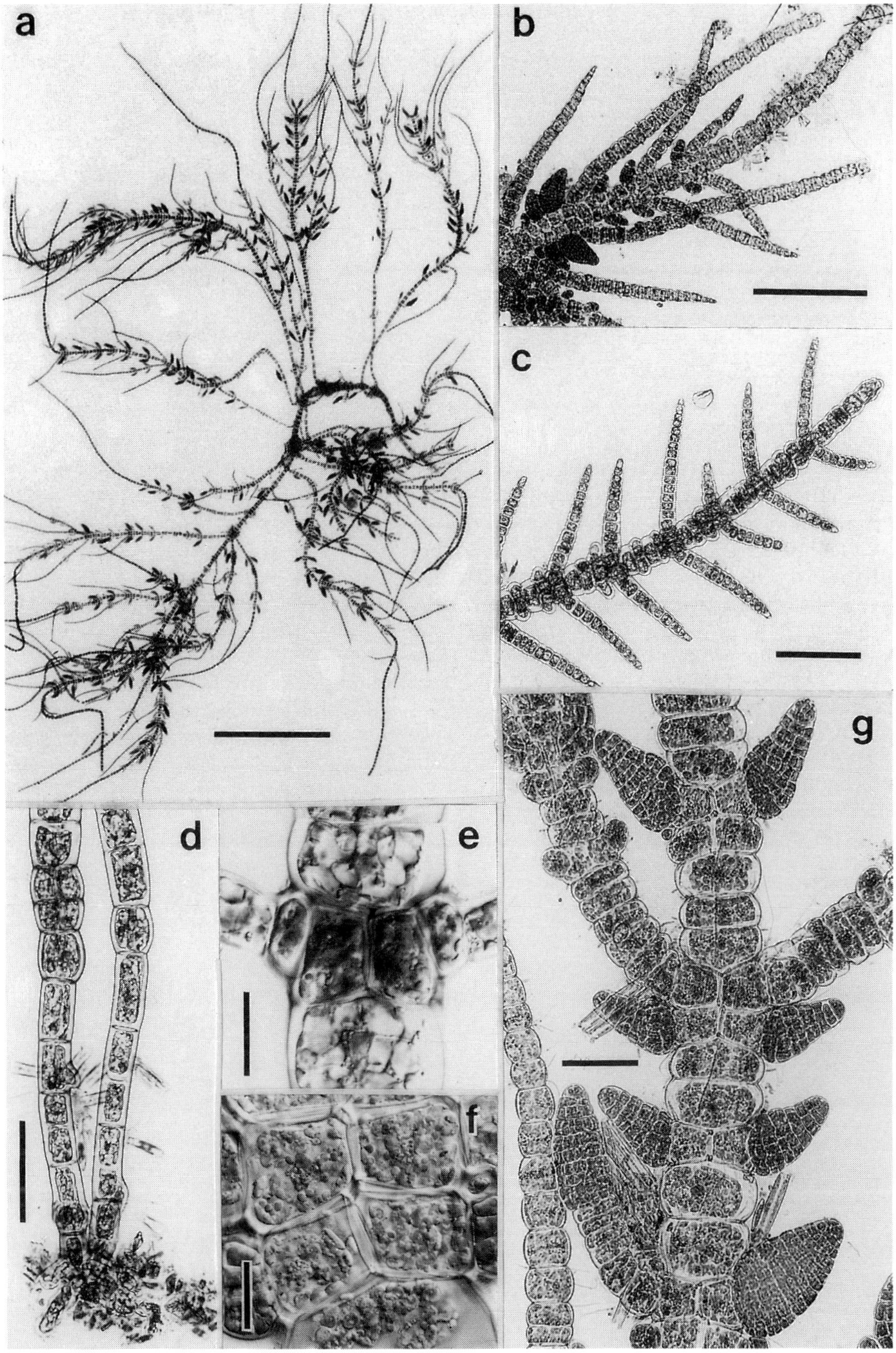

Fig. 31a, b, f, g *Geminocarpus geminatus* (Hooker f. & Harvey) Skottsberg: MELU, R 0507, 3364, 3365. **c–e** *G. austrogeorgiae* Skottsberg: MELU, R 2601, 2846. **a.** Plant habit. **b, c.** Apices of main axes. **d.** Base with rhizoidal processes. **e, f.** Detailed portions of main axes with opposite branching. **g.** Sessile plurangia on main axis. Scales: a = 1 mm; b = 200 µm; c = 100 µm; d, g = 50 µm; e, f = 20 µm.

NOTES: *Geminocarpus geminatus* is distinct from all other ectocarpalean taxa at Macquarie I. and shows minimal morphological variation throughout the Subantarctic. Macquarie specimens, however, are considerably shorter (≤2 cm) than type material from Cape Horn (to 10 cm) (Lamb & Zimmermann, 1977). *Geminocarpus geminatus* has also been reported from the Antarctic (Antarctic Peninsula—Lamb & Zimmermann, 1977; Bay of Alasheyev, Eastern Antarctica—Zinova, 1966; Victoria Land—Lucas, 1919), but these records probably represent *G. austrogeorgiae*. Both species are similar in being initially uniseriate and later partly multiseriate (Fig. 31c), but *G. austrogeorgiae* has narrower main branches (36 µm vs. 66 µm diameter) (Fig. 31e) and lacks the proliferous rhizoids (Fig. 31d) characteristic of the lower axes of *G. geminatus*. Other differences such as dark-brown colour, production of endophytic rhizoids from the basal cell of erect filaments and more elongate unangia in *G. austrogeorgiae* separate it from *G. geminatus*. Habits also are distinctive in that Macquarie plants entangle considerably and have obscured main axes, whereas Antarctic plants are more rigid, produce shorter lateral branches, do not become entangled into ropy strands and have distinct main axes.

Pilayella Bory de Saint-Vincent, 1823: 393

Type species: *P. littoralis* (Linnaeus) Kjellman, 1872: 99.

Thalli form long (to 25 cm) tufts or shorter (to 2 cm) mats of freely separating, largely uniseriate filaments that are epiphytic or epilithic and variously branched, with opposite or alternate branching most common. Prostrate axes give rise to upright filaments that are largely composed of barrel-shaped cells strongly constricted at their nodes. Growth is diffuse. Phaeophycean hairs are absent. Plastids are discoid or parietal and band-shaped, with each containing one pyrenoid. Spherical to subspherical unangia and plurangia generally form catenate series, which usually have an intercalary, but sometimes terminal position on vegetative branches.

Pedersen (1984) considers this genus to be the type for a new family in the Tilopteridales, the Pilayellaceae, which is proposed to accommodate branched parenchymatous plants that do not produce true hairs. In the present survey, *Pilayella* is maintained in the Ectocarpaceae because of its largely uniseriate thalli. Plants vary minimally, although some unusual features such as stellate plastid shape, the presence of pseudo-hairs (without basal meristems), and occasional localized meristems have been described (Lindauer *et al.*, 1961).

One species of *Pilayella*, the type, represents the genus throughout the Subantarctic.

Pilayella littoralis (Linnaeus) Kjellman, 1872: 99

Fig. 32

BASIONYM: *Conferva littoralis* Linnaeus, 1753: 1165.

TYPE LOCALITY: N. Europe.

TYPE: presumably in LINN, but specimen not seen.

SOUTHERN OCEAN DISTRIBUTION: Antarctic Peninsula (Asensi, 1966*b*); Campbell I. (Levring, 1945); Falkland Is (Cotton, 1915); Fuegia (Skottsberg, 1907, 1921); Îles Kerguelen (Zinova, 1958); Macquarie I.; Patagonia (Asensi, 1966*a*). (N.B. Papenfuss (1964*a*) indicated distribution at South Georgia, but the original published record for this locality has not been located for the present report.)

MATERIAL EXAMINED: Macquarie I.: collections MA-131, 134; Herbarium specimens: MELU, R 2150, 3344, 3369.

DESCRIPTION: Thalli to 8 cm tall, yellowish-brown, epilithic or epizoic, mat-forming, composed of largely uniseriate oppositely branched filaments forming thick skeins (Fig. 32a). Main axes mostly of uniform diameter (35–42 µm), tapering gradually toward apices, terminating in blunt apical

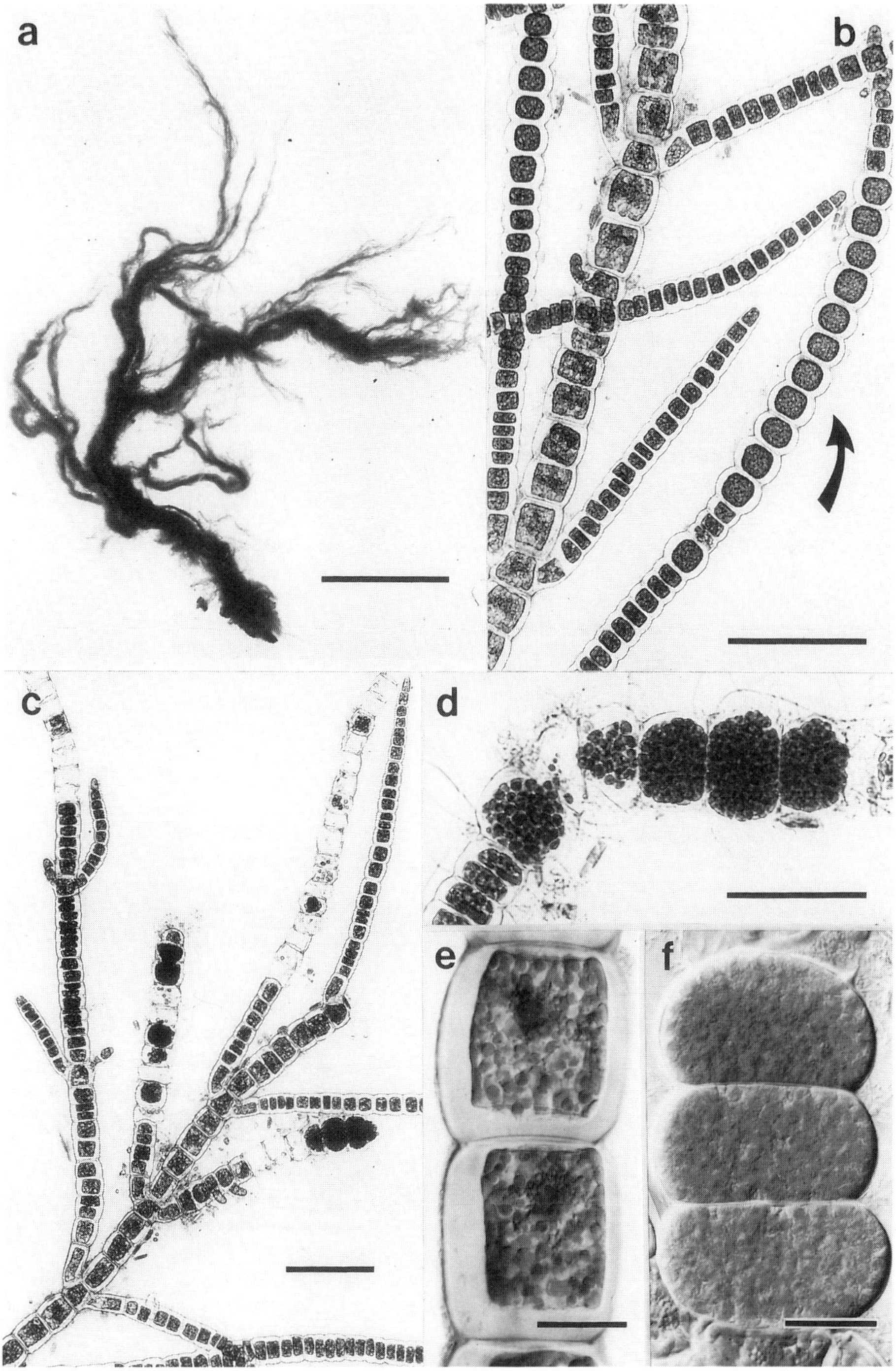

Fig. 32 *Pilayella littoralis* ((Linnaeus) Kjellman: MELU, R 2150, 3344. **a.** Rope-like thallus habit. **b.** Oppositely branched uniseriate axis with intercalary series of unilocular sporangia (arrow). **c.** Thallus with mature and empty unilocular sporangia. **d.** Detail of unilocular sporangia releasing spores. **e.** Detail of vegetative cells. **f.** Detail of sporangia. Scales: a = 1 cm; b, c : 100 μm; d = 50 μm; e, f = 20 μm.

cells. Growth diffuse, rarely localized at lateral branch bases (Fig. 32c). Lateral branches with variously bent, 2–3 times elongate, basal cells effecting narrow branching angles (Fig. 32c). Cells mostly barrel-shaped, 1–2 times as long as broad, with numerous discoid chloroplasts, each containing one or more pyrenoids (Fig. 32e). Unangia spherical (Fig. 32d) to ovoid (Fig. 32f) when mature, common in long, intercalary, catenate series on lateral branches (Fig. 32b, c). Plurangia not seen.

Pilayella littoralis is known from two collections made at the north end of the island. In one instance, a nearly pure stand of this species covered the surface of a small (*c.* 30 cm diameter) boulder in the mid-intertidal where *Durvillaea* was prevalent. In another intertidal locality, *Pilayella* grew epizoically atop limpet shells.

NOTES: Most plants appear to have diffuse meristems, although some localized growth at the bases of lateral branches approximates the trichothallic meristems of phaeophycean hairs. These branches, however, are highly pigmented, have additional meristematic zones throughout their lengths, and are very distinct from true hairs and pseudo-hairs.

Streblonema Derbès & Solier *in* Castagne, 1851: 100

Type species: *S. sphaericum* (Derbès & Solier) Thuret *in* Le Jolis, 1863: 73.

Plants are microscopic, chiefly endophytic, cryptic on host surfaces (although often visible as dull-colored spots when present on red or green algae), and primarily known indirectly through studies of their (larger) host algae. Vegetative filaments generally are twisted and composed of irregularly shaped cells. A completely endophytic system of branched, prostrate filaments gives rise to assurgent assimilatory filaments that may or may not emerge from host surfaces. Upright filaments, composed of cells with one or more discoid to band-shaped plastids, have apical or subapical meristems, whereas prostrate filaments have diffuse meristems. Phaeophycean hairs are produced by many species of *Streblonema*. Fertile branches are mostly emergent above host surfaces and commonly bear uniseriate or multiseriate plurangia.

Most species of *Streblonema* are described from the Northern Hemisphere, the majority from the North Atlantic. The genus has been previously documented from high-latitude southern waters only by *S. patagonicum* Skottsberg (1921). The newly described *S. codiatile* from Macquarie I. is the second record for the genus in the Subantarctic. Although *Streblonema* is not recorded from the Antarctic, I have observed some hydroid-tube-inhabiting brown plants with irregularly shaped cells that may belong to this genus (unpubl. pers. obs.). These endozoic filaments, however, could be the alternate (microscopic) stage of some larger (macroscopic) brown alga.

Streblonema codiatile R. W. Ricker sp. nov.

Fig. 33

HOLOTYPE: (Fig. 33a): MELU, R 1212; with isotypes in ADU, BM, Herb. Ricker, MELU, and UC.

TYPE LOCALITY: Aerial Cove, Macquarie I. (54°29′30″ S lat., 157°58′15″ E long.); on floor of protected cove in 4–5 m depth (Ricker, 2 December 1977).

DISTRIBUTION: Macquarie I.

MATERIAL EXAMINED: collections MA-46, 83, 111, 284, 300, 302, 309, 364; Herbarium specimens: MELU, R 1210, 1212, 1215, 1435.

DIAGNOSIS: Plantae in *Codio* endophyticae, filis erectis et dense intertextis prostratis quibus utriculum hospitis normaliter laxos pro cancellis rigidis continent. Color texturae infectae brunneolus ab assimilatoribus uniseriatis erectis dense pigmentiferis oriens. Fila erecta 130–175 μm longa, saepe, arcuata, simplicia aut ramificatione sparsa subdichotoma ad secundam,

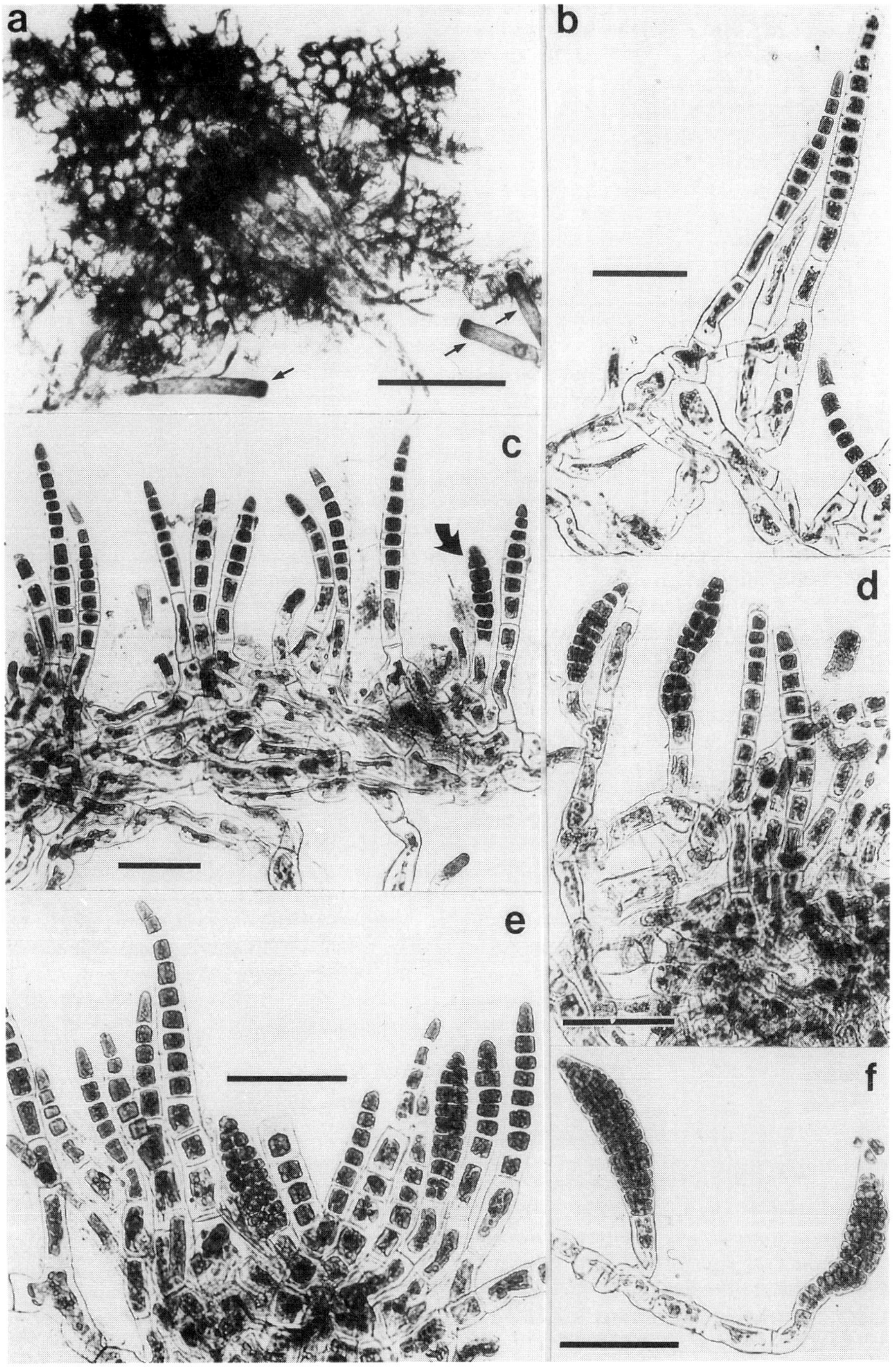

Fig. 33 *Streblonema codiatile* R. W. Ricker: MELU, R 1210, 1212, 1215, 1435. **a.** Photograph of holotype (R 1212) in water prior to drying as herbarium specimen. *Codium* utricles (arrows) normally penetrate mat lacunae. **b.** Upright filaments arising from prostrate axes. **c.** Habit of uniseriate upright filaments and young plurangium (arrow), interwoven prostrate axes. **d–e.** Habits of sporangia and upright filaments. **f.** Plurangia. Scales: a = 1 mm; b–f = 50 µm.

cellulis plurisque fere isodiametricis composita, cellulis centralibus interdum duplo latioribus quam longioribus (8–11 μm × 5–11 μm) et cellula basali 3–6 plo longiore quam latiore (6–11 μm × 25–40 μm). Fila prostrata ramosa, cellulis elongatis leviter pigmentiferis composita, saepe rhizoideis angustis lateraliter productis, structuram rigentem constituentia, propter divisiones cellularum longitudinales interdum partim biseriata. Plurangia uniseriata ad pluriseriata, oblongo-lanceolata ad lunata, 15–17 μm × 50–72 μm, terminalia vel lateralia pediculis unicellularibus insidentia, positionibus similibus assimilatoribus. Pili phaeophyceani atque unangia ignota.

Streblonema codiatile is named in recognition of its only known host, *Codium*.

Description: Plants endophytic in *Codium*, composed of erect and densely interwoven prostrate filaments that hold normally loosely associated host utricles in rigid lattice (Fig. 33a). Brownish colour of infected host tissue resulting from densely pigmented erect, uniseriate assimilators. Erect filaments 130–175 μm long, often arcuate, simple or with sparse subdichotomous to secund branching, composed mostly of cells with lengths nearly equalling breadths; mid-filament cells sometimes twice as broad (8–11 μm) as long (5–11 μm); basal cell of each upright elongated 3–6 times its diameter (6–11 μm × 25–40 μm). Prostrate filaments branched (Fig. 33b), composed of lightly pigmented, elongate cells that often laterally produce narrow rhizoids; both constituting rigid framework (Fig. 33c). Longitudinal cell divisions occasionally producing partially biseriate filaments. Plurangia uniseriate to pluriseriate, oblong-lanceolate to crescentic, 15–17 μm × 50–72 μm, terminal or lateral on single-celled stalks (Fig. 33d–f), developing in similar position to assimilators. Phaeophycean hairs and unangia unknown.

Notes: *Streblonema codiatile* is similar to *S. patagonicum* in the production of lightly pigmented, irregularly branched prostrate filaments, the position, shape, and dimensions of plurangia, the absence of any known unangia, and the occasional longitudinal cell divisions that produce partially biseriate filaments. It differs, however, in its host specificity (*S. patagonicum* is known only in *Cladosiphon zosterae*), production of short, arcuate assimilators and lack of hairs. Also, plastids of *S. codiatile* apparently fragment with age, beginning as single, parietal bands in young cells and later dividing into multiple strap-like bands in older cells. Plastids in cells of the prostrate filaments are clearly discoid and less numerous than those in the assimilators. *Streblomena patagonicum* apparently has only discoid plastids.

Howe (1914) described *S. cokeri*, which Skottsberg (1921) considered to be a near relative of *S. patagonicum*, from the coast of Peru. This species grows in an erect, branched species of *Codium* (in addition to growing in *Spatoglossum*, *Prionitis*, and *Lobocolax*) and is superficially similar to *S. codiatile*. *Streblonema cokeri* differs from *S. codiatile*, however, in having larger dimensions of filaments (to 1.8 mm long by 13–25 μm broad), in its habit with penicillate tufts of filaments on host utricle apices (vs. formation of a filamentous lattice around the utricles), and in its lack of longitudinal cell divisions, which occasionally occur in filaments of *S. codiatile*.

Ralfsiaceae Farlow, 1881: 17, 86

Members of this family are entirely encrusting (with the exception of *Analipus*) and composed of predominantly pseudoparenchymatous tissues derived from horizontally running filaments that give rise to more or less tightly compacted, usually laterally adjoined, vertical cell rows. Growth is intercalary or marginal. Cells have one or more plastids with or without a single pyrenoid/plastid. Unilocular and plurilocular structures, generally borne on separate plants, are produced laterally or terminally from, or intercalary in, vertical cell files. Life histories, where known, have isomorphic phases.

This family previously has been placed in its own order, the Ralfsiales, by Nakamura (1972), or variously treated as part of the Ectocarpales (Fritsch, 1945; Papenfuss, 1951; Nelson, 1982) or Chordariales (Kylin, 1947*a*; Lindauer *et al.*, 1961; Abbott & Hollenberg, 1976). Nakamura considered the following combination of characters to be diagnostic for the Ralfsiales: initial germination of spores into prostrate discs, cells with single, parietal, plate-like plastids that lack pyrenoids, and life histories with alternations of isomorphic somatic phases. Using several 'ralfsialean' taxa, Fletcher (1978) and Nelson (1982) both presented exceptions to each of Nakamura's criteria and preferred to consider the group (along with the Lithodermataceae, a family sometimes split from, but allied with, the Ralfsiaceae) as a single family in the Ectocarpales. Fletcher (1978) discussed differences between the pseudoparenchymatous crusts of the Ralfsiaceae and those truly parenchymatous crusts of the Cutleriaceae (Cutleriales) and Sphacelariaceae (Sphacelariales), and pointed out that encrusting plants in the Elachistaceae and Corynophlaeaceae (both families belonging to the Chordariales) produce more elaborate pustules compared to the relatively undifferentiated upright filaments of ralfsiacean plants. Possibly, as Skottsberg (1921) suggested when he placed various encrusting brown algae in the Myrionemataceae, a closer relationship exists between crusts in the Ralfsiaceae and Myrionemataceae. Members of the latter, however, generally have loosely organized erect filaments that dissociate freely.

Three genera (*Lithoderma, Petroderma,* and *Ralfsia*) are listed from the Southern Ocean (Papenfuss, 1964*a*) and two (*Lithoderma, Ralfsia*) occur at Macquarie I.

Lithoderma J. E. Areschoug, 1875: 22

Lectotype: *L. fatiscens* J. E. Areschoug 1875: 23.

Thalli are encrusting, up to *c.* 10 cm diameter, thin, firm, usually dark olive-brown to black, closely adherent to rock substrata, and composed of a monostromatic basal layer that gives rise to simple or sparingly branched, laterally adherent erect filaments. Unangia are nearly spherical and borne terminally on vertical cell files. Plurangia generally are unknown, but where known are linear-oblong to elongate-elliptical, usually uniseriate but sometimes biseriate, and borne laterally on vertical filaments.

Lithoderma is considered to be a problematical genus with close, but poorly understood, relationships to *Pseudolithoderma* Svedelius and *Sorapion* Kuckuck (Fletcher, 1978). Species numbers consequently are difficult to accurately assess. De Toni (1895) listed seven (Northern Hemisphere) species for *Lithoderma*, whereas Waern (1952), while describing an additional two new species, recognized only five species in the genus.

Skottsberg (1921), when erecting *L. piliferum* from the Falkland Is, discussed the possibility of this species belonging to either *Lithoderma* or *Pseudolithoderma*, the two being distinguished by their respective lateral or terminal positions of plurangia (Fritsch, 1945; Lund, 1959). Unfortunately, plurangia have not been seen in the Southern Ocean plants previously placed in *Lithoderma* or those presently reported from Macquarie I. Assuming that *Pseudolithoderma* is generically distinct from *Lithoderma*, confirmation of correct generic placement of these southern species awaits information on the disposition of plurangia.

Lithoderma piliferum Skottsberg var. ***debile*** R. W. Ricker var. nov.

Fig. 34a, b

Holotype (Fig. 34a): MELU, R 1964; with isotypes in ADU, BM, Herb. Ricker, and UC.

Type locality: 500 m south of Green Gorge, Macquarie I. (54°38′00″ S lat., 158°55′00″ E long.); growing at base of submerged rock wall in 11 m depth (Ricker & Horning, 12 January 1978).

Distribution: Macquarie I.

Material examined: Macquarie I.: collections MA-238, 278, 283, 284, 306, 309, 381; Herbarium specimens: MELU, R 1783, 1964, 3343, 3368.

Additional species examined:
L. antarcticum Skottsberg, 1953: 539.
Vestfold Hills, Antarctica: Tidegauge Pt., Heidemann Bay; plants epilithic in 5 m depth (Ricker, Butler, & Tucker, 14 December 1982): MELU, R 2722.

DIAGNOSIS: Thalli ad 130 μm crassi, encrustantes, leves, pallide brunnei ad fuligineos, substratis saxeis arcte adhaerentes, maculas 2–3 cm diametris formantes; filis arcte spissis et lateraliter conglutinatis cum cellulis apicalibus leviter elongatis et tholiformibus composti, nunc simplicibus cum cellulis rectangularibus et generaliter latoribus quam longioribus (6–12 μm × 6–9 μm), nunc dichotomiis singulis prope bases filorum cum cellulis quae dichotomiam subtendunt plerumque duplo latioribus quam longioribus et cum cellulis superioribus plerumque cubicis et in duobus ordinibus parallelis verticalibusque dispositis. Pili absunt. Structurae reproductivae ignotae.

The varietal name *'debile'* refers to the alga's characteristic lack of hairs.

DESCRIPTION: Thalli to 130 μm thick, encrusting, smooth, pale brown to sooty, tightly adherent to rock substrata (Fig. 34a), forming patches 2–3 cm diameter, composed of closely packed, laterally adjoined, simple or branched filaments (Fig. 34b). Single dichotomies present near bases of branched filaments, with cells subtending dichotomies commonly twice as broad as long; cells above usually cuboidal, arranged in two parallel vertical files. Unbranched filaments with angular cells sometimes nearly isodiametric, but generally broader than long (6–12 μm × 6–9 μm), terminated by slightly elongated dome-shaped cells. Hairs lacking. Reproductive structures unknown.

Crusts commonly cover small rocks in shallow tidepools in the mid-littoral and frequently are associated with *Hildenbrandia lecannellieri* on rocky surfaces in the mid- to lower littoral. Occurrence in the subtidal is seemingly less common, but crusts were seen on stones along the floors of *Macrocystis* forests in 13 m depth.

NOTES: Plants of the new variety have some features in common with both species described by Skottsberg but, in spite of their lack of hairs, more closely resemble *L. piliferum* from the Subantarctic in crust thickness and cell shapes and dimensions. Crusts of *L. antarcticum* (Fig. 34c) from the Antarctic have unbranched vertical cell rows as do some crusts from Macquarie I. The cells of *L. antarcticum*, however, are smaller, more numerous per file and result in slightly thicker crusts (to 170 μm) than those produced by the Macquarie alga. In the field, crusts of both species appear as thin, dark films, but *L. antarcticum* has a noticeably darker blackish-brown colour.

Ralfsia Berkeley *in* J. E. Smith & Sowerby, 1843: t. 2866

Type species: *R. deusta* Berkeley *in* J. E. Smith & Sowerby, 1843: t. 2866 (= *Ralfsia fungiformis* (Gunnerus) Setchell & Gardner 1924*b*: 11).

Thalli are firm, wholly encrusting, sometimes rhizoidally attached, and composed of a basal (hypothallial) layer, from a few to many cells thick, of horizontally running filaments that gives rise to an upper (epithallial) layer of assurgent, laterally adjoined assimilatory filaments. Growth of the hypothallium is along the margins. The thallus is composed of compact pseudoparenchyma. Cells of assimilators have single parietal plastids with or without pyrenoids. Hairs are present or absent and form singly or in small groups in depressions on the thallus surface. Unangia develop terminally on assimilators and may occur in groups to form raised sori on crust surfaces. Plurangia are terminal, intercalary, or lateral near the tips of pigmented vertical filaments. Life histories have isomorphic phases, with gametophytes producing isogametes.

Ralfsia comprises a heterogeneous assemblage of species, some of which are life history stages of heteromorphic members of the Scytosiphonaceae (Edelstein *et al.*, 1970) and do not belong to this isomorphic genus. Most 'species' still need to have their life histories studied in order to ascertain correct generic placement.

One species of *Ralfsia*, *R. australis* Skottsberg (1921), is recorded from the Subantarctic and occurs at Macquarie I.

Ralfsia australis Skottsberg, 1921: 20

Fig. 34d, e

Type locality: Falkland Is.

Type: presumably in S, but specimens were not seen.

Distribution: Antarctic Peninsula, McMurdo Sound (Moe, pers. comm.); Falkland Is (Skottsberg, 1921); Macquarie I.

Material examined: Macquarie I.: collections MA-83, 119; Herbarium specimens: MELU, R 2141, 2142, 2149.

Description: Thalli to 120 μm thick, encrusting, pale brown to orange-brown, drying to dark brown, lobed and sometimes overlapping producing uneven surfaces; easily detached from substratum as scaly, cartilaginous fragments. Plants composed of several cell-layers of prostrate, mostly colourless filaments bearing assurgent, pigmented filaments. Crusts attached rhizoidally. Rhizoids infrequent, produced from lowermost, horizontally running filaments composed of oblong cells (12–17 μm × 17–30 μm). Uppermost horizontally running filaments curving upwards, giving rise to laterally adhering assimilators composed of cells 1–1.5 times as long as broad. Assimilatory filaments with elongate to dome-shaped apical cells (5–10 μm diameter). Reproductive structures not seen.

Plants form small brown patches on rocky surfaces exposed to sun in the mid-littoral and on shaded rocks to 6 m depth in the sub-littoral.

Notes: Lacking reproductive structures, generic placement of this alga in *Ralfsia* may be incorrect. Because Macquarie plants largely agree with Skottsberg's (1921) description of *R. australis*, however, they are placed tentatively in this genus. The two differ apparently only in the shape of vegetative cells of assimilatory filaments, which in Macquarie thalli are more squat than those figured by Skottsberg (1921, fig. 8n) for Falkland plants. This may simply be a difference between reproductive and sterile thalli as Skottsberg's plants were bearing gametangia.

Chordariales

Although some authors (Fritsch, 1945; Russell & Fletcher, 1975; Clayton, 1981) consider the Chordariales a part of the Ectocarpales (*sensu lato*), the two orders are distinguished in the present account by their different life histories. As opposed to the alternation of isomorphic generations, as is characteristic of the Ectocarpales (*sensu stricto*), the Chordariales has heteromorphic generations with larger, morphologically distinctive sporophytes (producing either unilocular or both unilocular and plurilocular sporangia) and microscopic, filamentous gametophytes (bearing only plurilocular structures containing isogametes). Sporophytes in the Chordariales have apical, subapical, or diffuse meristems that give rise to either microscopic and filamentous (e.g., *Myrionema*) or macroscopic, pseudoparenchymatous thalli (e.g. *Papenfussiella*). The larger plants, being either uniaxial or multiaxial, consist of colourless medullary and pigmented cortical tissues. Cortical cells contain one to many plate-like, band-shaped, or discoid chloroplasts generally with pyrenoids. Phaeophycean hairs are present in many taxa.

The order comprises 10 families with some 60 genera (Wynne, 1982*b*), of which five families, including 19 genera and some 32 species, occur in high-latitude southern waters (Papenfuss, 1964*a*; Price, 1969; Ricker, pers. obs.). Four families (Elachistaceae, Myrionemataceae, Chordariaceae, Chordariopsidaceae), representing seven genera and eight species, occur at Macquarie I.

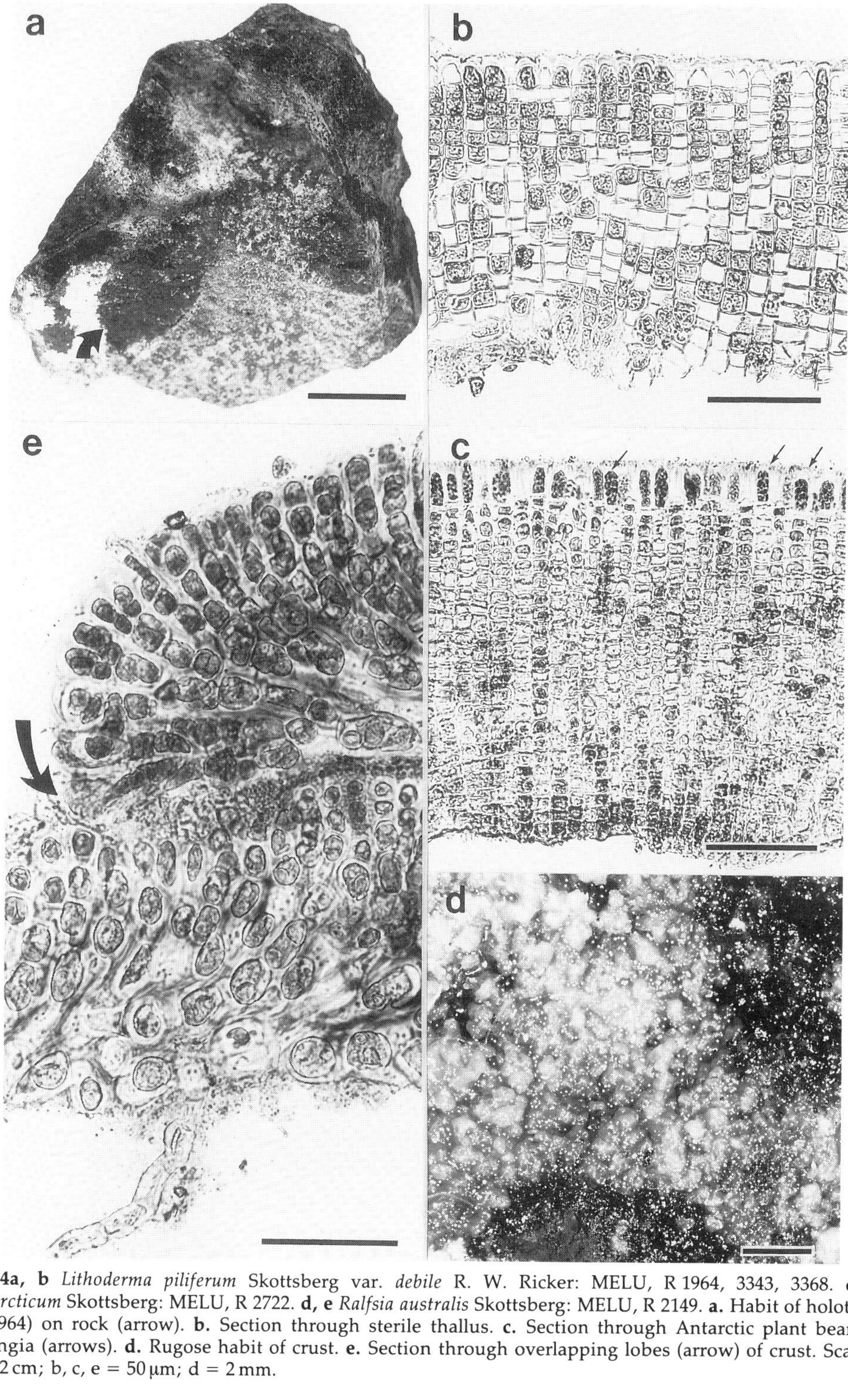

Fig. 34a, b *Lithoderma piliferum* Skottsberg var. *debile* R. W. Ricker: MELU, R 1964, 3343, 3368. **c** *L. antarcticum* Skottsberg: MELU, R 2722. **d, e** *Ralfsia australis* Skottsberg: MELU, R 2149. **a.** Habit of holotype (R 1964) on rock (arrow). **b.** Section through sterile thallus. **c.** Section through Antarctic plant bearing unangia (arrows). **d.** Rugose habit of crust. **e.** Section through overlapping lobes (arrow) of crust. Scales: a = 2 cm; b, c, e = 50 μm; d = 2 mm.

Myrionemataceae Nägeli 1847: 145, 252

Plants usually are endophytic or epiphytic and form discoid or irregularly shaped patches constructed with a monostromatic or distromatic prostrate layer, upper-surface cells of which give rise to either erect filaments, colourless hairs, or sporangia. Erect assimilatory filaments are short, simple or sparingly branched, monosiphonous and composed of cells containing several discoid plastids, each with one pyrenoid. Hairs and sporangia develop laterally on erect filaments. Unangia and plurangia are known, the latter either uniseriate or multiseriate, linear or ovate-lanceolate.

Papenfuss (1964*a*) listed two genera (*Hecatonema, Myrionema*) from the Southern Ocean. Only the latter occurs at Macquarie I.

Myrionema Greville, 1827: pl. 300

Type species: *M. strangulans* Greville, 1827: pl. 300.

Papenfuss (1964*a*) listed eight *Myrionema* species from southern high latitudes, seven of which are based on plants from the Southern Ocean. Five of these seven were described by Skottsberg (1907—*M. densum, M. incommodum, M. macrocarpum*; 1921—*M. fuegianum, M. patagonicum*), and the remaining two by Reinsch (1890—*M. inconspicuum, M. (?) paradoxum*). Reinsch's descriptions and illustrations of both species are equivocal, but Skottsberg (1921) has suggested that *M. inconspicuum* may be identical to *M. densum*. Baardseth (1941) recorded the Northern Hemisphere species *M. corunnae* Sauvageau from Tristan da Cunha, but this too may represent *M. densum*, which Skottsberg (1921) described as very similar to *M. corunnae*.

Species in this genus are separated from one another by differences in habitat (epiphytic vs. endophytic) or substratum (host), shape and persistence of prostrate (basal disc) systems, presence or absence of erect assimilatory filaments and phaeophycean hairs, shapes and sizes of assimilators, sporangia, and gametangia, and number of chloroplasts per cell.

Only one species of *Myrionema, M. incommodum*, grows at Macquarie I., where it is present in the cortical tissues of the brown alga *Adenocystis utricularis*.

Myrionema incommodum Skottsberg, 1907: 52

Fig. 35

Type locality: South Georgia.

Type: ? (not found at Skottsberg's Herbarium in Stockholm (S), where the bulk of his Antarctic algal collections is housed).

Distribution: Antarctic Peninsula (Skottsberg, 1953); Australia (Womersley, 1967); Falkland Is, Fuegia, South Georgia (Skottsberg, 1907, 1921); Macquarie I.; Patagonia (Asensi, 1966*a*).

Material examined: Macquarie I.: collection MA-213; Herbarium specimen: MELU, R 3326.

Description: Plants filamentous, forming dark-brown, sometimes tomentose patches to 1 mm across, endophytic in or epiphytic on cortical tissues of *Adenocystis utricularis* (Fig. 35a, b); endophyte anchored among host cortical cells by elongate basal cells. Thalli with pigmented, simple or irregularly branched assimilators (7–9 μm diameter) consisting of cells 2–3 times as long as broad (Fig. 35e); uniseriate plurangia (35–48 μm × 6–11 μm) with elongate single-celled stipes, borne laterally on assimilators (Fig. 35f), usually emergent from host cortex (Fig. 35d). Unilocular structures and phaeophycean hairs unknown.

The host alga *Adenocystis* grows abundantly in the lower intertidal, where it is recognized easily by its pale-brown, turgid sacs. On the lower halves of many sacs, dark brown patches may develop from infections of either *Gononema* or *M. incommodum*. Individuals of the former usually form more emergent, epiphytic growths and also produce basal masses of colourless pseudo-

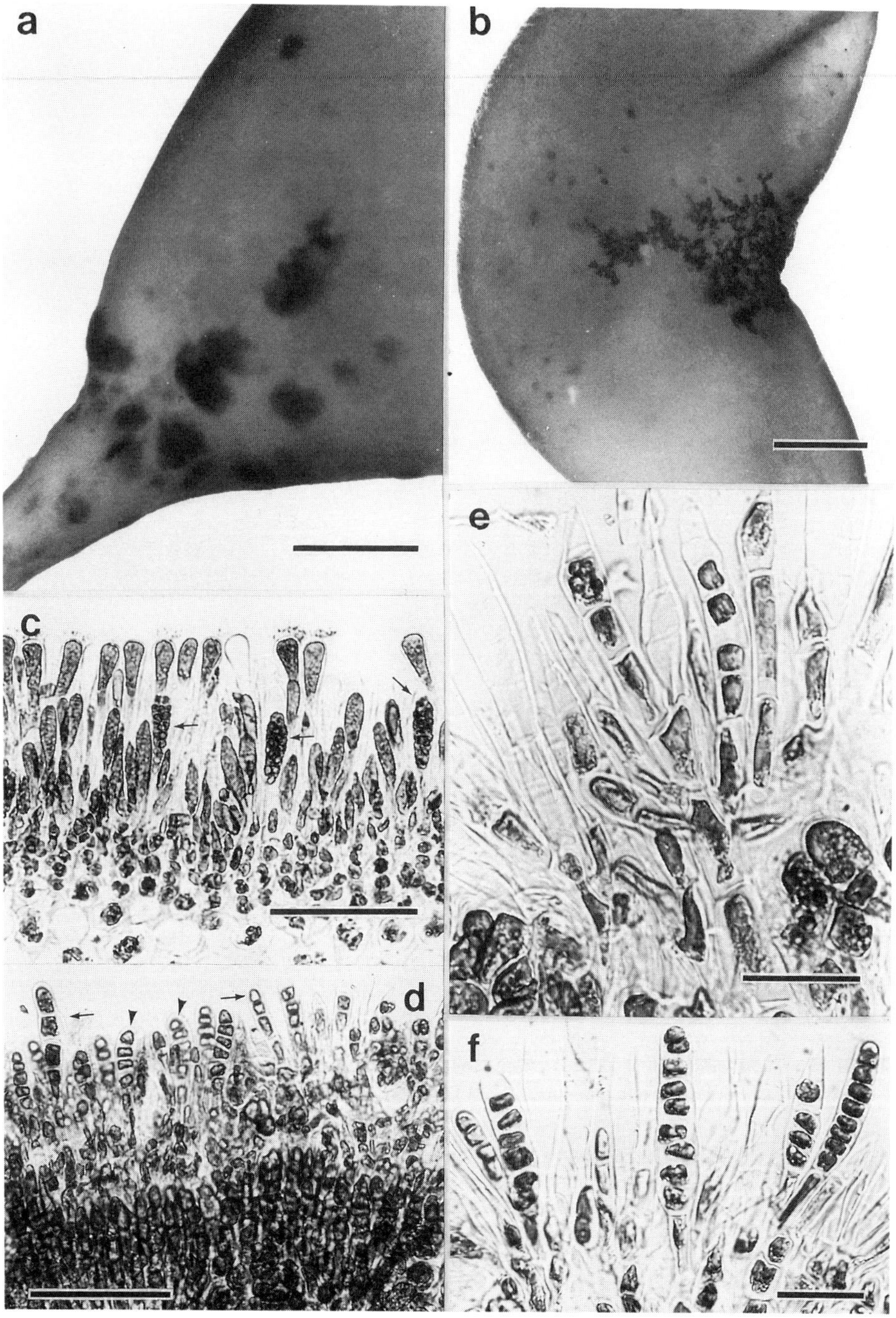

Fig. 35 *Myrionema incommodum* Skottsberg: MELU, R 3326. **a.** Habit of plants forming woolly patches on lower thallus of infected *Adenocystis*. **b.** Mid-thallus of *Adenocystis* with compact masses of infecting, endophytic filaments. **c.** Uninfected cortex of *Adenocystis* bearing unilocular sporangia (arrows). **d.** Infected *Adenocystis* showing *Myrionema* assimilators (arrows) and plurangia (arrowheads) between and above cells of host cortex. **e.** Detail of branching filaments bearing emptied plurangia. **f.** Stipitate plurangia. Scales: a = 1 mm; b = 2 mm; c, d = 50 μm; e, f = 20 μm.

parenchyma (Fig. 38b, c). Additionally, *Gononema* often grows only in cryptostomata in the host cortex, whereas *Myrionema* may spread throughout host cortical tissues. At Macquarie I., *M. incommodum* was seen only on fertile host thalli and no prostrate basal disc (as is typical for most *Myrionema* species) could be found. Uninfected fertile host tissues, bearing clavate unicellular paraphyses and spindle-shaped unilocular sporangia (Fig. 35c), were distinguished readily microscopically from the more cluttered cortical tissues of infected plants that contained plurangia produced by the endophyte (Fig. 35d).

NOTES: In contrast to Skottsberg's (1921) description of South American plants that produce weakly branched assimilators, colourless hairs and smaller, sometimes biseriate plurangia, plants from Macquarie I. have highly branched assimilators, no phaeophycean hairs and larger, strictly uniseriate plurangia. Also, Skottsberg's (1907, figs 58–60) initial illustrations of this species showed structures he believed to be unilocular sporangia, but the production of such sporangia has not been confirmed in the present survey. In spite of these differences, Macquarie plants are identified as *M. incommodum* because this is the only Southern Ocean species whose basal disc breaks down as the endophyte ramifies through fertile cortical tissues of *Adenocystis*. The extensive spread of *M. incommodum* through the cortex of *Adenocystis* appears to be typical for this species, and this in part led Skottsberg (1921) to consider Harvey's (1858, pl. 48, figs 3, 4) illustration of the anatomy of *Adenocystis* to be incorrect and actually representative of *M. incommodum*.

Myrionema macrocarpum is another species that also grows on *Adenocystis* (Skottsberg, 1907) and conceivably might be present at Macquarie I., but its vegetative structure (with a persistent basal disc) and swollen, pluriseriate plurangia were not found on host plants at this locality.

Elachistaceae Kjellman, 1890: 41

Members of this family usually are epiphytic and form basal cushions from which long assimilators emanate. The cushions are either loosely or compactly constructed of branched colourless filaments. Unilocular and plurilocular sporangia arise where the assimilator bases attach to the cushions, thereby forming a distinct (dark) zone separating the pseudoparenchymatous, colourless tissue below from the pigmented filaments above.

The Elachistaceae appears to be closely related to the Ectocarpaceae and has been placed by some authors (Fritsch, 1945; Clayton, 1981) in the Ectocarpales. It is presently grouped with the Chordariales primarily because its members develop pseudoparenchyma.

Wynne (1982*b*) attributes five genera (*Elachista, Halothrix, Herpodiscus, Leptonematella* (= *Leptonema* in Papenfuss, 1964*a*; see Wynne, 1969), *Portphillipia*) to the Elachistaceae, but Pedersen (1978) suggests that *Leptonematella* is more closely related to *Pogotrichum* in the Pogotrichaceae (Dictyosiphonales). *Halothrix* is known only from the Northern Hemisphere, *Herpodiscus* and *Portphillipia* from the Southern Hemisphere, and the remaining genera, in addition to *Gononema* (transferred in the present survey from the Ectocarpaceae), occur in both hemispheres. *Elachista* and *Gononema*, however, are the only genera representing the Elachistaceae at Macquarie I.

Elachista Duby, 1830: 972 (***nom. et orth. cons.***)

Type species: *E. scutulata* (J. E. Smith) Areschoug, 1843: 262.

Members of this genus typically are cushion-like, with basal masses of colourless pseudoparenchyma that bear two types of unbranched pigmented assimilators. Short (determinate) assimilatory filaments (known either as paraphyses (Kuckuck, 1929) or cortical filaments (Skinner, pers. comm.)) and long (indeterminate) assimilatory hairs with basal meristems arise terminally and laterally from branched, colourless medullary filaments that constitute the basal cushions. Unilocular and plurilocular structures develop at the juncture of assimilators and medullary filaments. Phaeophycean hairs are not produced.

Elachista species are distinguished from one another by their occurrence on specific hosts, and by differences in the structure of medullary tissues (based on branching patterns of the colourless filaments), sizes and shapes of assimilator cells and sporangia, presence or absence unangia and plurangia, and life history patterns.

Three species of *Elachista*, *E. antarctica* Skottsberg, *E. flaccida* (Dillwyn) Fries, *E. meridionalis* Skottsberg, are listed from the Southern Ocean (Papenfuss, 1964*a*), but only the first occurs at Macquarie I.

Elachista antarctica Skottsberg, 1953: 537

Fig. 36

TYPE LOCALITY: Melchoir Is, Antarctic Peninsula.

TYPE: ? (not found at Skottsberg's Herbarium in Stockholm (S), where the bulk of his Antarctic algal collections is housed).

DISTRIBUTION: Antarctic Peninsula (Skottsberg, 1953); Macquarie I.

MATERIAL EXAMINED:

1. Macquarie I.: collection MA-264; Herbarium specimens: MELU, R 1962, 2411–2415.
2. Antarctic Peninsula: epiphytic on *Palmaria* (Moe, March 1974): UC, Microscope Slide: no. 310a.

DESCRIPTION: Plants pulvinate, 1–2 mm diameter, light- to olive-brown, epiphytic on *Palmaria decipiens* (Fig. 36a); basal cushion hemispherical, with loosely to tightly compacted medulla (Fig. 36b, c) composed of reticulum of interconnected, irregularly to oppositely branched, colourless filaments comprising cells 2–3 times as long as broad (Fig. 36f); lower medullary cells with small knobs interlocking between host cells. Cushion surface covered by two types of assimilatory filaments, each containing numerous discoid plastids (Fig. 36g); cortical assimilators determinate, but of various lengths, straight to slightly curved, often with narrowed bases and inflated terminal cells; assimilative hairs to 5 mm long, composed of narrow cells (4–5 times as long as broad) immediately subtending basal meristems with broader, nearly isodiametric cells (Fig. 36d) that distally produce barrel-shaped cells 1–2.5 times as long as broad (20–25 µm × 30–50 µm) (Fig. 36e). Unilocular sporangia clavate to obpyriform (20–30 µm × 85–110 µm), sessile, borne laterally from basal cells of assimilators. Plurilocular structures and phaeophycean hairs unknown.

Some 30 plants, growing epiphytically on *Palmaria decipiens*, were collected from a tidepool in the lower intertidal. Because the epiphytes were observed on only one individual of the host alga, *E. antarctica* is presumed to be rare at Macquarie I.

NOTES: Plants from Macquarie I. are similar to individuals of the same species from Anvers I. along the Antarctic Peninsula, some 50 km south of the type locality in the Melchoir Is, in their host preference and production of a basal cushion that gives rise to long and short assimilatory filaments. Antarctic plants differed, however, in their longer (150–160 µm), somewhat angular unilocular sporangia and more compact medullary tissues.

In the present account, certain thallus structures are interpreted differently from Skottsberg's (1953) original concepts. Also, plants examined in the present survey differ in some morphological respects from the type description. Skottsberg noted that basal cushions give rise to numerous assimilatory filaments of differing lengths and he considered the shorter filaments,

---

Fig. 36 *Elachista antarctica* Skottsberg: MELU, R 1962, 2415; Moe, 310a. **a.** Habit of plants epiphytic on *Palmaria decipiens*. N.B. pigmented assimilatory filaments arising from basal cushions. **b.** Section through cushion and host from Macquarie I. **c.** Section through cushion and host (*'Leptosarca simplex'* = *Palmaria*) from Antarctic Peninsula. **d.** Bases of assimilators. **e.** Assimilator cells from mid-filament. **f.** Cushion-like medulla composed of branching filaments (arrow). **g.** Unilocular sporangia interspersed between long (hollow arrow) and short (solid arrow) assimilators. Scales: a = 1 mm; b = 300 µm; c = 200 µm; d, e, g = 25 µm; f = 100 µm.

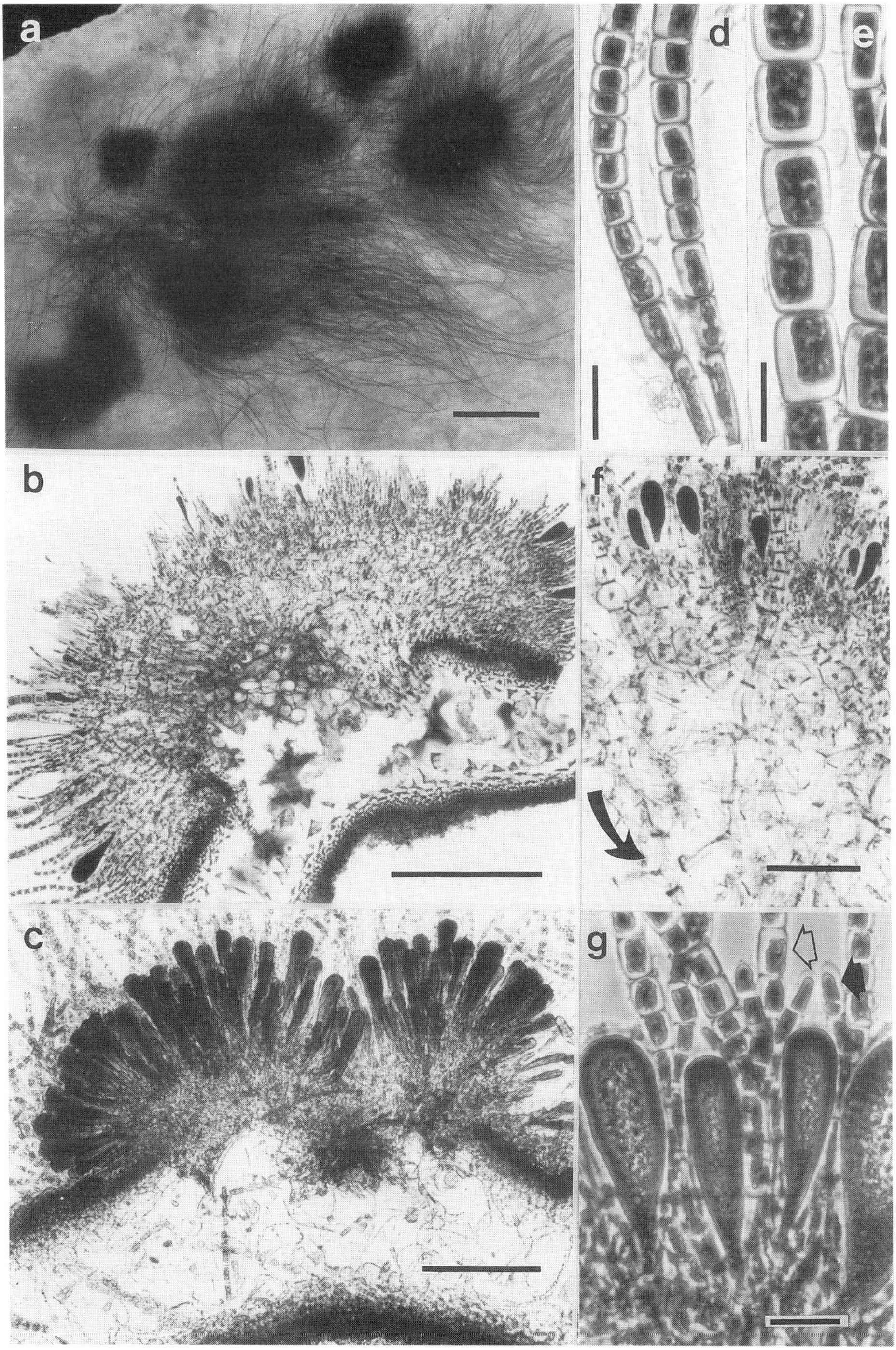
a
b
c
d
e
f
g

which he felt might be mistaken as paraphyses, to be underdeveloped long filaments. At present, *E. antarctica* is characterized as having two distinct types of assimilators (similar to other *Elachista* species (Setchell & Gardner, 1925)), the short ones known as cortical filaments and the long ones as assimilatory hairs. It appears that the latter are deciduous and lost something during the summer season, because some plants are heavily clothed in a tomentose covering, whereas others are nearly naked.

The medulla of Macquarie specimens is composed of branched filaments comprising oblong to more elongate cells. This differs considerably from Skottsberg's (1953, fig. 6a) illustration which shows a compact basal mass of nearly isodiametric cells not organized in distinct, branched filaments. Skottsberg also described the long, hair-like assimilators as having rounded apices and tapered upper portions composed of some 30–50 cells in the narrowed apical zone. Comparable filaments from Macquarie plants terminate in elongate apical cells with rounded tips, but the distal portions of the filaments are tapered to a lesser extent than those described by Skottsberg. It is unknown, however, how greatly this feature varies.

Gononema Kuckuck & Skottsberg *in* Skottsberg, 1921: 9

Lectotype: *G. pectinatum* (Skottsberg) Kuckuck & Skottsberg *in* Skottsberg, 1921: 10 (see Papenfuss (1964*a*) re. lectotypification).

Kuckuck & Skottsberg (*in* Skottsberg, 1921) described *Gononema* as being composed of basal masses of colourless tissue (similar to but not as extensive as in *Elachista*) and erect, branched, monosiphonous, pigmented filaments (similar to *Ectocarpus*). Additionally, members of the genus produce two types of plurangia, of which the longer generally are borne near the juncture of pigmented and non-pigmented cells, and the smaller on the erect filaments. When naming the genus, Skottsberg (1921) was referring to the extremely fertile filaments that its two constituent species (one formerly a member of *Ectocarpus* and the other *Elachista*) displayed.

Skottsberg (1921) placed this genus in the Ectocarpales because he considered the branching of filaments, shape of plurangia, and position of hairs very similar to other members of this order. In the present account, however, placement in the Elachistaceae is based upon the production of a compact, basal mass of colourless pseudoparenchyma. This feature alone is considered to be taxonomically more important at the family level than are the criteria discussed by Skottsberg. Although the Elachistaceae has been characterized by some (e.g., Newton, 1931) as lacking phaeophycean hairs, other concepts of the family (e.g., Setchell & Gardner, 1925; Lindauer *et al.*, 1961) have included hair-bearing members.

Kuckuck (*in* Skottsberg, 1921) intended to place *Ectocarpus tomentosoides* Farlow and *E. ellipticus* Saunders in *Gononema*, and Skottsberg (1921) indicated that *E. elachistaeformis* Heydrich from the Danish West Indies probably should follow similarly. Should these three *Ectocarpus* species ever be formally moved to *Gononema*, the genus would contain representatives in both tropical and high-latitude seas. At present, however, the genus is known only from cold waters in the Subantarctic and Arctic (following Pedersen's (1981) addition of two species, *G. aecidioides* and *G. alariae*, from Greenland). Both Southern Hemisphere taxa occur at Macquarie I.

Gononema pectinatum (Skottsberg) Kuckuck & Skottsberg *in* Skottsberg, 1921: 10

Fig. 37

Basionym: *Ectocarpus pectinatus* Skottsberg, 1907: 11.

Type locality: Falkland Is.

Fig. 37 *Gononema pectinatum* (Skottsberg) Kuckuck & Skottsberg: MELU, R 3383. **a.** Habit of tufted filaments growing epiphytically on thin black crust (possibly a lichen). **b.** Upper thallus with epiphytic protozoan (arrow). **c.** Base of phaeophycean hair (arrow). **d.** Mid-filament with pectinate branching and empty plurangia. **e.** Cell detail. **f.** Detail of branching. **g.** Plurangia. Scales: a = 1 mm; b = 50 μm; c, f, g = 20 μm; d = 40 μm; e = 10 μm.

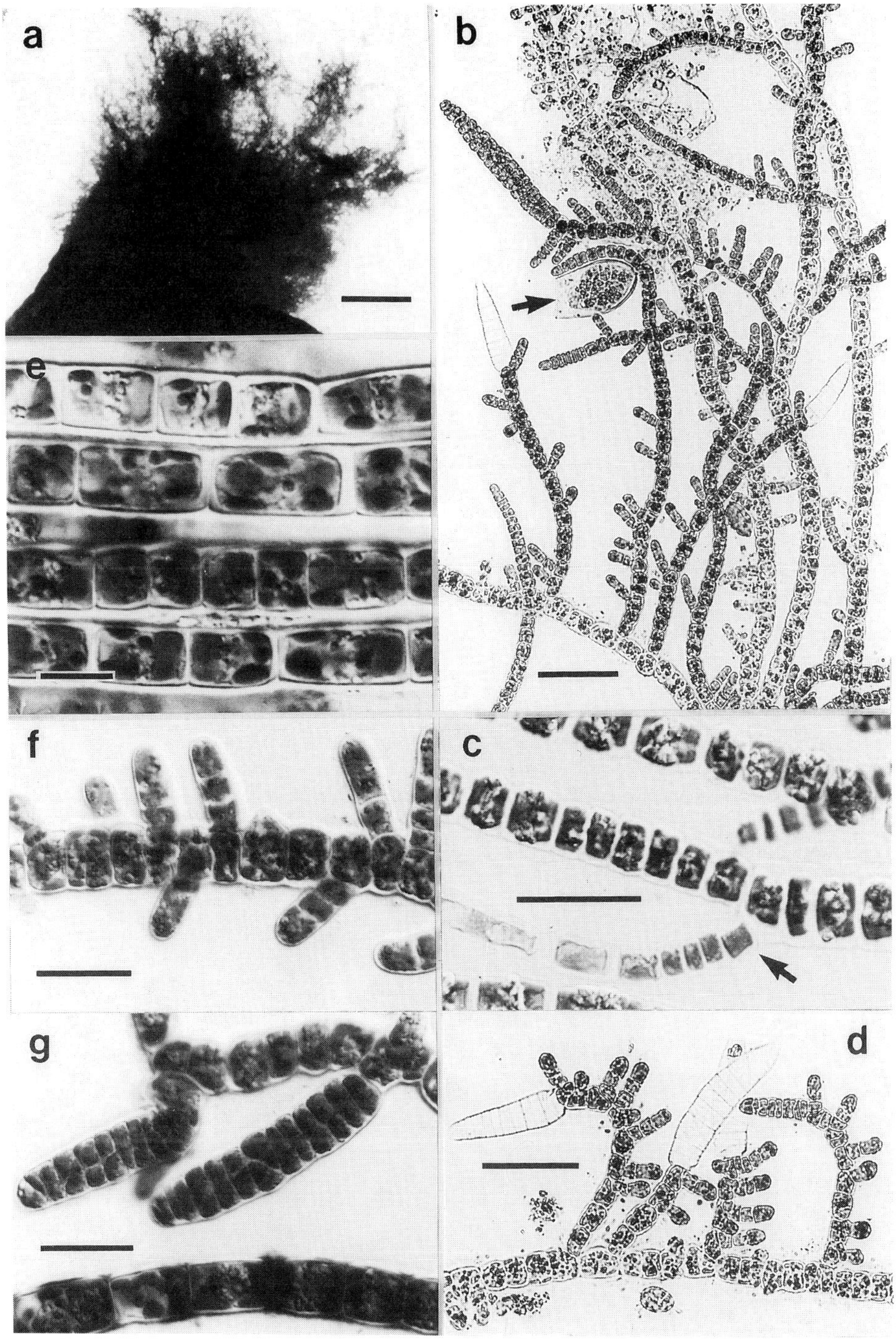
a
b
e
f
c
g
d

TYPE: ? (not found at Skottsberg's Herbarium in Stockholm (S), where the bulk of his Antarctic algal collections is housed).

DISTRIBUTION: Falkland Is (Skottsberg, 1907, 1921); Macquarie I.

MATERIAL EXAMINED: Macquarie I.: collection MA-134; Herbarium specimen: MELU, R 3383.

DESCRIPTION: Plants filamentous, forming tufts to 3 mm tall (Fig. 37a), epiphytic on thin, black crust (possibly lichen) covering rock. Thallus base a mass of colourless cells bearing erect, pigmented filaments. Upright axes (8–10 μm diameter) pectinately to alternately to irregularly branched (Fig. 37b), with numerous, short (2–4 cells) pectinately produced laterals (presumptive plurangia). Phaeophycean hairs borne laterally on branches (Fig. 37c). Growth intercalary. Cells mostly broader than long, sometimes more elongate, each containing several plate-like to band-shaped chloroplasts (Fig. 37e, f). Plurangia to 105 μm long, oblong-lanceolate, pluriseriate, sessile, borne singly at apices or laterally along main axes (Fig. 37b, d, g). Unangia unknown.

A single fertile clump of reddish-brown filaments was collected in late December from a shallow mid-intertidal pool.

NOTES: This species was described first by Skottsberg (1907) as a member of *Ectocarpus*, but later was moved to *Gononema* (Skottsberg, 1921) because it produced phaeophycean hairs laterally and bore numerous plurangia pectinately along main axes. In these respects as well as in the dimensions of filaments, diffuse growth, and absence of basal plurangia (which are produced by *G. ramosum*), plants from Macquarie I. agree well with Skottsberg's (1907, 1921) descriptions of *G. pectinatum*. In contrast, however, the plants originally described by Skottsberg grew epiphytically on *Cladophora*, and later collections (1921) came from cryptostomata of *Adenocystis*. Furthermore, the plurangia of the type species are pluriseriate (Skottsberg, 1907, figs 6–8) as a result of longitudinal divisions of the 2–4-celled lateral plurangium initials, whereas plurangia on Macquarie plants apparently lack longitudinal divisions and thus resemble determinate, vegetative laterals.

In both of Skottsberg's descriptions of *G. pectinatum*, few details appear regarding the basal thallus structure. In the present survey, a small mass of disorganized colourless cells was seen at the base of the erect, pigmented filaments. This tissue was much less extensive than the pseudoparenchyma observed at the base of *G. ramosum* (Fig. 38c).

Gononema ramosum (Skottsberg) Kuckuck & Skottsberg *in* Skottsberg, 1921: 11

Fig. 38

BASIONYM: *Elachista? ramosa* Skottsberg, 1907: 54.

TYPE LOCALITY: Fuegia.

TYPE: ? (not found at Skottsberg's Herbarium in Stockholm (S), where the bulk of his Antarctic algal collections is housed).

DISTRIBUTION: Falkland Is, Fuegia (Skottsberg, 1907, 1921); Macquarie I.; Patagonia (Asensi, 1966*a*).

MATERIAL EXAMINED: Macquarie I.: collections MA-215, 381; Herbarium specimen: MELU, R 2148.

DESCRIPTION: Plants filamentous, forming compact tufts (to 500 μm tall) of pigmented assimilators arising from basal, pseudoparenchymatous mass of colourless filaments embedded in cryptostomata of *Adenocystis utricularis* (Fig. 38a–c). Assimilator cells of equal length to width, or longer than wide (9–25 μm × 4–5 μm), filled with numerous band-shaped to discoid chloroplasts (Fig. 38d). Phaeophycean hairs 4–6 μm diameter, produced directly from thallus base (Fig. 38e). Plurangia lanceolate to elongate-linear (6–10 μm × 55–85 μm), uniseriate to biseriate, borne laterally and terminally on erect filaments (Fig. 38f–h). Unangia unknown.

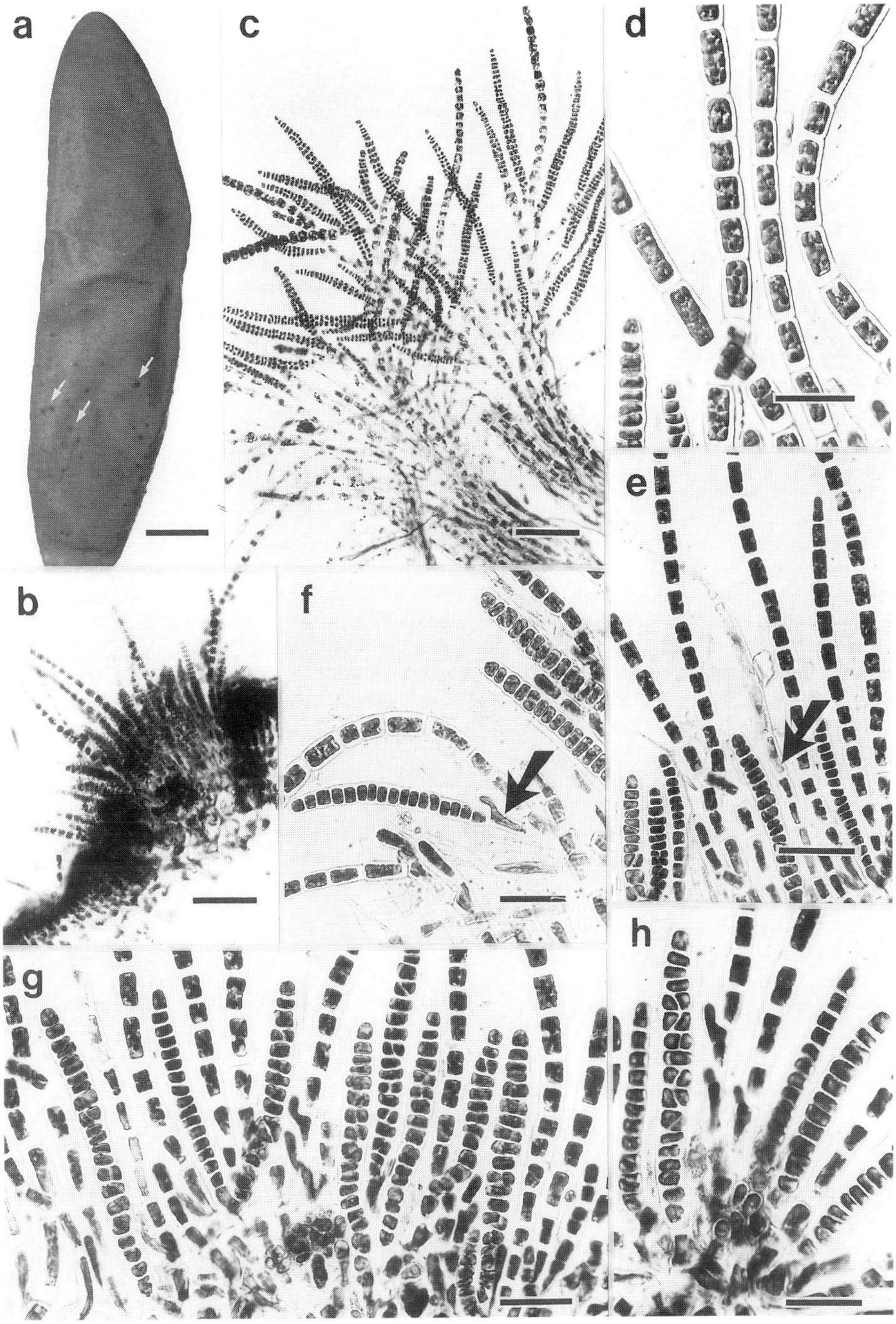

Fig. 38 *Gononema ramosum* (Skottsberg) Kuckuck & Skottsberg: MELU, R 2148. **a.** Habit of tufts (arrows) on *Adenocystis utricularis*. **b.** Section through host and endophyte. **c.** Habit of clustered filaments bearing plurangia. **d.** Detail of assimilatory filaments. **e.** Base of phaeophycean hair (arrow). **f.** Young plurangium apically positioned but being displaced laterally by continued growth of assimilator (as lateral bulge — arrow) from cell bearing plurangium. **g, h.** Plurangia. Scales: a = 1 cm; b, c = 50 µm; d, f–h = 20 µm; e = 30 µm.

Some 15 clumps of filaments, growing in separate cryptostomata on one individual of *Adenocystis*, were studied. The host alga was collected from a tide-channel, 1–2.5 m deep, that flowed into a large tidepool in the lower intertidal. In this habitat, *Adenocystis* grows in abundance alongside various encrusting coralline algae, *Acrosiphonia pacifica*, and *Codium subantarcticum*.

NOTES: Plants largely agree with Skottsberg's (1921) description of the species, except that plurangia are not produced on well-developed erect filaments. The only plurangia present on plants from Macquarie I. are equivalent to those that Skottsberg (1921) referred to as larger, basal gametangia. These apparently develop apically on young assimilatory filaments, but later are displaced laterally by continued development of the assimilatory filaments from the cells bearing plurangia (Fig. 38f).

In general, Macquarie plants lacked the long, plurangium-bearing filaments (similar to the erect axes of *E. pectinatum*) described by Skottsberg for mature individuals from the Falkland Is and Fuegia. In this respect, Macquarie specimens were more like the epiphyte on *Caepidium* that Skottsberg interpreted as an immature individual of *G. ramosum*.

Chordariaceae Greville, 1830: 44

Members of this family are distinguished by the presence of intercalary meristems and outer surface coverings of anticlinal, simple or branched, assimilatory filaments (Kylin, 1940). Plants typically have three tissue zones that consist of an unpigmented, filamentous medulla (sometimes with a central cavity), an intermediate, subcortical layer that comprises either rounded or elongate colourless cells arranged compactly or loosely in outwardly radiating filaments, and a cortical layer of anticlinal pigmented filaments. Assimilatory filaments are either short and compacted, thus producing a cartilaginous texture, or long and loosely united in a gelatinous matrix. If, as in some species, plants have both short and long assimilators, thallus consistency is determined by the more abundant or more persistent type of filament. Chordariacean taxa are separated from one another largely on differences in the above criteria as well as in the presence or absence of colourless hairs, monopodial vs. sympodial development, unaxial or multiaxial thalli, and shapes and sizes of cortical cells and sporangia (Kylin, 1940; Inagaki, 1958).

This primarily cold-water family contains 29 genera (Wynne, 1982*b*), of which 12 occur in high-latitude southern seas. Ten of these, comprising about 13 species distributed throughout the Southern Ocean, are listed by Papenfuss (1964*a*, including *Scytothamnus*, which was placed in the Dictyosiphonales). Two additional genera, *Haplogloia* and *Papenfussiella*, which were not listed by Papenfuss, plus *Scytothamnus* occur at Macquarie I.

Haplogloia Levring, 1939: 48

Type species: *H. andersonii* (Farlow) Levring, 1939: 50.

Levring (1939) established the genus *Haplogloia* to accommodate the Californian species *Mesogloia andersonii* Farlow. This species' uniaxial construction (evident in young apices) and lack of phaeophycean hairs ruled out its inclusion in either *Myriogloia* (multiaxial), where it had been placed by Setchell & Gardner (1925), or *Mesogloia* (bearing phaeophycean hairs), where it had been placed by Kuckuck (1929). Additional generic features recognized by Levring, but considered less important that the two mentioned above, included the occurrence of two types of assimilatory filaments (deciduous long and persistent short) and a characteristic shape of the short assimilators.

When Kylin (1940) established the genus *Papenfussiella*, he considered it to be a close relative of *Haplogloia*, but distinguished the former by its multiaxial construction and presence of only one type of assimilatory filament. Kylin further emphasized the taxonomic importance of assimilatory filament types when he established the northeastern Pacific species *H. Kuckuckii*, which he characterized as having long assimilators that are more readily lost than those in *H. andersonii*.

Oddly enough, in spite of the taxonomic importance he placed on assimilatory filaments, Kylin described a New Zealand alga with long and short assimilators and an uncertain axial construction as a species of *Papenfussiella (P. lutea).*

Inagaki (1958) described a third species of *Haplogloia, H. kurilensis,* from the Kurile Is. Although he was unable to determine whether the thallus was uniaxial or multiaxial, he assigned the species to *Haplogloia* on the basis of other anatomical features—the production of two types of assimilatory filaments and the relationship of cortex to medulla (i.e., lacking a subcortical zone of outwardly radiating filaments as occurs in *Myriogloia, Suringaria* and *Tinocladia*). He distinguished this species from others in the genus on the basis of the shapes and sizes of its assimilatory cells.

In the present account, axial construction is considered to be taxonomically more important at the generic level than the presence of one or more types of assimilatory filaments. On this basis, the newly described chordariacean alga from Macquarie I. is best assigned to *Haplogloia.* Both this species and *H. andersonii* (Santelices & Abbott, 1978) represent the genus in the Southern Hemisphere.

Haplogloia moniliformis R. W. Ricker sp. nov.

Fig. 39

HOLOTYPE (Fig. 39a): MELU, R 0606; with isotypes in ADU, BM, MELU, Herb. Ricker, and UC.

TYPE LOCALITY: North of Garden Bay, Macquarie I. (54°29′58″ S lat., 158°57′35″ E long.), growing on rocks and encrusting coralline algae in tidepool at upper edge of *Durvillaea* zone (Ricker, 20 December 1977).

DISTRIBUTION: Macquarie I.

MATERIAL EXAMINED: Macquarie I.: collections MA-114, 116, 134, 137, 271; Herbarium specimens: MELU, R 0604–0608, 3363.

DIAGNOSIS: Plantae ad 10 cm altae, gelatinosae, spadiceae ad brunneas, axibus teretibus irregulariter et profuse ramosis (ad 1.5 mm diametris) ex base parva discoidea orientibus, ramificatione radiali ad 3 ordines, ramis vestitis a strato piloso filorum pigmentiferorum (ad 200 μm longitudinibus) in cellulis subcorticalibus incoloribus portatorum. Cortex uno fili assimilatorii typo; assimilatores maturi plerumque arcuati, compositi 3–5 cellulis basalibus angustatis 1.5–2.5 plo longioribus quam latioribus quae catenas moniliformes cellularum accrescentium doliiformium vel rotundatarum (12–15 μm diametris) subtendunt. Apices juvenes fillo axiali singulo, cellulis axialibus quae 5–10 cellulas post basin assimilatoris terminalis sunt longitudinaliter sic divisis ut rhizoidea basipetalia producant quae cum rhizoideis ex cortice miscenur et medullam incolorem formant. Rami vetustiores axe centrali non distincto, medulla ex matrice gelatinosa composita quae aggregationem filorum incolorium interdum lateraliter connexorum laxam circumit. Sporangia unilocularia vel obpyriformia (60 μm × 33 μm) vel sphericalia vel ovoidea, inter corticem abundatia, in cellulis subcorticalibus inflatis vel rotundatis vel elongatis curvisque productia. Structurae pluriloculares necnon pili phaeophyceani ignoti.

Haplogloia moniliformis is named in recognition of the moniliform chains of cells present in the cortical assimilators of this species.

DESCRIPTION: Plants to 10 cm tall, gelatinous, light- to dark-brown, with terete, irregularly but profusely branched axes (to 1.5 mm diameter) arising from small discoid bases (Fig. 39a); branching radial (Fig. 39b), to three orders, with branches covered by hairy layer of pigmented assimilatory filaments (to 200 μm long) borne on colourless subcortical cells (Fig. 39f). Cortex with one type of assimilatory filament; mature assimilators usually arcuate, composed of 3–5 narrow (1.5–2.5 times elongate) basal cells that subtend moniliform chains of enlarged, barrel-shaped to isodiametric (12–15 μm diameter) cells (Fig. 39f). Young apices with single axial filament (Fig. 39c), with axial cells longitudinally dividing 5–10 cells behind base of terminal assimilator,

Fig. 39 *Haplogloia moniliformis* R. W. Ricker: MELU, R 0604–0606, 3363. **a.** Habit of holotype (R 0606) **b.** Detail of branching in upper thallus. **c.** L.S. young apex with terminal portion of apical filament missing from plane of section. **d.** L.S. mature apex. **e.** T.S. main axis. **f.** Cortical peel showing colourless subcortical cells (arrow) bearing assimilatory filaments. **g.** Unilocular sporangium. Scales: a = 1 cm; b = 2 mm; c, g = 20 μm; d, f = 50 μm; e = 200 μm.

giving rise to basipetal rhizoids that mix with cortical-derived rhizoids to form colourless medulla. Older branches without distinct axial filaments (Fig. 39d), with medulla consisting of gelatinous matrix surrounding loose aggregation of colourless, occasionally laterally connected filaments (Fig. 39e). Unilocular sporangia obpyriform (60 µm × 33 µm), spherical, or ovoid; abundant throughout cortex; produced directly on inflated, either rounded or elongate and curved, subcortical cells (Fig. 39g). Plurilocular structures and phaeophycean hairs unknown.

Most plants studied grew in a 40 cm wide belt on rocky surfaces beginning 20 cm below the water surface in a 1–1.5 m deep tidepool in the mid- to lower-intertidal. Aside from *Durvillaea antarctica* (which fringed the seaweed rim of this pool), some encrusting coralline algae and the band of *Haplogloia*, the pool was mostly barren.

NOTES: *Haplogloia moniliformis*, with its profuse branching and predominantly intertidal distribution, is distinguished readily from the less-branched, subtidal *Papenfussiella tristanensis*, which is the closest relative to *Haplogloia* at Macquarie I.

Aside from the remote location of *H. moniliformis* in the Southern Hemisphere, the new species differs from others in the genus by the formation of just one type of assimilatory filament. Because only a small number of plants (*c.* 15) of roughly equal age have been studied, however, developmental changes in the composition of cortical tissue (i.e., the types of assimilatory filaments present) cannot be documented. The particular type of assimilators that form a cortex, a feature considered by Kylin (1940) to be taxonomically important at the generic and specific level, may change throughout the growth of *Haplogloia* as it apparently does in some species of *Chordaria*. Searles (1980) recently described a broad range of morphologies in *Chordaria linearis* from extremely tomentose thalli (similar to those of *Haplogloia* and *Papenfussiella*) to hairless plants with compact cortical tissues (more typical of *Chordaria*). He was unable to determine whether these morphological differences represented developmental stages of young to older plants or were caused by variations between microenvironments where the plants grew.

Papenfussiella Kylin, 1940: 17

Type species: *P. gracilis* Kylin, 1940: 17.

Kylin (1940) erected the genus *Papenfussiella* for plants with multiaxial, terete fronds lacking a transition zone of outwardly radiating, colourless filaments between their medullary and cortical tissues. He described four new species (*P. gracilis* and *P. laxa* from South Africa; *P. lutea* from New Zealand; *P. tristanensis* from Tristan da Cunha) and made one new combination (*P. callitricha* from Greenland). Levring (1941) later described *P. mosleyi* from the Juan Fernandez Is and Inagaki (1958) moved a Japanese species [listed by Kylin (1940) as *Myriocladia kuromo* Yendo] into *Papenfussiella*.

The seven species in the genus have been separated on differences in thallus size, branching patterns, lengths of assimilatory filaments, and dimensions of sporangia and assimilatory filament cells. Based on an ecological study of *P. callitricha* (Rosenvinge) Kylin, however, Hooper & South (1977) have shown that thallus dimensions, branching pattern and sporangial size are variable depending upon whether the alga is from a rough-water or calm site. Consequently, the shapes and sizes of assimilatory cells and filaments remain the principal criteria for distinguishing taxa.

Of the five species that occur in the Cold Temperate to Subantarctic waters of the Southern Hemisphere, four (*P. gracilis, P. lutea, P. mosleyi* and *P. tristanensis*) are known only from their type localities. *Papenfussiella laxa*, first described from South Africa, has been reported subsequently from New Zealand (Lindauer *et al.*, 1961), but Nelson & Adams (1983) dispute this record and believe that a morphologically highly variable *P. lutea* is the only representative of the genus in New Zealand.

Papenfussiella from Macquarie I. has similarities with all five Southern Hemisphere species in the genus but is morphologically closest to *P. tristanensis*.

Papenfussiella tristanensis Kylin 1940: 19

Fig. 40

TYPE LOCALITY: Seal Bay, Tristan da Cunha.

HOLOTYPE: Herb. Baardseth, O (no. 215). (NB holotype illustrated by Kylin (1940, taf. 3, fig. 6).)

DISTRIBUTION: Macquarie I.; Tristan da Cunha (Kylin, 1940; Baardseth, 1941).

MATERIAL EXAMINED:
1. Macquarie I.: collections MA-11, 34, 35, 113, 114, 115, 116, 117, 137, 140, 227, 244, 271, 378; Herbarium specimens: MELU, R 0597, 0603, 0610, 0914, 1795.
2. Tristan da Cunha: holotype and isotypes. Seal Bay; Station 51, 5–20 m depth (leg. Baardseth, 4 January 1938; det. H. Kylin): O, no. 215. (N.B. there are five herbarium sheets kept in a single envelope that bears the number 215, but one specimen is labelled as the type.)

DESCRIPTION: Plants 40–50 cm tall (Fig. 40a), arising from small discoid bases. Thalli terete, slimy, with hairy surfaces (Fig. 40a, inset), olive- to golden-brown when wet, drying to light- or dark-brown; main axes (to 3 mm thick) percurrent, sparsely and irregularly branched to three orders, with lax, slightly thinner laterals that gradually taper apically. Fronds sometimes with central cavity, otherwise with loosely compacted inner mass of colourless, rhizoidal filaments of small (4–5 μm) diameter and monopodially developed, sub-apically derived filaments of larger (20–23 μm) diameter (Fig. 40b, c); without transition zone of branching filaments between medulla and cortex; outermost colourless (subcortical) cells giving rise directly to pigmented assimilatory filaments (Fig. 40d). Assimilators (to 1.5 mm long) with lightly pigmented narrow bases (4–6μm diameter) 3–5 cells long, distally swelling abruptly to barrel-shaped (11–15 μm diameter × 4–22 μm long) meristematic cells filled with numerous discoid plastids (Fig. 40e). Assimilator cells elongate (to 35 μm long) toward filament apices, the diameter generally remaining uniform throughout filament (except near apices where filaments taper gradually), constricting weakly at cross-walls except for basal, barrel-shaped cells. Assimilators terminated by cells with rounded apices. Basal cells of assimilatory filaments laterally bearing rhizoids (that may subsequently produce assimilatory filaments) and unilocular sporangia (Fig. 40f, g). Unilocular sporangia clavate to obpyriform (50 μm broad × 100 μm long), or ovate (25 μm × 40 μm) to nearly spherical (50 μm diameter); plurilocular structures unknown.

Papenfussiella tristanensis grew in depths of 5–15 m, where it was abundant on rocks and sometimes attached to encrusting coralline algae. Some 100 specimens were studied during the present survey, all of which originated in the subtidal and generally grew in lubricous clumps on boulders that were 30–50 cm in diameter. None of these plants bore epiphytes, although they often grew adjacent to members of the Delesseriaceae that commonly epiphytized other brown algae.

NOTES: Plants from Macquarie I. are morphologically similar to the holotype of *P. tristanensis* and anatomically comparable to Kylin's (1940) description and illustrations of Tristan da Cunha specimens. There is considerable overlap, however, in thallus size, assimilatory filament length, and sporangial and assimilatory cell dimensions between *P. gracilis*, *P. tristanensis*, and *P. mosleyi*. This probably reflects the need for a critical review of all species in the genus and a reappraisal of the criteria used to separate species. Using Kylin's (1940) illustrations, the degree of constriction between assimilatory cells appears to be taxonomically diagnostic, but in Macquarie plants this feature proved to be highly variable.

Two type of assimilatory filaments (arcuate and short; longer and hair-like) can be distinguished in Macquarie thalli, but the shorter ones are considered in the present account to be immature long assimilators. Inagaki (1958) described dimorphic assimilatory filaments in *P. kuromo*, but also listed this feature as diagnostic for the genus *Haplogloia*. Additionally, Inagaki noted some uniaxial apices (a feature that Levring (1939) considered to be diagnostic for *Haplogloia*) in the Japanese alga he placed in *Papenfussiella*. Similar confusion over the criteria

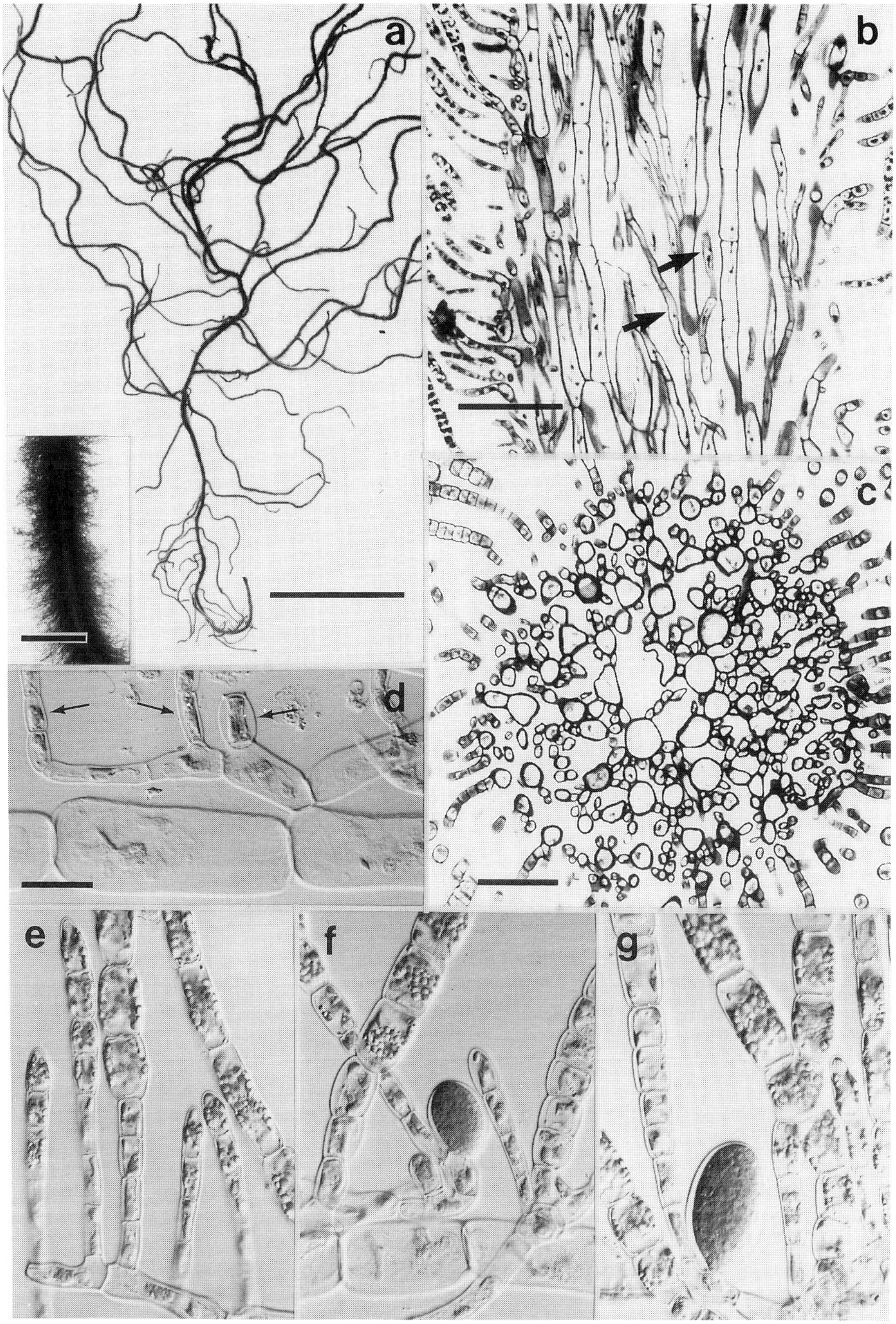

Fig. 40 *Papenfussiella tristanensis* Kylin: MELU, R 0597, 0914, 1795. **a.** Habit. Inset: detail of hairy thallus surface. **b.** L.S. main axis showing rhizoids (arrows) in medulla. **c.** T.S. main axis showing larger diameter medullary cells and smaller diameter rhizoids. **d–g** Cortical peels. **d.** Colourless subcortical filament bearing three assimilatory filaments (arrows). **e.** Base of assimilatory filaments. **f.** Subcortical filament bearing assimilators and rhizoid (arrow). **g.** Unilocular sporangium. Scales: a = 1 cm; inset = 3 mm; b = 100 μm; c = 50 μm; d–g = 20 μm.

used to separate *Haplogloia* and *Papenfussiella* was generated by Kylin (1940) when he described *P. lutea*. Macquarie plants have been placed in *Papenfussiella* rather than *Haplogloia* because of their multiaxial construction.

Scytothamnus Hooker f. & Harvey, 1845: 531

Type species: *S. australis* (J. Agardh) Hooker f. & Harvey, 1845: 531.

Scytothamnus contains three species, all of which are restricted to temperate or colder waters of the Southern Hemisphere. The type, *S. australis*, was moved by Price (1969) from the Dictyosiphonales to the Chordariales because of growth from a subapical meristem consisting of a group of uniseriate filaments that produce a pseudoparenchymatous thallus. The remaining two members, *S. fasciculatus* and *S. hirsutus*, were considered by Skottsberg (1921) to be anatomically similar to one another and unlike the type species, in that both were believed to develop from single apical cells and form true parenchyma as typically occurs in the Dictyosiphonaceae. *Scytothamnus hirsutus* has not been studied since Skottsberg (1921) originally described it, but *S. fasciculatus* has been examined more recently by Asensi (1975). Through developmental studies of culture- and field-grown plants, Asensi reached similar conclusions to those of Price regarding the filamentous nature of plants and consequently retained *S. fasciculatus* with *S. australis* in the Chordariales.

Scytothamnus australis apparently is widely distributed along cold temperate shores and the northern edge of the Subantarctic Zone (Levring, 1941; Lindauer *et al.*, 1961; Womersley, 1967), whereas *S. fasciculatus* grows farther south, has a broad Subantarctic distribution and is the only representative of the genus at Macquarie I.

Scytothamnus fasciculatus (Hooker f. & Harvey) Cotton, 1915: 170

Fig. 41a–f, i, j

Basionym: *Dictyosiphon fasciculatus* Hooker f. & Harvey, 1845: 178.

Synonym: *Scytothamnus australis sensu* Skottsberg, 1907: 48.

Type locality: Auckland Is.

Type: Herb. Hookerianum (BM), but specimens were not seen. Nelson & Adams (1983) indicate 'Harvey Herbarium (TCD)' as the location for the type, but this differs from Cotton's (1915) account that places the type material in Herb. Hookerianum.

Distribution: Argentina (Asensi, 1975); Auckland Is (Hooker f. & Harvey, 1845); Australia (Womersley, 1967); Campbell I. (Levring, 1945); Chile (Levring, 1960); Falkland and Kerguelen Is (Hooker f., 1867); Fuegia, South Georgia (Skottsberg, 1907); Macquarie I.; New Zealand (Lindauer *et al.*, 1961). (N.B. In addition, Papenfuss (1964*a*) indicated distributions at the South Shetland Is

Fig. 41a–f, i, j *Scytothamnus fasciculatus* (Hooker f. & Harvey) Cotton: MELU, R 1892, Lindauer no. 5142. **g, h** *Scytothamnus australis* (J. Agardh) Hooker f. & Harvey: MELU, K 5239. **a.** Habit of thallus in water. **b.** Branching detail upper thallus. **c.** Surface view of branch with embedded sporangia (arrowhead) and sporangial cavities (arrow). **d.** T.S. branch with mature unilocular sporangium (small arrow) and cavity of emptied sporangium (large arrow). **e.** L.S. branch with embedded sporangia and hollow medulla. **f.** Detail of cortex in T.S. showing nearly isodiametric subcortical cells. **g.** T.S. branch showing cortical cells in files. **h.** L.S. branch showing filamentous medulla. **i.** L.S. (rehydrated herbarium material) *S. fasciculatus* from New Zealand showing compact subcortical and elongate medullary cells. **j.** L.S. branch showing compact subcortical and elongate medullary cells. Scales: a = 1 cm; b = 2 mm; c = 500 μm; d, f, i, j = 25 μm; e = 100 μm; g, h = 50 μm.

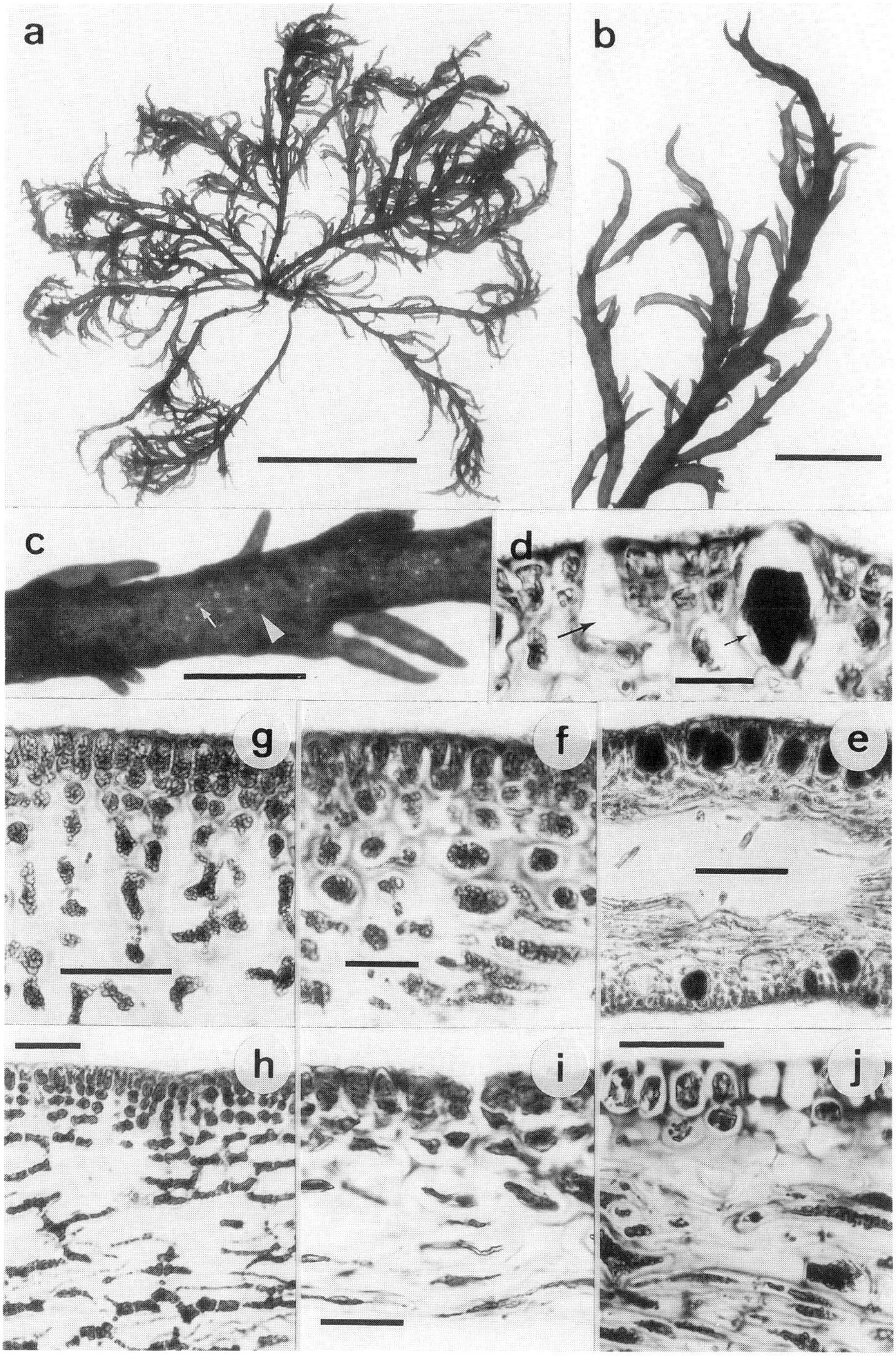
a
b
c
d
g
f
e
h
i
j

and Victoria Land, but the original published records for these localities have not been located for the present report.)

MATERIAL EXAMINED:
Scytothamnus fasciculatus
1. Macquarie I.: collection MA-133; Herbarium specimen: MELU, R 1892; Garden Cove, intertidal rocks (Taylor, December 1950): MELU, R 3354–3359 (= Taylor collection no. 562).
2. New Zealand: Stewart I. (E. W., 10 December 1944): ex Herb. V. W. Lindauer, specimen no. 5142. Lonneker's Nugget, Stewart I., upper eulittoral rocks (Kraft, 7 November 1972): MELU, K 5270.

ADDITIONAL SPECIES EXAMINED:
Scytothamnus australis (J. Agardh) Hooker f. & Harvey
1. Australia: Airey's Inlet, Victoria, mid-littoral rocks (Kraft, 26 January 1975): MELU, K 5239.
2. New Zealand: Opito Bay, Bay of Islands, common in mid-intertidal on boulders (Kraft, 8 October 1972): MELU, K 5000.

DESCRIPTION: Plants 2–3 cm tall, bushy, gregarious, with several individuals growing from common discoid base (Fig. 41a); thalli light- to olive-brown, cartilaginous, irregularly branched with slightly to strongly curled branches (Fig. 41b). Main axes and radially-borne, lanceolate laterals terete (Fig. 41c), composed of pigmented cortex 1–2 cell layers thick with slightly elongate (8 μm × 12 μm) surface cells (Fig. 41d, f); subcortex 3–5 cell layers of compacted, isodiametric (15 μm diameter) to oblong cells (12 μm × 25 μm) not arranged in distinct files (Fig. 41f); medulla filamentous, with central cavity (Fig. 41e, j). Phaeophycean hairs 5–8 μm diameter, abundant in some plants, absent in others, produced in cortical depressions (cryptostomata ?). Unilocular sporangia 25–28 (–50) μm × 40–45 (–70) μm, pyriform to oval, embedded in and scattered throughout cortex of upper branches, not arranged in distinct sori (Fig. 41c, d). Plurangia unknown.

During the present survey, I collected a clump of ten plants from a shallow tidepool in the mid- to lower-intertidal at the north end of the island. Previously, in 1950, Taylor had collected *Scytothamnus* (growing adjacent to *Rhizoclonium*, *Bostrychia*, and *Chaetangium*) from this end of the island but at another locality and from the upper intertidal.

NOTES: Lindauer *et al.* (1961) described some small plants (1–2 cm tall) that were fertile and similar to those presently reported from Macquarie I. In general, however, plants documented from most localities (e.g., Argentina—Asensi, 1975; Australia—Womersley, 1967; New Zealand—Lindauer *et al.*, 1961) are larger (to 25 cm tall). A large New Zealand specimen, when sectioned, revealed a comparable anatomy to that of the small plants from Macquarie I. (Fig. 41i). This particular construction, consisting of a shallow cortex, compact (seemingly non-filamentous) subcortex comprising rounded cells, and a filamentous medulla with a central cavity, is the same as that illustrated by Asensi (1975) for South American plants. It is also noticeably different from the more filamentous organization of the type species, *S. australis*, which often is morphologically similar to the smaller plants of *S. fasciculatus* from Macquarie I. *Scytothamnus australis* has a clearly filamentous cortex, with cortical files 3–4 cells long (Fig. 41g, h), and medulla without a central cavity. Anatomical similarities between *S. fasciculatus* and members of *Dictyosiphon* suggest that this species might be better moved back to its original position in *Dictyosiphon*—a move that Pedersen (1984) has recently proposed. Additionally, he has proposed the combination *Dictyosiphon hirsutus* (Skottsberg) Pedersen for *S. hirsutus* Skottsberg on the basis of this alga and *D. foeniculaceus* (Hudson) Greville having similar generic features.

Chordariopsidaceae Kylin, 1940: 53

This monotypic family was established by Kylin (1940) to accommodate wiry, somewhat cartilaginous plants with uniaxial thalli that supposedly develop from three-sided apical cells. Such a construction and type of meristem distinguish this family from the Chordariaceae, whose members are morphologically similar but differ anatomically in their multiaxial construction and subapical or diffuse (intercalary) meristems.

Chordariopsis Kylin, 1940: 54

Type species: *C. capensis* (C. Agardh) Kylin, 1940: 54.

Known by only the type species, *Chordariopsis* was segregated from *Chordaria* because of the former's development from an apical cell and the presence of a single, central axis (Kylin, 1940). These two features, however, are often difficult to see (especially in older branches that have become hollow) and other anatomical differences may better separate both genera. Sterile thalli of *Chordariopsis* have compact cortical tissues (in contrast to the assimilatory filaments that cover *Chordaria*) and fertile thalli are covered by a layer of unicellular, rarely two-celled, clavate paraphyses that are unlike the multicellular paraphyses of *Chordaria* that generally have swollen terminal cells (Kylin, 1940). Pyriform, unilocular sporangia develop among the club-shaped paraphyses. Plurilocular structures are unknown.

Chordariopsis capensis (C. Agardh) Kylin, 1940: 54

Fig. 42

Basionym: *Chordaria flagelliformis* f. *capensis* C. Agardh, 1824: 257.

Synonyms: see Nelson & Adams, 1983. *Chordaria dictyosiphon* (Harvey) Kützing *sensu* Kenny & Haysom, 1962: 258.

Type locality: Cape of Good Hope, South Africa.

Type: Herb. Agardh (LD), but specimens were not seen.

Distribution: Auckland Is (Nelson & Adams, 1983); Falkland Is (Cotton, 1915); Fuegia (Skottsberg, 1907); Îles Kerguelen (Askenasy, 1888); Macquarie I. (Zinova, 1958); South Africa (C. Agardh, 1824); South America (pers. obs.).

Material examined:

1. Macquarie I.: collections MA-213, 238, 278; Herbarium specimens: MELU, R 1988, 2409, 2410; Garden Cove, intertidal (Kenny, 3 February 1949): MELU, R 3335, 3336.
2. South Africa: Platboombai Cape Peninsula, intertidal (Hommersand, 9 December 1977): MELU, R 3631; Sea Point, Cape Town, intertidal (Hommersand, 5 December 1977): MELU, R 3632.
3. South America: Isla San Pedro, Gulfo de Penas, 3–4 m depth (Searles, Leister, & Brauner, 6 November 1972): Searles, no. 72-32-9.

Description: Plants to 15 cm tall, dark-brown, with terete, cartilaginous fronds attached by small, discoid holdfasts. Thalli pseudoparenchymatous, with whip-like, lax to weakly curled laterals (to 6 cm long) arising perpendicularly from all sides of percurrent main axes (Fig. 42a); lateral branches (≤1 mm diameter) mostly simple, sometimes once-branched at wide angles. Construction of distal axes either distinctly filamentous (main indeterminate branches; Fig. 42d) or compact and with parenchyma-like tissue (lateral determinate branches; Fig. 42e, f). Sterile thallus parts (e.g., at branch apices) consisting of pigmented, cortical files 2–3 cells long, with outermost cells *c.* twice as long as broad (Fig. 42e, f); subcortex lacking or limited to single layer of colourless, oblong cells; medulla with branched, colourless filaments generally of equal diameter; rhizoidal filaments absent (Fig. 42d). Older axes usually with central cavity. Sexual thalli covered by 1-(2-)celled clavate paraphyses (50–70 μm long) surrounding spindle-shaped, unilocular sporangia (35–40 μm × 20 μm); vegetative cortical files completely absent where paraphyses present (Fig. 42b, c). Colourless hairs 5–7 μm diameter, rare, sometimes branched near bases, often occurring in pairs among paraphyses. Plurilocular structures unknown.

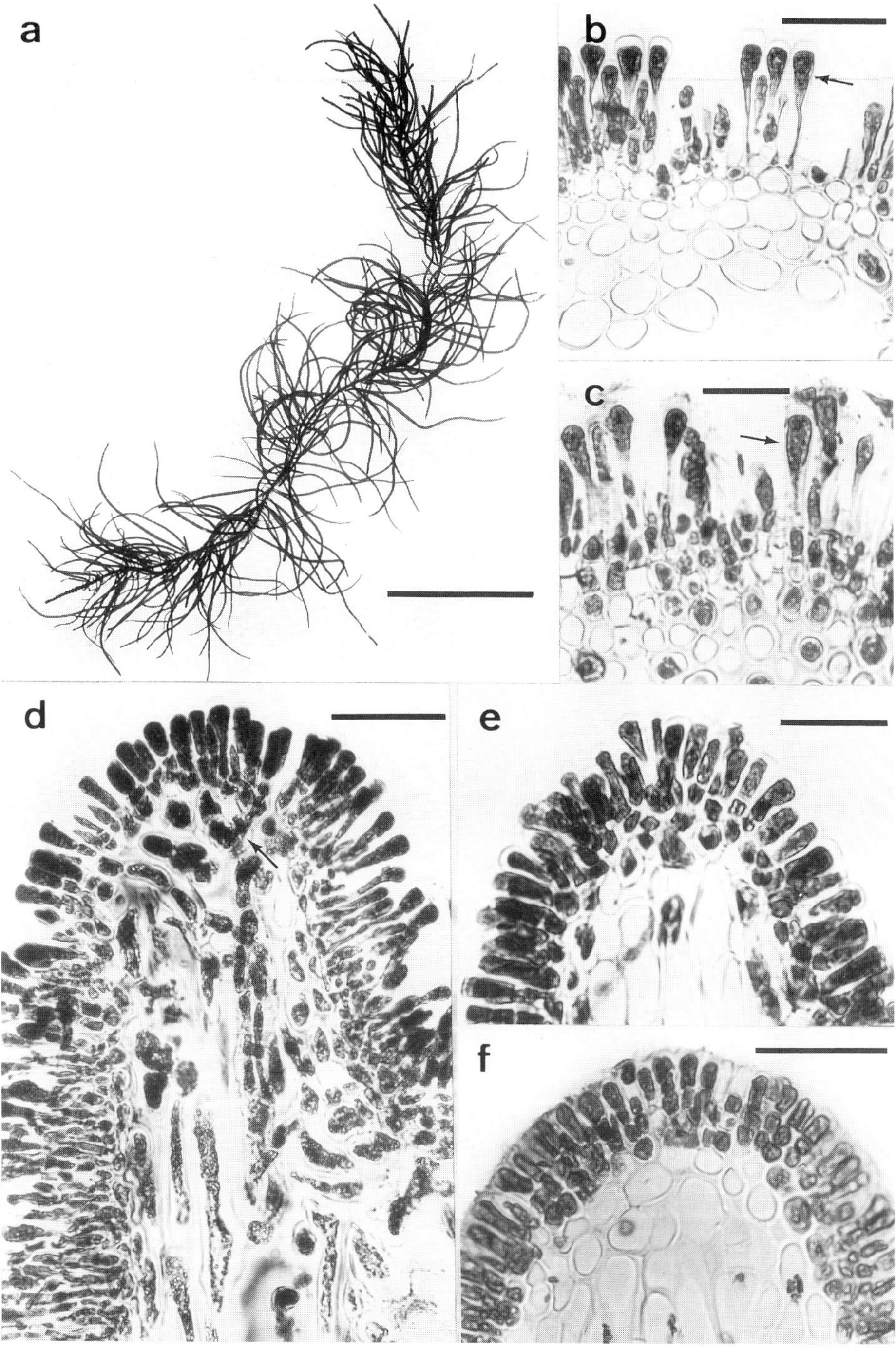

Fig. 42 *Chordariopsis capensis* (C. Agardh) Kylin: MELU, R 1988, 2409, 2410, 3631. **a.** Habit of thallus in water. **b.** T.S. fertile axis showing unicellular paraphyses (arrow). **c.** T.S. fertile axis of South African plant with unicellular paraphyses (arrow). **d.** L.S. apex of main (indeterminate) axis showing medulla with branched filaments (arrow). **e, f.** L.S. apex of (determinate) lateral branch showing compact cellular medulla. **e.** Macquarie specimen. **f.** South African specimen. Scales: a = 2 cm; b, d, f = 50 µm; c, e = 25 µm.

Some 25 plants, of which all are fertile, were collected from the intertidal at Macquarie I. Plants grow epilithically alongside *Iridaea* and *Plocamium* in the mid- to lower-intertidal in shallow tidepools *c.* 30 cm above the upper edge of the *Durvillaea* zone.

NOTES: Nelson & Adams (1983) recently reported *C. capensis* from the Auckland Is and discussed its distribution throughout the Subantarctic. Their brief account on the Auckland plants agrees with Kylin's (1940) description of the species, although it lacks discussion on thallus development, type of meristem and construction of the medulla.

In the present account, Macquarie, South African and South American plants have been compared to confirm the occurrence of the same entity at all three localities. The plants are believed to be *C. capensis,* in spite of certain consistent differences between them and the type description. Most notably, the uniaxial construction and development from a single, three-sided, apical cell are not apparent as was illustrated and described by Kylin (1940, fig. 30a) for South African plants. In contrast, the presently described algae have two different apical anatomies, depending upon whether apices of (indeterminate) main axes or (determinate) lateral branches were sectioned. Distal portions of determinate branches (Fig. 42e, f), although lacking a definite central axis and single apical cell, better resemble Kylin's illustrations (that display a parenchyma-like medullary tissue) than do the indeterminate branches that have a distinctly filamentous organization. Another difference is the presence of phaeophycean hairs in Macquarie plants and absence of such in Kylin's description of the type.

In some morphological and anatomical respects, the presently described plants are similar to the genus *Chordaria.* They are placed in *Chordariopsis,* however, because of their distinctive, unicellular, clavate paraphyses, lack of *Chordaria*-like vegetative assimilators, and absence of rhizoidal filaments in the medulla (as commonly present in *Chordaria*).

Sphacelariales

Members of the Sphacelariales show apical growth, heterotrichy and darkening of cell walls when placed in bleach [eau de Javelle (Reinke, 1890)]. The chemical reactions causing cell walls to blacken in bleach are not known, but apparently they occur only in sphacelarialean species (Prud'homme van Reine, 1974).

Prominent apical cells, distinctive for all members of the group, divide transversely to form primary segments which in turn divide similarly into secondary segments. Secondary segments are usually of equal height and are designated as upper (superior) and lower (inferior). Longitudinal divisions of secondary segments occur in most species, thus resulting in a true parenchyma with cells arranged in tiers along axes. Continued transverse and longitudinal divisions of secondary segments occur in some species and result in small cortical cells. Cortication of lower branches is generally further increased by descending, rhizoidal filaments.

Plants may be entirely encrusting or endophytic, but more often consist of erect, stiff branches that arise from basal crusts. Cells of prostrate and erect axes have single, centrally positioned nuclei (that often are surrounded by tannins) and peripherally distributed lenticular or discoid plastids.

An alternation of isomorphic somatic phases occurs in some Sphacelariales. Gametophytes produce isogametes, anisogametes, or oogametes; sporophytes produce biflagellate meiospores in unangia. Asexual recycling of either phase occurs through biflagellate zoospores produced in plurangia. Additionally, many species reproduce vegetatively by means of specialized deciduous branchlets (propagules).

Most of the Sphacelariales were originally separated into two groups by Reinke (1890), based on two fundamental patterns of branch development. In one group, the Acroblastae, laterals are initiated directly from the apical cells. In the second group, the Hypacroblastae, laterals arise from subapical cells. Sauvageau (1903) subsequently introduced many other terms for types of branch development and used developmental differences to distinguish five groups within the order. Oltmanns (1922), using Sauvageau's concepts and terminology, separated the Sphacelariales into three families (Cladostephaceae, Sphacelariaceae and Stypocaulaceae), to which Fritsch (1945) later added a fourth (Choristocarpaceae). Prud'homme van Reine (1982) recently removed the

Choristocarpaceae from the Sphacelariales because members of the former do not produce transversely dividing primary segments, nor do their cell walls blacken in bleach. Wynne (1982*b*), however, continues to recognize this family in the Sphacelariales.

Prud'homme van Reine (1982) uses the various patterns of branch development in the order to define two subgroups of Reinke's (1890) Acroblastae (Acrohomoblastic, Acroheteroblastic), and three subgroups of his Hypacroblastae (Dichoblastic, Hemiblastic, Meriblastic). Prud'homme van Reine's concepts and terminology are employed when discussing the taxonomic placement of Macquarie's two sphacelarialean taxa, with *Sphacelaria bornetii* belonging to the Sphacelariaceae and *Alethocladus corymbosus* to the Stypocaulaceae. These two families, along with the Cladostephaceae, are widely represented in the Subantarctic, but only the Stypocaulaceae extends into the Antarctic where *Alethocladus corymbosus* recently has been collected on the Antarctic Peninsula (Moe, pers. comm.).

Sphacelariaceae Decaisne, 1842: 329, 341

The Sphacelariaceae contains plants with predominantly upright, branched axes, although one genus, *Onslowia* (Searles & Leister, 1980), is composed of creeping, endophytic filaments. All laterals arise from subapical cells (hypacroblastic branching; Prud'homme van Reine, 1982). Branches are either long (indeterminate), resembling main axes, or short (determinate). The latter are never borne in whorls as in the Cladostephaceae and are frequently modified for bearing reproductive structures (unangia or plurangia). Hairs, present in some species, develop as acroblastic laterals from lenticular initials derived directly from apical cells (Fritsch, 1945). Sexual reproduction is by isogametes or anisogametes, and life histories usually have isomorphic or slightly heteromorphic somatic phases (van den Hoek & Flinterman, 1968). In some taxa reproduction is known only vegetatively through propagules which, when present, are often morphologically characteristic for the species (Abbott & Hollenberg, 1976).

Of the six genera credited to this family (Fritsch, 1945; Searles & Leister, 1980), only *Sphacelaria* has been reported from the Subantarctic. A single species of this genus grows at Macquarie I.

Sphacelaria Lyngbye ***in*** Hornemann 1818: t. 1600

Type species: *S. reticulata* Lyngbye *in* Hornemann 1818: t. 1600 (see Prud'homme van Reine (1982: 59–61) re. typification).

Thalli are 1 mm to 10 cm long, tuft-forming, and mostly composed of upright, slightly rigid filaments arising from either discoid bases one to several cell-layers thick, or from prostrate masses of free filaments that creep across or penetrate host surfaces. Uprights are initially uniseriate but soon become multiseriate. Transverse divisions of primary segments precede any longitudinal cell divisions. Secondary segments divide longitudinally in all erect axes. Branch primodia, which subsequently become apical cells of lateral branches, arise laterally either from the entire length of superior (secondary) segments (hemiblastic branching) or from portions of superior segments (meriblastic branching). Branching can be either sparse and irregular or densely pinnate, alternate, opposite, or unilateral. Colourless hairs with basal meristems and generally basal sheaths are present in some species. Unangia and plurangia occur on either the same or separate plants.

Sphacelaria bornetii Hariot, 1887*a*: 57

Fig. 43

Type locality: Orange Bay, Fuegia.

Type: presumably in PC (Koster, 1969), but specimens were not seen.

Distribution: Fuegia (Hariot, 1887*a*, 1889); Macquarie I.

MATERIAL EXAMINED: Macquarie I.: collection MA-255; Herbarium specimens: MELU, R 1912, 3385.

DESCRIPTION: Thalli with rigid, erect, sparsely branched filaments to 2 mm tall, developing from a dense prostrate mat epiphytic on *Ballia callitricha* (Fig. 43a). Filaments initially uniseriate with elongate, blunt-tipped, apical cells (Fig. 43b), soon becoming multiseriate, with tiers of 4–5 segments per level. Axes uncorticated, except near bases of erect filaments where several uniseriate basipetal rhizoids may develop; tiers and segments remaining visible throughout mature filaments. Secondary transverse septa in secondary segments begin forming ten to fifteen tiers below apices, at which point lateral branches arise (Fig. 43f). Basal cells of laterals generally extending across entire lengths of superior (secondary) segments (Fig. 43e; hemiblastic branching), less frequently from portions of secondary segments (Fig. 43d, g; meriblastic branching). Branching irregular to alternate; sterile plants bearing only indeterminate branches similar in shape to, but slightly thinner (16–20 μm diameter) than, main axes (25–29 μm diameter). Reproductive thalli with long (indeterminate) and short (determinate) branches, the latter usually adaxially bearing series of 2–5 unangia (Fig. 43c, h, i). Unangia cylindrical (13–35 μm) to ovoid (23 × 30 μm), sessile. Plurangia similarly shaped, but larger (30–35 × 55–70 μm), stipitate on one-celled stalks, borne either directly on main axes and indeterminate branches or terminally on short lateral branches that sometimes bear multiple plurangia in small fascicles. Propagules unknown.

Although the host alga, *B. callitricha*, is common around Macquarie I. in shaded pools of the mid-littoral down to 20 m depths, *S. bornetii* was found in only one tide pool. It produced a bristly mat composed of thalli with stiff branches that are easily distinguished from the ectocarpalean thalli, which form lax tufts of branches, that also grow epiphytically on *Ballia*.

NOTES: *Sphacelaria* has more species than any other genus in the Sphacelariales, and although most of these species are characterized by their propagule morphology, the identity of the present alga is quite certain despite its lack of these structures. Macquarie specimens agree completely with Hariot's (1889) illustration and emended description of the South American type collection, which also lacked propagules and reached 2 mm lengths. His first account of this entity (1887*a*) erroneously described the plants as 2 cm tall. Womersley (1967) described *S. carpoglossi* from Southern Australia and stated that it is related to *S. bornetii* from Cape Horn. Macquarie plants have secondary transverse septa, however, and are clearly different from the Australian species which lacks such septa.

Stypocaulaceae Oltmanns, 1922: 95

This family has plants with mostly erect, branched thalli whose branches arise directly from segmentation of the apical cell (acroblastic branching), similar to hair formation in *Sphacelaria*. Most genera are coarsely filamentous, and some (*Phloecaulon, Ptilopogon*) have tissues differentiated into a medulla, cortex, meristoderm, and an investment of cortical rhizoids. These are the largest and most complex plants in the order. In contrast, the genus *Alethocladus* is less corticated and more *Sphacelaria*-like in construction, producing branches with regular tiers of cells below short uniseriate axes. Hairs, usually sheathed, can be present or absent except in *Alethocladus*, where they are never present. Reproductive structures (sporangia, antheridia, oogonia), where known, are borne singly or in groups in the axils of normal vegetative or specialized branches. Reproduction is mostly isogamous or anisogamous, but oogamy is known in *Halopteris*. No species produce propagules.

Of all the family members (*Alethocladus, Halopteris, Ptilopogon* and *Phloecaulon*), *Halopteris* is distributed most widely, growing in both Northern and Southern Hemispheres. The first three genera are listed from the Subantarctic (Papenfuss, 1964*a*), but only *Alethocladus* is known from Macquarie I., where it appears to be very rare.

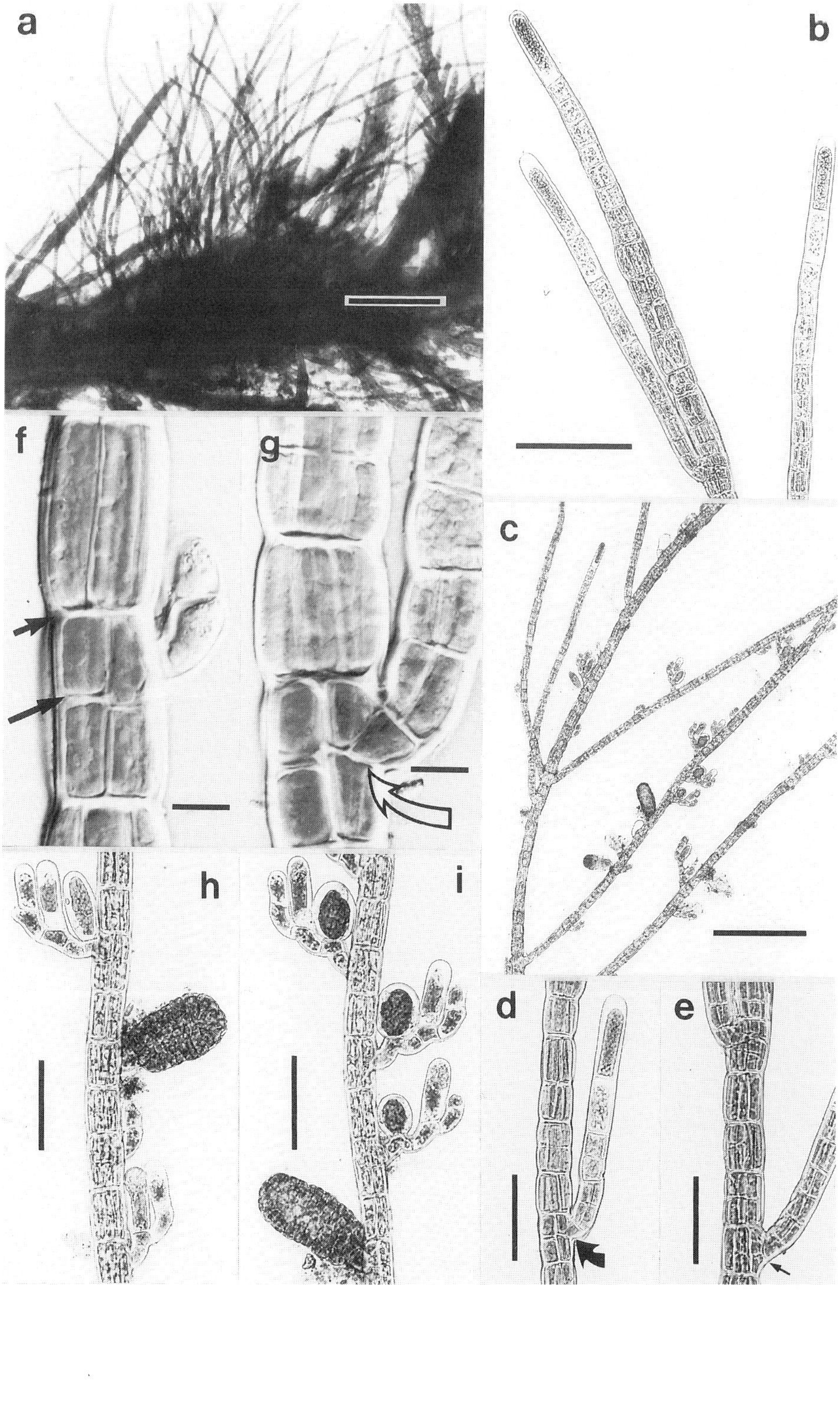
a
b
c
d
e
f
g
h
i

Alethocladus Sauvageau, 1903: 288

Type species: *A. corymbosus* (Dickie) Sauvageau, 1903: 288.

Plants grow to 11–12 cm tall, form tufts of branched, erect axes with tiered, *Sphacelaria*-like construction, and have habits similar to some wiry species of *Halopteris*. Axes of *Alethocladus*, however, generally are less robust than those of *Halopteris*. The type of basal attachment is unknown. Lower branches sometimes are corticated by rhizoids produced from cortical cells. Branching is irregular to distichous-alternate, with branches arising from lenticular primordia directly cut from apical cells. Primordia usually give rise only to lateral branches (acrohomoblastic branching), but not infrequently divide into smaller upper cells, which generally remain dormant (in branch axils), and larger lower cells, which give rise to lateral branches (acroheteroblastic branching; pers. obs.). Cells contain numerous discoid plastids. Hairs and reproductive structures are unknown.

The genus is monotypic and known only from the Subantarctic and Antarctic. It is distinct from the closely related genus *Halopteris*, which produces hairs and sporangia from the small cells located in branch axils.

Alethocladus corymbosus (Dickie) Sauvageau, 1903: 288

Fig. 44

Basionym: *Sphacelaria corymbosa* Dickie, 1876*a*: 50.

Synonyms: see Papenfuss, 1964*a*.

Type locality: Îles Kerguelen.

Type: K *in* BM, no. H 306/67 92 KEW.

Distribution: Antarctic Peninsula (Moe, pers. comm.); Îles Crozet (Hemsley, 1884); Îles Kerguelen (Dickie, 1876*a*); Macquarie I.

Material examined:

1. Macquarie I.: collections MA-303, 309; Herbarium specimen: MELU, R 1960.
2. Îles Kerguelen: syntype material from Observatory Bay, on *Mytilus*; (Eaton, 16 November 1874): K *in* BM (no. H 306/67 92 KEW).

Description: Plants to 1.5 cm tall, wiry, with prominent main axes terminated by large apical cells; branching distichous-alternate, to three orders (Fig. 44a), with branches arising from primordia directly derived from apical cells. Primordia abutting primary septa medially along their bases (Fig. 44b–d), usually developing into lateral branches (acrohomoblastic branching) without producing small axillary cells (Fig. 44e) as typically occur in *Halopteris*. Lower axes corticated by uniseriate and multiseriate rhizoids. Reproductive structures unknown.

Plants were found loose among collections of *Microzonia velutina* taken from subtidal rocks in 5–10 m depth. Only two collections from Caroline Cove yielded specimens of this species and consequently the distribution of *A. corymbosus* around the island is uncertain.

Fig. 43 *Sphacelaria bornetii* Hariot: MELU, R 1912, 3385. **a.** Epiphytic habit on *Ballia callitricha*. **b.** Apices. **c.** Branching of upper axes. **d.** Lateral branch origin from upper portion of transversely divided secondary segment (arrow). **e.** Lateral branch borne from secondary segment that subsequently divided transversely into three cells (arrow). **f.** Detail of primary (short arrow) and secondary (long arrow) transverse divisions. **g.** Detail of lateral branch arising from upper portion of superior segment divided by secondary transverse division (arrow). **h, i.** Unangia in series of 2–3 and solitary stipitate plurangia. Scales: a = 500 μm; b = 100 μm; c = 200 μm; d, e, h, i = 50 μm; f, g = 10 μm.

Although obvious reproductive structures were not seen, some branches loosely entangled in the upright axes had rhizoidal portions that might function as propagules for vegetative reproduction.

NOTES: Dickie's (1876*a*) original description of the species is brief and based on sterile plants, but he reported *Alethocladus* growing on *Mytilus* shells and worm tubes, and its habit as caespitose with erect axes to 5–8 cm tall. Sauvageau (1903) described the plants as tufts that may become many centimeters tall. The record of *Alethocladus* from Macquarie I. is based on three small

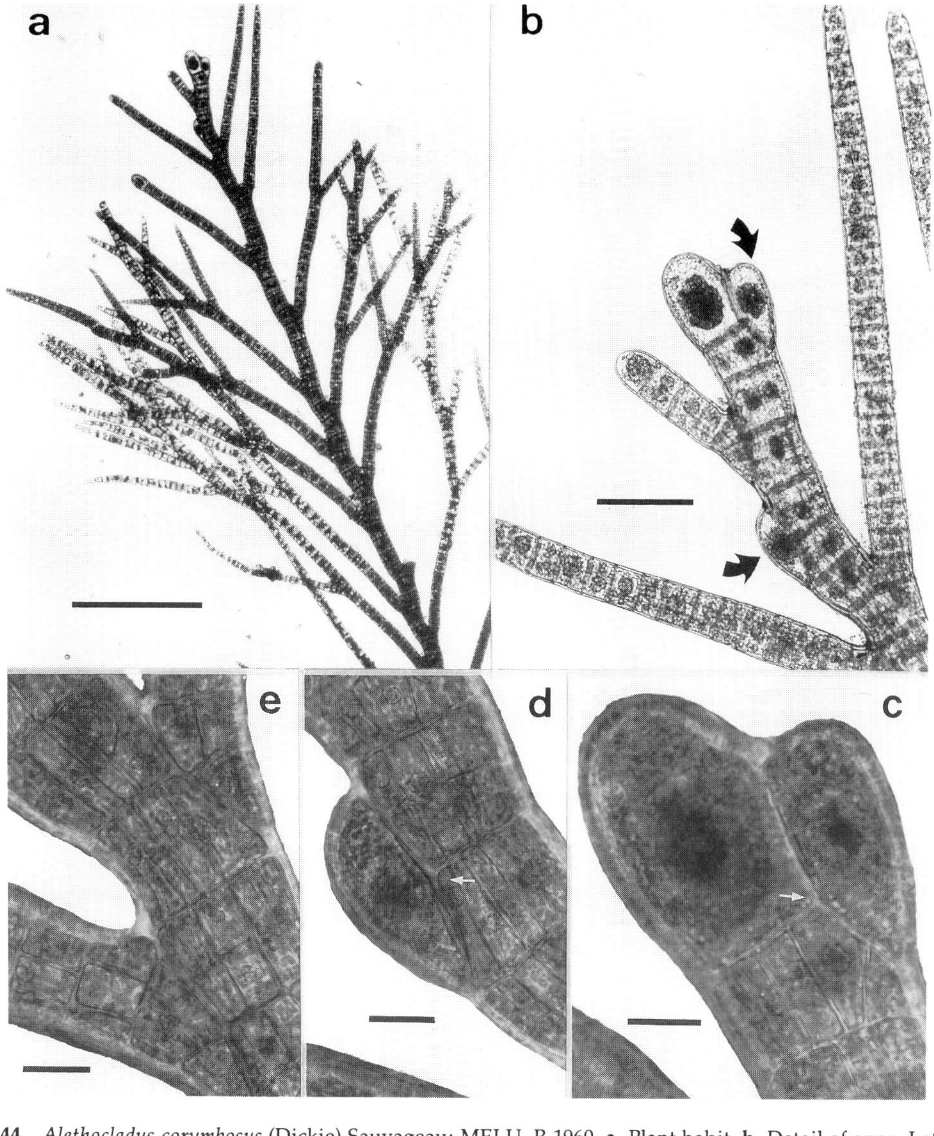

Fig. 44 *Alethocladus corymbosus* (Dickie) Sauvageau: MELU, R 1960. **a.** Plant habit. **b.** Detail of apex. Lateral branch primordia (arrows) abutting primary transverse septa. **c, d.** Details of lateral branch primordia showing branch initials extending across primary septa (arrows). **e.** Detail of mature branch bases. Scales: a = 500 μm; b = 100 μm; c–e = 25 μm.

specimens that differ somewhat from the previously published descriptions of the species. Branches do not form pronounced corymbs and one plant had some colourless, hair-like processes terminating multiseriate rhizoids, although these lacked basal meristems characteristic of true hairs.

Moe's recent discovery of this alga along the Antarctic Peninsula suggests that the species is widely distributed throughout southern high latitudes and not wholly endemic to the Subantarctic as reported by Papenfuss (1964*a*). Askenasy (1888) reported *Sphacelaria funicularis* from Fuegia and Papenfuss (1964*a*) has questioned whether this might actually be a record for *Alethocladus*. Askenasy's description, however, is of richly fertile plants whose branching is comparable to that of *Halopteris*. This would be an unusual and notable record for *Alethocladus* should it be verified.

Dictyosiphonales

Members of this order have dissimilar generations involving alternations of macroscopic sporophytes with microscopic, filamentous gametophytes that reproduce by isogametes formed in plurilocular structures. Sporophytes have apical, subapical, or diffuse meristems, form parenchyma and arise from discoid or rhizoidal bases. Thalli are simple or branched and either solid, cylindrical or flattened, or tubular to saccate. Cortical cells are pigmented and filled with numerous discoid or lenticular plastids, each usually containing one pyrenoid. Phaeophycean hairs commonly are produced either singly or in tufts from cortical cells. Medullary cells generally are large and colourless. Sporophytes bear only unilocular or both unilocular and plurilocular sporangia. Unilocular structures, although initially superficial, often become secondarily embedded in the cortex. Plurilocular organs mostly remain superficial, where they are either scattered or aggregated in sori.

Papenfuss (1951) viewed the Dictyosiphonales as containing six families, the Chnoosporaceae, Dictyosiphonaceae, Giraudyaceae, Myriotrichaceae, Punctariaceae and Striariaceae, but he later (1964*a*) recognized a seventh when he removed *Scytosiphon* from the Punctariaceae and placed this genus in the earlier-established Scytosiphonaceae (Farlow, 1881). Nakamura & Tatewaki (1975) presented life history data for some members (*Colpomenia, Endarachne, Petalonia, Scytosiphon*) of the Scytosiphonaceae which justify movement of the family from the Dictyosiphonales to a separate order (Scytosiphonales). Because some of Papenfuss' original six families have been seen as heterogeneous assemblages, three additional families have been created. The Delamareaceae was established by Zinova (1953) initially to accommodate *Delamarea*, which Papenfuss (1951) had previously placed in the Dictyosiphonaceae. Zinova (1954) further expanded the family by adding a new genus from the Northern Hemisphere (*Stschapovia*) and including the Southern Hemisphere *Cladothele* (previously in the Dictyosiphonaceae) and North Atlantic *Coelocladia* (a member of the Striariaceae). Pedersen (1976) subsequently described the Coelocladiaceae for *Coelocladia*, and later (1978) established another family, the Pogotrichaceae, for *Pogotrichum*, considered by Papenfuss (1951) as part of *Litosiphon* in the Punctariaceae.

The Delamareaceae, Punctariaceae and Striariaceae, together including seven genera and eight species, represent the Dictyosiphonales in high-latitude southern waters. Papenfuss (1964*a*) had included the cold-water genus *Scytothamnus* in the tropical family Chnoosporaceae, but this genus since has been moved by Price (1969) to the Chordariaceae.

Two families (Delamareaceae, Punctariaceae), each with one genus and species, occur at Macquarie I.

Delamareaceae Zinova, 1953: 128

Members of this family are unique in having distinctly enlarged cortical cells. The family is characterized mostly on the features of *Delamarea*, however, and the other constituent genera, *Cladothele* and *Stschapovia*, require further study concerning their systematic positions (Pedersen, 1974).

One genus, *Cladothele*, from this family is represented in the Subantarctic Zone.

Cladothele Hooker f. & Harvey, 1845: 293

Type species: *C. decaisnei* Hooker f. & Harvey, 1845: 293.

Plants have terete, irregularly branched thalli arising from small discoid bases. Prominent apical cells divide to produce solid axes composed internally of large colourless cells and externally of smaller, but outwardly swollen and inflated, pigmented cortical cells ('papillae' of Hooker f. & Harvey (1845)). Reproductive organs, unangia and plurangia, generally are borne on separate plants but sometimes occur on the same thallus (Skottsberg, 1921). Plurangia are convex and develop singly or in confluent, sometimes irregularly shaped patches. Unangia are superficial, bulbose and solitary. Asensi (1976) studied the life history of *C. decaisnei* and observed a protonemal stage producing plurilocular sporangia (the products of which recycled this filamentous phase) and also gave rise directly (through buds) to typical upright thalli. He was unable, however, to observe an alternation of karyological phases.

Hooker f. & Harvey (1845) described this genus as 'certainly related to *Codium*' because of *Cladothele*'s surface 'papillae' that grossly resemble *Codium* utricles. Skottsberg (1921) examined these inflated cortical cells and found them quite unlike those in any other dictyosiphonalean genus, including *Stictyosiphon* (where Murray (1891) had placed *Cladothele*), which he distinguished from *Cladothele* based on the former having a trichothallic meristem and the latter an apical one. He also recognized a close morphological resemblance between *Cladothele* and the Northern Hemisphere genus *Delamarea*.

One species of *Cladothele*, *C. striarioides*, grows in tidepools at Macquarie I.

Cladothele striarioides (Skottsberg) Zinova, 1954: 241

Fig. 45

Basionym: *Cladothele decaisnei* var. *striarioides* Skottsberg, 1921: 38.

Type locality: not indicated.

Type: S, but specimens were not seen.

Distribution: Auckland Is (Lindauer *et al.*, 1961); Fuegia (Skottsberg, 1921); Macquarie I.; Patagonia (Asensi, 1966*a*).

Material examined: Macquarie I.: Garden Cove, intertidal rock pool (Taylor, 4 February 1951): MELU, R 3352, 3353 (= Taylor collection no. 511).

Description: Plants to 7 cm tall, dark brown when dry, with terete, irregularly branched main axes (to 500 μm diameter; Fig. 45a, b) developed from distinct apical meristems. Monosiphonous portions, to at least six cells long, terminating branches before longitudinal divisions of axial cells give rise to pluriseriate construction. Hairs infrequent, present only on young branches, rarely borne from apical cells. Branches solid, with monostromatic cortex composed of outwardly swollen pigmented cells containing numerous discoid plastids aggregated along externally facing walls; medulla comprising irregularly shaped, large, colourless cells. Plurangia abundant on older branches, generally protuberant from vegetative thallus surfaces, occurring singly or in scattered patches, ranging greatly in shape and size (e.g., 76 × 53 μm, 68 × 30 μm, 84 × 40 μm, 65 × 30 μm, 84 × 53 μm, 68 × 42 μm). Unangia unknown.

This species is known from only one intertidal collection, where plants, apparently surrounded by or epiphytic upon *Codium subantarcticum*, were growing in a tidepool.

Notes: No material of this species was collected during the 1977–78 expedition to Macquarie I. and the data presented here come partly from my own observations of Taylor's dried herbarium collections, but largely from notes made by Mrs Iona Christianson on wet-preserved plants (also gathered by Taylor).

Cladothele from Macquarie I. agrees with Skottsberg's (1921) description of South American plants. Individuals of this genus, whether wet or dried, are readily distinguished from all other brown algae at Macquarie I. that lack the 'papillate' cortical cells. Such swollen cells in *C. striarioides*, when dry, separate slightly from one another to render a distinctive, tessellated pattern on branch surfaces (Fig. 45c).

Punctariaceae (Thuret) Kjellman, 1880: 9

Members of this family usually have unbranched thalli that are either solid, flattened and blade-like, or hollow and saccate. Papenfuss (1964*a*) listed five genera and species, *Adenocystis utricularis, Asperococcus echinatus, Cladochroa chnoosporiformis, Corycus lanceolatus* and *Punctaria plantaginea*, from the Subantarctic Zone. Of these, only *Adenocystis* also occurs in the Antarctic, where apparently it is restricted to rocky shores along the Antarctic Peninsula (Lamb & Zimmermann, 1977), and it is also the only member of the Punctariaceae that grows at Macquarie I.

Adenocystis Hooker f. & Harvey ***in*** Harvey & Hooker f., 1845: 179

Type species: *A. lessonii* (Bory de Saint-Vincent) Hooker f. & Harvey *in* Harvey & Hooker f., 1845: 179 (= *A. utricularis* (Bory de Saint-Vincent) Skottsberg, 1907: 39).

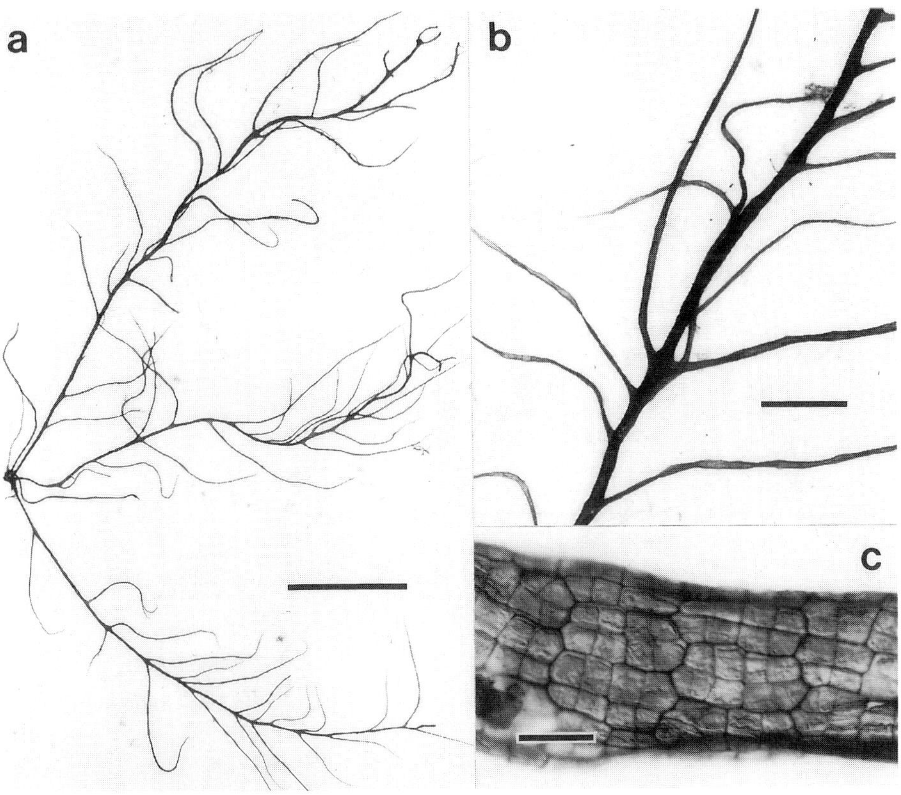

Fig. 45 *Cladothele striarioides* (Skottsberg) Zinova: MELU, R 3352, 3353. **a.** Habit. **b.** Branching pattern. **c.** Detail of rehydrated branch surface. Scales: a = 5 mm; b = 1 mm; c = 50 μm.

Adenocystis contains one species with two known forms (Skottsberg, 1921), both of which grow in the Subantarctic Zone. The genus forms unbranched, turgid sacs that are attached by short stipes to small discoid holdfasts. Fronds appear punctate, with small whitish tufts of colourless filaments arising from microscopic hair-pits (cryptostomata) scattered across thallus surfaces. Unilocular sporangia develop on the saccate macroscopic plants and give rise to meiospores that develop into filamentous or discoid microthalli, which are dioecious gametophyes. Gametes are isomorphic and produced either in swollen irregularly shaped cells of discoid thalli or in plurilocular structures on filamentous microthalli (Müller, 1984).

The combination of habit, vegetative anatomy (with cryptostomata) and production of only unilocular sporangia on its saccate fronds separates *Adenocystis* from other unbranched, cylindrical brown algae (Wynne, 1973). *Adenocystis* is morphologically very similar to, and occurs in some localities with *Utriculidium durvillaei.* The former, however, bears only unilocular sporangia on its saccate fronds and has cortical cells that contain numerous discoid plastids, whereas the latter bears only plurilocular organs and has cortical cells that contain one plastid. Some authors (e.g., Delépine & Asensi, 1978) have suggested that both taxa might be alternate phases of one species, although Naylor's (1955) and Müller's (1984) life history studies have not supported this conclusion. Of the five unbranched, saccate, brown algae (*A. utricularis, Asperococcus echinatus, Corycus lanceolatus, Colpomenia sinuosa* and *Utriculidium durvillaei*) listed by Papenfuss (1964*a*), *Adenocystis* is distributed most widely throughout the Subantarctic and is the only one present at Macquarie I.

Adenocystis utricularis (Bory de Saint-Vincent) Skottsberg, 1907: 39

Fig. 46

Basionym: *Asperococcus utricularis* Bory de Saint-Vincent 1826*b*: 594.

Synonyms: see Papenfuss, 1964*a*.

Type locality: Falkland Is.

Type: Herb. Bory de Saint-Vincent, PC (Womersley, 1967), but specimens were not seen.

Distribution: Circumpolar—probably growing on all ice-free, rocky Subantarctic shores. Presently reported from: Antarctic Peninsula, Falkland Is, Fuegia, South Georgia (Skottsberg, 1907); Argentina (Kühneman, 1970); Auckland, Campbell, and Chatham Is, New Zealand (Lindauer *et al.*, 1961); Chile (Levring, 1960); Îles Crozet (Levring, 1944); Îles Kerguelen (Dickie, 1879; Reinbold, 1907); Macquarie I. (Zinova, 1958; Kenny & Haysom, 1962); South Orkney Is (Gepp & Gepp, 1905*a*, 1912); South Shetland Is (Gain, 1912); Stewart I. (Adams *et al.*, 1974); Tasmania (Cribb, 1954). (NB Lindauer *et al.* (1961) additionally indicated distribution at Marion I. and Victoria Land, but the original published records for these localities have not been located for the present report.)

Material examined: Macquarie I.: collections MA-83, 111, 133, 134, 137, 213, 215, 238, 278, 283, 381, 397; Herbarium specimens: MELU, R 0058, 0059, 0062, 2510, 3579.

Description: Plants to 9 cm tall, with unbranched, clavate to obovate-fusiform, saccate fronds connected by solid stipes (≤5 mm long) to small discoid bases (Fig. 46a, b); generally pale brown when sterile, darker brown when fertile. Thallus surfaces with small tufts of colourless filaments (Fig. 46f) arising from minute pits (cryptostomata; Fig. 46e) distributed over entire frond. Thallus

Fig. 46 *Adenocystis utricularis* (Bory de Saint-Vincent) Skottsberg: MELU, R 0062, 2510, 3579. **a, b.** Habit of saccate thalli. (N.B. four plants in 'a' epiphytic on *Mesophyllum* (R 3579).) **c.** Section through sterile cortex. **d.** Section through fertile cortex showing clavate-shaped unicellular paraphyses and spindle-shaped unilocular sporangia (arrow). **e.** Section through cortical hair pit. **f.** Surface view cortical hairs. **g.** Sporeling apex with terminal hair tuft. Scales: a = 2 cm; b = 1 cm; c = 10 cm; d, e = 50 cm; f = 500 μm; g = 25 μm.

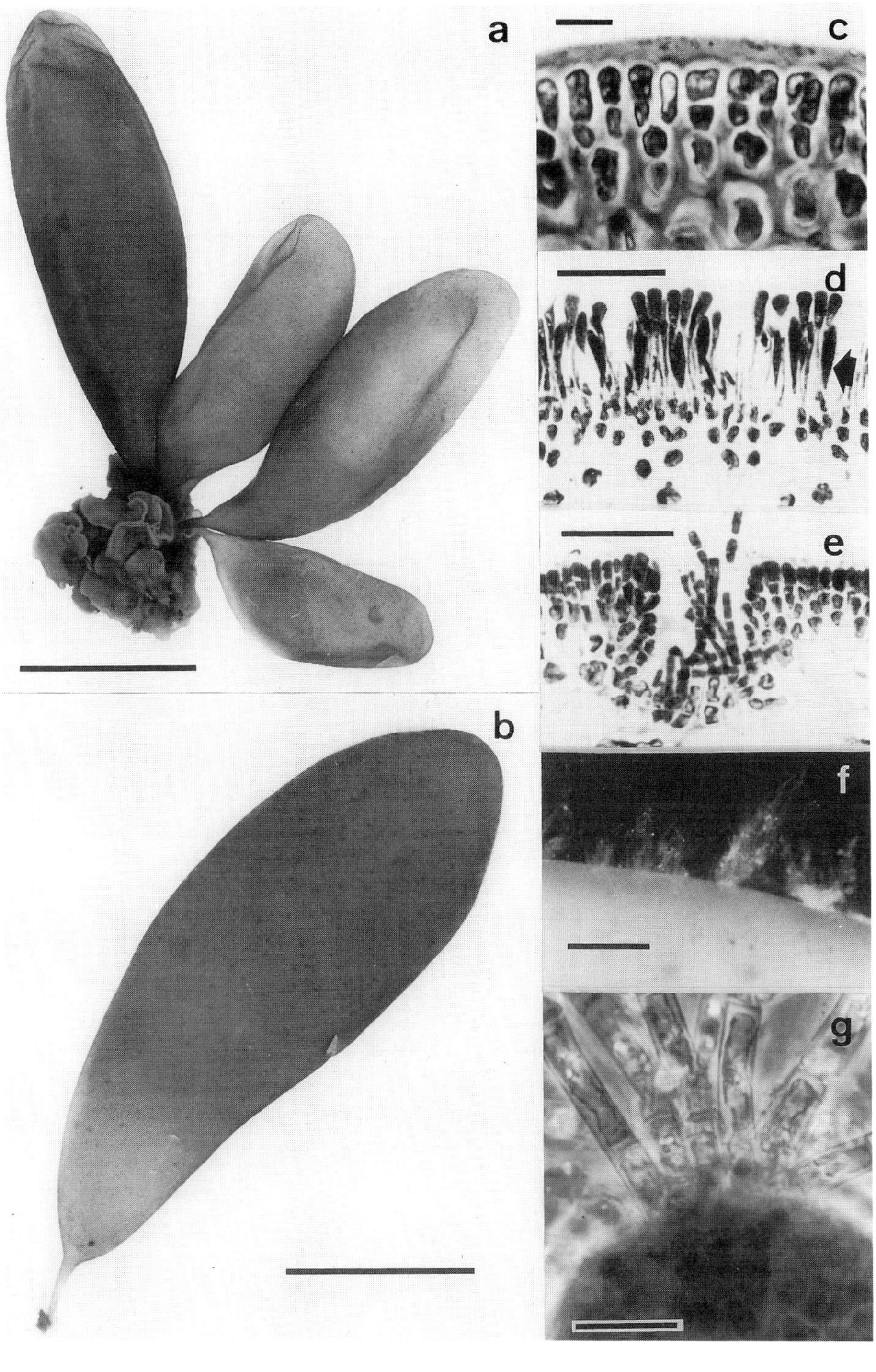
a
b
c
d
e
f
g

with 2–3 cell layers of pigmented cortical cells (3.5–5.5 μm × 9–14.5 μm diameter; Fig. 46c), 1–2 cell layers of colourless, isodiametric (20–25 μm diameter) to periclinally oblong (18 μm × 40 μm) subcortical cells, and medulla with central cavity surrounded by colourless filaments (11 μm × 100 μm). Reproductive plants with indistinct, confluent sori covering most of thallus; sori comprising clavate, unicellular paraphyses (to 50 μm long) surrounding spindle-shaped unilocular sporangia (to 30 μm long; Fig. 46d). Plurilocular structures unknown.

Extremely young individuals of *Adenocystis* have solid and cylindrical thalli that develop from single, or groups of (Fig. 46g), hairs borne in apical hair-pits. Thalli do not become vesiculate until their overall height roughly equals the stipe length of mature plants, at which time a bulbous, hollow knob develops distally. Such sporelings may form dense patches on rocky surfaces in the lower intertidal in early summer, later growing into extensive clusters of turgid, water/mucilage-filled sacs. Individual thalli occasionally become punctured and flaccid, but these usually fill with sediment and retain their inflated shape.

More than 200 individuals, ranging from 1 mm to 9 cm tall, have been studied from tidepool or lower intertidal collections. Most of these plants grew epilithically, but some were epiphytes on *Chaetangium* and *Mesophyllum schmitzii*. *Adenocystis* often was infected by either or both of the two brown endophytes *Gononema ramosum* and *Myrionema incommodum*, the former usually growing in or adjacent to cryptostomata and the latter occurring diffusely through the host cortex. Both endophytes were visible macroscopically as dark-brown, irregularly shaped patches on host thalli. Other algae growing alongside *Adenocystis* include: *Acrosiphonia pacifica*, *Rhizoclonium ambiguum*, *Scytosiphon lomentaria* and *Polysiphonia anisogona*.

NOTES: Skottsberg (1907) reported one exceptionally deep (to 20 m) collection of *A. utricularis* from Fuegia. This contrasts with the strictly intertidal habitat occupied by this species at Macquarie I. Also, in contrast to Levring's (1945) observations of Stewart I. plants that apparently become fertile at a certain height (5 cm tall), thallus size for Macquarie plants is unrelated to the alga's reproductive status. At Macquarie I. some individuals only 2 cm tall produced unilocular sporangia, whereas others to 7 cm tall remained sterile and could be recognized by their pale brown colour.

Scytosiphonales

Chadefaud (1960) recognized two classes of brown algae which he distinguished by the presence (Phaeosporeae; = Phéosporées) or absence (Cyclosporeae; = Cyclosporées) of an alternation of generations. Kylin (1933) had already divided the Phaeophyta into three classes, the Isogeneratae, Heterogeneratae and Cyclosporeae, based on differences in life history patterns. The Isogeneratae (with isomorphic generations) and Heterogenerate (with heteromorphic generations) together are equivalent to the Phaeosporeae. However, Kylin's tripartite classification has been abandoned by some authors because it is based on heterogeneous assemblages of orders, the members of which develop differently and produce either parenchymatous or non-parenchymatous tissues or both. Nakamura (1972) divided the Phaeosporeae into three subclasses: the Haplostichidae with both generations of plants forming non-parenchymatous tissues, the Polystichidae with both generations forming parenchyma, and the Haplopolystichidae having an alternation between non-parenchymatous and parenchymatous phases.

Three orders, the Dictyosiphonales, Scytosiphonales and Laminariales, constitute the Haplopolystichidae. Of these, the Scytosiphonales is the only one having macroscopic gametophytes (Nakamura & Tatewaki, 1975). Additionally, members of this order are defined by the presence of a single plate-like cloroplast per pigmented cell, as contrasted with taxa in the Dictyosiphonales or Laminariales whose cells contain numerous discoid plastids.

The order comprises two families, the Chnoosporaceae and Scytosiphonaceae, which are well represented in both hemispheres. Only the latter, however, is present at Macquarie I.

Scytosiphonaceae Farlow, 1881: 15, 62

Plants grow diffusely or have intercalary, basal meristems giving rise to unbranched, tubular, bladelike, or globose thalli emanating from discoid bases. Genera and species in this family are separated mostly on their vegetative morphologies and, in some instances, on the presence or absence of vegetative structures such as hairs, hair-pits (cryptostomata), or paraphyses borne among plurilocular structures on fertile thalli.

Until Nakamura & Tatewaki (1975) demonstrated the fusion of zooids from plurangia borne on separate thalli in several genera, life histories were conceived as having macroscopic sporophytes that produced only plurilocular structures (e.g., Lindauer *et al.*, 1961). From these observations, they concluded that the macroscopic thalli are actually heterothallic gametophytes that usually produce isogametes. In some instances, as in *Petalonia fascia*, gametes develop directly without fusing (Nakamura & Tatewaki, 1975). Sporophytes consist of either crusts or microscopic filaments that produce unilocular sporangia.

This family contains eight genera with roughly 20 species, of which five species are listed from high-latitude southern waters (Papenfuss, 1964*a*). Two genera, each with one species, grow at Macquarie I.

Petalonia Derbès & Solier, 1850: 265 (***nom. cons.***)

Type species: *P. debilis* (C. Agardh) Derbès & Solier, 1850: 265 (= *P. fascia* (O. F. Müller) Kuntze, 1898: 419; see Setchell & Gardner (1925) re. synonymy).

J. Agardh (1896) recognized four species for *Petalonia (Phyllitis)*, but Setchell & Gardner (1925) considered the genus monotypic. Parke & Dixon (1968) record three species (*P. fascia*, *P. zosterifolia*, *P. filiformis*) from British shores, and Nakamura & Tatewaki (1975) distinguished the first two by differences in their life histories. Gametophytes are unbranched, foliose, and usually solid but with occasional cavities. Sterile fronds have two distinct tissue layers including a small-celled, pigmented, monostromatic cortex sometimes bearing tufts of colourless hairs, and a medulla composed of several layers of large colourless cells. Fertile thalli produce uniseriate plurilocular gametangia in extensive sori often covering both thallus surfaces.

The type species of the genus occurs at Macquarie I. and throughout the Subantarctic Zone.

Petalonia fascia (O. F. Müller) Kuntze, 1898: 419

Fig. 47

Basionym: *Fucus fascia* O. F. Müller, 1778: 768.

Synonyms: *see Papenfuss, 1964a.*

Type locality: North Atlantic (Norway).

Type: presumably in BM, but specimens were not seen.

Distribution: Occurrence in temperate to colder waters of both hemispheres; reported from the following Southen Hemisphere localities: Argentina (Kühnemann, 1970); Australia (Womersley, 1967); Chile (Levring, 1960); Falkland Is, South Georgia (Skottsberg, 1907); Fuegia (Hariot, 1889); Heard I. (Chittleborough, 1956); Îles Kerguelen (Reinbold, 1907); Juan Fernandez Is (Levring, 1941); Macquarie I.; New Zealand (Lindauer *et al.*, 1961); Tristan da Cunha (Baardseth, 1941).

Material examined: Macquarie I.: collections MA-111, 129, 134, 137, 213, 238; Herbarium specimens: MELU, R 1668, 1671, 3271. Garden Cove, upper sub-littoral ('u.s.l.') pools (Wollaston, 5 December 1960): ADU, A 25,568.

DESCRIPTION: Plants to 10 cm tall, with flattened and foliose, pale to dark orange-brown, mostly undivided fronds (Fig. 47a), the larger blades sometimes becoming laciniate. Sterile thalli to 75 μm thick, consisting of a monostromatic cortex with angular, nearly isodiametric (4–5 μm diameter) cells; medulla with 3–4 layers of large and small, colourless cells (Fig. 47b). Fertile thalli 110–120 μm thick, covered nearly completely on both surfaces by gametangial sori. Medulla colourless, generally with two cell-layers. Plurilocular gametangia uniseriate, to 42 μm long (Fig. 47c). Sporophytes and unilocular structures unknown.

Approximately 20 specimens were studied from collections made in the lower intertidal to upper sub-littoral. Plants generally grew in clusters, with each blade arising from a distinctly separate discoid base, on either rocks or limpet shells. Collections of fertile plants were made in early summer (December). Other algae associated with *P. fascia* included *Acrosiphonia pacifica*, *Ulva rigida*, and *Scytosiphon lomentaria*.

Scytosiphon C. Agardh, 1820: 160 (*nom. cons.*)

Type species: *S. filum* var. *lomentaria* (Lyngbye) C. Agardh, 1820: 162 (= *S. lomentaria* (Lyngbye) Link, 1833: 232).

Plants generally are gregarious, growing in tufts of simple, tubular individuals, each with a cylindrical or compressed thallus that is either constricted at regular to irregular intervals or smooth throughout. Fronds develop from intercalary meristems near thallus bases.

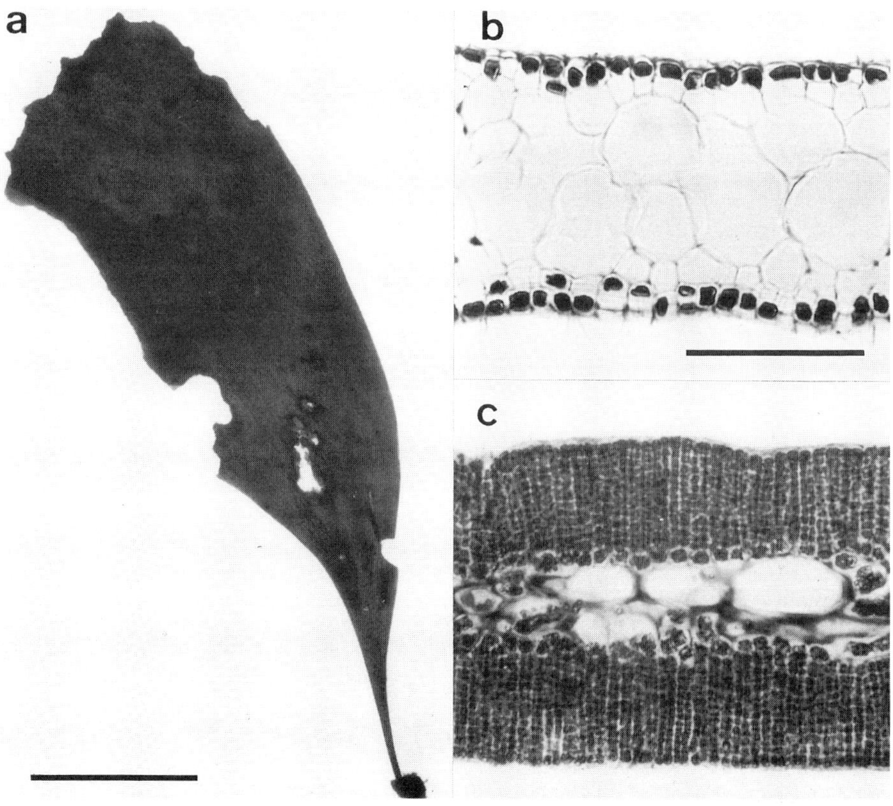

Fig. 47 *Petalonia fascia* (Müller) Kuntze: MELU, R 1668, 1671. **a.** Habit of fertile thallus (R 1668). **b.** T.S. sterile blade. **c.** T.S. fertile blade bearing plurilocular gametangia. Scales: a = 5 mm; b, c = 50 μm.

Wynne (1973) presented a key to the four known species of *Scytosiphon* (*S. crispus* Skottsberg, *S. lomentaria*, *S. dotyi* Wynne, *S. pygmaeus* Reinke) and gave geographic distributions for each. Two species are listed from the Subantarctic Zone by Papenfuss (1964*a*), *S. crispus* being known only from the Falkland Is and *S. lomentaria* appearing to be widespread throughout the Subantarctic Zone.

Scytosiphon lomentaria (Lyngbye) Link, 1833: 232

Fig. 48

BASIONYM: *Chorda lomentaria* Lyngbye, 1819: 74.

TYPE LOCALITY: North Atlantic (Denmark).

TYPE: presumably in C (Koster, 1969), but specimens were not seen.

DISTRIBUTION: Widespread in temperate to colder waters of both hemispheres; reported from the following Southern Hemisphere localities: Argentina (Kühnemann, 1970); Auckland Is (Laing, 1909); Australia (Womersley, 1967); Chile (Levring, 1960); Falkland Is, Fuegia, South Georgia (Skottsberg, 1907); Îles Kerguelen (Hooker, 1846); Juan Fernandez Is (Levring, 1941); New Zealand (Lindauer *et al.*, 1961); Tristan da Cunha (Baardseth, 1941).

MATERIAL EXAMINED: Macquarie I.: collections MA-83, 111, 119, 134, 137, 213, 238; Herbarium specimens: MELU, R 1670, 2462, 3608–3626.

DESCRIPTION: Thalli to 30 cm long, tubular, to 7 mm diameter, light to dark-brown, unbranched, usually with constrictions along axes, gradually tapering basally and apically; growing singly, but more often in tufts (Fig. 48a, b); composed of 2–3 layers of pigmented, isodiametric (5 μm diameter) to oblong (5 μm × 9 μm), cortical cells (Fig. 48f) often bearing clusters of colourless hairs; medulla with central cavity surrounded by vertically elongate cells, which in thallus transections appear irregularly shaped, and of large and small diameter (Fig. 48e). Fertile fronds with plurilocular gametangial sori usually becoming confluent, nearly or entirely covering frond surfaces (Fig. 48c); old thalli degenerating (Fig. 48b, d) following release of gametes, producing irregular thallus surfaces with bulges (remaining, filled gametangia) and depressions (already emptied and degenerated gametangia). Sori containing uniseriate, plurilocular gametangia (to 95 μm long), and few unicellular, clavate-shaped paraphyses (Fig. 48g). Sporophytes and unilocular structures unknown.

Scytosiphon lomentaria, which is sometimes partly covered by *Ectocarpus* on its upper thallus and *Rhizoclonium* on its base, grows on rocks and limpet shells in the lower intertidal alongside *Durvillaea antarctica*, *Adenocystis utricularis* and *Petalonia fascia*. Thalli become fertile at Macquarie I. in early December and are highly eroded, following the release of gametes, by the end of the summer.

NOTES: *Scytosiphon lomentaria* from Macquarie I. agrees well with individuals of the same species reported from the Northern Hemisphere (Abbott & Hollenberg, 1976; Kornmann & Sahling, 1977) and other Southern Hemisphere localities (Lindauer *et al.*, 1961). One observation based on Macquarie plants, which apparently has not been noted in other accounts of this species, deals with the appearance of older thalli that usually become dwarfed, full of holes and heavily covered with epiphytes. These senescing plants develop warty, irregular surfaces, lose their distinctive constrictions along axes and may superficially agree with descriptions of the Northern Hemisphere genus *Coelocladia*.

Desmarestiales

The two families Arthrocladiaceae and Desmarestiaceae, both comprising essentially cold-water members, constitute the order. The former, known singly by the North Atlantic species *Arthrocladia villosa* (Hudson) Duby, has been reported recently for the first time from the

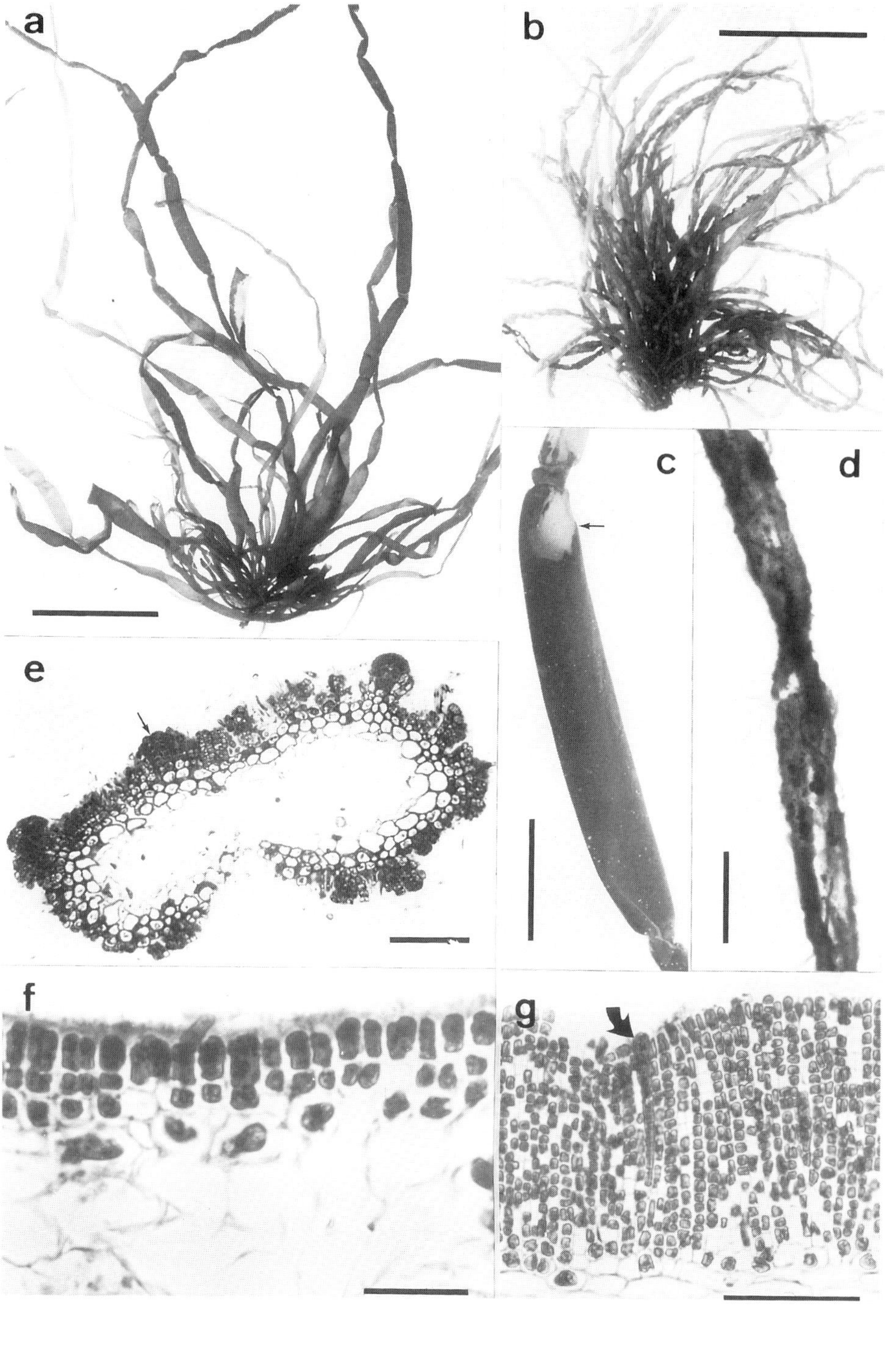
a
b
c
d
e
f
g

Southern Hemisphere (southern coast of Australia; Skinner & Womersley, 1983). The Desmarestiaceae, however, is well represented in both Northern and Southern Hemispheres by *Desmarestia*, and in the cold Antarctic waters by the additional genera *Himantothallus* and *Phaeurus*. Because three of the four constituent genera occur around Antarctica, Moe & Silva (1981) consider this region to be the center of distribution for the order.

Life histories involve an alternation of heteromorphic somatic phases, with sporophytes being macroscopic (up to 10 m long) and having uniaxial construction, trichothallic meristems, and unilocular sporangia produced in either sori (*Desmarestia*, *Himantothallus*) or catenate series on branched filaments (*Phaeurus*). Gametophytes are microscopic and either monoecious (Kornmann, 1962*a*) or dioecious (Schreiber, 1932; A. Chapman & Burrows, 1971). Sexual reproduction, known for *Arthrocladia* (Müller & Meel, 1982) and *Desmarestia* (Kornmann, 1962*a*), is oogamous.

Desmarestiaceae (Thuret) Kjellman 1880: 10

Sporophytes are terete or flattened, branched or unbranched, and filamentous or foliaceous, with membranous, cartilaginous, or coriaceous blades. Some species are annuals, but most Antarctic and Subantarctic representatives appear to be perennials which produce massive holdfasts and tough thickened fronds. Such coarse thalli usually have large amounts of secondary pseudoparenchymatous tissues derived from periclinal divisions of surface (meristoderm) and axial cells. Primary meristems are trichothallic, but often unrecognizable in mature plants because either the terminal filaments are missing or the actively dividing regions have been covered by cortication.

The Desmarestiaceae is the only group of brown algae aside from the Laminariales (kelps) to produce cells seemingly modified for solute transport ('trumpet hyphae'). The trumpet hyphae of the Desmarestiaceae differ from those of the kelps, however, in their unique covering of ensheathing filaments (B. Jönsson, 1901; Gepp & Gepp, 1905*a*, 1907).

Development of unilocular sporangia has been observed in *Desmarestia* (Skottsberg, 1907; Moe & Silva, 1977*b*), *Himantothallus* (Moe & Silva, 1981) and *Phaeurus* (Lamb & Zimmermann, 1977). There appears to be a great range of sporangial characters that eventually may be used widely for taxonomy at the infrageneric level. At present, however, sporangia have not been reported for most species (Moe & Silva, 1977*b*), and taxonomy is usually based solely on vegetative characteristics (A. Chapman, 1972).

Desmarestia occurs throughout the Subantarctic Zone and is the only genus representing the family at Macquarie I.

Desmarestia Lamouroux, 1813: 43 (***nom. cons.***)

Type species: *D. aculeata* (Linnaeus) Lamouroux, 1813: 45.

Desmarestia, with nearly 40 described species, has a broader distribution than other genera in the order. Its members arise from solid or hapteroid, conical holdfasts and sometimes produce extremely robust (to 5 m long) thalli which are morphologically highly variable.

A physiological feature unique to some oppositely branched species of *Desmarestia* (including *D. willii*, but apparently not *D. chordalis*) is the production of sulfuric acid in vacuoles (Bold & Wynne, 1978). When collecting such acid-producing plants, it is best to keep them separate from other algae and, if possible, preserve them immediately to prevent self-digestion.

Papenfuss (1964*a*) listed 11 species from the Antarctic and Subantarctic Zones. Of these *D. chordalis* and *D. willii* grow at Macquarie I., where they are abudant in subtidal communities.

Fig. 48 *Scytosiphon lomentaria* (Lyngbye) Link: MELU, R 1670, 2462, 3608, 3626. **a.** Habit of fertile plants. **b.** Habit of senescent plants. **c.** Surface view of gametangial sorus. Arrow indicates edge of darkly pigmented sorus and pale sterile tissue. **d.** Surface view of senescent frond with mostly emptied gametangia. **e.** T.S. old frond (d) showing clusters of filled gametangia (arrow) among patches of discharged gametangia. **f.** Section through vegetative cortex. **g.** Section through fertile cortex with plurilocular gametangia and infrequent unicellular paraphyses (arrow). Scales: a = 2 cm; b, c = 1 cm; d = 1 mm; e = 100 μm; f = 25 μm; g = 50 μm.

Desmarestia chordalis Hooker f. & Harvey, 1845: 249

Figs 49, 50, 51

SYNONYMS: *Desmarestia rossii* Hooker f. & Harvey 1845: 249–250.

TYPE LOCALITY: Îles Kerguelen.

LECTOTYPE: Herb. Hookerianum (BM), no. 17 890/64.

DISTRIBUTION: Apparently restricted to the Subantarctic Zone*; presently reported from: Falkland Is (Skottsberg, 1907; Cotton, 1915); Fuegia (Skottsberg, 1907); Heard I. (Dickie, 1876*d*; pers. obs.); Îles Kerguelen (Hooker f. & Harvey, 1845); Macquarie I. (Zinova, 1958); South Orkney Islands (Gepp & Gepp, 1905*a*).

MATERIAL EXAMINED:

1. Macquarie I.: collections MA-11, 14, 34, 39, 46, 56, 98, 113, 114, 115, 116, 119, 137, 140, 222, 227, 237, 248, 271, 283, 364, 397; Herbarium specimens: MELU, R 0194, 3517; Microscope slides: R/M 0316, 0317, 0319, 0322, 0324.
2. Heard I.: Atlas Cove (17 April 1953); Herbarium specimens: MELU, R 3518, 3519.
3. Îles Kerguelen: Christmas Harbour, 4–10 m depth (J. D Hooker, specimen no. 679): BM, no. 17 890/64.

DESCRIPTION: Plants 1–2 m tall, dark-brown, with flattened stipes arising from massive conical holdfasts composed of fused haptera (Fig. 49c, d). Thalli with percurrent main axes, opposite to sub-opposite pinnate branching to four orders. Final-order branches cord-like, to 30 cm long × 2 mm broad, with bluntly rounded to mucronate apices. Immature thalli (rossii-stage) yellow-brown to olivaceous, with delicate pinnately branched axes bearing lanceolate pinnules 1–2 cm long, arising from discoid to slightly lobed, solid, conical holdfasts (Fig. 49a, b); sporelings and some juveniles with trichothallic hair-tufts along margins or apices of laterals and main axes (Fig. 49a). Longitudinal (Fig. 50b, d) and transverse (Fig. 51b, d) sections of mature (chordalis-stage) thalli showing meristoderm of elongate, rounded, pigmented cells surrounding cortex, 4–8 cells deep, and medulla of intermixed large and small diameter filaments. Some large, elongated, medullary cells (trumpet hyphae), having flared ends with pitted transverse walls ('sieve plates'; Fig. 51f), surrounded by monostromatic layer of small-celled filaments (ensheathing filaments; Figs 50e, 51e). Identical sectional views of young and old plants showing comparable construction but with shapes of meristoderm cells differing—immature thalli (rossii-stage) having rounded and nearly isodiametric cells (Figs 50a, c; 51a, c), and older thalli (chordalis-stage) having elongate cells (Figs 50b, d; 51b, d). Reproductive structures unknown.

More than 80 plants have been studied, all of which grew either on rocky or coralline-encrusted surfaces in the subtidal (5–22 m deep). Aside from *Macrocystis*, *D. chordalis* is the largest and most conspicuous subtidal brown alga at Macquarie I. It is also one of two (*D. willii* being the other) dominant overstory species that further shade the largely red-algal understory community in *Macrocystis* forests.

The cord-like fronds of *D. chordalis* often support creeping hydroid colonies, the brown alga *Ectocarpus constanciae*, and the red algae *Delesseria lancifolia* and *Schizoseris dichotoma*. The massive holdfasts of *D. chordalis* are epiphytized usually by encrusting coralline algae and some delesseriaceous species.

NOTES: *Desmarestia chordalis* and *D. rossii*, first described by Hooker f. & Harvey (1845) as separate species because of differences in branching, were suspected by Levring (1944) to actually constitute only one species. Macquarie specimens, covering a broad range in ages, show a continuum from delicate, pinnately branched young plants (rossii-stage) to older thalli with thick

*Skottsberg (1906) reported *Desmarestia 'rossii'* from the Antarctic Peninsula and Victoria Land. Subsequent studies in these areas (Moe, pers. comm.) and other Antarctic localities (pers. obs.), however, have not confirmed the presence of this species in the Antarctic. It seems likely that Skottsberg probably misidentified *D. menziesii*, a common Antarctic species, as *D. rossii*.

Fig. 49 *Desmarestia chordalis* Hooker f. & Harvey: MELU, R 0194, 3517. **a.** Habit of juvenile thallus showing filamentous (trichothallic) structure of laterals (Slide R/M 0316). **b.** Habit of rossii-stage. **c.** Habit of mature (chordalis-stage) thallus. **d.** Detail of holdfast from mature (chordalis-stage) thallus. Scales: a = 4 mm; b = 5 cm; c = 20 cm; d = 2 cm.

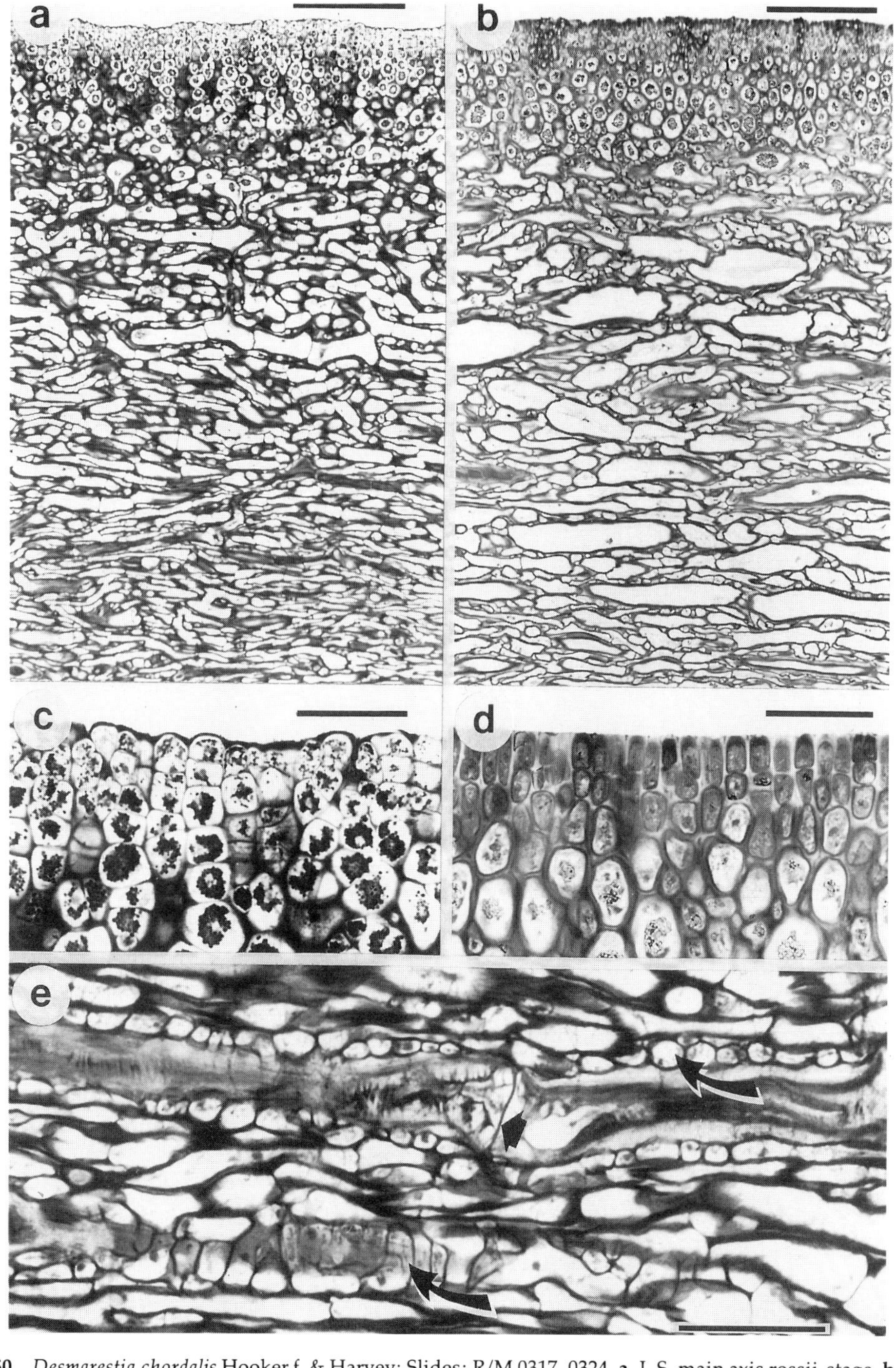

Fig. 50 *Desmarestia chordalis* Hooker f. & Harvey: Slides: R/M 0317, 0324. **a.** L.S. main axis rossii-stage. **b.** L.S. main axis chordalis-stage. **c.** Detail rossii-stage cortex in L.S. **d.** Detail chordalis-stage cortex in L.S. **e.** L.S. chordalis-stage medulla showing ensheathing filaments (curved arrows) and 'sieve plate' (short arrow) separating two trumpet-hyphal cells. Scales: a, b = 200 μm; c–e = 50 μm.

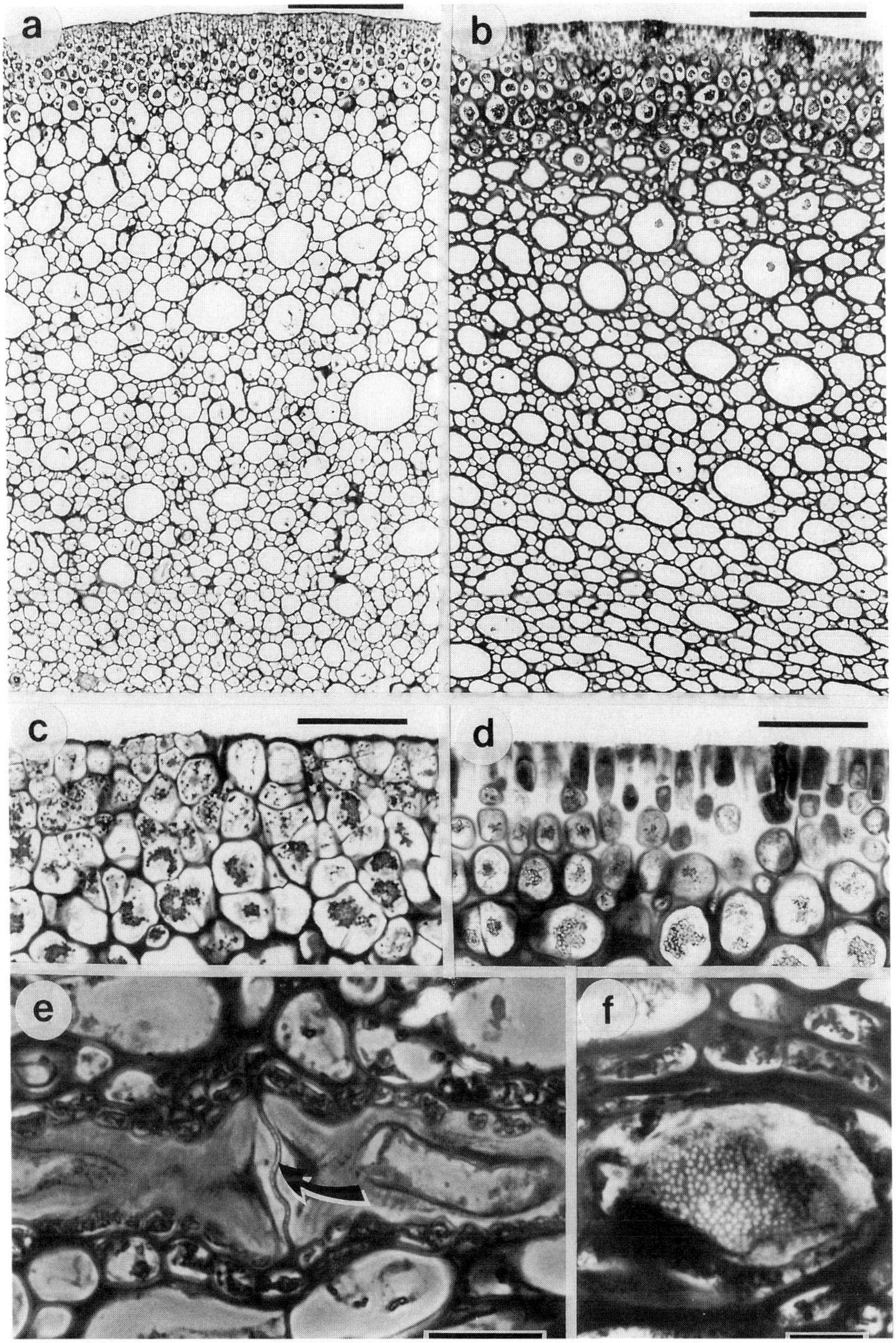

Fig. 51 *Desmarestia chordalis* Hooker f. & Harvey: Slides: R/M 0317, 0319, 0322. **a.** T.S. main axis rossii-stage. **b.** T.S. main axis chordalis-stage. **c.** Detail rossii-stage cortex in T.S. **d.** Detail chordalis-stage cortex in T.S. **e.** L.S. chordalis-stage medulla with detail of septum (arrow) separating trumpet-hyphal cells. **f.** Surface view 'sieve plate' showing pits in septum. Scales: a, b = 200 µm; c, d = 50 µm; e = 20 µm; f = 10 µm.

masses of coarse, cord-like fronds (chordalis-stage), thus confirming Levring's suspicions.

Either name, *'chordalis'* or *'rossii'*, could be validly applied to this entity, but the name *D. chordalis* is used in this report because of its original application to the mature thallus form.

Desmarestia chordalis is the most robust *Desmarestia* species in high-latitude southern waters. It has a comparable habit and community role to that of its Antarctic equivalent, *D. menziesii* J. Agardh. The two species apparently do not overlap geographically, but some specimens of *D. chordalis* from Heard I. appear similar to *D. menziesii* by producing small, lanceolate pinnules. Such spine-like pinnules are characteristic of the slightly less robust *D. menziesii.*

Desmarestia willii Reinsch, 1888*a*: 191

Fig. 52

SYNONYMS: see Papenfuss, 1964*a*.

TYPE LOCALITY: South Georgia.

LECTOTYPE: Herb. Reinsch (M), no. 67.83/53.

DISTRIBUTION: Auckland Is (Laing, 1909); Îles Crozet (Levring, 1944); Îles Kerguelen (Reinbold, 1908; Levring, 1944); Falkland Is (Skottsberg, 1907; Cotton, 1915); Fuegia (Hariot, 1889; Skottsberg, 1907); Macquarie I. (Zinova, 1958; Kenny & Haysom, 1962); Marion I. (Dickie, 1876*b*); South Georgia (Reinsch, 1888*a*). (N.B. Papenfuss (1964*a*) additionally indicated distributions at the Prince Edward and South Shetland Is and Victoria Land, but the original published records for these localities have not been located for the present report. Gain's (1912) report of *D. willii* from Peterman I. (part of the Palmer Archipelago) is not included in Papenfuss' list and apparently also was overlooked by Lamb & Zimmermann (1977).)

MATERIAL EXAMINED:

1. Îles Kerguelen: Deutsche Tiefsee Expedition (Apstein, 26–28 December 1898): HBG, no. 206.
2. Macquarie I.: collections MA-11, 14, 34, 46, 56, 83, 113, 114, 115, 116, 117, 134, 137, 140, 215, 222, 227, 237, 244, 248, 271, 283, 284, 295, 300, 302, 303, 306, 309, 364, 367, 377, 378, 397; Herbarium specimens: MELU, R 0819, 0821, 0822, 3513; Microscope slides: R/M 0323, 0395.
3. South Georgia: Lectotype (Will, 1883): M, 67.83/53; possible syntype: M, 67.83/52.

DESCRIPTION: Thalli to 1 m long, yellowish-brown, with branched percurrent main axes (2–3 mm diameter), arising from lobed discoid holdfasts (Fig. 52a, b). Branching oppositely pinnate (Fig. 52d) to four orders, with terete to slightly compressed branches. Internode lengths variable; pinnae either compact or loose, with pinnules sometimes short and spicate but generally longer (to several cm), delicate, thread-like. Trichothallic meristems visible on upper branches of some small plants (Fig. 52c), generally absent from larger specimens, where cortication extends to blunt-tipped apices. Main axes with circular (Fig. 52e) or oval cross-sectional outline; axial cell centrally positioned, thick-walled, ensheathed by single layer of smaller cells (Fig. 52g); medulla with large and small diam colourless cells (Fig. 52e); cortex with 1–2 layers of pigmented, rounded (15–40 µm diameter) cells (Fig. 52f). Reproductive structures unknown.

More than 50 plants have been studied, all of which grew on similar rocky substrata at comparable subtidal depths as did *D. chordalis. Desmarestia willii*, however, generally occurred more frequently below depths of 15 m than did *D. chordalis.*

In contrast to the coarse, strap-like plants of *D. chordalis* that often were covered partly by encrusting coralline algae or members of the Delesseriaceae, thalli of *D. willii* formed large, soft bushes and were unusually distinctive in their general lack of epiphytes.

NOTES: *Desmarestia willii*, often confused with *D. viridis* (O. F. Müller) Lamouroux from the Northern Hemisphere, was distinguished particularly from this species by Reinsch (1888*a*). The two taxa supposedly differ in cross-sectional outline of stipes, cortical thicknesses, organization of certain medullary cells, orders of branching and degree of cortication at branch apices.

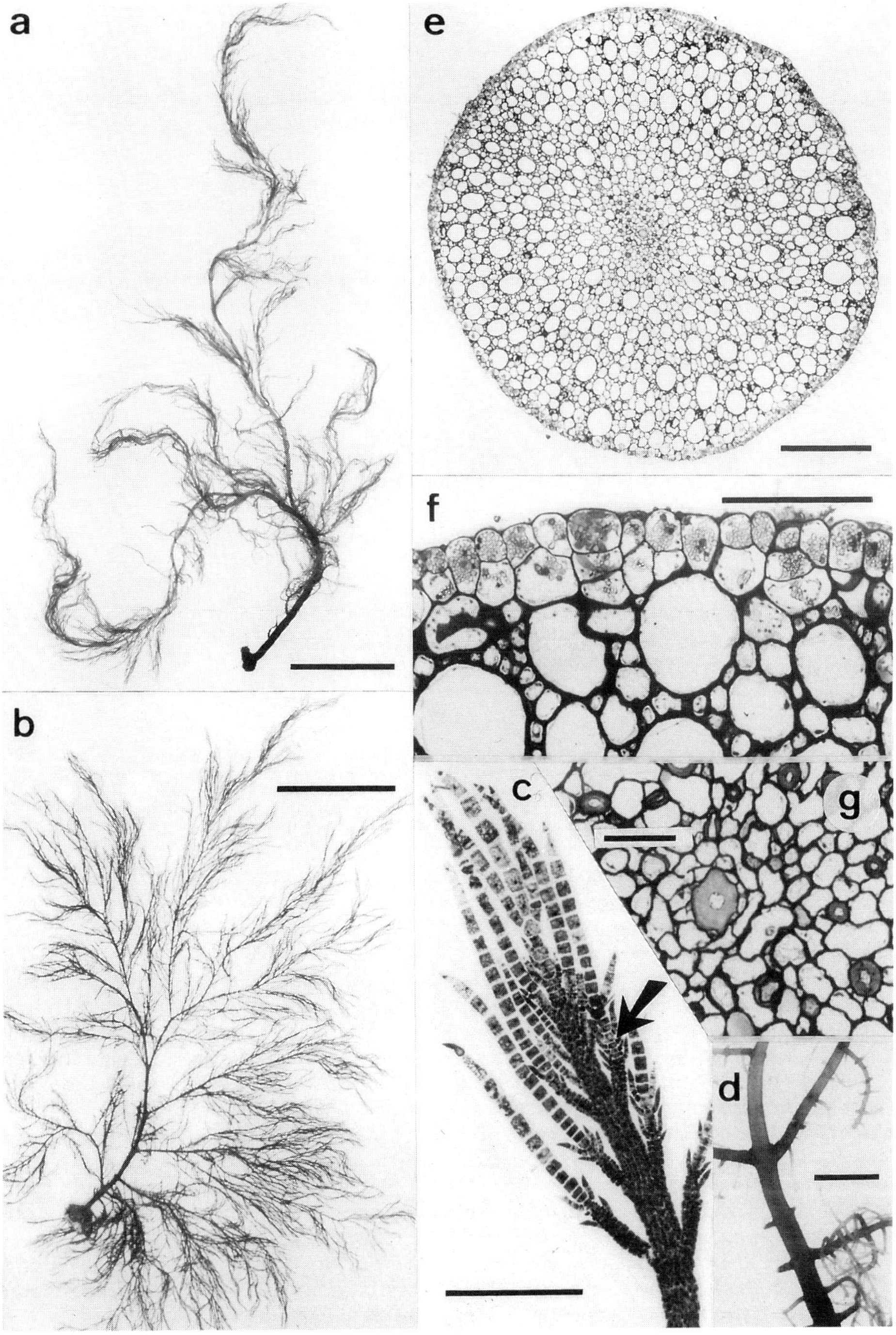

Fig. 52 *Desmarestia willii* Reinsch: MELU, 0819; Slides: R/M 0323, 0395. **a.** Habit of juvenile thallus. **b.** Habit of mature thallus. **c.** Trichothallic meristems (arrow) on upper branch. **d.** Detail of opposite branching along main axis. **e.** T.S. main axis, mature thallus. **f.** Detail of cortex from 'e'. **g.** T.S. main axis showing detail of axial filament surrounded by smaller, thick-walled medullary cells ('vascular strands'). Scales: a = 2 cm; b = 5 cm; c = 250 μm; d = 5 mm; e = 400 μm; f = 100 μm; g = 50 μm.

Desmarestia willii originally was described as having stipes with elliptical, cross-sectional outlines, a monostromatic cortex and medullary tissues that appear as a reticulum composed of large and small cells. Also, some thick-walled 'vascular strands' (= Strange) in the medulla were said to be distributed around the central, oval-shaped axial cell with its fine mesh of ensheathing tissue. In the present study, *D. willii* has been seen with pinnate branching to four orders and with cortication extending completely to branch apices. In contrast, *D. viridis* was distinguished by Reinsch as having stipes with circular, cross-sectional outlines, a (3–)4-layered cortex, medullary tissues without 'vascular strands' organized around the central axis, pinnate branching to only three orders, and cortication not extending to the terminal filaments.

Features such as stipe shape and the degrees of cortication and branching show a wide range of variation, which suggests that *D. viridis* and *D. willii* are less distinct than Reinsch (1888*a*) indicated. Specimens from Macquarie I. are similar in habit to both species but are presently referred to the Southern Hemisphere *D. willii* until the exact relationship between both taxa is clarified.

Cutleriales

Three genera, *Cutleria, Microzonia* and *Zanardinia,* of the single family Cutleriaceae constitute the Cutleriales. The order is represented most widely by *Cutleria,* which grows in tropical to temperate seas in the Northern and Southern Hemispheres (Allender & Kraft, 1983). In contrast, *Zanadinia* is known mainly from the Mediterranean (Fritsch, 1945) and *Microzonia* is restricted to cold-temperate and Subantarctic waters of the Southern Hemisphere (O'Donnell, 1954; Asensi, 1971).

Cutleriaceae Hauck, 1883/1884: 318, 403

Plants generally grow from marginal, trichothallic meristems that produce strap- to fan-shaped thalli composed of true parenchyma. Although a trichothallic meristem has been reported for *Microzonia* (O'Donnell, 1954; Asensi, 1971), its presence has not been confirmed through the present studies, which have revealed growth from a marginal zone of dome-shaped apical cells. Possibly two different entities are being placed in *Microzonia,* or the trichothallic filaments are deciduous and missing from plants studied for the present survey.

Life histories include alternations of heteromorphic (*Cutleria*) and isomorphic (*Zanardinia*) somatic phases (Yamanouchi, 1912, 1913). Reproduction is anisogamous, with gametophytes bearing large (female) and small (male) plurilocular gametangia on either the same or different thalli. Gametangia in *Cutleria* are borne laterally on branched filaments, whereas those in *Zanardinia* are borne terminally on simple filaments. Sporophytes of *Cutleria* and *Zanardinia* bear sori with dense aggregations of sessile, unilocular sporangia that contain 4, 8, 16, or 32 zoospores (Fritsch, 1945). Unilocular sporangia in *Microzonia* are borne laterally and terminally on branched filaments and are believed to contain 16–32(–64?) spores (O'Donnell, 1954).

Syringoderma is similar to *Microzonia* in having fan-shaped blades developed from mariginal zones of apical cells, cortical-cell rhizoids that thicken blades, and unilocular sporangia produced on uniseriate, branched filaments (Walker & Henry, 1978; Henry, 1984). Four species of *Syringoderma* have been described, with *S. australe* Levring (1940) from high latitudes in the Southern Ocean, and *S. abyssicola* (Setchell & Gardner) Levring (1940), *S. floridana* Henry (1984) and *S. phinneyi* Henry & Müller (1983) from the Northern Hemisphere. This genus was originally placed in the Dictyotales, later believed to be more similar to the Sphacelariales (Delépine, 1968; Wynne, 1972), and recently moved to its own order, the Syringodermatales (Henry, 1984). The taxonomic relationship between *Microzonia* and *Syringoderma* is uncertain at present and needs to be clarified through detailed developmental and life history studies of the former to ascertain whether or not it should also be placed in the Syringodermatales.

Microzonia J. Agardh, 1894: 18

Type species: *M. velutina* (Harvey) J. Agardh, 1894: 19.

Originally described by Harvey (1855) as a species of *Zonaria* (probably because of its fan-shaped thallus), *Z. velutina* was moved by J. Agardh (1894) to the new genus *Microzonia* because of differences in thallus construction and the development of unilocular sporangia on uniseriate, branched filaments (vs. sessile development of sporangia in *Zonaria*). O'Donnell (1954) later transferred *Microzonia* from the Dictyotales to the Cutleriales because of the sporangial filaments and presence of a trichothallic meristem. The disposition of unilocular sporangia in *Microzonia*, however, differs from that in either *Cutleria* or *Zanardinia*.

Microzonia is monotypic and has been previously recorded from New Zealand (Lindauer *et al.*, 1961) and Argentina (Asensi, 1971). The occurrence of this alga at Macquarie I. constitutes the southernmost record of the genus.

Microzonia velutina (Harvey) J. Agardh, 1894: 19

Figs 53, 54

Basionym: *Zonaria velutina* Harvey, 1855: 218–219.

Synonyms: see Lindauer, *et al.*, 1961.

Type locality: East coast New Zealand.

Holotype: not designated; syntypes in Herb. Harvey, TCD (unnumbered specimens).

Distribution: Argentina (Asensi, 1971); Macquarie I.; New Zealand (Harvey, 1855; Lindauer *et al.*, 1961); Stewart I. (Adams *et al.*, 1974).

Material examined:

1. Macquarie I.: collections MA-298, 302, 303, 306, 309, 364, 377, 378; Herbarium specimens: MELU, R 1150, 1710, 1961, 1965; Microscope slides: R/M 0044, 0207.
2. New Zealand: Port Cooper, Banks' Peninsula (D. Lyall, March 1849): TCD, unnumbered specimen; Milford Sound, S.W. Coast N. Zealand (D. Lyall, March 1851): TCD, unnumbered specimen; South bank mouth of Owahanga River, east coast North I. (Parsons, 24 January 1973): CHR 319738; Oaro, 20 km south of Kaikoura, east coast South I. (Parsons, 24 October 1973): CHR, 319921.

Description: Plants to 3 cm long × 1–2 cm broad, flabellate, light- to dark-brown, decumbent (Fig. 53a, b), attached by basal rhizoids or rhizoidal mat covering most of ventral blade-surface. Thalli usually in imbricate clusters, with individual fronds splitting into thin stipe-like straps giving rise apically to young fan-shaped blades (Fig. 53d). Growth from marginal zone of apical cells (Figs 53c, 54); fronds monostromatic at margins (Fig. 53f, g), becoming tristromatic (to 39 μm thick) 1–5 cells behind apices (Fig. 54), thickened further by rhizoidal cortication toward blade bases (Fig. 53g–i). External, acropetal and basipetal rhizoids developing mostly from cortical cells on ventral blade-surface. Lower thallus (to 45 μm thick) generally consisting of two layers of rectilinear (mostly elongated), medullary cells surrounded by monostromatic cortex; upper and lower cortical surfaces dimorphic (Fig. 53i); dorsal cortical cells roughly half the size of ventral cells. Unilocular sporangia ovate (20–25 μm × 40–45 μm), lateral and terminal on branched monosiphonous filaments produced from ventral blade-surface (Fig. 53k). Sporangial filaments in dense dark-brown rows forming felty cushion covering all but thallus margins (Fig. 53j). Gametangia unknown.

Microzonia velutina grows subtidally (4–15 m deep) on either rocks or *Mesophyllum schmitzii* in shaded and unshaded sites on the floors of *Macrocystis* forests. Generally, numerous fronds overlap like shingles and form extensive mats to *c.* 1 m diameter. In most cases, sand covers the basal parts of thalli (often including the reproductive portions) and leaves only young blades

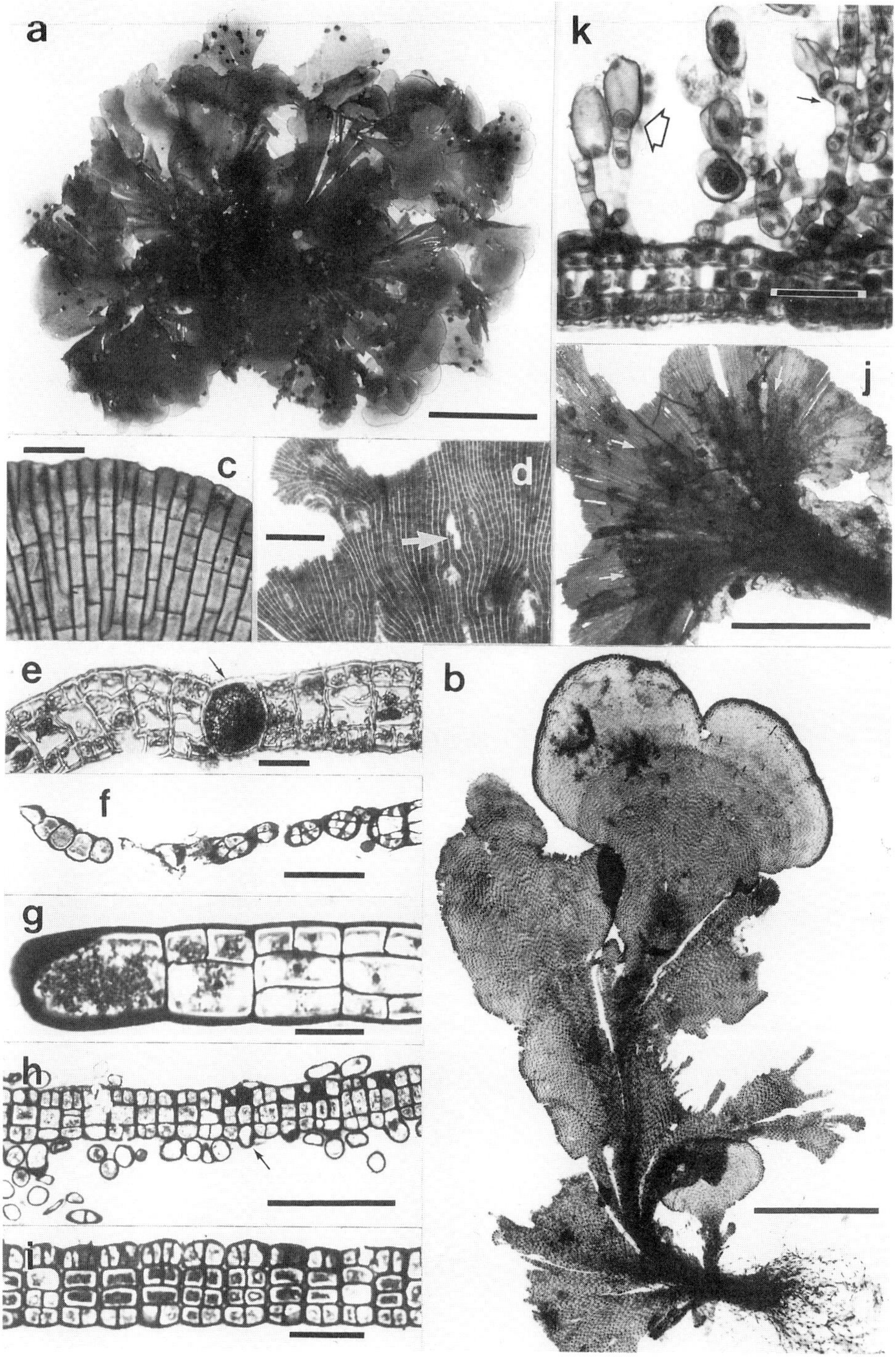
a
k
j
c
d
e
b
f
g
h
i

emergent. Plants often host bryozoans and epiphytic algae on their dorsal blade-surfaces. *Antithamnionella ternifolia* and *Melobesia membranacea* are common epiphytes, with *Dilophus decumbens* and *Hymenena decumbens* less frequent colonizers. Internally, thalli of *M. velutina* are infected sometimes by *Eurychasma*-like fungal infections (Fig. 53e), similar to those described by Jenneborg (1977) for *Syringoderma* and a variety of other algae.

NOTES: *Microzonia velutina* from Macquarie I. and type specimens from New Zealand lack marginal hair-fringes. Expansion of blades apparently occurs through division of marginal apical cells as in some Dictyotales (e.g., *Chlanidophora, Stypopodium, Zonaria*).

Macquarie thalli generally are thinner, more delicate, and a lighter shade of brown than are the darker (drying to black), coriaceous, intertidal (type) specimens collected from New Zealand. Lindauer *et al.* (1961), however, described deep-water plants with light-coloured, more membranous and fragile fronds that differed from the coarse thalli of intertidal plants.

Dictyotales

Members of the Dictyotales are parenchymatous and grow from apical meristems consisting of either single, lenticular cells, or marginal rows of elongate, tubular cells. Thalli are flattened, strap-like or flabellate, and either erect, procumbent, or decumbent. Life history patterns include an alternation of isomorphic generations, with gametophytes and sporophytes bearing reproductive organs in superficial sori. Gametangia (producing eggs and antheridia in plurangia) and sporangia are sometimes interspersed with paraphyses.

The Dictyotales generally abound in tropical and subtropical waters (Fritsch, 1945; Bold & Wynne, 1978; Wynne, 1982*b*), although roughly two-thirds of the 17 genera currently recognized for the order (Papenfuss, 1977; Allender, 1980) grow along the cold-temperate shores of southern Australia (Womersley, 1967). Lindauer *et al.* (1961) also indicated a wealth of dictyotalean algae from temperate New Zealand waters. Areas where cold and warm waters mix also show considerable diversity of Dictyotales (Earle, 1969; Allender & Kraft, 1983). The cold waters of the Subantarctic Zone, however, are largely devoid of these algae (Papenfuss, 1964*a*), and the Antarctic Zone is completely without members of this order.

The order consists of two families, the Dictyotaceae and the monotypic, recently created Dictyotopsidaceae (Allender, 1980). A single member of the Dictyotaceae represents the order at Macquarie I.

Dictyotaceae Lamouroux *ex* Dumortier, 1822: 72, 101

The Dictyotaceae comprises some 16 genera, of which 15 occur in the Southern Hemisphere and 14 are represented in Australian waters. Twenty-four genera have been referred to this family at one time or another, but now some are placed in separate orders or designated as synonyms of existing dictyotacean genera (Papenfuss, 1977). With *Microzonia* moved to the Cutleriales (O'Donnell, 1954) and *Syringoderma* placed in its own order (Henry, 1984), the only record of a dictyotacean alga from southern high latitudes is Zinova's (1958) report of *Dictyota dichotoma* (Hudson) Lamouroux from Macquarie I. Other dictyotacean genera (*Distromium, Padina, Zonaria*) have been reported from islands along the northern edge of the Subantarctic Zone (Juan

Fig. 53 *Microzonia velutina* (Harvey) J. Agardh: MELU, R 1150, 1710, 1961; Slide R/M 0207. **a.** Imbricate cluster of several fronds in water. **b.** Thallus habit (single frond; slide R/M 0207). **c.** Surface view thallus margin. **d.** Surface view upper blade with lacunae. (N.B. filaments (arrow) growing into holes.) **e.** T.S. thallus containing *Eurychasma*-like fungal infection (arrow). **f–i.** T.S. frond, beginning 1 mm behind apex and progressing basipetally every 5 mm. 'f' and 'g' showing blade margins, 'h' and 'i' near mid-thallus. Note rhizoidal cortication in 'h' and outgrowth of rhizoid (arrow) from cortical cell. **j.** View of ventral surface of blade showing pubescent area (arrows) where sporangia are borne. **k.** T.S. frond showing ventral surface bearing branched filaments (arrow) laterally and terminally producing unilocular sporangia. (N.B. sporangia show internal proliferation within old sporangial walls (hollow arrow).) Scales: a = 1 cm; b = 1 mm; c, i, k = 50 µm; d = 300 µm; e, g = 25 µm; f, h = 100 µm; j = 5 mm.

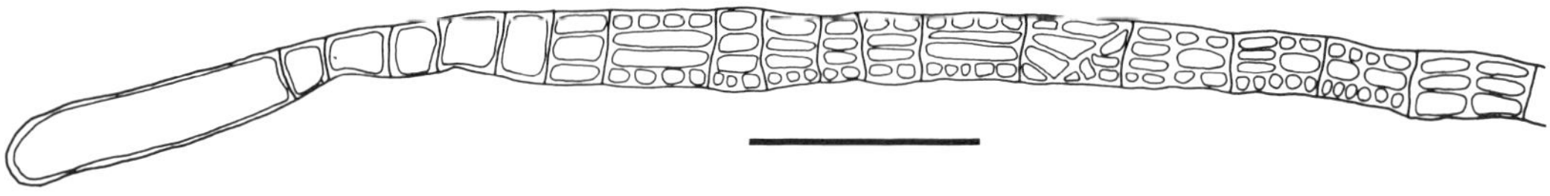

Fig. 54 *Microzonia velutina* (Harvey) J. Agardh: MELU, Slide R/M 0044. Camera lucida drawing: Longitudinal section from apex toward base. Scale: 200 μm.

Fernandez Is—Levring, 1940, 1941; Stewart I.—Adams *et al.*, 1974; Tristan da Cunha—Baardseth, 1941), but *Dictyota* at Macquarie I. represents the highest latitude record, in both Northern and Southern Hemispheres, for any member of the Dictyotaceae.

Dictyota dichotoma is a broadly misunderstood concept that has been used to receive many species (Earle, 1969), but none with decumbent habits or anatomy similar to the new species of *Dilophus* described in the present survey. Because of superficial similarities between *Dictyota* and *Dilophus*, and confirmation of only the latter genus at Macquarie I., Zinova's report of *Dictyota dichotoma* is suspected to actually represent the new species of *Dilophus*.

Dilophus J. Agardh, 1882: 106

Lectotype: *D. gunnianus* J. Agardh, 1882: 106.

J. Agardh (1882) used differences in frond anatomy to separate *Dilophus* (which has a medulla with multiple cell layers) from the superficially similar *Dictyota* (which has a medulla with one layer), and later (1894) used thallus morphology, the number of cell layers in the medulla, and the arrangement of medullary cells to characterize 14 species of *Dilophus*. J. Feldmann (1937) further distinguished members of the two genera by their different attachment structures, *Dilophus* having wiry stolons and *Dictyota* basal rhizoids. Womersley (1967) considered the morphological and anatomical features utilized by Agardh to be too variable and recommended re-evaluation of the genus.

Four species, mostly from Japan and the central-eastern Pacific, have been added to the genus since Agardh's (1894) monograph. These, plus the newly described alga from Macquarie I., increase the number of species in *Dilophus* to 19.

Dilophus decumbens R. W. Ricker sp. nov.

Fig. 55

HOLOTYPE: MELU, R 1143 (MA-295), with isotypes in ADU, BM, Herb. Ricker, and UC.

TYPE LOCALITY: Green Gorge, Macquarie I. (54°38′00″ S lat., 158°55′00″ E long.); growing on 0.3–1 m diameter boulders on floor of *Macrocystis* forest in 17 m depth (Ricker, 13 January 1978).

DISTRIBUTION: endemic to Macquarie I.

MATERIAL EXAMINED: Macquarie I.: collections MA-34, 83, 244, 248, 271, 283, 295, 298, 302, 303, 309, 364, 367, 377, 378; Herbarium specimens: MELU, R 0096, 0097, 0841, 1141–1148.

DIAGNOSIS: Plantae ad 5 cm altae axibus decumbentibus ligulatis brunneis dichotome vel subdichotome ramosis ad 1 cm latitudine, superioribus thalli partibus late rotundatis aut valde attenuatis, cum laminis ex singulis cellulis apicalibus lenticularibus evolventibus. Ventrales laminarum paginae cum agellis rhizoideorum ad bases et prope apices locatis, dorsalibus ex

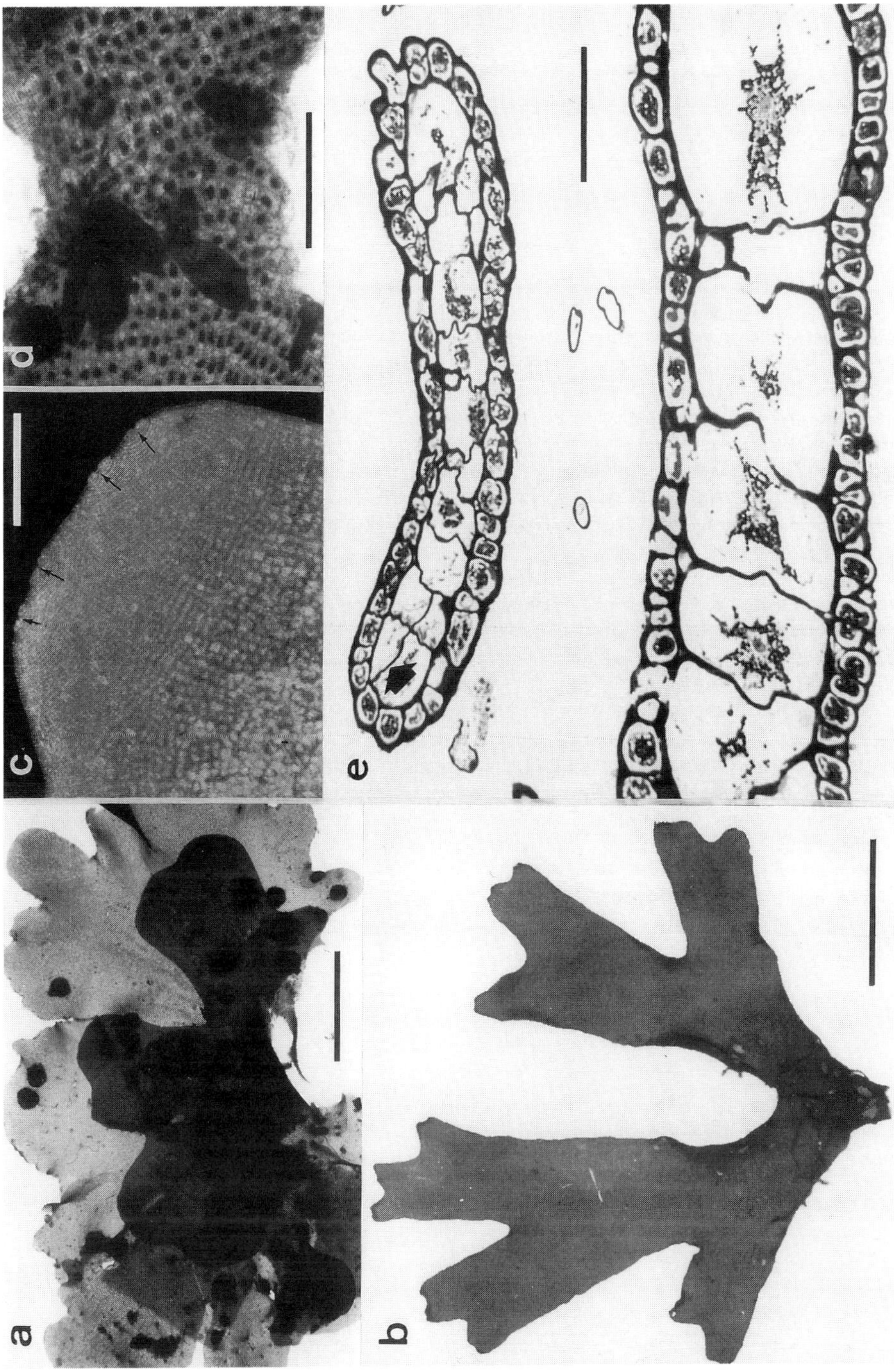

Fig. 55 *Dilophus decumbens* R. W. Ricker: MELU, R 0096, 0841. **a.** Habit of plant epiphytic on *Hymenena decumbens*. **b.** Habit of epilithic plant. **c.** Upper frond with apical cells (arrows). **d.** Bladelets from dorsal surface near thallus base. **e.** T.S. main blade (below) and bladelet (above) showing two-layered medulla (arrow) at margin. Scales: a = 3 mm; b = 1 cm; c = 500 μm; d = 1 mm; e = 100 μm.

cellulis corticalibus laminulas (ad 700 μm longitudine) interdum ferentibus. Frondes 160–200 μm crassae, pro parte maxima tristromaticae, praeter margins ubi medulla strata cellularum 2–3 irregulariter composita continere potest. Transectiones laminarum medullam incolorem ex cellulis vel quadratis (115 μm diametris) vel rectangularibus (100 μm × 200 μm) et corticem monostromaticum pigmentosum ex cellulis rectilinearibus vel triangularibus vel quadratis vel rectangularibus (20–25 μm × 30–40 μm) compositam patefaciunt. Structurae reproductivae ignotae.

The species name refers to the alga's characteristic decumbent habit. This feature, in addition to blade dimensions and a remote location, distinguishes *D. decumbens* from other species in the genus.

Description: Plants to 5 cm tall, with axes decumbent, ligulate, to 1 cm broad, deep brown, branched dichotomously to subdichotomously (Fig. 55a, b). Upper thallus broadly rounded or markedly attenuated; blades developing from single, lenticular, apical cells (Fig. 55c). Ventral blade-surfaces with patches of rhizoids at bases and near apices; dorsal blade-surfaces sometimes bearing bladelets (to 700 μm long) from cortical cells (Fig. 55d). Fronds 160–200 μm thick, mostly tristromatic, except at margins where medulla may comprise 2–3, irregularly organized layers of cells (Fig. 55e). Blade transections revealing colourless medulla, composed of cuboidal cells nearly isodiametric (115 μm diameter) to elongate (100 μm × 200 μm), and monostromatic pigmented cortex with rectilinear (mostly pyramidal and cuboidal) cells 20–25 μm × 30–40 μm (Fig. 55e). Reproductive structures unknown.

Over 200 specimens have been examined, all of which were collected between 5–20 m depths on shaded floors of *Macrocystis* forests. Plants grew either on boulders, generally forming mats several cm in diameter, or on various plants such as *Macrocystis* holdfasts, *Hymenema decumbens* (Fig. 55a) and *Microzonia velutina*.

Notes: Macquarie I. appears to be unique among other Subantarctic islands of comparable latitude in having a representative of the Dictyotales. This representative combines some features that supposedly are diagnostic for *Dilophus* (a medulla with multiple cell layers at least in part) and others that are diagnostic for *Dictyota* (attachment with basal rhizoids). The Macquarie alga is aligned with *Dilophus* on the basis of its blade anatomy. Although J. Feldmann (1937) characterized *Dilophus* as having stoloniferous attachments, which certainly is true for some species (e.g., *D. intermedius* (Zanardini) Allender & Kraft, *D. marginatus* J. Agardh; Allender & Kraft, 1983), not all species placed in this genus attach by stolons. Some species from southern Australia (e.g., *D. fastigiatus* (Sonder) J. Agardh) attach by means of basal rhizoids resembling those produced by *D. decumbens* (pers. obs.).

Laminariales

Members of the Laminariales have large sporophytes (the phase often referred to as 'kelps'), usually composed of highly differentiated tissues. Thalli consist of a holdfast, stipe, and blade, comprising an epidermis, cortex, and medulla. Distinctive anatomical features of most species in the order include cortical mucilage ducts, medullary cells elongated into trumpet-shaped hyphae, and sieve plates (Fritsch, 1945). Growth from a localized meristem, positioned in the 'transition zone' at the juncture of the stipe and blade, is also characteristic for laminarialean species. Additional growth (in girth) comes from a superficial meristem (meristoderm). Sporophytes reproduce only through unilocular sporangia, the products of which develop into gametophytes. Sporangia generally are borne on both faces of specialized fronds (sporophylls).

Gametophytes in the Laminariales are microscopic, dioecious filaments that are not recognizable in the field. Female thalli are sparsely branched and consist of large cells, some of which function as oogonia. Male plants are highly branched, composed of small cells and produce antheridia, in which each locule produces a single biflagellate sperm (Wynne, 1982*b*).

The order is generally conceived as having four families, the Alariaceae, Chordaceae, Laminariaceae and Lessoniaceae. Two of these, the Alariaceae and Lessoniaceae, have been recorded from the Subantarctic (J. D. Hooker, 1846; Skottsberg, 1907), although the presence of the former is doubted by Papenfuss (1964*a*; dubious record of *Ecklonia* from Fuegia). The transfer of *Phaeoglossum* and *Phyllogigas* into the Desmarestiales (Moe & Silva, 1977*a*, 1981) leaves the Antarctic without any representatives of the Laminariales.

Lessoniaceae Setchell & Gardner, 1925: 591, 621

This family is distinguished from other families in the order by the pattern of blade division. In the Lessoniaceae, blades divide longitudinally through their basal meristems and give rise to numerous blades, each with a separate basal meristem located at the region of transition from stipe to blade. Blade splitting in the most closely related family, the Laminariaceae, does not extend into the transition zone. Consequently, each individual member of the Laminariaceae has only one basal meristem regardless of the number of blades present.

Subantarctic members of the Lessoniaceae include five species of *Lessonia* and one species of *Macrocystis*, although Papenfuss (1964*a*) suggested that *Lessonia* in the Southern Ocean might actually consist of only three species. Kenny & Haysom (1962) reported *Lessonia variegata* J. Agardh and *Macrocystis pyrifera* from Macquarie I., but the first species was not encountered during the present study. Because *Macrocystis* juveniles superficially resemble those of *Lessonia* and also develop similarly (Fritsch, 1945; Lindauer *et al.*, 1961), it is possible that the two taxa were earlier confused. None of the Kenny & Haysom collections examined during the present survey contained *Lessonia*, although numerous *Lessonia*-like juveniles of *Macrocystis* were present.

Macrocystis C. Agardh, 1820: 46

Lectotype: *Macrocystis pyrifera* (Linnaeus) C. Agardh, 1820: 47.

The genus *Macrocystis* consists of three species and several forms (Womersley, 1954; Nicholson, 1979). It occurs widely throughout the Subantarctic Zone but also disjunctly in the Northern Hemisphere along the west coast of North America. Primarily a cold-temperate genus, it may have had a Northern Hemisphere origin, subsequently moved southward (possibly during periods of glaciation), and spread rapidly once plants reached the southern tip of South America, where they could be transported vast distances reasonably quickly by the prevailing West Wind Drift currents (Nicholson, 1979). Recession of the ice age and consequent broadening of the tropics may have led to the Northern and Southern Hemisphere populations becoming isolated.

Two species of *Macrocystis*, *M. angustifolia* Bory de Saint-Vincent and *M. pyrifera*, grow in the Southern Hemisphere, but only *M. pyrifera* extends into high-latitude southern waters, where it has a circumpolar distribution and is a common feature of most, if not all, Subantarctic islands.

Macrocystis pyrifera (Linnaeus) C. Agardh, 1820: 47

Fig. 56

Basionym: *Fucus pyriferus* Linnaeus, 1771: 311.

Synonyms: see Papenfuss, 1964*a*.

Type locality: oceano Aethiopico (Southern Atlantic Ocean; see discussion by Womersley, 1954).

Type: LINN. (N.B. type not studied during present survey).

Distribution: This species grows throughout the Subantarctic, extends northwards along the west coast of South America to roughly 14° S lat., and also occurs along the west coast of California (Womersley, 1954; Nicholson, 1979). Records in the Subantarctic include: Auckland and Campbell Is (Laing, 1909); Chile (Levring, 1960); Falkland Is (Skottsberg, 1907; Cotton, 1915); Fuegia

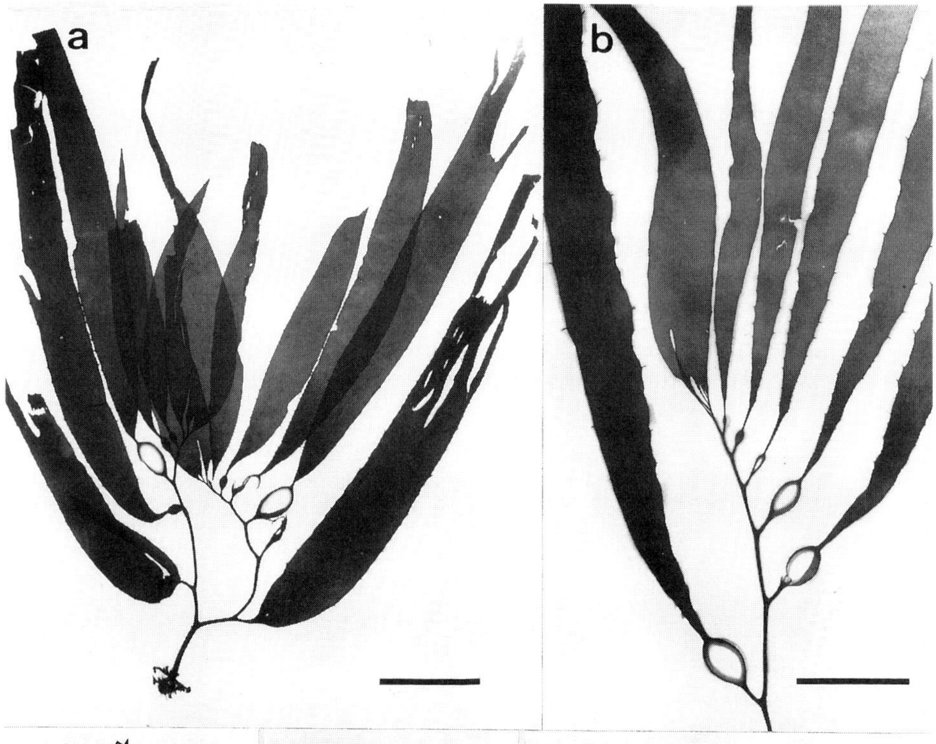

Fig. 56 *Macrocystis pyrifera* (Linnaeus) C. Agardh: MELU, R 0024, 0153, 1904. **a.** Habit of young sporophyte. **b.** Upper branch detail. **c.** Cut-away view of holdfast showing central axis. **d.** T.S. sterile frond. **e.** T.S. sporophyll showing unicellular paraphyses (hollow arrow) and unilocular sporangia (solid arrow). **f.** Surface-view of sporophyll with sterile margin. Scales: a, b = 5 cm; c = 2 cm; d = 100 μm; e = 25 μm; f = 1 mm.

(Harvey & Hooker f., 1845); Îles Crozet (Reinbold, 1908); Îles Kerguelen (Askenasy, 1888; Levring, 1944); Macquarie I. (Zinova, 1958; Kenny & Haysom, 1962); Marion I. (Dickie, 1876*b*); South Georgia (Reinsch, 1890; Skottsberg, 1907); South Sandwich Is (Zinova, 1966); St. Paul (Reinbold, 1907); Stewart I. (Adams *et al.*, 1974); Tristan da Cunha (Baardseth, 1941).

Material examined: Macquarie I.: collections MA-11, 14, 34, 35, 39, 46, 54, 56, 83, 95, 98, 111, 113, 114, 115, 116, 117, 140, 215, 227, 244, 248, 271, 295, 298, 300, 302, 303, 306, 309, 364, 377, 378, 381; Herbarium specimens: MELU, R 0024, 0153, 1376, 1904, 2146.

Description: Plants to 25 m long, consisting of hapteroid bases, terete, cartilaginous, upright axes, lateral and terminal blades. Holdfasts conical, with solid core surrounded by numerous outwardly radiating dichotomously forked haptera (Fig. 56c). Central axis (arising from holdfast) dichotomously branched four or more times in lower thallus, producing at least eight (usually entangled) cord-like uprights. Juvenile thalli with dichotomously branched terete axes bearing terminal, lanceolate to strap-like , entire or once-divided blades; blade division acropetal. Mature thalli usually bearing terminal, falcate blades repeatedly split acropetally (Fig. 56a, b) resulting in lateral fronds along entire length of upright axes. Mature fronds strap-like, to 90 cm long × 3 cm broad, 270–300 μm thick, with smooth or corrugated surfaces, with or without marginal teeth; blades usually subtended by ovoid to pyriform, gas-filled vesicles (pneumatocysts). Vesicles entirely lacking in some thalli, but where present, first appearing 5–15 cm behind terminal falcate blades. Sterile blades each with monostromatic superficial meristoderm, large-celled colourless cortex, and colourless filamentous medulla (Fig. 56d); reproductive fronds (sporophylls) with sterile margins (Fig. 56f), otherwise covered on both blade faces with clavate unicellular paraphyses 45–60 μm long, and fusiform unilocular sporangia 45–50 μm long (Fig. 56e). Gametophytes not seen.

Two distinct forms of this species were present at Macquarie I. Long, spindly plants growing at depths of 20–30 m in offshore kelp beds along the east coast of the island were invariably sterile. Plants in comparable offshore beds along the west coast, although seemingly less abudant, were similar but slightly more robust. Specimens from *c.* 5 m depth in Caroline cove, which is partially protected by offshore rocks, however, were 5–6 m long, densely branched, and richly fertile with basal sporophylls during mid-summer (January).

Macrocystis generally was devoid of epiphytes excepting those that often covered parts of its holdfast. These included *Dilophus decumbens*, encrusting coralline algae, several members of the Ceramiaceae (the most common of which were *Ballia callitricha* and *Pterothamnion simile*), and members of the Delesseriaceae.

Notes: The upright portions of *M. pyrifera* from Macquarie I. show considerable morphological variation, but the holdfast morphology remains relatively constant, as apparently is true throughout the entire range of the species (Womersley, 1954).

Durvillaeales

Although some authors (Fritsch, 1945; Clayton, 1981) treated *Durvillaea* as a member of the Fucales because of overall reproductive similarities, Petrov (1965) considered the genus sufficiently distinct to warrant its own order (Durvilleales). Hay (1979*b*) recognized the Durvillaeales because of *Durvillaea*'s unique combination of diffuse growth and conceptable-borne reproductive structures. *Durvillaea*, the sole genus in the order, has a *Laminaria*-like habit but lacks the localized intercalary meristem characteristic of laminarialean plants. It also produces gametangial conceptacles similar to those in the Fucales and would clearly be misplaced in the Laminariales. Placement in the Fucales, however, is debatable because *Durvillaea* has a diffuse meristem, rather than an apical one as typically occurs in fucalean species (Fritsch, 1945; Nizamuddin, 1968).

Durvillaeaceae (Oltmanns) De Toni, 1891: 173

Oltmanns (1889) recognized *Durvillaea*, *Sarcophycus* (now a synonym of *Durvillaea*), and *Ecklonia* (a member of the Laminariales) as distinct from other members of the Fucaceae and placed the first genus (and the latter two with question) in his group Durvilleae. The name Durvillaeaceae first appeared as a *nomen nudum* (De Toni, 1891) and was later validated (De Toni, 1895) with a Latin diagnosis mostly describing the positioning and development of oogonia.

Durvillaea alone, comprising four species, constitutes the Durvillaeaceae (Hay, 1979*b*). Diffuse vegetative growth and a thallus that is differentiated into a meristoderm, cortex, and medulla characterize members of this strictly Southern Hemisphere family (Naylor, 1949). Hyphae, which arise as outgrowths from the cortical and medullary cells, traverse and course longitudinally through the blades. Plants are dioecious and have conceptacles scattered over the entire frond. Antheridia are mostly borne on branched filaments, and oogonia are either attached directly to the conceptacle wall or borne on branched filaments.

Durvillaea Bory de Saint-Vincent, 1826*a*: 192

Lectotype: *Durvillaea antarctica* (Chamisso) Hariot, 1892: 1432.

The morphology and anatomy of *Durvillaea* are best known through the studies of Herriott (1923) and Naylor (1949) on *D. antarctica*, other species having received much less attention. Hay (1979*a*, 1979*b*) reviewed the genus and some of the criteria used to separate the four recognized species, *D. antarctica*, *D. chathamensis* Hay, *D. potatorum* (Labillardiere) Areschoug, and *D. willana* Lindauer. One species, *D. antarctica*, usually produced buoyant laminae that have honeycomb-like medullary tissues. This species is also the most widely distributed in the genus, being common throughout the Subantarctic. In contrast, the three species with non-inflated blades have restricted distributions, with *D. chathamensis* confined to the Chatham and Antipode Is., *D. potatorum* to southeastern Australia and *D. willana* to southern New Zealand.

This genus superficially resembles some members of the Desmarestiales and Laminariales in having a large and roughly conical holdfast, a terete to slightly compressed stipe, and a broadly expanded blade that is dissected to various degrees. At least nine species have been accredited to *Durvillaea*, but some simply represent degrees of variation in one species (e.g., blade thickness and habitat dissimilarities). Hay (1979*b*) separated *Durvillaea* species by differences in holdfast morphology, stipe dimensions and branching pattern, lamina morphology and anatomy, and habitat.

Durvillaea does not grow in the Antarctic, but three species have been listed from the Subantarctic (Papenfuss, 1964*a*). One of these, *D. caepaestipes*, is a synonym of *D. antarctica*, which is a second species recognized by Papenfuss (Hay, 1979*b*). The third species, *D. potatorum*, was considered by Nizamuddin (1968) as endemic to Australian waters. Rabenhorst (1878) reported it from the Auckland Is, but Papenfuss (1964*a*) questioned this distribution record. The Stewart and Chatham Is., both in the New Zealand Subantarctic, have *D. willana* Adams *et al.*, 1974) and *D. chathamensis* (Hay, 1979*b*) respectively, in addition to the circumpolar species *D. antarctica*. Only *D. antarctica*, however, grows as far south as Macquarie I. (Bennett (1971), Delépine (1966), Kenny & Haysom (1962), Simpson (1976) and Zinova (1958, 1964, 1973)).

Durvillaea antarctica (Chamisso) Hariot, 1892: 1432

Fig. 57

BASIONYM: *Fucus antarcticus* Chamisso *in* Choris, 1822: 7.

Fig. 57 *Durvillaea antarctica* (Chamisso) Hariot: MELU, R 1398, 1399. **a.** Habit of plants encircling intertidal boulder. **b.** T.S. lamina with chambered medullary tissue; section showing raised profile of pustule. **c.** Epidermal peel (including meristoderm, cortex, and outer layer of medulla) showing honeycomb-like construction of medulla. **d.** Pustule surface from intertidal lamina. **e.** Section through meristoderm (arrows) and cortex. **f.** Section through pustule with refractile, possibly fungal, bodies (darkly stained here) dispersed in cortex. **g.** Section through meristoderm, cortex, and outer medulla. Scales: a = 20 cm; b, d = 15 mm; c = 10 mm; e = 50 µm; f, g = 100 µm.

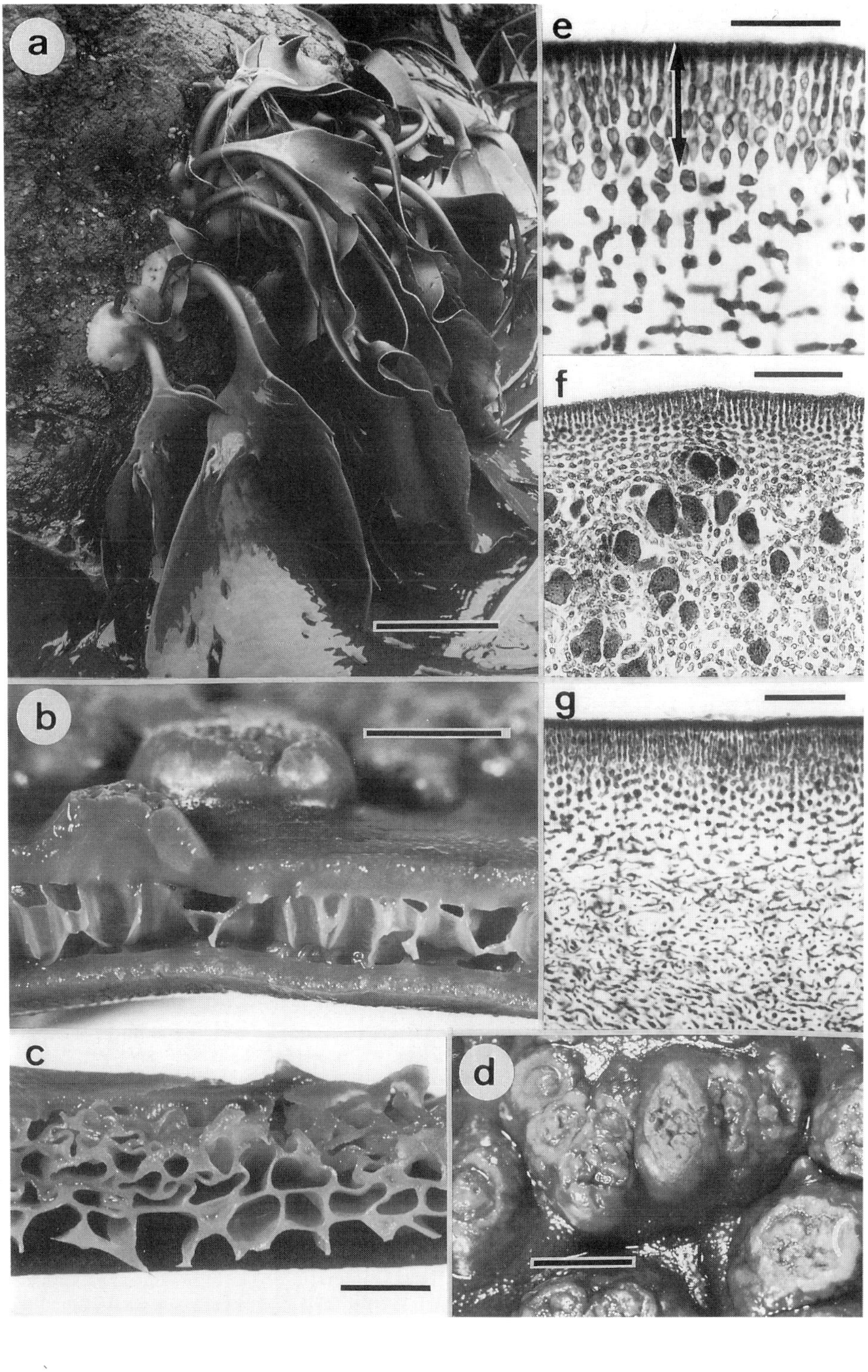
a
b
c
d
e
f
g

SYNONYMS: see Hay, 1979*b*.

TYPE LOCALITY: Cape Horn.

TYPE: plate 7 accompanying Chamisso's (1822) description in *Voyage pittoresque autour de monde* (see Hay, 1979*b*).

DISTRIBUTION: Circumpolar throughout the Subantarctic; known as far north as Coquimbo (*c*. 29° S) in Chile and as far south as Macquarie I. (54°30′ S). Delépine (1966) and Hay (1979*a*) map the alga's distribution throughout the Subantarctic.

MATERIAL EXAMINED: Macquarie I.: collections MA-11, 16, 39, 53, 95, 97, 112, 129, 134, 137; Herbarium specimens: MELU, R 1398, 1399.

DESCRIPTION: Thalli either broad and flat (Fig. 57a), thick or thin, or highly dissected into thong-like straps of rubbery to leathery consistency. Calm-water plants and juveniles usually solid-bladed; rough-water plants with chambered laminae (Fig. 57b, c). Massive conical holdfasts commonly fusing to form single discs to 0.5+ m diameter bearing multiple fronds. Blades subtended by stout stipes 10–20 cm long. Frond anatomy with outermost tissue (meristoderm) 5–6 cell-layers thick composed of elongate tear-shaped cells (Fig. 57e), outer cortex with rounded to irregularly shaped cells, inner cortex with periclinally elongated cells (Fig. 57g), and innermost medulla with colourless filaments. Reproductive structures not observed.

The pattern of meristoderm, cortex and medulla may be disrupted in spots by perturbations such as physical battering of fronds and subsequent regrowth, invertebrate grazing (although invertebrates on *Durvillaea* were rarely seen) and infection. Large pustules (Fig. 57b, d), common on many intertidal plants, have a thickened meristoderm of small tear-shaped cells (8–12 cell layers deep) outside a disorganized cortex rich with tiny refractile bodies (possibly of fungal origin; Fig. 57f).

Although not studied during the present survey, reproduction in *D. antarctica* has been documented by Hay (1979*b*). Plants are dioecious, with male and female thalli at times arising from common, coalesced holdfasts.

Plants generally grow in the lower intertidal but also extend to 15 m depth in an extremely rough subtidal zone near Anchor Rock. The subtidal algal community in the locality is unusual because of *Durvillaea*'s presence, and the flora here approximates one typical of most intertidal regions around the island.

NOTES: *Durvillaea antarctica*, as characterized by Hay (1979*b*, *c*), is a strictly intertidal species distinct from *D. chathamensis*, which extends into the subtidal. *Durvillaea antarctica* forms a zone above mean low water to about mid-tide level, whereas *D. chathamensis* grows lower down at the low-tide mark and into the sub-littoral. Nizamuddin (1968) describes *D. antarctica* (using the name *D. caepaestipes*) as 'sub-littoral and sub-Antarctic' based on a supposed record by Lindauer, Chapman & Aiken that, however, is not apparent in their report (1961). The occurrence of *D. antarctica* from the subtidal at Macquarie I., therefore, is the first verified extension of this species into the subtidal. The overlap in the vertical distributions for *D. antarctica* and *D. chathamensis* thereby eliminates one distinction between these two species. It seems prudent, however, to continue thinking of *D. antarctica* as a primarily intertidal species because of its dominance in this zone and general absence from most subtidal habitats.

Hay (1979*b*) suggested that the internal gas-containing chambers of the honeycomb-like tissue that make the fronds of *D. antarctica* buoyant may aid dispersion over vast expanses of open ocean. Although these buoyant fronds might also be expected to serve as rafts for the dispersal of attached algae, *Durvillaea* from Macquarie I. only rarely bore epiphytes, these generally being cosmopolitan members of the Ectocarpales (e.g., *Ectocarpus siliculosus*, Fig. 30).

In New Zealand waters an endophytic brown alga, *Herpodiscus durvillae* South, infects *D. antarctica* and forms large galls comparable in external morphology to those on Macquarie plants (South, 1974), but this endophyte has not been collected at Macquarie I. Cantacuzène (1930) and Andrews (1976) describe a range of fungal and pathological infections in numerous brown algae, but none seem responsible for the galls on Macquarie *Durvillaea*.

Rhodophyta

Bangiales

This order traditionally has contained the three families Bangiaceae, Boldiaceae and Erythropeltidaceae (Bold & Wynne, 1978; Dixon, 1982), but Garbary *et al.* (1980*a*) recently placed the latter two families, plus the Compsopogonaceae, in a new order, the Erythropeltidales. This order, however, is predated by the Compsopogonales, which Kylin (1956) erected for the Compsopogonaceae. Separation of the above-mentioned families is based upon the presence of pit connections in the sporophytic (conchocelis) stage of members of the Bangiaceae and a complete lack of pit connections in all life history phases of species in the remaining three families. Differences in spore production also have been used to split the Bangiales into two orders. In the present account, in which only gametophytes have been studied, a traditional concept of the order is upheld. The Bangiales is represented at Macquarie I. and throughout the Southern Ocean by two families, the Bangiaceae and Erythropeltidaceae.

Bangiaceae Nägeli, 1847: 136

Members of this family have dissimilar generations involving alternations of erect filamentous, cylindrical, or foliose gametophytes with branched filamentous sporophytes. Gametophytes show diffuse growth and are constructed of cells that lack pit connections and have 1–2 stellate plastids, each one central pyrenoid. Thalli are largely undifferentiated except at their bases where the cells develop rhizoidal processes. Plants reproduce asexually by transformation of vegetative cells into monospores or aplanospores. Sexual reproduction has been demonstrated by Hawkes (1978), who was able to show fusion of gametes that develop from undifferentiated vegetative cells.

Sporophytes have apical or intercalary meristems, primary pit connections and cells with one to several plastids of various types, each with one or more pyrenoids. Recycling of the sporophyte stage occurs by monospores that show unipolar germination. Otherwise, the sporophyte produces specialized lateral (conchosporangial) branches that give rise to conchospores showing bipolar germination and eventually developing into gametophytes.

The family is well represented throughout the world, as well as in high-latitude southern waters, by the genera *Bangia* and *Porphyra*, but only species of the latter occur at Macquarie I.

Porphyra C. Agardh, 1824: xxxii (***nom. cons.***)

Type species: *P. purpurea* (Roth) C. Agardh, 1824: 191 (= *P. umbilicalis* (Linnaeus) Kützing, 1843: 383—see Silva (1952) re. synonymy).

Taxonomic distinctions within this genus are based upon the membranous blades of the gametophyte (Kurogi, 1972; Conway *et al.*, 1975; Garbary *et al.*, 1980*b*). Criteria used to separate infrageneric groups include: monostromatic vs. distromatic fronds; blade shape, colour, and thickness; vegetative cell size and shape; number of plastids per cell; presence of only one (dioecious) or both (monoecious) types of reproductive organs on a single individual; location (marginal or submarginal) and arrangement (intermixed with vegetative cells or composed strictly of reproductive ones) of reproductive organs on blade surfaces; division patterns for carpogonia and spermatangia and subsequent organization of tiers of carposporangia and spermatia; and various ecological features such as seasonality, habitat preference, and geographic distribution.

Most *Porphyra* species have two somatic phases (Garbary *et al.*, 1980*b*), with the smaller sporophytic (conchocelis) phase, which forms delicate pink patches of microscopic filaments in shells, often being overlooked in the field. The only Subantarctic record of this stage comes from shells collected at the Falkland Is (Cotton, 1915).

Porphyra has many species distributed around the world, but the greatest concentration of these is along the eastern and western sides of the North Pacific (Conway *et al.*, 1975). Papenfuss

(1964*a*) listed nine species from the Antarctic and Subantarctic Zones. In the present survey, four are reported from Macquarie I., two of which are new. Insufficient material of a fifth entity has prevented its identification at the specific level.

Porphyra columbina Montagne, 1842*b*: 14

Fig. 58

SYNONYMS: see Womersley & Conway, 1975.

TYPE LOCALITY: Auckland Is.

TYPE: Herb. Montagne, PC; unnumbered specimen which has been designated by H.B.S. Womersley (5.12.1952) as the type for the species.

DISTRIBUTION: Argentina (Pujals, 1963); Auckland Is (Montagne, 1842*b*, 1845); Australia (south-eastern corner, including Tasmania) (Womersley & Conway, 1975); Fuegia (Levring, 1960); Heard I. (pers. obs.); Macquarie I. (Zinova, 1958, 1973); Stewart I. (Adams *et al.*, 1974). (N.B. Levring (1953) additionally indicated distributions at Campbell I. and parts of Australia, but the original published records for these localities have not been located for the present report.)

MATERIAL EXAMINED:
1. Auckland Is: specimen designated as the type (no collection data available).
2. Heard I.: north of Wharf Pt., Atlas Cove, rocks in eulittoral (Smith, 6 February 1983): MELU, R 3717.
3. Macquarie I.: collections MA-119, 120, 130, 133, 137, 238, 278; Herbarium specimens: MELU, R 1883–1888, 2015, 2016, 2030–2032, 2113; Hasselborough Bay, reef (Haysom, 29 March 1950): MELU, R 3714 (= Haysom specimen no. MI/49/x64); Garden Bay, upper littoral rocks (Taylor, 8 January 1951): MELU, R 3712, 3713 (= Taylor collection no. 546).

DESCRIPTION: Plants to 15 cm tall, often arising from cuneate bases and producing round to reniform, ovate to linear to irregularly lobed, monostromatic fronds with ruffled margins (Fig. 58b, c). Blades olive, green-brown, green-red, orange-brown, plum, or silvery-purple in colour, with surfaces composed of perpendicularly arranged rows of cells. Vegetative cells rectilinear, 2–4 times elongate (4–9 μm × 10–14 μm), generally in tetrads or sometimes triads (Fig. 58d). Fronds monoecious, with spermatangia developing marginally in extensive bands and carposporangia sub-marginally in scattered patches intermingled with vegetative cells, or with all three cell types (male, female, sterile) intermixed along blade margins. Spermatangia in groups of 64–128, carposporangia in groups of 16–32 (Fig. 58e).

At Macquarie I., *P. columbina* is commonly found on rocky surfaces in the high intertidal, where it is exposed to drying winds and direct sunlight. Consequently, plants become partially desiccated, but are frequently wetted by Macquarie's pounding surf and ubiquitous sea-spray. *Porphyra columbina* is often the only macrophyte present on high-intertidal boulders (Fig. 58a), but it also grows alongside *Acrosiphonia* and *Bostrychia*. In some instances, plants are covered by patches of an unknown filamentous alga, which possibly is a member of the Phaeophyta.

NOTES: Of all species in the genus *P. columbina* is most widely distributed throughout the Subantarctic and certainly is the most prominent at Macquarie I. Although its habit is extremely variable, this species characteristically grows in the high-intertidal to supratidal, has distinctive blade surfaces comprising tetrads of cells arranged in rows, and upon drying, unlike other species at Macquarie I., adheres weakly to paper. In the Antarctic Zone, *P. endiviifolium* (A. & E. S. Gepp) Y. Chamberlain, which lacks aligned tetrads of cells, occupies a similar habitat after the ice has broken out in mid- to late-summer (pers. obs.).

Plants from Macquarie I. are considerably smaller than those 1 m long specimens described by Levring (1953) from Australia. Womersley & Conway (1975) describe this species from Australia

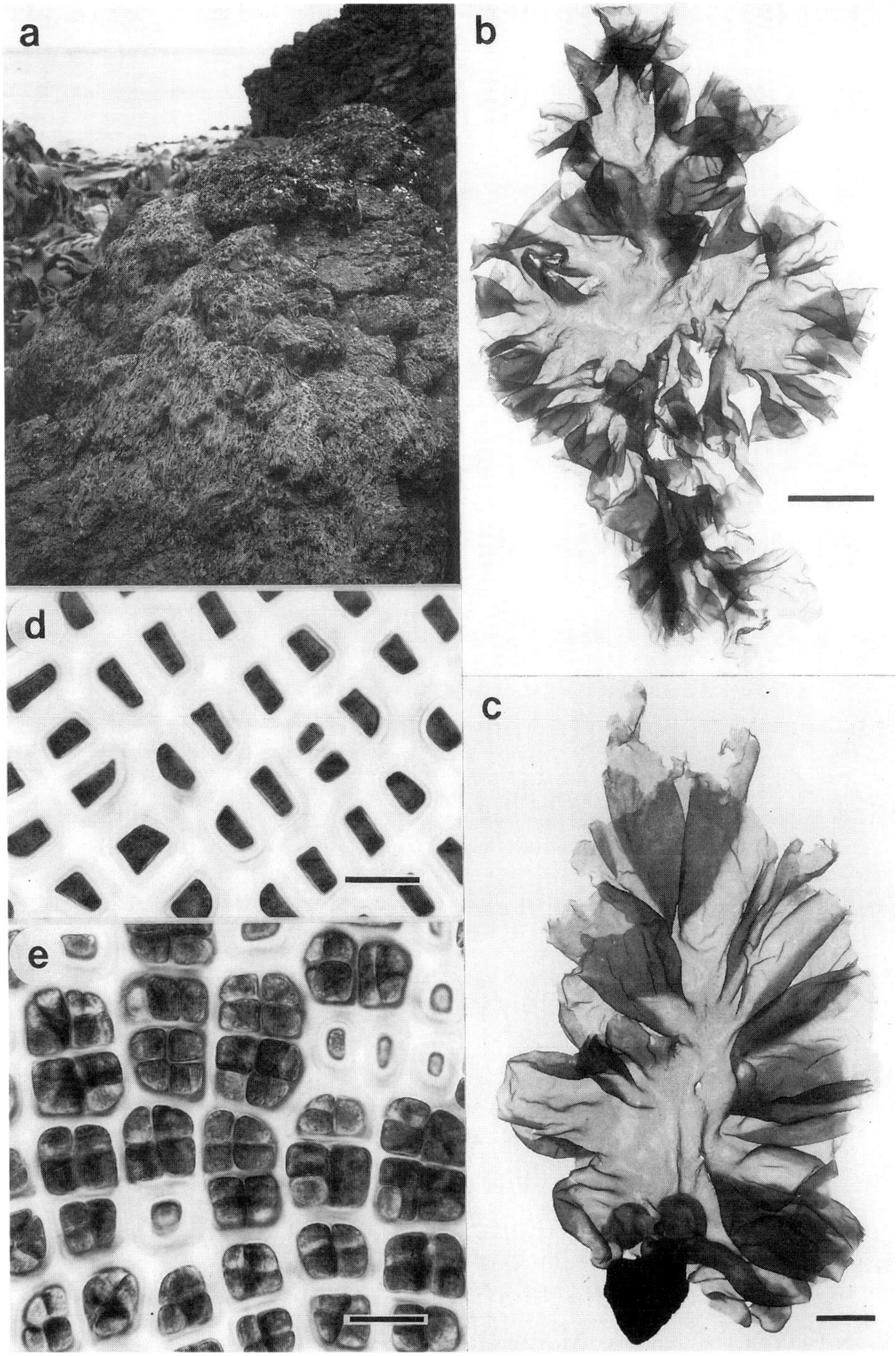

Fig. 58 *Porphyra columbina* Montagne: MELU, R 2030, 2032. **a.** Upper intertidal boulder covered with *Porphyra*. **b, c.** Habits of monoecious thalli. **d.** Surface view of vegetative cells from mid-thallus. **e.** Surface view of carposporangia. Scales: b = 2 cm; c = 1 cm; d, e = 20 μm.

as 'essentially a winter form', whereas at Macquarie I. it is present at least during the summer (the duration of our collecting from November to February) and probably persists throughout the year.

Kenny & Haysom (1962) reported *P. umbilicalis* (Linnaeus) Agardh from Macquarie I., but no specimens matching descriptions (Newton, 1931; Kornmann & Sahling, 1977) of this typically intertidal species were present in their collections. Possibly their report actually refers to a form of the only known intertidal species at Macquarie I., *P. columbina*. Simpson's (1976) reference to and illustrations of '*P. umbilicalis*' most certainly represent *P. columbina*.

Porphyra ionae R. W. Ricker sp. nov.

Fig. 59

HOLOTYPE (Fig. 59a): MELU, R 3715; with isotype in BM.

TYPE LOCALITY: Hasselborough Bay, Macquarie I. (54°29′50″ S lat., 158°57′00″ E long.), drift specimens (Taylor, 1950).

DISTRIBUTION: Macquarie I.

MATERIAL EXAMINED: Macquarie I.: Hasselborough Bay, drift collection (Taylor, 1950): MELU, R 3715, 3716 (= Taylor collection no. 552).

DIAGNOSIS: Plantae ad 40 cm latae et 25 cm altae, laete roseae, monostromaticae, in partibus sterilibus 70–80 µm crassae, lobis irregularibus late expansae, marginibus imperturbatis, exsiccatis impolitis vel nitidis charta firme adhaerentibus. Cellulae vegetativae aspectu laminae superficiali 11–19 µm diametro, aspectu laminae transectionali 25–36 µm altitude et 8–18 µm latitudine, quaeque chloroplasto parietali laminato (?) cum pyrenoide centrali, in thallo inferiore elongatae basin rhizoidealem formare. Thalli monoecii, carpogoniis et spermatangiis et saepe cellulis vegetativis additionalibus intermixtis in agellis diffusibus vel soris parvis rotundis super totas laminarum paginas extensis. Carposporangia in fasciculis sedena. Spermatangia in quoque fasciculo numero ignoto.

Porphyra ionae is named in recognition of Mrs Iona Christianson, to whom I am grateful for the loan of these specimens as well as other Macquarie I. algae.

DESCRIPTIONS: Plants to 40 cm across by 25 cm tall (Fig. 59a), with monostromatic fronds bright rose-pink in colour, 70–80 µm thick in sterile portions, broadly expanded with irregular lobes and unruffled margins, adhering firmly to paper upon drying with a mat to glossy finish. Vegetative cells 11–19 µm diameter in blade surface view (Fig. 59b), 25–36 µm tall by 8–18 µm broad in blade section (Fig. 59c), each with one laminate (?), parietal plastid with a central pyrenoid. Cells of lower thallus elongated to form rhizoidal base (Fig. 59d). Thalli monoecious, with carpognia and spermatangia and often additional vegetative cells intermixed in diffuse patches or small round sori spread over entire blade surfaces (Fig. 59e). Carposporangia developed in packets of 16; spermatangial packets with an undetermined number of spermatangia.

Two specimens of this species were obtained from drift collections along the northern end of Macquarie I. It has not been collected since, and nothing is known about its abundance or distribution in this vicinity. Because of its conspicuously large size and absence from any intertidal collections, the species is presumed to have originated in the subtidal.

NOTES: *Porphyra ionae* differs from the intertidal species *P. columbina* in habit, cell organization, and presumably habitat. The disordered arrangement of cells on blade surfaces of *P. ionae* also distinguishes it from the subtidal species *P. woolhousiae*, which it resembles in blade colour. The latter two species can also be distinguished by differences in blade thickness and shape, and the distribution of reproductive structures across blade surfaces.

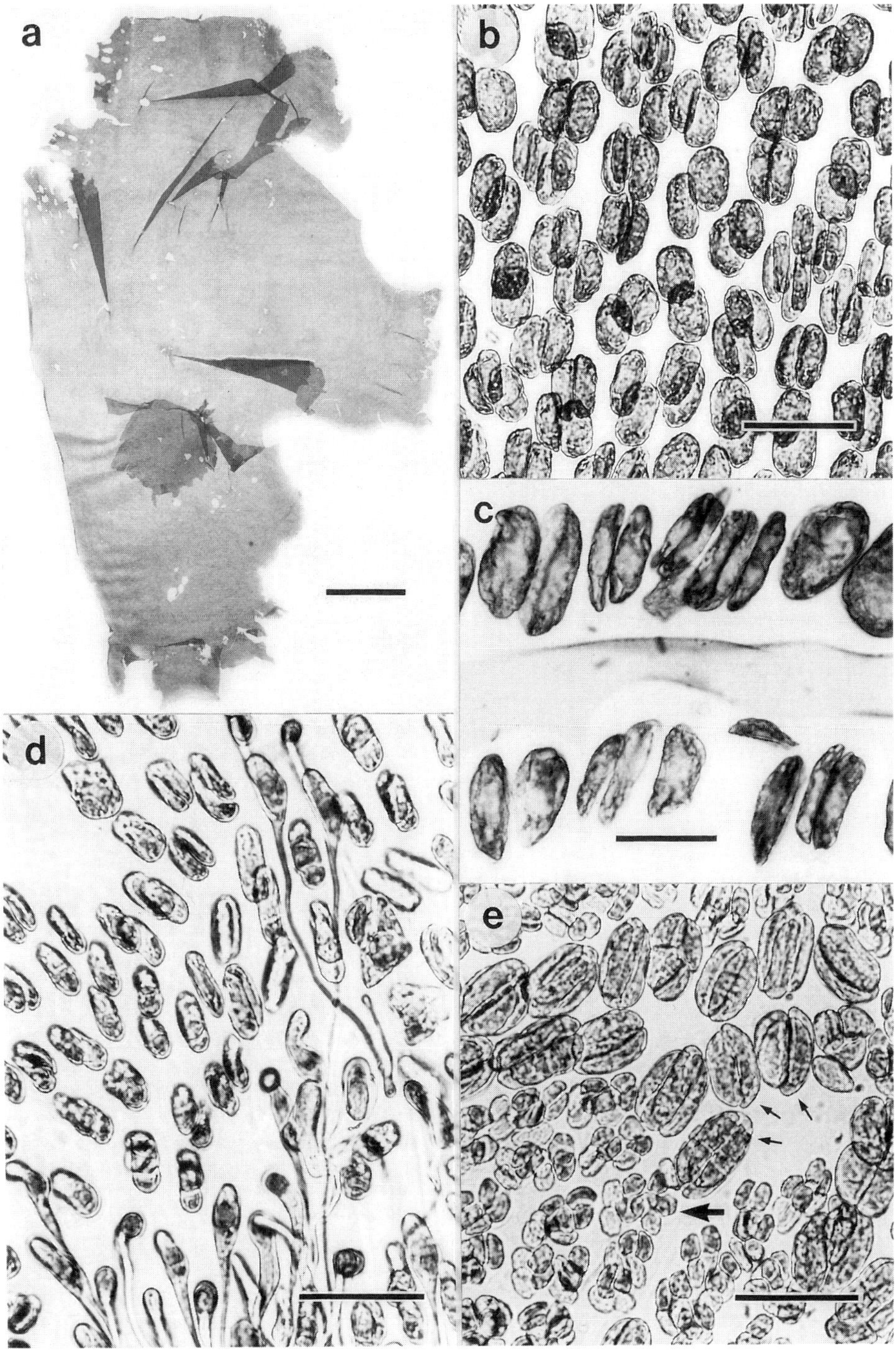

Fig. 59 *Porphyra ionae* R. W. Ricker: MELU, R 3715, 3716. **a.** Habit of holotype (R 3715). **b, d, e.** Blade surface views. **b.** Vegetative cells in upper thallus. **c.** Section through (sterile) mid-thallus. (N.B. photograph shows two overlapping monostromatic fronds.) **d.** Lower thallus with rhizoid-bearing cells. **e.** Carposporangia (small arrows) and spermatangia (large arrow). Scales: a = 5 cm; b, d, e = 40 μm; c = 20 μm.

Porphyra plocamiestris R. W. Ricker sp. nov.

Fig. 60

Holotype (Fig. 60a): MELU, R 1733.

Type locality: Goat Bay, Macquarie I. (54°29′25″ S lat., 158°57′35″ E long.), growing epiphytic on *Plocamium secundatum* Kützing in *c.* 10 m depth (Ricker, 30 November 1977).

Distribution: Antarctic Peninsula (Moe, pers. comm.); Macquarie I.

Material examined: Macquarie I.: collections MA-14, 39, 46, 113, 117, 244, 248; Herbarium specimens: MELU, R 0811, 1459, 1460, 1733, 3458, 3459.

Diagnosis: Plantae epiphyticae purpurascentes ad roseas foliaceae rotundae (1.0–1.5 cm diametris) vel elongate (ad 14 cm longitudinibus) laciniatescentes monostromaticae ad 30 μm crassae ex base cordata ad cuneatum orientes. Thalli steriles aspectu cutis cellulas irregulariter positas rotundatas ad rectilineares 8–14 μm diametris praebentes quasque chloroplasto cum pyrenoide centrali. Laminae monoeciae, paginis totis nunc cum tegimene continuo carpogoniorum et spermatangiorum et cellularum sterilium, nunc cum soris unisexualibus irregulariter dispositis et per zonas cellarum sterilium separatibus. Carpogonia parietibus acutis (trichogynibus?) in ambabus laminarum paginis. Carposporangia in fasciculis octona. Spermatangia in fasciculis octona ad sedena.

Porphyra plocamiestris is named in recognition of the genus *Plocamium*, which most commonly hosts this new species.

Description: Plants epiphytic, purplish to rose in colour, foliaceous, monostromatic, to 30 μm thick, with rounded (1–1.5 cm diameter) to elongate (to 14 cm long) fronds becoming lacinate, arising from a cordate to cuneate base (Fig. 60a, b). Sterile thallus surfaces (Fig. 60c) composed of irregularly positioned, rounded to rectilinear cells, 8–14 diameter, each with a plastid and central pyrenoid. Blades monoecious, with entire surfaces covered by either a continuous patch of intermixed carpogonia, spermatangia, and sterile cells, or by irregularly arranged unisexual sori separated by zones of sterile cells (Fig. 60d, e, i). Carpogonia with pointed end walls (trichogynes?) at both blade surfaces (Fig. 60e, g). Carposporangia in packets of 8 (Fig. 60f, h), spermatangia in packets of 8–16 (Fig. 60h, i).

Porphyra plocamiestris grows subtidally at depths of 4–20 m, where it is epiphytic on various branched red algae, although most commonly on either *Plocamium hookeri* or *P. secundatum*. The abundance of this alga was not noted, but it occurs along both east and west coasts of Macquarie I.

Notes: At Macquarie I., *P. plocamiestris*, *P. ionae*, and *P. woolhousiae* grow in the subtidal. The first two species produce reproductive structures across entire blade surfaces in contrast to the marginal arrangement in *P. woolhousiae*. *Porphyra plocamiestris*, with its smaller dimensions overall and more regular cell arrangement on blade surfaces, is easily distinguished from *P. ionae*.

Moe (pers. comm.) collected plants from the Antarctic Peninsula that are similar, if not identical, to *P. plocamiestris*. These too grow mostly on *Plocamium*, although on a species not present at Macquarie I., *P. cartilagineum* (Linnaeus) Dixon.

Fig. 60 *Porphyra plocamiestris* R. W. Ricker: MELU, R 1733, 3458: **a.** Habit of holotype (R 1733) on *Plocamium secundatum*. **b.** Plant habit on *Plocamium hookeri* (R 3458). **c–f.** Blade surface views. **c.** Vegetative cells in mid-thallus. **d.** Spermatangial packets. **e.** Carpogonia displaced on their sides. **f.** Carposporangia. **g, h.** Sectional views of fertile thalli. **g.** Carpogonia. **h.** Carposporangia (hollow arrow) and spermatangia (solid arrow). **i.** Camera lucida drawing of blade surface showing spermatangial packets (arrow). Scales: a = 1 cm; b = 2 cm; c–h = 20 μm; i = 40 μm.

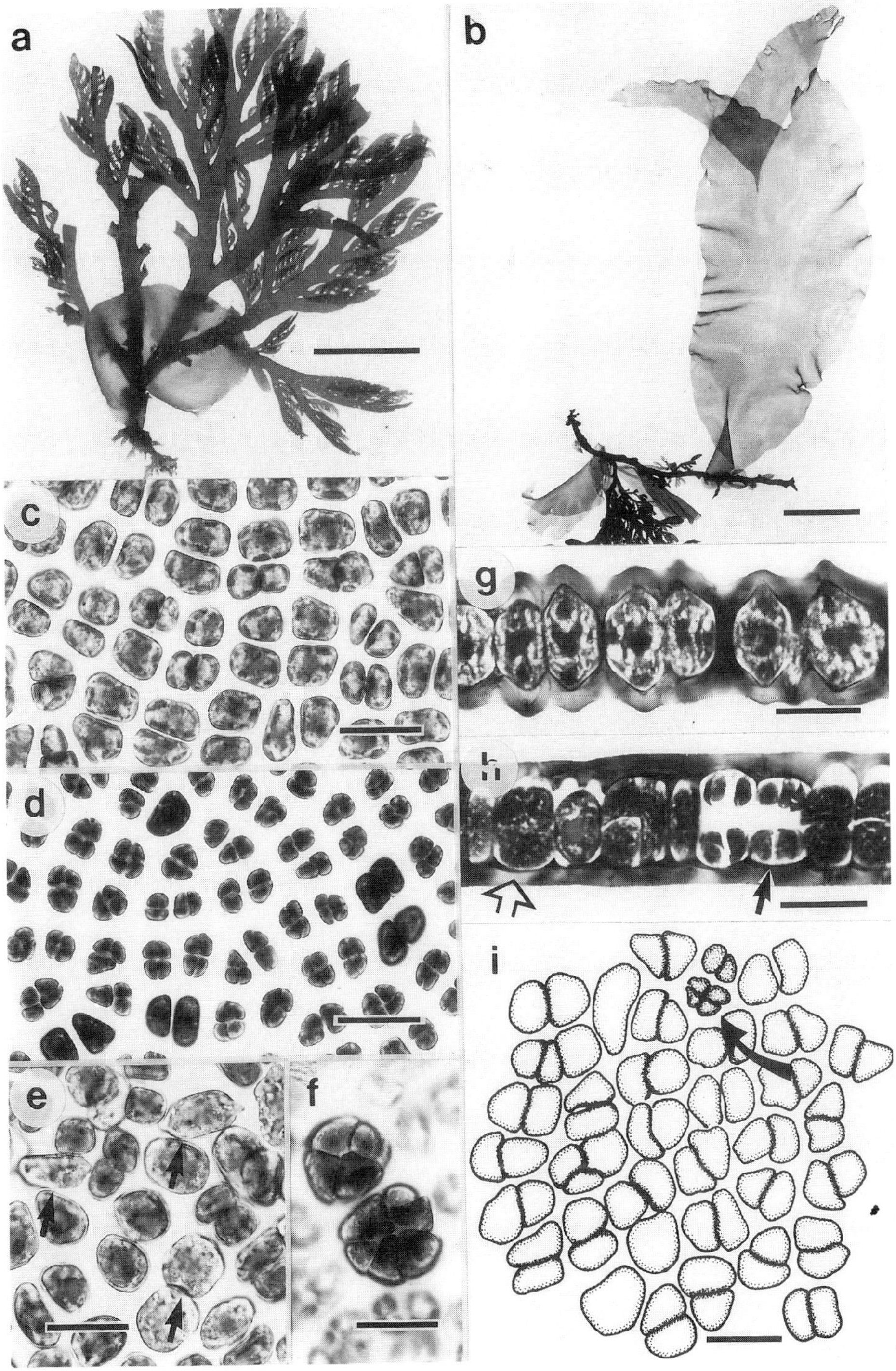
a
b
c
d
e
f
g
h
i

Porphyra woolhousiae Harvey, 1863: pl. cclxv

Fig. 61

TYPE LOCALITY: Tasmania.

TYPE: Herb. Harvey, TCD (no. 3); NB type specimen illustrated by Harvey (1863, pl. cclxv, fig. 1) and Womersley & Conway (1975, fig. 1b).

DISTRIBUTION: Falkland Is (Skottsberg, 1923); Îles Kerguelen (Hamel, 1928); Macquarie I.; New Zealand (Levring, 1955); South America (Pujals, 1963); Tasmania (Harvey, 1863).

MATERIAL EXAMINED:
1. Macquarie I.: collections MA-11, 271, 378; Herbarium specimens: MELU, R 0001, 0130, 0135–0137, 0158, 0504, 1729, 1732, 3442–3457.
2. Tasmania: type specimen, presented by Miss Woolhouse of Sheffield. TCD. no. 3.

DESCRIPTION: Plants epilithic or epizoic, with fronds lanceolate (45 cm long × 9 cm broad) to expansive and irregularly laciniate (19 cm long × 22 cm broad) (Fig. 61a), monostromatic, pink to rose to burgundy in colour, sometimes perforate, often with slightly ruffled margins, arising from umbilicate (Fig. 61b), occasionally cuneate, bases. Growth diffuse. Sterile blade surfaces composed of paired, 8–10 μm diameter, rounded to somewhat angular cells in upper thallus (Fig. 61c) and solitary, isodiametric (10–14 μm diameter) to twice as long as broad cells in lower thallus (Fig. 61d); basal cells mostly with rhizoidal extensions (Fig. 61e). Fronds monoecious, with spermatangia usually borne in continuous colourless band along thallus margins, but occasionally with pigmented vegetative cells or dark-red packets of carposporangia intermixed; thallus margins dissolving at maturity to release gametes. Female reproductive structures mostly submarginal (Fig. 61f), forming red speckles adjacent to colourless band. Carposporangia appearing to be in groups of 8–16, spermatangia in groups of 32–64 (Fig. 61g).

Porphyra woolhousiae grows in the subtidal to 11 m depth, where it sometimes covers boulder surfaces to the exclusion of other macrophytes. Plants also were collected alongside *Ulva angusta* and *Iridaea cordata* on boulder outcrops in a sand plain at 10 m depth.

NOTES: The identity of this species is not wholly certain. Macquarie individuals differ considerably from the type specimen, which grows as an epiphyte, has marginal and diffuse growth, and has blade surfaces composed of rectilinear (polygonally shaped) cells. Womersley & Conway (1975) described plants of *P. woolhousiae* from Australia that are much smaller than the type specimen but have similar growing regions and comparable cell shapes on blade surfaces. A marginal meristem is considered by Womersley & Conway to be taxonomically important and unique to *P. woolhousiae* among Australian species in the genus.

Macquarie plants, in spite of lacking marginal meristems and having blade surfaces composed of rounded cells, are referred to *P. woolhousiae* because of their growth in the subtidal, blade morphology and colour, monostromatic construction, and reproductive strategy with monoecious fronds developing spermatangia in marginal bands which deliquesce upon maturation.

Porphyra sp.

Fig. 62

MATERIAL EXAMINED: Macquarie I.: collection MA-14; Herbarium specimen: MELU, R 1459.

DESCRIPTION: Plant to 3 mm tall, with pale purple to pink, oval-lanceolate frond arising from reniform base with overlapping lobes epiphytic on *Plocamium hookeri* (Fig. 62a). Blade to 12 μm thick along thallus margin (Fig. 62b), with surface largely consisting of disordered mixture of vegetative cells and presumptive carpogonia (Fig. 62e); sterile thallus margins composed of rounded cells 7–10 μm diameter (Fig. 62c); base composed largely of rhizoidal cells (Fig. 62d).

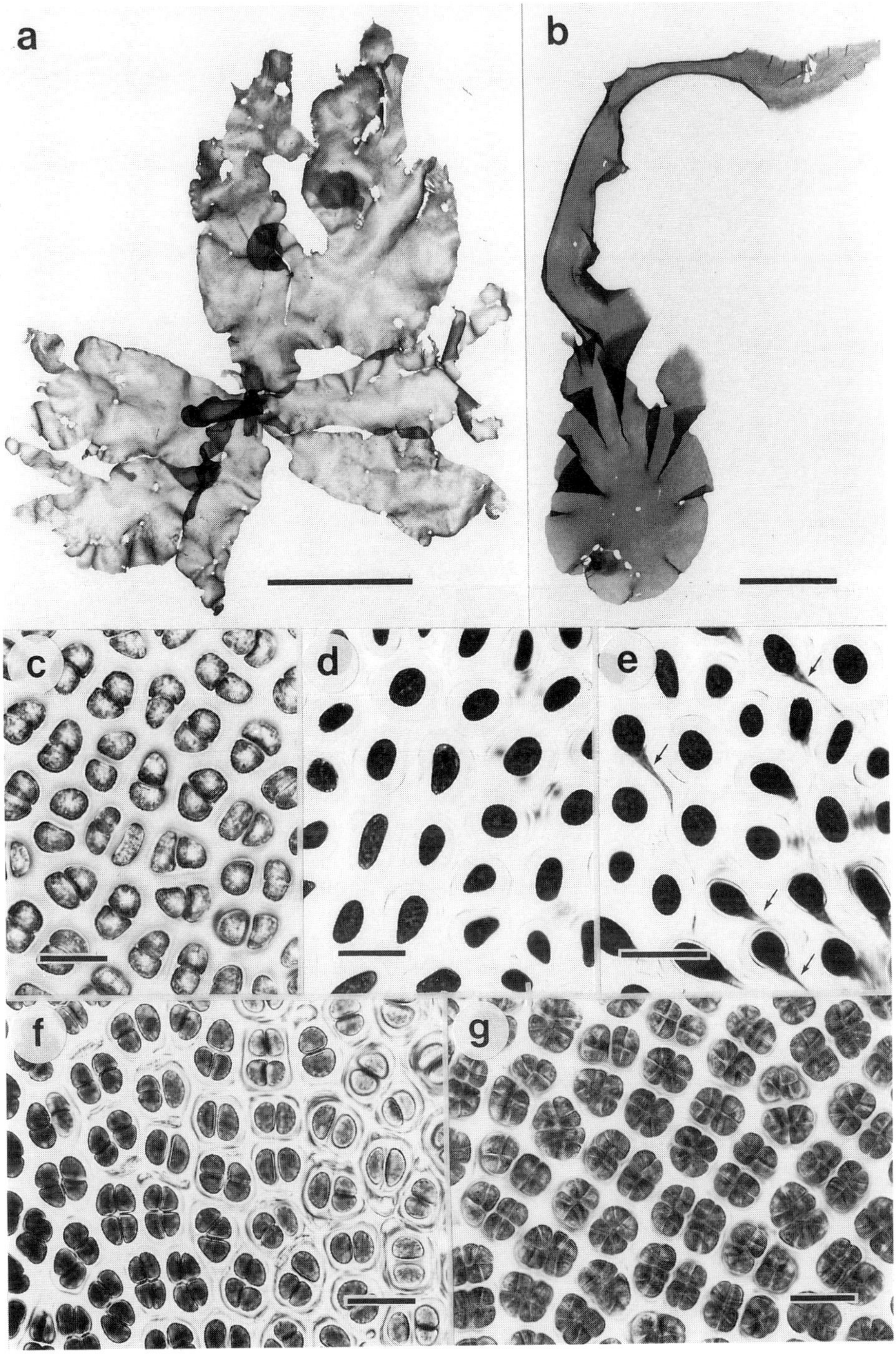

Fig. 61 *Porphyra woolhousiae* Harvey: MELU, R 0001, 1729, 3442. **a, b.** Thallus habits. **c–g.** Blade surface views. **c.** Upper thallus. **d.** Lower thallus. **e.** Rhizoidal cells (arrows) at thallus base. **f.** Carposporangia near frond margin. **g.** Spermatangia near frond margin. Scales: a = 4 cm; b = 5 cm; c–g = 20 cm.

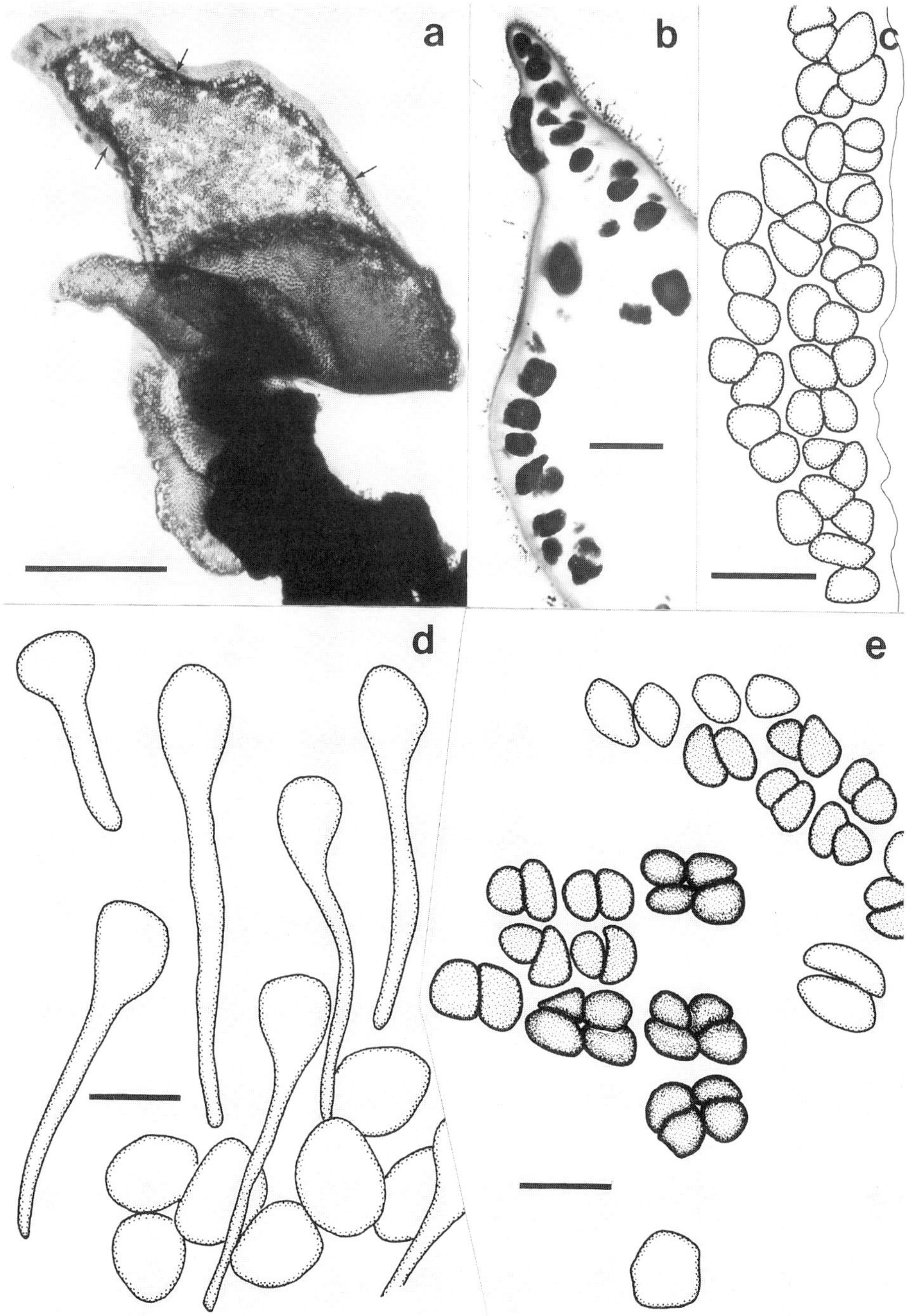

Fig. 62 *Porphyra* sp.: MELU, R 1459. **a.** Habit of thallus epiphytic on *Plocamium*. (N.B. epiphyte is inflated from dark band (arrows) inwards.) **b.** Section through frond margin. **c–e.** Camera lucida drawings. **c.** Blade margin. **d.** Rhizoids at thallus base. **e.** Surface view of presumptive carposporangia (shaded). Scales: a = 1 mm; b–e = 20 μm.

This plant was collected from 10 m depth, where its host commonly grows on small boulders and rock walls. Because only one specimen of this entity was collected at Macquarie I., details on its distribution, as well as its identity, remain unknown.

NOTES: In order to study the construction of this alga, the entire plant was embedded in plastic for sectioning, thereby leaving no representative specimen. Consequently, herbarium vouchers include a photograph of the plant habit and a microscope slide with a section through mid-thallus.

Erythropeltidaceae Skuja, 1939: 33

Members of this family have dominant phases that are either crustose, filamentous, or foliose. Plants have differentiated basal systems, but do not produce rhizoids. Thalli are composed of cells that contain either a single stellate plastid or one to several parietal plastids with or without pyrenoids. The dominant phase may alternate with a filamentous 'pseudoconchocelis' stage that functions in various ways depending upon the particular genus involved (Garbary *et al.*, 1980a). Reproduction is by monospores produced by an unequal division of an undifferentiated vegetative cell. The smaller portion remains in the parent thallus and the larger monospore is shed.

Five genera, *Erythrocladia, Erythrotrichia, Porphyropsis, Membranella* and *Smithora,* all of which grow along the west coast of North America, are known for the family. The first three occur in the Subantarctic, with *Porphyropsis* singly representing the family at Macquarie I.

Porphyropsis Rosenvinge, 1909: 68

Type species: *P. coccinea* (J. Agardh *ex* J. E. Areschoug) Rosenvinge 1909: 69.

Plants initially develop a monostromatic crust that thickens into a multi-layered basal cushion. Small vesicles arise from the pulvinate basal system and eventually rupture to form delicate monostromatic fronds comprising cells widely separated in an extensive wall matrix. Each vegetative cell contains one parietal, lobed plastid without a pyrenoid. Reproduction is by monosporangia produced by an unequal division of an undifferentiated cell.

Two species are described for the genus, the type previously known from only the Northern Hemisphere and *P. minuta* Womersley & Conway endemic to Australia. *Porphyropsis minuta* is superficially similar to *P. coccinea,* but is distinguished from it partly by differences in vegetative organization. More important, however, is the different mode of monospore formation in which entire cells along blade margins of *P. minuta* are converted into monospores. This is unlike monospore development resulting in small residual cells remaining in the blade following monospore release—the method considered typical for the genus as well as other members of the Erythropeltidaceae (Garbary *et al.*, 1980b).

Porphyropsis coccinea (J. Agardh ***ex*** J. E. Areschoug) Rosenvinge 1909: 69

Fig. 63

BASIONYM: *Porphyra coccinea* J. Agardh *ex* J. E. Areschoug 1850: 407.

TYPE LOCALITY: Atlantic Ocean along northern coast of Europe.

TYPE: presumably in L (Koster, 1969), but specimens were not seen.

DISTRIBUTION: This species is widespread throughout the Northern Hemisphere, occurring in the eastern and western parts of both the Atlantic and Pacific Oceans (Rosenvinge, 1909; Kylin, 1956; Abbott & Hollenberg, 1976), but is known only from two Southern Hemisphere localities, Chile (Santelices & Abbott, 1978) and Macquarie I.

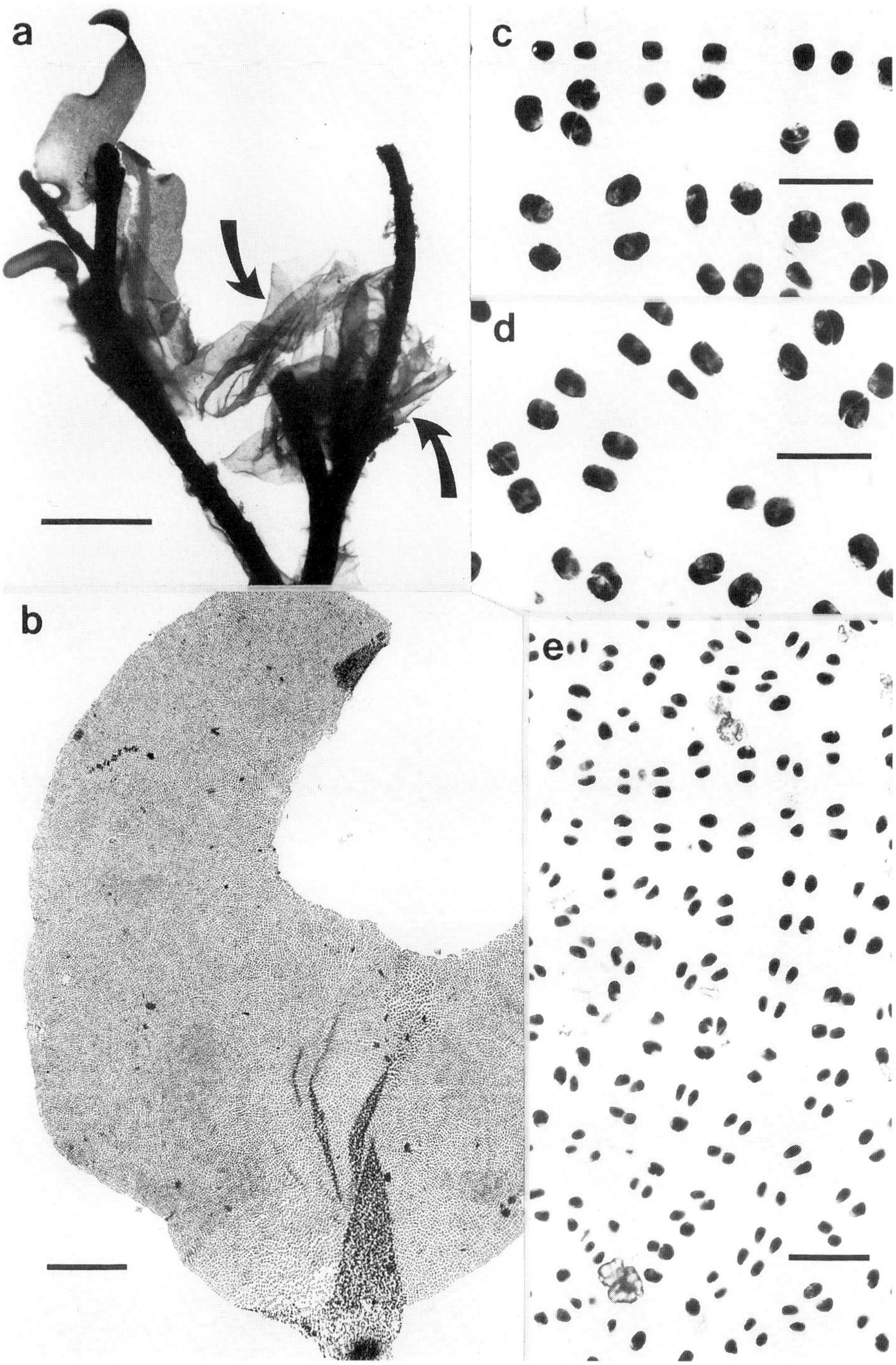

Fig. 63 *Porphyropsis coccinea* (J. Agardh *ex* J. E. Areschoug) Rosenvinge: MELU, R 0495, 0587, 0843. **a.** Habit of plants (arrows) epiphytic on *Lophurella hookeriana.* **b.** Habit of individual blade. **c–e.** Blade surface views. **c.** Cells along blade margin. **d.** Cells of mid-thallus. **e.** Mid-thallus. Scales: a = 2 mm; b = 300 μm; c, d = 20 μm; e = 30 μm.

MATERIAL EXAMINED: Macquarie I.: collections MA-14, 46, 114, 116, 117, 244, 302, 303, 309; Herbarium specimens: MELU, R 0495, 0587, 0843, 1459, 1987.

DESCRIPTION: Plants to 7 mm tall, with delicate, membranous, pink fronds arising from a basal cushion epiphytic on various red and brown algae. Blades round to oval, largely undifferentiated (Fig. 63b), composed of rounded cells, 4–7.5 µm diameter, often in diads or tetrads aligned in irregular rows (Fig. 63e); cells widely separated by extensive wall matrix (Fig. 63c, d). Reproductive structures unknown.

Plants grow subtidally to 20 m depth as epiphytes on the fan-shaped brown alga *Microzonia velutina* and the branched red alga *Lophurella hookeriana*. Because of their small size, inconspicuous habit and delicate construction, few individuals were preserved for study and no field observations were made regarding the distribution of this species around Macquarie I.

NOTES: Plants from Macquarie I. have features similar to both species in the genus. In size alone, Macquarie specimens are closer to the Australian *P. minuta*, but this species has blade surfaces with cells arranged in distinct rows oriented perpendicular to one another (Womersley & Conway, 1975, fig. 7c). Blade surfaces of Macquarie specimens, as well as those of individuals of the type species from British Columbia (Garbary *et al.*, 1980b), are more disordered, with cells sometimes arranged in irregular rows. Additionally, Womersley & Conway (1975) described the colour of *P. minuta* as 'always a greenish-brown-purple, never rose-red as in *P. coccinea*'.

Acrochaetiales

The highly debated status of the Acrochaetiaceae as one of 12 families in the Nemaliales (Dixon, 1973), or as the only family constituting the Acrochaetiales (Chemin, 1937*a*), has been reviewed by Kraft (1981) and Woelkerling (1983). Because acrochaetioid algae lack auxiliary cells and develop gonimoblast tissue directly from all or part of the carpogonium, these plants usually have been considered as members of the Nemaliales. Such placement prevails in most recent treatments of the family. Recognition of the Acrochaetiales, as suggested by Garbary (1978), emphasizes the phylogenetic importance of this group and results in orders that are seemingly monophyletic rather than polyphyletic as the Nemaliales *sensu lato* appears to be.

Acrochaetiaceae Fritsch *ex* W. R. Taylor, 1957: 209, 210

The Acrochaetiaceae generally is considered to comprise either the single, morphologically variable genus *Audouinella* (Woelkerling, 1971; Garbary, 1979), or as many as seven of the eight genera recognized by Kylin (1956).* Papenfuss (1945, 1947) distinguished four genera, *Acrochaetium, Rhodochorton, Audouinella* and *Chromastrum* on the basis of differences in their chloroplast structures, a feature that alone has proved too variable for generic distinctions (Woelkerling, 1971). Drew (1928) considered differences in cell structure and the production of either monosporangia or tetrasporangia inadequate for separating acrochaetioid genera and united them all in *Rhodochorton*. Woelkerling (1971) similarly conceived the group as having one large genus, but showed that the name *Audouinella* predates *Rhodochorton*. He maintained *Colaconema* Batters as a form genus for asexual species of uncertain affinities. Although use of a single genus to represent a broad range of forms is appealing, this type of classification de-emphasizes reportedly stable morphological differences that characterize smaller groups of species (Stegenga, 1979). On the basis of life history and morpho-systematic studies, Stegenga & Vroman (1977) considered it best to reflect such differences by recognition of 5(–7) genera (*Acrochaetium, Audouinella, Chromastrum, Colaconema, Kylinia, Liagorophila, Rhodochorton*).

In the present account, *Acrochaetium* and *Rhodochorton* are reported from Macquarie I., even though the defining features for these genera (as treated by Stegenga, 1979) were not evident. As the generic concepts within the Acrochaetiaceae are still being debated, each species growing at Macquarie I. has been placed arbitrarily in the initial genus that it was assigned to in this family.

* *Conchocelis* is known now to be a phase in the life history of *Porphyra*.

By a similarly conservative approach, the family can be viewed as having four genera, *Acrochaetium*, *Audouinella*, *Kylinia* and *Rhodochorton*, with a total of seven species in Antarctic and Subantarctic waters (Papenfuss, 1964*a*).

Acrochaetium Nägeli *in* Nägeli & Cramer, 1858: 532

Type species: Not indicated.

Acrochaetium has been characterized variously and in the past was considered to be the largest acrochaetioid genus (Kylin, 1956). Stegenga (1979) described it as having usually one (but sometimes a few) parietal plastid (each with a pyrenoid), virtually isomorphic generations, and a well-developed carposporophyte. Abbott & Hollenberg (1976) reported stellate plastids in addition to parietal laminate ones, and plastids that lacked or had up to two pyrenoids. Reproduction is commonly by monosporangia, less so by tetrasporangia.

In early accounts, species within the genus were often distinguished on substrate specificity, i.e., growth as epiphytes or endophytes in particular hosts, but because plants are able to grow and reproduce in culture free from host tissues, the taxonomic value of this character now is considered less important. Stegenga (1979) has suggested that vegetative cell and monosporangial dimensions are probably better criteria for species distinctions.

Papenfuss (1964*a*) listed four species from the Subantarctic and none from the Antarctic. Of these, *A. daviesii* is present at Macquarie I.

Acrochaetium daviesii (Dillwyn) Nägeli, 1861: 405

Fig. 64

Basionym: *Conferva daviesii* Dillwyn, 1809: 73.

Synonyms: see Woelkerling, 1971 (as *Audouinella daviesii*).

Type locality: Bantry Bay, Ireland.

Lectotype: BM (Dixon & Irvine, 1977).

Distribution: This species is considered by Woelkerling (1971) to be cosmopolitan, occurring in both Northern and Southern Hemispheres, and is reported from the following Southern Ocean localities: Campbell I. (Levring, 1945); Fuegia (Hariot, 1889); Macquarie I.; Stewart I. (Adams *et al.*, 1974); Tristan da Cunha (Baardseth, 1941).

Material examined: Macquarie I.: collection MA-120; Herbarium specimen: MELU, R 1947.

Description: Plants filamentous, to 2 mm tall, tuft-forming (Fig. 64a), with thalli alternately to irregularly branched (Fig. 64b, c). Filaments mostly uniform diameter throughout, sometimes tapering minimally toward apices, composed of elongate cells (15–)23–28(–53) μm × 13–18 μm, each with single, plate-like, parietal plastid (Fig. 64e, f) sometimes containing prominent central pyrenoid. Growth apparently apical and diffuse. Tetrasporangia borne singly or in small clusters (Fig. 64d, g). Sexual structures not seen.

Acrochaetium daviesii, which grew as a dense mat on driftwood, is known from Macquarie I. through only one collection made at Hasselborough Bay (at the northern end of the island). Abundance and distribution of this alga around the island is unknown.

Notes: This species is best known for producing monosporangia, but tetrasporangia have been reported previously on plants from California (Abbott & Hollenberg, 1976). Only tetrasporangia have been observed on Macquarie thalli. Although *A. daviesii* from Macquarie I. has a similar

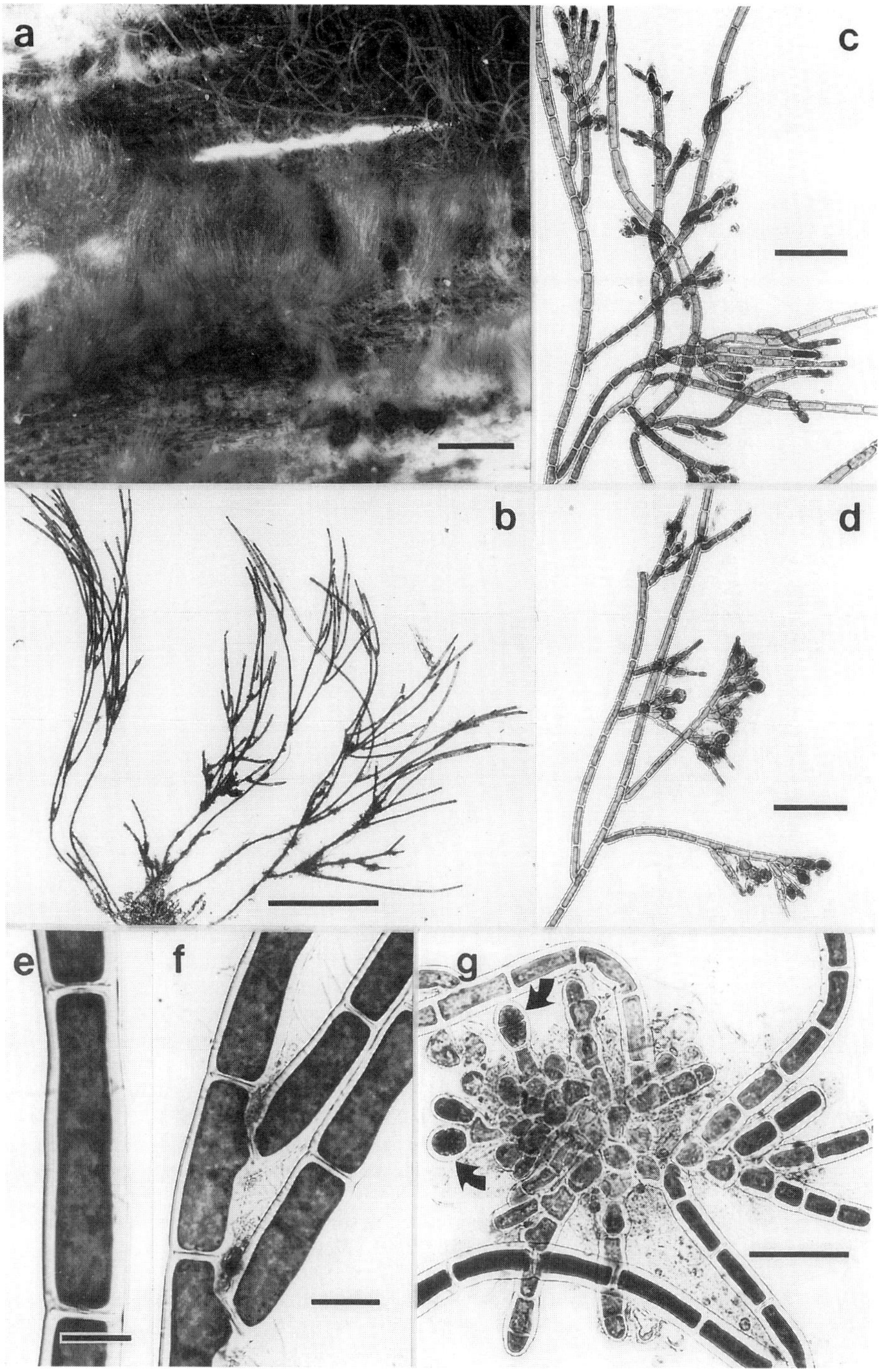

Fig. 64 *Acrochaetium daviesii* (Dillwyn) Nägeli: MELU, R 1947. **a.** Habit of tuft-forming plants associated with *Rhizoclonium* on driftwood. **b.** Plant habit. **c.** Branching of upper axes. **d.** Portion of tetrasporophyte with tetrasporangia. **e, f.** Detail of cells. **g.** Developing tetrasporangia (arrows). Scales: a = 2 mm; b = 500 μm; c, d = 100 μm; e, f = 20 μm; g = 50 μm.

habit to *Rhodochorton pupureum*, the two species can be separated by differences in their plastids and production of tetrasporangia.

Rhodochorton Nägeli, 1861: 355

Lectotype: *R. rothii sensu* Nägeli (see Farr *et al.*, 1979).
Plants are filamentous with prostrate portions consisting of intertwined filaments or basal discs that give rise to erect, usually sparsely branched filaments. Stegenga (1979) considered the primary characteristic for the genus to be its 'development of the tetrasporic phase directly from the fertilized carpogonium or from an intercalated carposporophyte-like structure', and never from free carpospores. Most other generic descriptions emphasize certain morphological and cytological characteristics, although expression of such features differs somewhat between the various accounts. Generally, cells contain one to serveral parietal plastids, sometimes appearing fragmented into numerous discs, with or without pyrenoids. Sporophytes usually produce cruciately divided tetrasporangia, but sometimes bisporangia.

The genus is represented worldwide, extending also into the Antarctic, by *R. purpureum*. Papenfuss (1964*a*) listed this species as the single representative of the genus in southern high latitudes, but a second species, *R. concrescens*, is presently reported along with *R. purpureum* from Macquarie I.

Rhodochorton concrescens Drew, 1928: 167

Fig. 65

Type locality: Carmel Bay, Monterey County, California.

Type: UC, no. 294561 (= Gardner, no. 4828).

Southern Ocean distribution: Antarctic Peninsula (Moe, pers. comm.) Australia, New Zealand (Levring, 1953); Macquarie I.

Material examined: Macquarie I.: collections MA-94, 255, 367; Herbarium specimens: MELU, R 1912, 3312.

Description: Plants to 200–300 μm tall, epiphytic, with prostrate and erect portions (Fig. 65a–d). Prostrate system encrusting, often monostromatic, but sometimes thicker, giving rise to erect uniseriate axes. Filaments simple or branched, 6.5–7.7 μm diameter throughout, composed of cells with equal length to breadth or to three times elongate; apical cells with rounded apices. Plastids parietal, lacking pyrenoids, one per cell, often appearing lobed or fragmented into numerous discs, (Fig. 65g, h). Tetrasporangia (18.5 μm × 22 μm) cruciately divided and terminating either main axes or one-celled lateral branches (Fig. 65e, f). Sexual reproductive structures unknown.

Rhodochorton concrescens is known at Macquarie I. only as an epiphyte on the red algal host *Ballia callitricha* and grows in intertidal pools and in the subtidal to 10 m depths. *Sphacelaria bornetii*, also epiphytic on *Ballia*, sometimes occurs intermixed with *R. concrescens*.

Notes: This species was initially described from the Pacific coast of North America (Drew, 1928), but has been reported since from several Southern Hemisphere localities (Levring, 1953). Although Drew (1928) considered the extensive basal disc to be the most distinctive feature of this species, she did not give details on its thickness or construction. Plants described by Abbott & Hollenberg (1976) from Californian waters have prostrate portions to several cells thick. Mac-

Fig. 65 *Rhodochorton concrescens* Drew: MELU, R 1912, 3312. **a.** Habit of plants epiphytic on *Ballia callitricha*. **b.** *Ballia* with basal crust of epiphyte bearing erect filaments. **c, d.** Prostrate and erect portions of thallus. **e, f.** Detail of erect branches with developing tetrasporangia. **g, h.** Cell details. Scales: a = 100 μm; b–d = 50 μm; e–h = 20 μm.

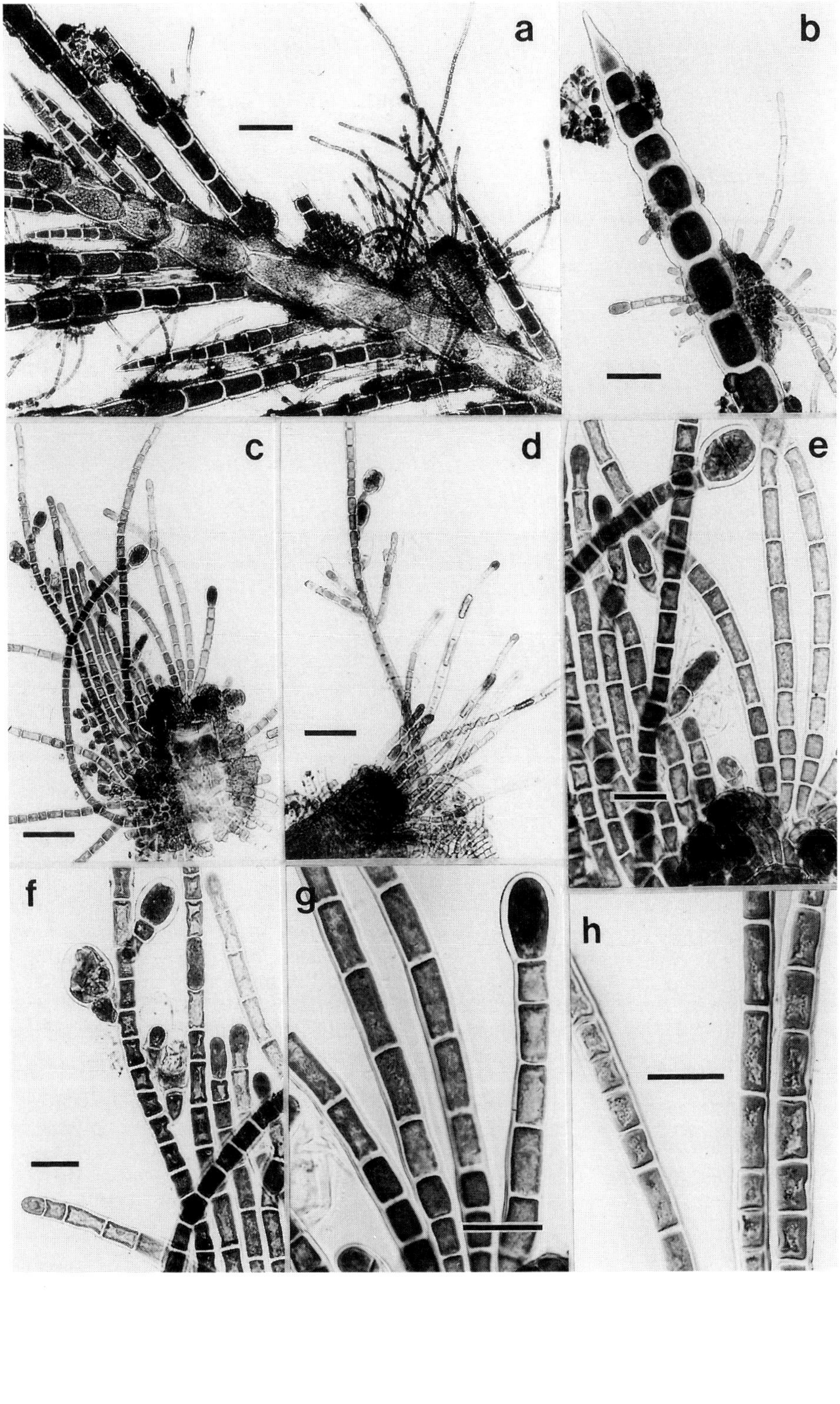
a
b
c
d
e
f
g
h

quarie thalli sometimes produce polystromatic crusts, but usually develop extensive monostromatic crusts. In nearly all other respects, however, Macquarie and Californian plants are similar.

Plectoderma minus Reinsch (1875) from South Georgia is poorly known but, based on illustrations by Reinsch (1890, pl. 13, figs 6, 7), appears to be similar, if not identical, to *R. concrescens*. Should the two taxa prove to be conspecific, the specific epithet *minus* would have priority.

Rhodochorton purpureum (Lightfoot) Rosenvinge, 1900: 75

Figs 66, 67

BASIONYM: *Byssus purpurea* Lightfoot, 1777: 1000.

SYNONYMS: see Woelkerling, 1973 (as *Audouinella purpurea*).

TYPE LOCALITY: Ruined Abbey, Island of Iona, Scotland.

TYPE: possibly destroyed (see Conway & Knaggs, 1966).

DISTRIBUTION: Cosmopolitan in temperate to polar marine waters and also known from terrestrial and freshwater sites (Knaggs, 1967; Woelkerling, 1973). Reported in southern high latitudes from the Antarctic Peninsula (Skottsberg, 1953), Argentina (Pujals, 1963), Australia (Levring, 1953; as *Rhodochorton rothii*), Chile (Levring, 1960), Kerguelen (Dickie, 1879) and Macquarie I.

MATERIAL EXAMINED: Macquarie I.: collections MA-44, 45, 120, 121, 312, 313, 352; Herbarium specimens: MELU, R 2287, 2451–2455.

DESCRIPTION: Plants epilithic, dark-purple, usually forming dense turf of stiff sparsely branched filaments with opposite to secund to irregular branching (Figs 66a, c, 67b, c). Filaments mostly uniform diameter throughout and having elongate (to 72 μm long) apical cells with rounded apices (Fig. 66e). Cells in lower branches 12–19 μm wide × 50–55 μm long, mostly containing numerous discoid plastids (3–4 μm diameter) lacking pyrenoids (Fig. 67d–f), but sometimes with plastids appearing as single, parietal reticulum surrounding one large (3.3–4 μm diameter) nucleus (Fig. 66f). Tetrasporangia cruciately divided and borne in clusters (Figs 66b, d, g). Sexual reproductive structures not seen.

This species grows in the high intertidal, often covering the damp undersides of boulders or lining the walls of caves with a burgundy red felt. In some instances, where it grows at the base of *Bostrychia* mats (Fig. 67a), it forms a much less compact turf with branches intertwining with the coarser axes of *Bostrychia*. *Rhodochorton purpureum* is common in many localities around Macquarie I., ranging from the mid-intertidal to caves 3 m above the upper edge of the *Durvillaea* belt.

NOTES: Although not studied here, the life history of this species is well known through the work of West (1970, 1972) and Stegenga (1978). Macquarie plants have been assigned to *R. purpureum* because of their occurrence in a habitat characteristic for this species and their production of distinctive tetrasporangial clusters that resemble those on European Atlantic tetrasporophytes (Stegenga, 1978, fig. 1). Cell dimensions of Macquarie and European plants are comparable and, as indicated by Stegenga (1978), cells in sporangial clusters are typically smaller than normal vegetative cells (Figs 66g, 67g).

Fig. 66 *Rhodochorton purpureum* (Lightfoot) Rosenvinge: MELU, R 2451–2455. **a.** Portion of upper thallus. **b.** Upper branches of tetrasporophyte. **c.** Branching of upper axes. **d.** Branches bearing clusters of tetrasporangia. **e, f.** Cell details. **g.** Detail of tetrasporangial cluster with immature, apically produced tetrasporangia. Scales: a, b = 500 μm; c, d = 200 μm; e, f = 20 μm; g = 50 μm.

a
b
c
d
e
f
g

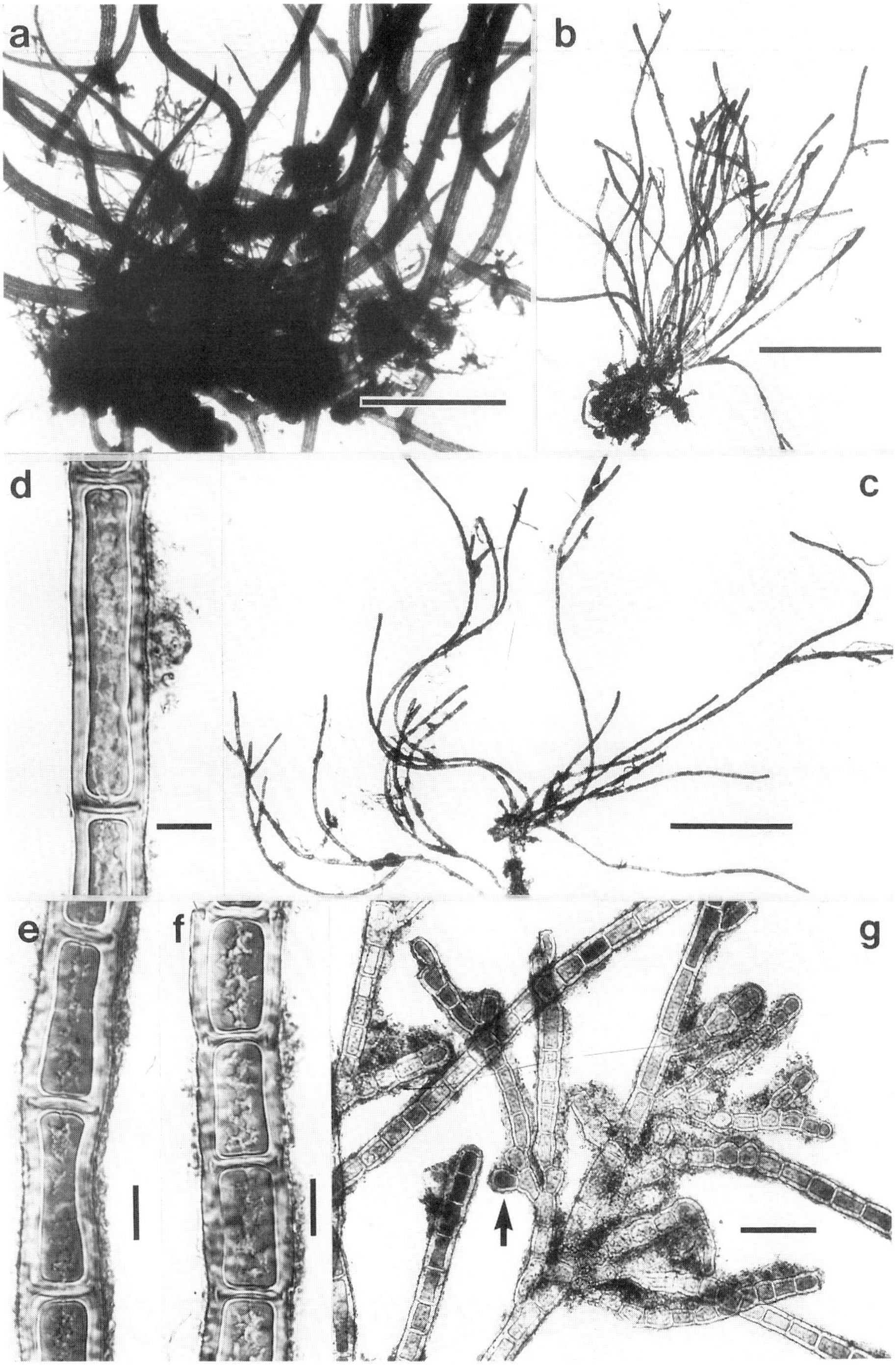

Fig. 67 *Rhodochorton purpureum* (Lightfoot) Rosenvinge: MELU, R 2287. **a.** Habit of filaments growing at base of *Bostrychia vaga* mat. **b, c.** Thallus habits. **d–f.** Cell details. **g.** Thallus with immature spore (arrow). Scales: a = 1 mm; b, c = 500 μm; d–f = 10 μm; g = 50 μm.

Rhodochorton variabile Drew, 1928: 174

Fig. 68

TYPE LOCALITY: Cypress Point, Monterey County, California.

TYPE: UC, no. 266281 (= Gardner, no. 4659).

SOUTHERN OCEAN DISTRIBUTION: Macquarie I.

MATERIAL EXAMINED: Macquarie I.: collection MA-11; Herbarium specimen: MELU, 1275.

DESCRIPTION: Plants 1–2 mm tall, forming epiphytic turf (Fig. 68a) composed of uniseriate, sparsely to much branched filaments with secund branching (Fig. 68b, c). Filaments mostly uniform diameter (8–9 μm) throughout composed of cells 2–4 times elongate (to 26 μm) with apical cells having rounded apices. Plastids parietal, band-shaped, lacking pyrenoids, one per cell. Tetrasporangia ovoid (14 μm × 17 μm), terminating 1–2-celled lateral branches (Fig. 68d, f) or sessile along main axes (Fig. 68g). Spermatangia-like structures borne in clusters (Fig. 68e). Female reproduction structures not seen.

Rhodochorton variabile is known from only one collection taken from 5 m depth, where it and *Ectocarpus siliculosus* grew intermixed as epiphytes on the course brown algal host *Desmarestia chordalis*. The distribution and abundance, as well as the morphological variability of this species at Macquarie I. is unknown.

NOTES: Drew (1928) described the limits of this species as 'vague', and 'the forms so numerous that it is difficult to give a comprehensive description'. Abbott & Hollenberg (1976) consider it to be synonymous with *Acrochaetium pacificum* Kylin (1925), which Drew (1928) had also placed in *Rhodochorton*, and Garbary *et al.* (1982) have included it as a variety of *Audouinella plumosa* (Drew) Garbary (1979). Macquarie plants are tentatively assigned to this species because of their obvious secund branching, a characteristic considered by Drew as outstanding for this species.

Bonnemaisoniales

Feldmann & Feldmann (1942) recommended that the Bonnemaisoniaceae, together with the Naccariaceae, be raised to ordinal status, but the order has not been widely accepted. Rather, both families generally are placed in the Nemaliales (Dixon & Irvine, 1977; Bold & Wynne, 1978; Kraft, 1981; Dixon, 1982). On the basis of life histories and reproduction, Chihara & Yoshizaki (1972) and West & Hommersand (1981) support recognition of the Bonnemaisoniales. Additional supportive evidence for the order comes from ultrastructural work by Pueschel & Cole (1982) on pit connections in red algae, in which members of the Bonnemaisoniaceae and Naccariaceae form a group distinct from other families in the Nemaliales.

Papenfuss (1964*a*) recognized the Bonnemaisoniales and included the Chaetangiaceae as one of its constituent families. In the present account, however, this family is retained in the Nemaliales, with the Bonnemaisoniaceae and Naccariaceae constituting the Bonnemaisoniales. Both families are represented in the Northern and Southern Hemispheres, but only the Bonnemaisoniaceae has representatives in southern high-latitude waters.

Bonnemaisoniaceae Schmitz, 1892: 20

Members of the Bonnemaisoniaceae have uniaxial thalli, three-celled carpogonial branches, gonimoblasts with large fusion cells bearing swollen terminal carposporangia, and spermatangia produced in superficial sori on small fertile branches or portions of main axes. Plants have erect gametangial thalli with fronds either terete and radially branched or variously flattened and bilaterally branched. The tetrasporangial phase, where known, is prostrate or erect, either encrusting or filamentous (West & Hommersand, 1981), but sometimes (as for *Delisea pulchra*;

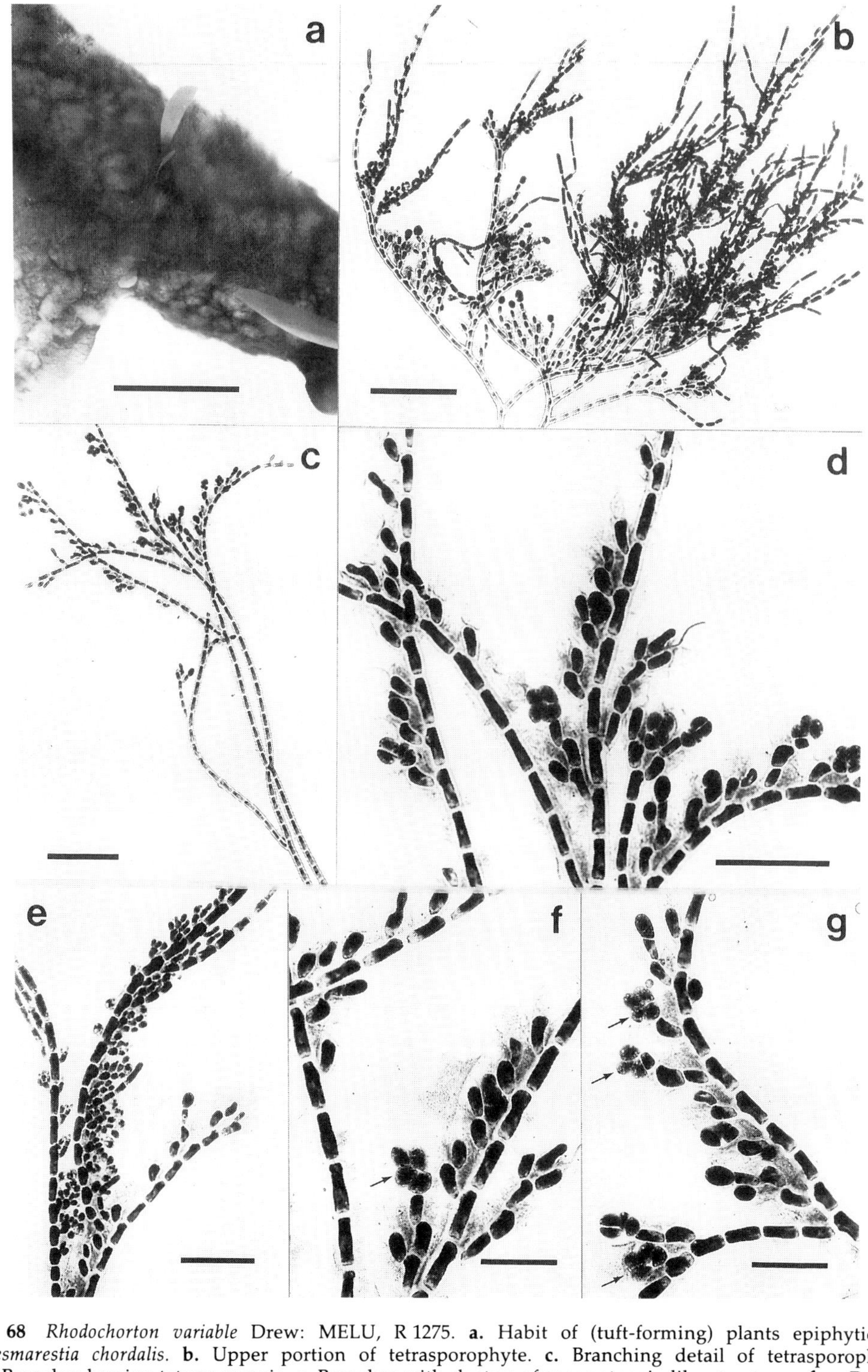

Fig. 68 *Rhodochorton variable* Drew: MELU, R 1275. **a.** Habit of (tuft-forming) plants epiphytic on *Desmarestia chordalis*. **b.** Upper portion of tetrasporophyte. **c.** Branching detail of tetrasporophyte. **d.** Branches bearing tetrasporangia. **e.** Branches with clusters of spermatangia-like structures. **f, g.** Details of tetrasporangia (arrows). Scales: a = 5 mm; b, c = 100 μm; d, e = 50 μm; f, g = 30 μm.

Levring, 1953) morphologically similar to the gametangial phase.

The family comprises six genera, *Asparagopsis*, *Bonnemaisonia*, *Delisea*, *Leptophyllis*, *Pleuroblepharidella* and *Ptilonia*, with members of two growing in high-latitude southern waters. Both *Delisea pulchra* (Greville) Montagne and *Ptilonia magellanica* (Montagne) J. Agardh are distributed widely throughout the Southern Ocean and might be expected to occur at Macquarie I. Of these, only *D. pulchra* is present in this vicinity. *Ptilonia* is represented at Macquarie I. by *P. willana* Lindauer, which was previously known only from New Zealand.

Delisea Lamouroux, 1819: 41

Lectotype: *D. elegans* Lamouroux, 1819: 42 (see Silva, 1957).

Chihara & Yoshizaki (1971, 1978) characterized *Delisea* as having: (1) uniaxial thalli; (2) primary axial rows producing two laterally directed pericentral cells, one developing into a short, simple (major determinate) branch or a long (indeterminate) branch, the other remaining poorly developed and mostly contributing to thallus cortication; (3) cortical gland cells; (4) spermatangia derived from cortical cells; (5) three-celled carpogonial branches arising from pericentral cells of main axes on fertile long (indeterminate) branches; (6) sterile (pericarp) cells developing from the supporting cell and the basal cell of the carpogonial branch; (7) a nutritive cell tuft produced from the hypogynous cell; and (8) post-fertilization development with a connection forming between the carpogonium and hypogynous cell, fusion of the basal portion of the nutritive tuft with the hypogynous cell, and production of the primary gonimoblast from the carpogonium.

Delisea can be separated from other genera in the family because its members produce branches alternately along the edges of flattened axes, have cystocarps sessile on indeterminate branches, and form spermatangial sori on unspecialized axes (Bonin, 1981). The genus has many similarities with *Ptilonia* and the two are believed by Chihara & Yoshizaki (1978) to be congeneric, but both genera are recognized in the present account (for reasons given in the description of *Ptilonia*).

Of the six species attributed to *Delisea*, (i.e., *D. compressa*, *D. flaccida*, *D. elegans*, *D. hypneoides*, *D. plumosa*, *D. pulchra*), all occur in temperate to colder waters of the Southern Hemisphere, but only *D. pulchra* extends into the Northern Hemisphere. All species have similar anatomy and reproductive characteristics, but can be separated by differences in thallus morphology and habit. The principal criteria distinguishing species are width of main axes and length of spinose (major determinate) branches, position and number of cystocarps per indeterminate branch, and arrangement of indeterminate branches along main axes.

Delisea pulchra (Greville) Montagne, 1844: 158

Figs 69, 70g

Basionym: *Bowiesia pulchra* Greville, 1830: lvii.

Synonyms: The following taxa, excluding *Bonnemaisonia prolifera*, are discussed by Bonin (1981):
- *Bonnemaisonia prolifera* Reinsch 1888*b*: 153
- *Calocladia pulchra* (Greville) J. Agardh 1841: 21
- *Delisea fimbriata* (Lamouroux 1819: 42 (see Silva (1957) re. nomenclatural problems with the name *D. fimbriata*)
- *Delisea japonica* Okamura, 1908: 139
- *Delisea serrata* Kützing 1849: 771

Type locality: reported from the southern seas of Australia (J. Agardh, 1863), but suspected by J. D. Hooker (1847) to be part of a collection of Macquarie I. algae sent to D. Fraser in Sydney, New South Wales.

Lectotype: Greville collection, E; N.B. type specimen illustrated by Bonin (1982, figs 8a, b).

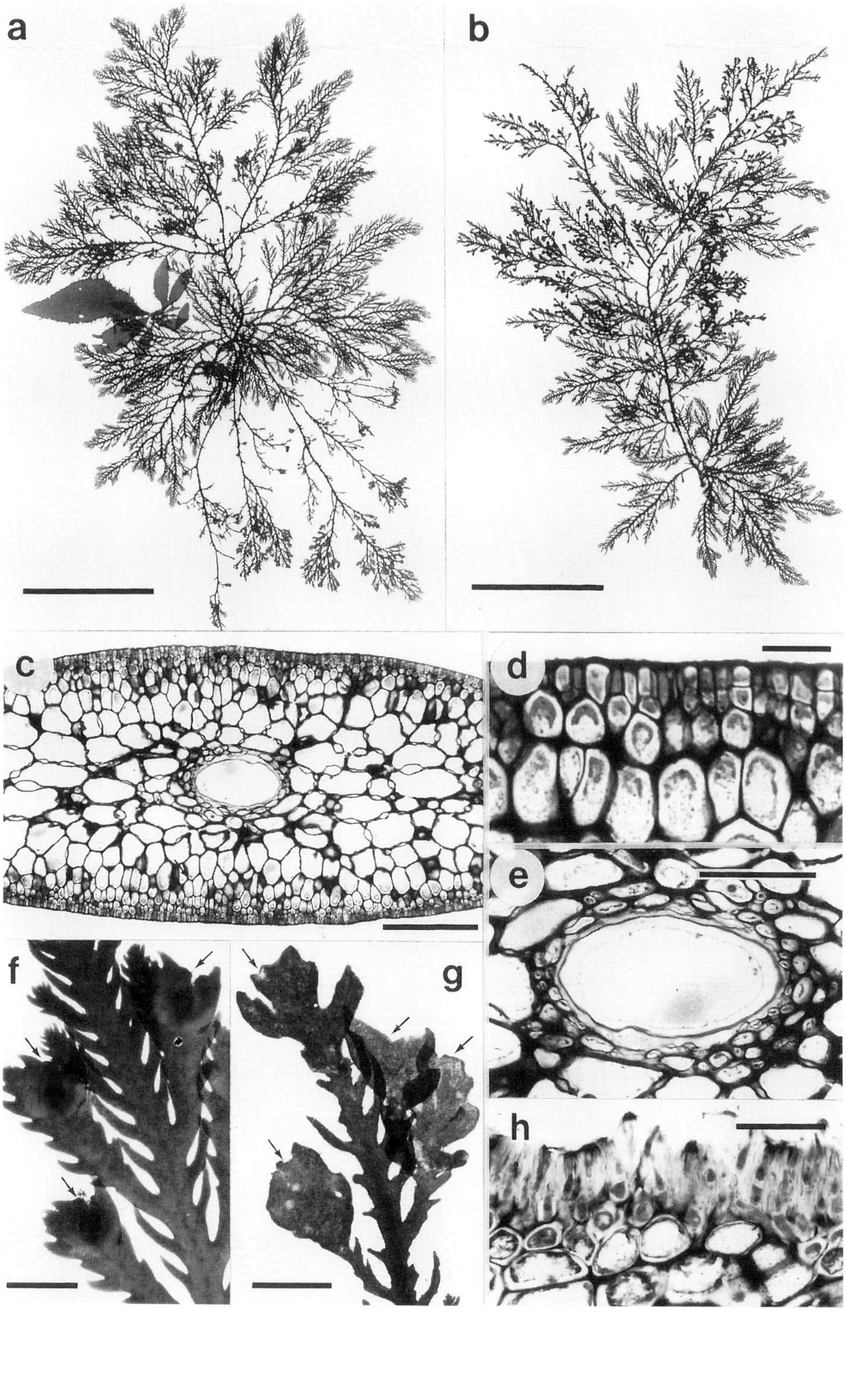
a
b
c
d
e
f
g
h

Southern Ocean distribution: Antarctic Peninsula (Skottsberg, 1953); Heard I. (Dickie, 1876*d*); Îles Kerguelen (Dickie, 1876*c*, 1879); Macquarie I.; South Georgia (Reinsch, 1888*b*).

Material examined:
1. Heard I.: Atlas Cove (Chittleborough, 11 September 1949): MELU, R 3729 (female).
2. Macquarie I.: collections MA-302, 303, 309, 364, 378; Herbarium specimens: MELU, R 0154, 0180–0182, 1703, 3282–3285; UC, R 0140.

Bonnemaisonia prolifera
South Georgia: Lectotype. 'Nordstrand der Landzünge' (Will, 3 July 1883): M, 67.83/17 (= Will, no. 80) (male).

Description: Plants to 20 cm tall, bushy, burgundy to red-brown to purple-black, arising from discoid holdfasts, usually with several coarse main axes alternately branched in one plane producing either spine-like (major) determinate or longer indeterminate laterals and indistinct (minor) determinate laterals (Fig. 69a, b). Axes 550–600 μm thick by 1–2 mm wide, comprising cortex with 1–2 layers of isodiametric to twice elongate pigmented cells and less frequent, distinctly larger, yellowish-brown gland cells (Fig. 69c, d) and medulla with numerous layers of large and small colourless cells surrounding axial filament. Transection of frond showing axial filament oval in outline, 100 μm × 175–190 μm, ensheathed by rhizoids (Figs 69e, 70g). Cystocarps sessile, singular, subterminal on indeterminate branches, with ostioles directed in the plane of thallus branching (Fig. 69f); carposporophytes with large basal fusion cells bearing terminal, elongate, clavate carposporangia. Spermatangial sori colourless, produced on both surfaces of flattened and expanded, irregularly lobed terminal portions of indeterminate axes (Fig. 69g, h). Tetrasporophytes not seen.

Delisea pulchra, known only as an infrequent member of the understory community in *Macrocystis* forests, grows in the subtidal to 20 m depths. Plants are commonly associated with *Rhodophyllis acanthocarpa*, which grows either on rocks or as an epiphyte on *Delisea* (Fig. 69a). Female and sterile thalli are considerably more abundant than male gametophytes.

Notes: Bonin's (1981) observations of New Zealand sporophytes show that tetrasporangia are produced in sori covering both sides of expanded upper portions of indeterminate branches, structures somewhat similar to those bearing spermatangial sori. Tetrasporangia were not seen in Macquarie material. *Delisea* species are partly separated on differences in the size and shape of major determinate (tooth-like) branches. Depending upon the reproductive status of plants, however, the marginal teeth of *D. pulchra* are either broad or narrow, short or long, terete or flattened. Those of male gametophytes generally are winged slightly, whereas those developing on female and sterile thalli usually are terete.

Ptilonia (Harvey) J. Agardh, 1852: 773

Type species: *P. magellanica* (Montagne) J. Agardh, 1852: 774.

Kylin (1956) used only morphological criteria to distinguish *Ptilonia*, which has thalli with entire or serrate margins, from the closely related *Delisea*, the fronds of which have either alternately fringed or toothed margins. Chihara & Yoshizaki (1978) considered this insufficient reason for recognizing two genera and, consequently, transferred all species to one genus, *Delisea*. Bonin (1981) disagreed with this and continued to recognize both genera through differences in the development of lateral branches, the general morphology of spermatangial sori, the development of carpogonial branches, and the location and orientation of cystocarps. In *Ptilonia*, major determinate axes do not emerge from the thallus to form dentate margins as exist in *Delisea*;

Fig. 69 *Delisea pulchra* (Greville) Montagne: MELU, R 0181, 0182, 3284, 3285. **a.** Habit of male thallus with epiphytic *Rhodophyllis acanthocarpa*. **b.** Habit of female thallus. **c–e.** T.S. mature axis. **d.** Detail of cortex. **e.** Detail of axial filament showing ensheathing rhizoids. **f.** Habit of subterminal cystocarps (arrows). **g.** Habit of spermatangial sori (arrows). **h.** Section through spermatangial sorus. Scales: a, b = 5 cm; c = 200 μm; d, h = 25 μm; e = 100 μm; f, g = 2 mm.

carpogonial branches are produced from pericentral cells of major determinate branches, and cystocarps, which usually have their ostioles directed away from the plane of thallus branching, are either terminal or subterminal on major determinate branches. Based on observations made during the present survey, spermatangial sori are comparable in both genera. Additional personal observations also suggest a biochemical difference between *Ptilonia* and *Delisea* species. *Ptilonia* species generally are delicate and begin decomposing within several hours if not preserved soon after removal from the sea. In contrast, *Delisea* species are much tougher and can withstand considerably longer intervals between collection and fixation. The Australian species *P. australasica* Harvey is notably different, and in this respect more like *Delisea*.

The genus contains six species, *P. magellanica*, *P. australasica*, *P. mooreana*, *P. subulifera*, *P. willana* and *P. okadai*, of which the first five occur in the Southern Hemisphere.

Ptilonia willana Lindauer, 1947: Exsiccatae No. 218

Fig. 70a–f

SYNONYM: *Delisea willana* (Lindauer) Chihara *in* Chihara & Yoshizaki, 1978: 387.

TYPE LOCALITY: near Twilight Bay, Port Pegasus, Stewart I.

LECTOTYPE: AKU, no. 91 (Lindauer set no. 6703); (Bonin, 1981: 52, fig. 10a).

DISTRIBUTION: Macquarie I.; South island of New Zealand and various islands (including Antipodes, Auckland, Chatham, Snares, Solander, and Stewart Is) in the New Zealand Subantarctic (Bonin, 1981).

MATERIAL EXAMINED: Macquarie I.: collections MA-98, 248, 303, 309; Herbarium specimens: MELU, R 1900–1903, 3287, 3288.

DESCRIPTION: Plants to 12 cm tall, bushy, arising from discoid holdfasts, usually with one main blade to 11 mm broad having a visible percurrent central axis bearing laterals alternately to irregularly in one plane (Fig. 70a). Thalli flattened, pinkish-brown when fresh, reddish-brown to greenish upon fixation in 4% formalin, with margins entire, finely serrate, or sometimes bearing small spicate branchlets (Fig. 70d) or irregularly shaped proliferations (Fig. 70b). Blades to 265 μm thick, with cortex monostromatic, pigmented, composed of isodiametric (7–11 μm diameter) cells usually arranged in single layer (Fig. 70f); medulla multilayered, composed of small and large colourless cells, with corticating rhizoids surrounding axial filament (Fig. 70e). Cystocarps sessile, subapical on indeterminate branches or marginal proliferations, each with one ostiole directed away from the plane of branching (Fig. 70b–d). Spermatangia and tetrasporangia not seen.

Ptilona is an understory species in *Macrocystis* forests and grows on rocky surfaces to 18 m depth. It is rare, but occurs at both the northern and southern ends of the island.

NOTES: Bonin (1981) studied this species from New Zealand and characterized it similarly to the above description. Additionally, she observed plants to be dioecious, with spermatangial sori produced on indeterminate branch tips, and she also noted that fresh plants emit a distinctive burnt rubber odor and have a slippery texture. Such characteristics were not noticed in Macquarie thalli, which were fixed shortly after collection and not studied while living.

Fig. 70a–f *Ptilonia willana* Lindauer: MELU, R 1900, 3287, 3288. **g** *Delisea pulchra* (Greville) Montagne: MELU, R 3284. **a.** Habit of female thallus. **b.** Thallus base with lateral proliferations containing cystocarps. **c.** Habits of cystocarps (arrows). **d.** Detail of spinous lateral proliferation (small arrow) and presumptive cystocarp bladelet (large arrow). **e–f.** T.S. main axis. **e.** Axial cell (ax). **f.** Detail of cortex. **g.** Optical section showing central axis with ensheathing filaments. Scales: a = 1 cm; b = 3 mm; c, d = 2 mm; e = 50 μm; f = 20 μm; g = 100 μm.

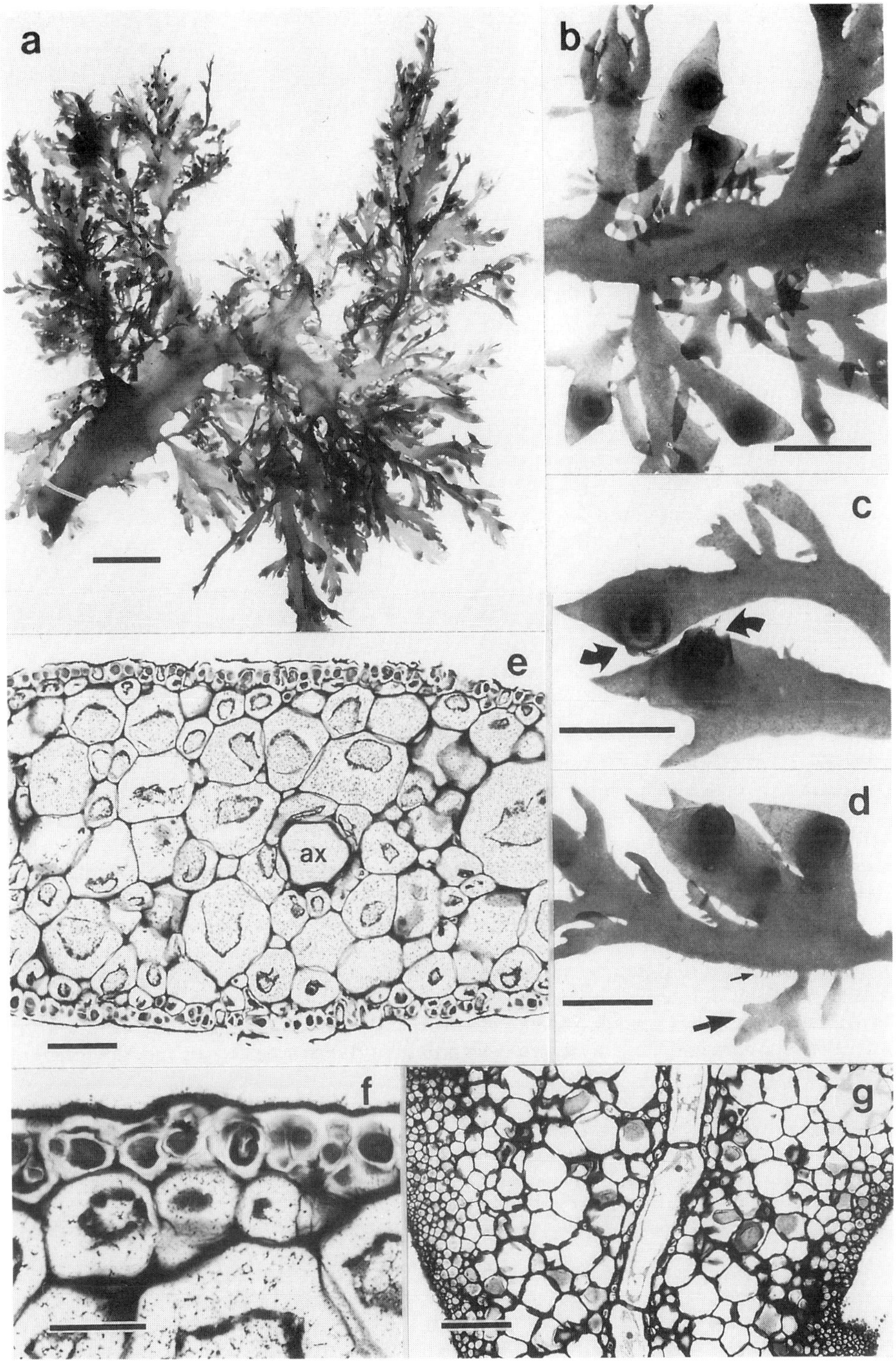
a
b
c
d
e
ax
f
g

The northern range of *P. willana* overlaps with the southern range of *P. mooreana* at Wellington on the South Island of New Zealand (Bonin, 1981). Female gametophytes of both species can be distinguished easily by differences in their respective sessile and stipitate cystocarps (V. J. Chapman, 1969). In this same way *P. willana* differs from the type species, *P. magellanica*.

Nemaliales

In its broadest concept, as advocated by Dixon (1982), the Nemaliales contains 13 families which have been united in one order because their members have gonimoblasts that develop directly from the carpogonium or from a cell in the carpogonial branch that functions as the generative auxiliary cell. Aside from this one unifying feature, the order comprises a heterogenous assemblage of families including a broad range of thallus morphologies and several types of life histories (Dixon, 1982). Such diversity has led some workers (e.g., Garbary, 1978; West & Hommersand, 1981) to recognize various nemalialean families as separate orders.*

In the present account, the order is conceived more narrowly than by Dixon and is represented at Macquarie I. by one species in the Chaetangiaceae.

Chaetangiaceae Kützing, 1843: 389, 391

Desikachary (1963) considered the Chaetangiaceae sufficiently distinct from other families of the Nemaliales to warrant separate ordinal status, but this has not been accepted widely (Dixon, 1982). Members have multiaxial thalli that may or may not be calcified. The gametophytic and sporophytic generations are either isomorphic or slightly to markedly heteromorphic with gametophytes being the dominant phase. Sporophytes are filamentous and *Acrochaetium*-like or encrusting (Dixon & Irvine, 1977; Delépine *et al.*, 1979).

This family comprises eight marine genera (Wynne & Kraft, 1981) mostly occurring in warm temperate to tropical waters, *Chaetangium* being the only representative in high-latitude southern waters.

Chaetangium Kützing, 1843: 392

Type species: *C. ornatum* (Linnaeus) Kützing, 1843: 392.

Gametophytic thalli are erect, 2–10 cm tall, uncalcified, soft to firm and cartilaginous, and simple to dichotomously (sometimes proliferously) branched with either terete or largely flattened axes that are solid or hollow and composed of loosely to tightly interwoven, small-diameter medullary filaments surrounded by a compact filamentous cortex. Gonimoblasts arise from cortical filaments and develop in upper portions of the thallus or on specialized reproductive proliferations. Gametophytes are dioecious, with both male and female thalli being vegetatively similar. Spermatangia are borne in conceptable-like structures that are distributed over the upper thallus (Martin, 1939). Sporophytic thalli are encrusting and give rise to cruciately divided tetrasporangia (Delépine *et al.*, 1979).

This exclusively Southern Hemisphere genus has approximately 12 species (De Toni, 1897, 1924; Levring, 1955), with *C. fastigiatum* being distributed most widely throughout the Southern Ocean.

Chaetangium fastigiatum (Bory de Saint-Vincent) J. Agardh, 1852: 460

Fig. 71

Basionym: *Halymenia fastigiata* Bory de Saint-Vincent, 1826*b*: 594.

*Acrochaetiaceae—to the Acrochaetiales (J. Feldmann, 1953); Batrachospermaceae, Lemaneaceae, & Thoreaceae—to the Batrachospermales (Pueschel & Cole, 1982); Chaetangiaceae—to the Chaetangiales (Desikachary, 1963); Bonnemaisoniaceae & Naccariaceae—to the Bonnemaisoniales (J. Feldmann, 1952); Gelidiaceae & Gelidiellaceae—to the Gelidiales (Kylin, 1923; Fan, 1961).

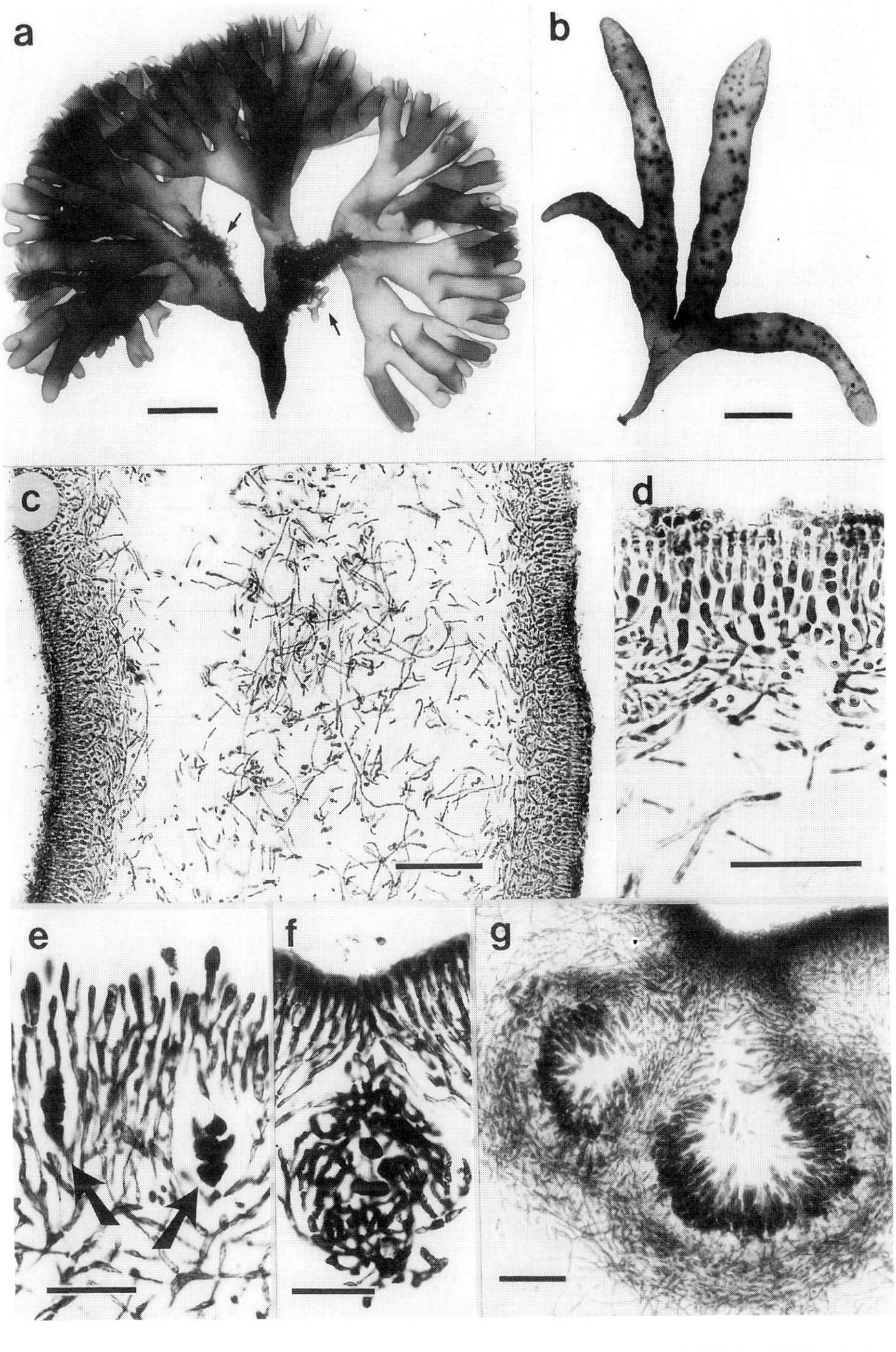

Fig. 71 *Chaetangium fastigiatum* (Bory de Saint-Vincent) J. Agardh: MELU, 3393, 3401. **a.** Habit of plant with proliferous branching (arrows) on lower thallus. **b.** Habit of female gametophyte. **c.** T.S. through sterile thallus. **d.** Detail of cortex. **e.** Carpogonial branches (arrows). **f.** Early development of involucre. **g.** Section through mature cystocarp. Scales: a = 6 mm; b = 5 mm; c, g = 100 µm; d = 50 µm; e, f = 20 µm.

SYNONYMS: see Papenfuss (1964*a*) and Levring (1953).

TYPE LOCALITY: Falkland Is.

TYPE: Herb. Agardhianum (LD), no. 32591. (Type specimens were not seen during the present survey, but fragments of Bory de Saint-Vincent's type material of *Halymenia* no. 23 were studied by Skottsberg (1923).)

DISTRIBUTION: Argentina (Pujals, 1963); Auckland Is (Montagne, 1842*b*); Campbell I. (Levring, 1945); Chile, Falkland Is, Fuegia (Skottsberg, 1923); Îles Crozet (Levring, 1944); Îles Kerguelen (Dickie, 1879); Macquarie I. (Zinova, 1958); New Zealand (Levring, 1955; Chapman, 1969); Peru (Acleto, 1973); South Georgia (Kylin & Skottsberg, 1919); Stewart I. (Adams *et al.*, 1974); Western Australia (De Toni & Forti, 1922).

MATERIAL EXAMINED:

1. Heard I.: Transect 1, north of Wharf Point, Atlas Cove, on rocks in the eulittoral (Smith, 6 February 1983): MELU, R 3735.
2. Macquarie I.: collections MA-111, 119, 120, 130, 133, 213, 238; Herbarium specimens: MELU, R 0091, 1303, 2023–2026, 3393, 3401.
3. New Zealand: Shag Point, low to mid-eulittoral reef flat (Kraft, 1 November 1972): MELU, K 5568.

Chaetangium pseudosaccatum Levring, 1955: 423
New Zealand: Shag Point, low-eulittoral reef flat (Kraft, 1 November 1972): MELU, K 5567.

DESCRIPTION: Plants to 3 cm tall, red-brown to purplish, arising from discoid holdfasts, with dichotomously (to seven times) branched axes to 3 mm diameter, mostly saccate and turgid with mucilaginous filling, but sometimes compressed and cartilaginous (Fig. 71a, b). Lower portions of main axes occasionally with proliferous branching (Fig. 71a). Thalli multiaxial and entirely filamentous (Fig. 71c), with cortex composed of anticlinal files (4–7 cells long) of pigmented cells arising from subcortex of colourless, interwoven, periclinal filaments (Fig. 71d); medulla with loosely organized colourless filaments in gelatinous matrix. Carpogonial branches three-celled, outwardly directed, borne at juncture of subcortex and cortex, and terminated by pyriform carpognia with narrow (not inflated) trichogynes (Fig. 71e). Young gonimoblasts with loosely organized filamentous pericarps (Fig. 71f); mature cystocarps embedded in medulla, but readily visible from thallus surface (Fig. 71b), with compact pericarps comprising tightly interwoven filaments surrounding filamentous gonimoblasts bearing masses of inwardly directed clavate carposporangia (Fig. 71g). Male gametophytes and tetrasporophytes not seen.

Chaetangium fastigiatum grows on rocky surfaces in the intertidal, usually forming extensive low carpets on small boulders. The thallus commonly is infected by what appears to be an undescribed ascomycete species (Kohlmeyer, pers. comm.). Some of the ascocarps (appearing as black spots on the host thallus) are in turn attacked by a hyperparasite similar to species of the oomycete genera *Atkinsiella* and *Petersenia*.

NOTES: Most plants studied from Macquarie and Heard Is have terete inflated (saccate) branches. Chapman (1969) described New Zealand *C. fastigiatum* as morphologically variable, having compressed and cartilaginous fronds that range from flattened to almost terete, but not saccate. The New Zealand species *C. pseudosaccatum* Levring (1955) has tubular to saccate fronds that are simple or sparsely branched and smaller (8–12 mm tall) than the more branched ones of *C. fastigiatum*. Plants from Macquarie I. are somewhat intermediate in form between these two species, but are aligned presently with *C. fastigiatum* on the basis of thallus size and degree of branching. Macquarie thalli are morphologically very different from New Zealand *C. fastigiatum* cited in the present survey and it may be that the two represent different species. In this case, all entities currently known as *C. fastigiatum* would need to be re-examined in order to accurately assess the geographic distributions of both taxa. On the basis of occurrence at Heard and Macquarie Is, however, it seems likely that the entity with saccate fronds may also occur at other Subantarctic localities.

Post-fertilization development, as briefly described in the present survey, is the same as previously documented by Delépine & Lambert (1981). These authors reported 3–5-celled carpogonial branches, development of the gonimoblast from the hypogynous cell, and involucres composed of branched filaments surrounding young carposporophytes (comparable to that presently illustrated in Fig. 71f), in contrast to Levring's (1945) description of cystocarps without involucral filaments. Delépine, Delesalle, & Lambert (1979) germinated carpospores in culture and followed their development into sterile discoid crusts anatomically resembling field-grown sporophytic crusts that produced cruciately divided tetrasporangia.

Cryptonemiales

Plants are uniaxial or multiaxial, pseudoparenchymatous, loosely or compactly constructed, either calcified, slimy and gelatinous, or cartilaginous and firm, and encrusting or foliose with erect or prostrate fronds. Although the Cryptonemiales includes morphologically diverse species that have life history patterns with alternations of either isomorphic or heteromorphic somatic phases, it is unified and distinguished from the closely related red algal order Gigartinales by the production of auxiliary cells prior to fertilization in specially derived (accessory) branches. Auxiliary cells, however, are often difficult to see, let alone interpret as to position and origin (Kraft, 1981). In some families, the position of the auxiliary cells in relation to the surrounding vegetative cells (e.g., formation of different ampullar types in the Halymeniaceae (= Cryptonemiaceae)—Chiang, 1970) or carpogonial branches (e.g., procarpic vs. non-procarpic species in the Kallymeniaceae—Norris, 1957) has been used to characterize levels of evolutionary development.

As presently conceived, the order has 14 families containing some 130 genera (Kraft, 1981), although recent work by Pueschel & Cole (1982) suggests that two of these families, the Corallinaceae and Hildenbrandiaceae, should be elevated independently to orders. Papenfuss (1964*a*) listed seven families, the Choreocolacaceae, Corallinaceae, Dumontiaceae, Halymeniaceae (= Cryptonemiaceae), Hildenbrandiaceae, Kallymeniaceae and Peyssonneliaceae (= Squamariaceae, totalling 28 genera and 74 species, from the Antarctic and Subantarctic Zones. These numbers of taxa have since been reduced through various taxonomic studies (e.g., Hooper & South, 1974) and major reviews of families (e.g., Corallinaceae, Adey, 1970). Eight genera and ten species represent the Corallinaceae, Halymeniaceae, Hildenbrandiaceae and Kallymeniaceae at Macquarie I.

Corallinaceae Lamouroux, 1812: 185

This is by far the largest cryptonemialean family. It is represented throughout the world by approximately 50 extant and 30 exclusively fossil genera with 400–500 species (Kraft, 1981). Dixon (1982), in contrast, totalled only some 30 genera, a difference that probably reflects the extreme variation in taxonomic opinion concerning the Corallinaceae. Members of this family are multiaxial, have life history patterns with an alternation of isomorphic somatic phases, produce reproductive structures in conceptables, and (most) have cell walls impregnated with calcium carbonate (in the form of calcite crystals; Johansen, 1981). Although many schemes have been proposed for grouping the constituent species, modern systematics within the family is founded upon Hauck's classification of two subfamilies, with encrusting and non-articulated species placed in the Melobesieae, and erect, articulated ones with non-calcified joints in the Corallineae (Littler, 1972).

Papenfuss (1964*a*) listed 14 corallinaceous genera (not counting Reinsch's *Plectoderma*, which has uncertain affinities) and 49 species from the Southern Ocean, with roughly one quarter of these being erect and articulated. In a recent survey of this family at several Antarctic and Subantarctic localities, Zaneveld & Sanford (1980) recorded 14 encrusting species and no articulated ones. At Macquarie I. alone, they reported 6 genera and 13 species*, although some of their identifications are dubious (e.g., *Pseudolithophyllum aequabile*, which the authors describe as having uniporate sporangial conceptacles, is the basionym for *Antarcticophyllum aequabile* (Foslie) Mendoza (1976*a*), which has multiporate sporangial conceptacles).

Levring (1945) reported the upright, articulated species *Corallina officinalis* at Macquarie I., but this distribution remains unconfirmed. Rather, the coralline flora in this vicinity apparently lacks erect species, but is well represented by encrusting taxa. Of the five species documented presently, only two, *Mesophyllum schmitzii* and *Pseudolithophyllum discoideum*, were reported previously by Zaneveld & Sanford (1980). Although not included in the present survey and not examined anatomically, some thick encrusting plants from Macquarie's intertidal had morphologies similar to species of *Antarcticophyllum* (Lemoine) Mendoza (1976*a*) and produced multiporate conceptacles (pers. obs.).

Melobesia Lamouroux, 1812: 186

Type species: *M. membranacea* (Esper) Lamouroux, 1812: 186.

Melobesia has been circumscribed variously in the past (cf. Suneson, 1943). It is conceived presently as circumscribed by Adey (1970). Its species have been confused with those of the genus *Heteroderma* and sometimes have been referred to under the generic name *Epilithon* Heydrich (Lebednik, 1977). Plants are encrusting, epiphytic or epizoic, generally two cells thick in sterile parts, with a monostromatic epithalium and monostromatic hypothallium, and up to 12 cells thick near conceptacles, with a perithallium becoming multilayered. Epithallial cells are mostly rounded. Secondary pit connections are absent, but lateral cell fusions common. Gametophytes are dioecious, with sexual conceptacles uniporate. Sporangial conceptacles are multiporate, with each sporangium possessing a densely staining (with Toluidine Blue 0) apical plug.

This genus is absent from the Antarctic and has been known previously in the Subantarctic by *Melobesia neglectum* (Foslie) Adey (1970). On the basis of Lemoine's (1915, pl. 9, fig. 2) illustration of a coaxial hypothallium, however, this species does not belong in the genus *Melobesia*, or in *Lithothamnium* as indicated by Zaneveld & Sanford (1980). Levring (1945) reported *Melobesia leptura* Fosile (1906) from Campbell Is, but Zaneveld & Sanford (1980), when studying plants collected from 165 m depth off the northeast side of Macquarie I., accepted Foslie's second opinion on this species as a member of *Heteroderma*, i.e., *H. leptura* (Foslie) Foslie (1909). The presence of *M. membranacea* at Macquarie I. therefore represents the southernmost range of the genus.

Melobesia membranacea (Esper) Lamouroux, 1812: 186

Fig. 72

BASIONYM: *Corallina membranacea* Esper, 1788–1830: pl. *Corallina* 12 (publication date of plate undetermined).

SYNONYMS: see Rosenvinge, 1917 (as *Epilithon membranacea*).

TYPE LOCALITY: on Rhodophyta, west coast of France.

TYPE: presumably in ER (Koster, 1969), but specimens were not seen.

**Heteroderma leptura* (Foslie) Foslie, *Leptophytum asperulum* (Foslie) Adey, *L. coulmanicum* (Foslie) Adey, *L. foecundum* (Kjellman) Adey, *Lithophyllum atalayense* Lemoine, *Lithothamnium macquariensis* Sanford, *L. neglectum* (Foslie) Foslie, *L. schmitzii* (Hariot) Heydrich, *L. zaneveldii* Sanford, *Phymatolithon lenormandii* (Areschoug) Adey forma *macquariensis* Zaneveld & Sanford, *Pseudolithophyllum aequabile* (Foslie) Adey, *P. almanense* (Lemoine) Zaneveld & Sanford, *P. discoideum* (Foslie) Lemoine.

SOUTHERN OCEAN DISTRIBUTION: Macquarie I.

MATERIAL EXAMINED: Macquarie I.: collections MA-34, 83, 95, 244, 271, 300, 303, 309, 378; Herbarium specimens: MELU, R 0117, 0588, 0589, 0622, 0628, 0794, 1166, 1186, 1217; Microscope slides R/M 0242–0250, 0619, 0621.

DESCRIPTION: Crusts pink, to 10 µm thick and with 2–3 cell-layers in vegetative parts, thicker (to 95 µm with 7–10 cell-layers) near conceptacles, coalescing with adjacent (gametangial and sporangial) crusts to form patches to 10 cm across (Fig. 72a, b). Epithallium monostromatic, with cells small (to 5 µm diameter), triangular to rounded in sectional views of crusts, cut off by oblique walls and incompletely covering underlying cells (Fig. 72c). Perithallium non-existent to one cell thick in sterile tissues, increasing to eight cell-layers near conceptacles; hypothallium monostromatic, with cells similar to those of perithallium (Fig. 72c, d). Secondary pit connections absent; lateral fusions present between adjacent cells in hypothallial and perithallial tissues (Fig. 72d). Male conceptacles to at least 145 µm inner diameter × 45 µm inner height, with spermatangia borne on roof and floor. Female conceptacles to at least 200 µm inner diameter × 60 µm inner height, with center of floor supporting numerous two-celled carpogonial branches; periphery of conceptacle with maturing carposporangia (Fig. 72f). Sporangial conceptacles to at least 145 µm inner diameter × 85 inner height, with sporangia zonately divided and 25–30 µm broad by 65–75 µm long (Fig. 72g).

Melobesia grows in the subtidal to 18 m depth and is a common epiphyte on the leafy brown alga *Microzonia velutina* and on the red algae *Plocamium hookeri*, *P. secundatum* and *Phyllophora appendiculata*. Although numerous plants were collected from many localities around the island, only a few specimens were studied anatomically to produce the above description.

NOTES: *Melobesia* is thinner and more delicate than other encrusting corallines presently documented from Macquarie I., although *Heteroderma leptura* (Foslie) Foslie, which also produces thin crusts, was reported previously from this locality by Zaneveld & Sanford (1980). Both genera, however, are readily distinguished by the multiporate tetrasporangial conceptacles of the former and uniporate ones of the latter.

Melobesia membranacea is the only species in the genus presently recorded from high-latitude waters in the Southern Ocean. Other Southern Ocean records for the genus exist (e.g., Stewart I.—Adams *et al.*, 1974; Tristan da Cunha—Baardseth, 1941), but species have not been identified.

Mesophyllum Lemoine, 1928: 251

Lectotype species: *M. lichenoides* (Ellis & Sollander) Lemoine, 1928: 252 (see Hamel & Lemoine (1953) re. lectotypification).

Lemoine (1928) erected *Mesophyllum* for plants with multiporate tetrasporangial conceptacles and a coaxial hypothallium, features that she considered separately as typical of the genera *Lithothamnium*, which has a non-coaxial hypothallium, and *Lithophyllum*, which has uniporate tetrasporangial conceptacles (Lebednik, 1978). *Mesophyllum* has been refined further since its inception (Suneson, 1937, 1943), and the present account follows the circumscription by Adey (1970). Members of this genus generally form thin leafy crusts, but some are thick and encrusting and others branched. Although similar to *Melobesia* in having a monostromatic epithallium, lateral cell fusions, no secondary pit connections, uniporate sexual conceptacles, and multiporate tetrasporangial conceptacles containing sporangia with apical plugs, thalli of *Mesophyllum* are considerably thicker and have a multi-layered (coaxial) hypothallium with cells of adjacent, periclinal filaments aligned vertically to form arching tiers. The meristem is restricted to the uppermost perithallial cells and appears as a row of elongate cells.

The total number of species belonging to *Mesophyllum* is uncertain (Woelkerling, pers. comm.), but of the 43 species placed in the genus by Adey (1970), most occur in the Southern Hemisphere. Two of these [*M. aucklandicum* (Foslie) Adey and *M. fuegianum* (Foslie) Adey], plus another two that were added to the genus by Mendoza [1977; *M. kühnemannii* Mendoza and *M. schmitzii*

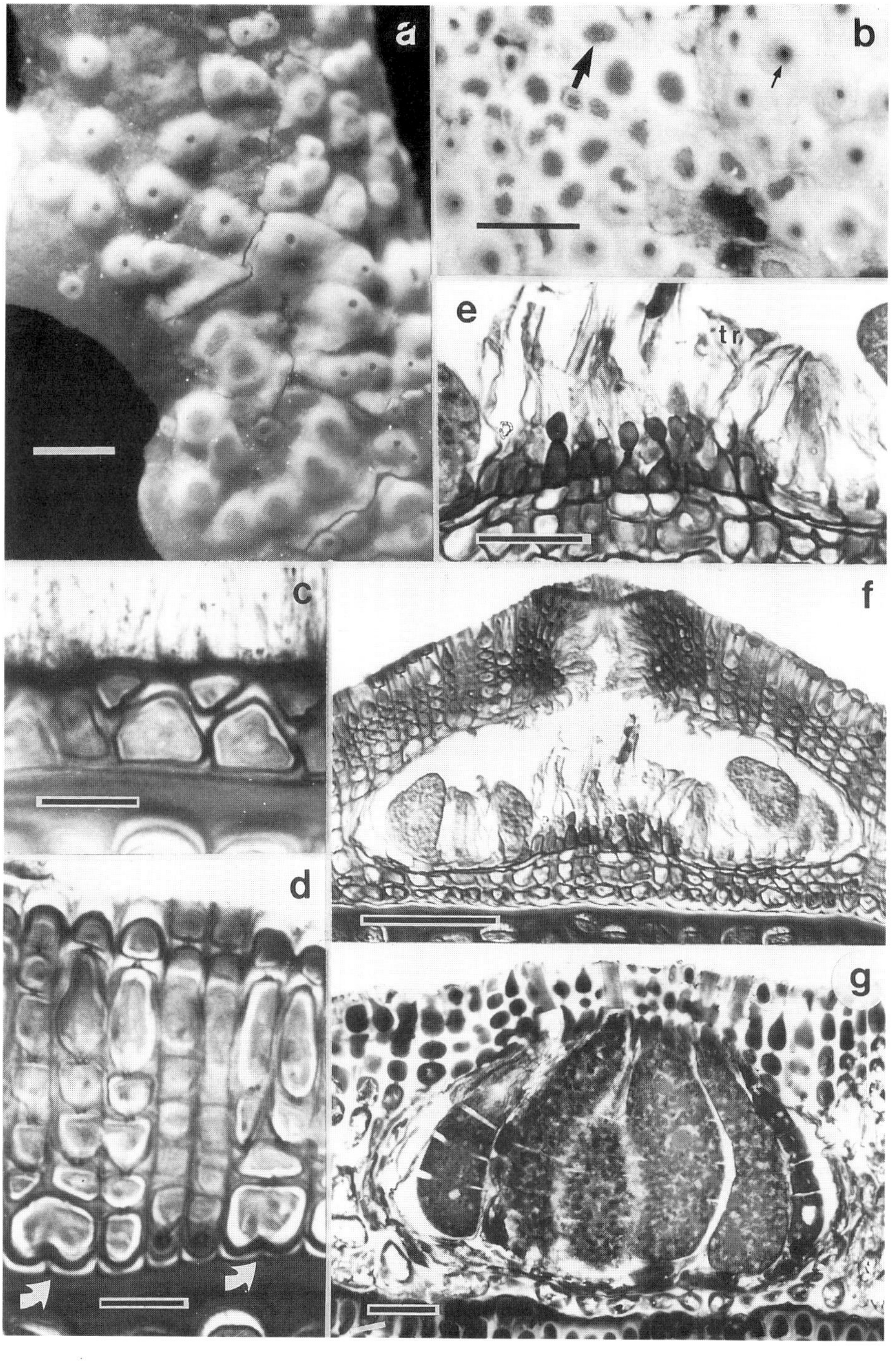
a
b
e
tr
c
f
d
g

(Hariot) Mendoza] plus *M. patena*, which is presently transferred from *Melobesia*, grow in the Subantarctic.

The genus is represented at Macquarie I. by *M. patena* and *M. schmitzii*.

Mesophyllum patena (Hooker f. & Harvey) R. W. Ricker comb. nov.

Fig. 73

BASIONYM: *Melobesia patena* Hooker f. & Harvey *in* Harvey, 1847: 111.

SYNONYMS: *Synarthrophyton patena* (Hooker f. & Harvey) Townsend, 1979: 252. (N.B. see Townsend (1979) for full synonymy.)

TYPE LOCALITY: Flat Point (near Castlepoint), New Zealand.

TYPE: TCD (Colenso, 1331). (N.B. specimen annotated by Johansen (September, 1967) as type is illustrated by Chapman & Parkinson (1974, pl. 72).)

DISTRIBUTION: The geographic range of this species is uncertain and needs to be confirmed through anatomical examinations of the specimens upon which records are based. As indicated by Townsend (1979), however, *M. patena* is reported from the following localities: Auckland Is, Australia (along parts of the southern coast of South Australia to Victoria and from Tasmania), Campbell I., Chatham I., Falkland Is, Fuegia, Macquarie I. (based on Levring's (1945) report of *Lithothamnion patena* from this locality), New Zealand, and South Africa.

MATERIAL EXAMINED:

1. Macquarie I.: collections MA-34, 255, 367; Herbarium specimens: MELU, R 0836, 2145; Microscope slides R/M 0370–0375, 0603.
2. New Zealand: Lectotype. Flat Point (Colenso): TCD, 1331 (male).

DESCRIPTION: Crusts epiphytic, pink, 140–400 μm thick in sterile parts, thicker near conceptacles, with smooth undulating surfaces (Fig. 73a, b), and composed of distinct tissue layers with cells lacking secondary pit connections. Epithallium monostromatic, with cells mostly oval in section and having rounded upper and lower walls, but sometimes flattened and slightly angular on lower surfaces (Fig. 73c); perithallium with several layers of spheroidal to somewhat elongate and sometimes rectilinear cells; hypothallium weakly to strongly coaxial (Fig. 73d, g). Lateral fusions of two or more cells common throughout perithallial and hypothallial tissues (Fig. 73d, e), absent in epithallium. Plants apparently dioecious, with gametangial conceptacles raised, uniporate, and with roofs developed from the overgrowth of surrounding tissues. Spermatangial conceptacles to 265 μm inner diameter × 90 μm inner height, with spermatangia produced along floor, walls and roof (Fig. 73f); spermatangial mother cells simple but sometimes (along conceptacle floors) appearing dendroid. Tetrasporangial conceptacles multiporate, 300–360 μm inner diameter × 135–145 μm inner height, with tetrasporangia zonately divided, oval to elongate, 35–45 μm broad × 85–100 μm long (Fig. 73g); conceptacle roof with numerous lateral cell fusions; pores filled with darkly staining (using Toluidine Blue 0) mucilage (Fig. 73h).

Plants grow as epiphytes on *Ballia callitricha*, where they wrap around host branches and form oval to rounded discs. The host alga occurs in the lower intertidal, but is most abundant in the subtidal, and is common along all sides of Macquarie I. *Mesophyllum patena*, however, has been collected from only one location along the sublittoral fringe, immediately below the lower limit of

Fig. 72 *Melobesia membranacea* (Esper) Lamouroux: MELU, R 0628, 1217; Slides R/M 0242–0250, 0619. **a.** Habits of gametangial and tetrasporangial plants epiphytic on *Plocamium*. **b.** Surface view of uniporate sexual conceptacles (small arrow) and multiporate tetrasporangial conceptacles (large arrow). **c–g.** Sectional views. **c.** Vegetative thallus. **d.** Tissue near female conceptacle showing fusions of adjacent hypothallial cells (arrows). **e.** Detail of degenerating carpogonial branches with loose cluster of trichogynes (tr). **f.** Female conceptacle. **g.** Tetrasporangial conceptacle. Scales: a, b = 500 μm; c, d = 10 μm; e, g = 20 μm; f = 50 μm.

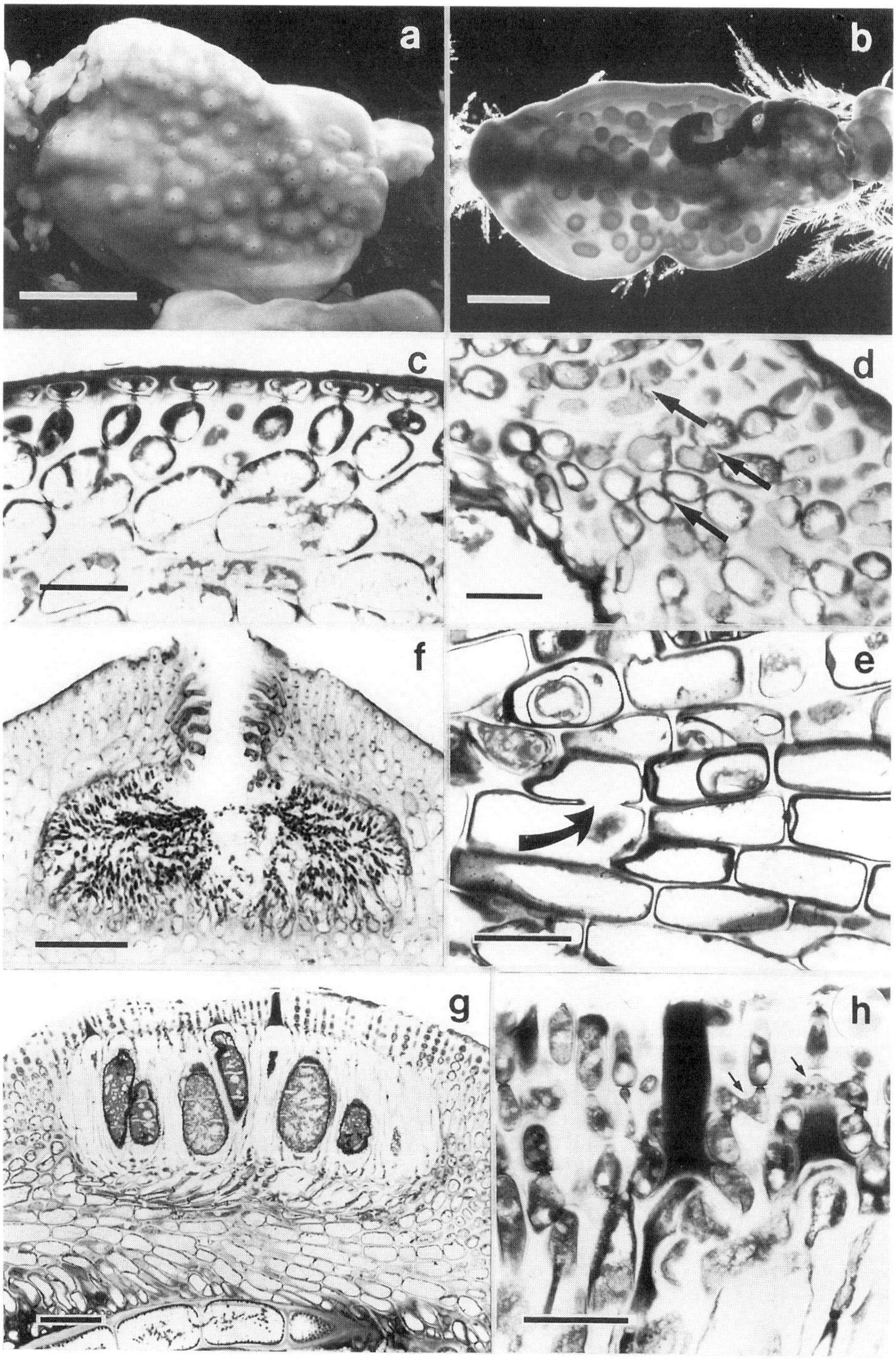
a
b
c
d
f
e
g
h

the *Durvillaea* belt, and in two subtidal habitats to depths of 10 m. Its distribution around the island is unknown and it is presumably uncommon.

NOTES: Considerable taxonomic and nomenclatural confusion over this species has been created by the various descriptions and names applied to it. Townsend (1979) selected *Melobesia patena* as the basionym for the new genus *Synarthrophyton* and based her description on observations of Australian plants without having studied type material. She circumscribed *Synarthrophyton* as having rounded epithallial cells, a coaxial hypothallium, secondary pit connections, lateral cell fusions restricted to conceptacle roofs, multiporate tetrasporangial and uniporate sexual conceptacles, procarps with a sterile cell, and spermatangia borne in dendroid clusters only along the conceptacle floor. Although some of these reproductive features were not seen in the present studies, examinations of type fragments of *M. patena* revealed an abundance of lateral cell fusions in perithallial and hypothallial tissues and the production of spermatangia from all inner surfaces of male conceptacles. Study of additional plants from Macquarie I., which have vegetative constructions identical with the type of *M. patena*, revealed information on tetrasporophytes, but failed to confirm details of the procarp structure. The present observations suggest placement of *M. patena* in *Mesophyllum* and do not warrant erection of a new genus (i.e., *Synarthrophyton*) for this species. An Australian entity such as described by Townsend would be distinct from *Mesophyllum* and not be conspecific with *M. patena*.

Hooker f. & Harvey (*in* Harvey, 1847) first described *Melobesia patena* with another new species, *M. antarctica*, which they considered as a possible variety of the former. *Melobesia antarctica* had been referred to earlier by Hooker f. & Harvey (*in* Hooker, 1846–47) as *M. verrucata* var. *antarctica*. Based on comparisons of Macquarie individuals and type material of *M. patena* with Lemoine's (1913) description and illustration of *Lithothamnium antarcticum* (Hooker f. & Harvey) Heydrich (1901), a later name for *M. antarctica*, both taxa are closely related if not identical. Heydrich (1907) obviously realized the close relationship between these species when he proposed the name *Lithothamnion patena* f. *antarctica* to accommodate *M. antarctica*. Should the two be conspecific, the epithet '*patena*' would take priority over '*antarcticum*' (following Article 57.2 of the International Code of Botanical Nomenclature), and the various names listed by Lemoine (1913) would become synonyms of *Mesophyllum patena*.

Mesophyllum schmitzii (Hariot) Mendoza, 1977: 28

Fig. 74

BASIONYM: *Lithophyllum schmitzii* Hariot, 1895: 98.

SYNONYMS: see Mendoza, 1977.

TYPE LOCALITY: Fuegia.

TYPE: presumably in PC (Mendoza, 1977), but specimens were not examined for the present survey.

DISTRIBUTION: Argentina (Mendoza, 1977); Falkland Is (Lemoine, 1915); Fuegia (Hariot, 1895); Macquarie I. (Zaneveld & Sanford, 1980); South Orkney Is (Holmes, 1905; Foslie, 1912). (N.B. Papenfuss (1964*a*) additionally indicated distribution at South Georgia, but the original published record for this locality has not been located for the present survey.)

Fig. 73 *Mesophyllum patena* (Hooker f. & Harvey) R. W. Ricker: MELU, R 2145; Slides R/M 0370, 0374, 0603; TCD, 1331. **a.** Habit of gametophyte. **b.** Habit of tetrasporophyte. **c–h.** L.S. crusts. **c.** Detail of monostromatic epithallium. **d, e.** Lateral fusions (arrows) of perithallial cells. (N.B. 'd' is from the type material of *Melobesia patena.*) **f.** Spermatangial conceptacle. **g.** Tetrasporangial conceptacle. **h.** Lateral fusions (arrows) of cells forming roof of tetrasporangial conceptacle. Scales: a, b = 2 mm; c–e, h = 20 μm; f, g = 50 μm.

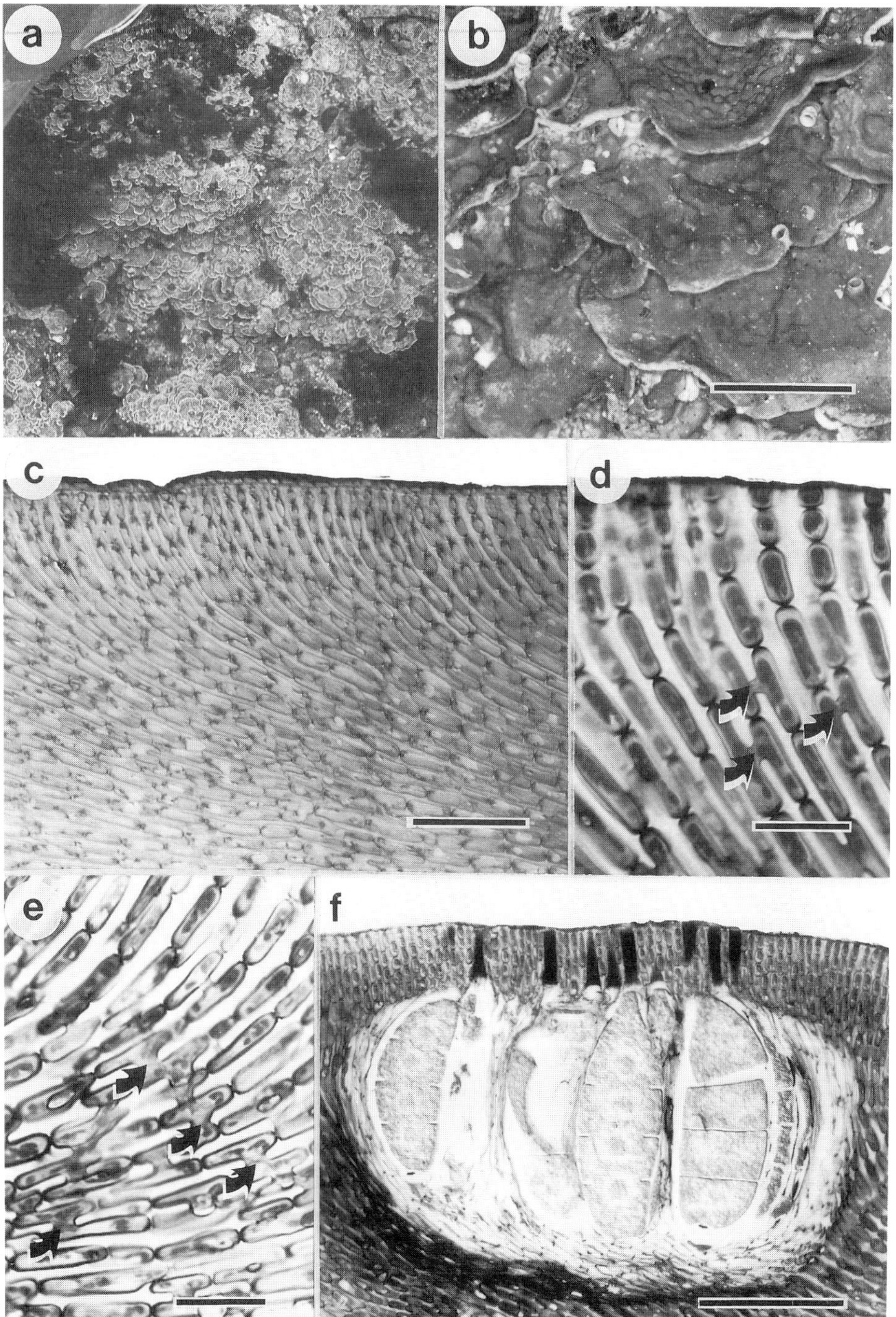
a
b
c
d
e
f

MATERIAL EXAMINED: Macquarie I.: collections MA-46, 83, 381; Herbarium specimens: MELU, R 1477–1480, 1491–1493, 1495–1500, 3741; Microscope slides R/M 0276–0282.

DESCRIPTION: Crusts pink to purplish-pink to dull red, leafy, with smooth to slightly convoluted surfaces and thickened whitish margins, 150–400 μm thick, 2–3 cm across, adnate proximally and free distally, forming imbricate clusters to >1 m across (Fig. 74a, b), and composed of distinct tissues with hypothallium being most extensive (Fig. 74c). Epithallium monostromatic, with rounded to angular cells covered by thick cuticle (Fig. 74d); perithallium 3–5 cells thick, with meristem not localized directly beneath the epithallium but showing progressive cell elongation into upper portion of largely non-coaxial hypothallium; hypothallial cells tubular, usually 5 μm diameter and of various (17–45 μm) lengths, forming occasionally branched files, with cells of adjacent, parallel filaments sometimes synchronously dividing to produce small sections of arching tiers. Secondary pit connections absent; lateral fusions of two or more cells common in perithallial and hypothallial tissues (Fig. 74d, e). Sporangial conceptacles multiporate, *c.* 360 μm inner diameter × 200 μm inner height, elevated above frond surfaces (Fig. 74f), with tetrasporangia zonately divided, oval to more elongate, *c.* 60 μm broad × 175 μm long, with densely staining (with Toluidine Blue 0) apical plugs. Gametangial conceptacles prominent as raised uniporate cones; sexual reproductive structures not studied.

Plants form extensive clumps of overlapping fronds on unshaded, rocky surfaces in mid- to lower-intertidal pools and in the subtidal to at least 6 m depths. In the intertidal, *M. schmitzii* is often associated with *Acrosiphonia pacifica*, *Adenocystis utricularis*, various members of the Ectocarpaceae and *Schizoseris condensata*. In the subtidal, it has been collected growing adjacent to or intermixed with *Microzonia velutina* on large boulders along the floor of kelp forests where the overhead *Macrocystis* canopy is missing.

NOTES: Mendoza (1977) transferred Hariot's (1895) *Lithophyllum schmitzii* to *Mesophyllum* because the type material, as well as additional plants from South America, had a coaxial hypothallium. Macquarie plants lack a coaxial hypothallium, although small portions sometimes show arching tiers of cells, but are tentatively referred to *M. schmitzii* because of anatomical similarities between crusts collected from the island during the present survey and those described and illustrated earlier by Zaneveld & Sanford (1980) as *Lithothamnium schmitzii* (Hariot) Heydrich. The habit of Macquarie thalli is comparable to that illustrated by Lemoine (1913, pl. 1, fig. 3) for *L. schmitzii* but is also strongly similar to that of the type specimen of *L. chathamense* Foslie (illustrated by V. J. Chapman & Parkinson (1974, pl. 71a), which purportedly is endemic to the Chatham Is.

Various anatomical differences distinguish Macquarie plants from the present concept of *Mesophyllum*. These include a non-coaxial hypothallium, growth from a diffuse region of cell elongation, and an epithallium composed of cells that are rounded toward the crust surface but angular on their inner-facing surfaces. Johansen (1976) described *Mesophyllum* species as having either a coaxial or non-coaxial ('multistratose') hypothallium and listed seven other features in common with the genus *Phymatolithon*, which has a non-coaxial hypothallium. The precise relationship between these two genera is clouded, but Woelkerling (pers. comm.) considers the latter to have a meristem not restricted to the subepithallial cells, in which cellular elongation is progressive into the hypothallium. Using this criterion, Macquarie plants should be aligned with the genus *Phymatolithon*. They are maintained presently in *Mesophyllum*, however, until both genera are more clearly circumscribed.

Pseudolithophyllum Lemoine, 1913: 45

Lectotype: *P. discoideum* (Foslie) Lemoine, 1913: 46 (see Hamel & Lemoine (1953) re. lectotypification).

Fig. 74 *Mesophyllum schmitzii* (Hariot) Mendoza: MELU, R 1491, 1496, 1497; Slides R/M 0276–0282. **a.** Habit of plants growing in lower intertidal pool. **b.** Overlapping gametangial crusts with uniporate conceptacles (arrows). **c.** L.S. crust. **d.** Detail of multi-layered perithallium showing lateral cell fusions (arrows). (N.B. epithallium is monostromatic.) **e.** Detail of hypothallium showing lateral cell fusions (arrows). **f.** Tetrasporangial conceptacle. Scales: b = 5 mm; c, f = 100 μm; d, e = 20 μm.

Members of this genus are encrusting, coarse, sometimes branched, and composed of a single- to multi-layered epithallium, well-developed perithallium, and generally a single-layered, but sometimes thicker, hypothallium. Secondary pit connections are lacking and lateral cell fusions abundant. Large hyaline cells, otherwise known as either megacells (Abbott & Hollenberg, 1976), heterocysts, or trichocytes (Adey, 1970), are absent. Sexual and asexual conceptacles are uniporate and lack apical plugs.

Adey (1970) recorded 22 species for the genus, although two, *P. aequabile* (Foslie) Adey and *P. subantarcticum* (Foslie) Adey, have been transferred by Mendoza (1976*a*) into a new genus, *Antarcticophyllum*, which Lemoine (1913) earlier considered to be a subgenus of *Lithophyllum*. *Pseudolithophyllum* is absent from the Antarctic, but represented by *P. consociatum* (Foslie) Lemoine, *P. discoideum* and *P. falklandicum* (Foslie) Adey in the Subantarctic (Lemoine, 1913). Only the first two species, however, occur at Macquarie I.

Pseudolithophyllum consociatum (Foslie) Lemoine, 1913: 48

Fig. 75

BASIONYM: *Lithophyllum consociatum* Foslie, 1905: 1.

SYNONYMS: see Lemoine, 1913.

TYPE LOCALITY: Îles Kerguelen.

TYPE: TRH.

DISTRIBUTION: Îles Kerguelen (Foslie, 1905); Macquarie I.

MATERIAL EXAMINED:
1. Îles Kerguelen: Type specimen.
2. Macquarie I.: collections MA-55, 134, 238, 278; Herbarium specimens: MELU, R 1481, 1482, 1485–1490; Microscope slides R/M 0342, 0343, 0347.

DESCRIPTION: Crusts epilithic, pinkish-purple, coarse, to 1 cm thick, either smooth or with knobby excrescences, and composed of distinct tissues, with perithallium most extensive (Fig. 75a, c, d). Epithallium composed of 3–5 layers of isodiametric to barrel-shaped cells, with upper cells appearing to be periodically shed (Fig. 75b); perithallium with upper layer of elongate (to 26 μm long) meristematic cells and numerous underlying layers of oval to rectangular cells *c.* twice as long as broad (10–13 μm long × 5–6 μm diameter) arranged in simple vertical files; hypothallium monostromatic, conforming to contour of substratum, often difficult to see, and composed of rectangular cells *c.* 7–8 μm broad × 20 μm long giving rise to two vertical rows of smaller (perithallial) cells. Secondary pit connections absent; lateral fusions of two or more cells common throughout perithallium (Fig. 75e, f), rare in epithallium (Fig. 75b), not observed in hypothallium. Sporangial conceptacles uniporate, *c.* 300 μm inner diameter × 120 μm inner height, with mound of cells centrally positioned on conceptacle floor (Fig. 75g). Tetrasporangia zonately divided, oval, *c.* 55 μm broad × 80 μm long. Gametangial conceptacles uniporate, flush with or slightly sunken below crust surfaces; sexual reproductive structures not studied.

Plants are common around Macquarie I. and form extensive crusts on rocky surfaces usually in mid-littoral pools *c.* 1 m above the upper limit of the *Durvillaea* zone, but also in the subtidal to depths of at least 3–5 m. In the intertidal, *P. consociatum* often grows adjacent to crusts of

---

Fig. 75 *Pseudolithophyllum consociatum* (Foslie) Lemoine: MELU, Slide R/M 0347; TRH, type specimen. **a.** Thallus habit. **b.** Section through epithallium showing (rare) lateral cell fusion (arrow). (N.B. elongate cells subtending epithallium are meristematic.) **c, d.** Sections through tetrasporophytes showing extensive perithallium. **e, f.** Lateral cell fusions (arrows) in perithallial cells. (N.B. 'd' and 'f' are from the type material.) **g.** Tetrasporangial conceptacle. Scales: a = 5 mm; b = 20 μm; c, d = 200 μm; e, f = 10 μm; g = 50 μm.

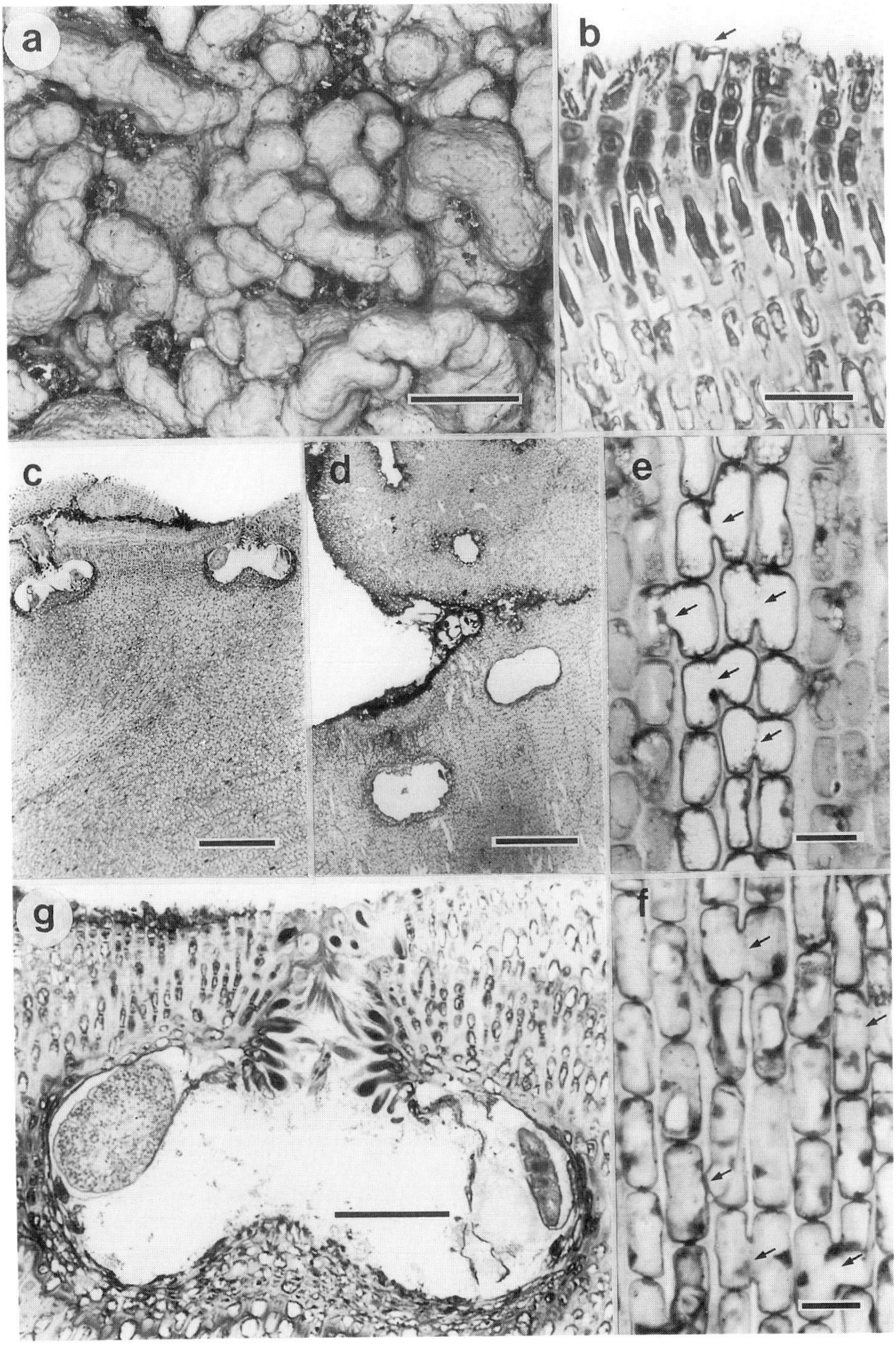
a
b
c
d
e
f
g

Hildenbrandia lecannellieri and is associated frequently with various filamentous (e.g., *Polysiphonia anisogona*) and foliaceous (e.g., *Iridaea cordata* and *Delesseria lancifolia*) red algae and the brown algae *Adenocystis utricularis* and *Sytosiphon lomentaria*. In the subtidal, it grows alongside the green alga *Codium subantarcticum* on unshaded rocky reefs.

NOTES: Lemoine (1913) described the habits of young plants of *P. consociatum* as discoid with smooth surfaces, older thalli with knobby excrescences, and illustrated (Fig. 14) perithallial construction with simple, vertical files of rectangular cells. She observed the perithallial cells of adjacent vertical files to be aligned oppositely to form horizontal cell layers. This feature supposedly distinguished *P. consociatum* from the type species, *P. discoideum*, which apparently lacks such layers and has cells of adjacent files offset but interconnected by numerous fusions.

Macquarie plants are constructed similarly to the type specimen of *P. consociatum** (see Fig. 75c–f), but do not show the horizontal layering of cells that Lemoine considered distinctive for this species. Some sections of the perithallium do have cells aligned horizontally, but this is not pronounced throughout the crust. Additionally, perithallial cells in *P. consociatum* are interconnected by numerous lateral fusions, as was initially described for *P. discoideum*. This feature, however, probably is common to all species in the genus.

Pseudolithophyllum consociatum can be identified by its usually knobby surface and extremely thick and compact perithallium consisting of rectangular cells. This contrasts with the closely related *P. discoideum*, which produces crusts with generally smooth surfaces and thinner, less compacted perithallial tissues comprising oval to barrel-shaped cells.

Pseudolithophyllum discoideum (Foslie) Lemoine, 1913: 46

Fig. 76

BASIONYM: *Lithophyllum* (?) *discoideum* Foslie, 1900: 73.

SYNONYMS: see Mendoza, 1976c.

TYPE LOCALITY: Fuegia.

TYPE: presumably in TRH (Koster, 1969), but specimens were not seen.

DISTRIBUTION: Argentina, South Georgia (Mendoza, 1976c); Falkland Is (Lemoine, 1915); Fuegia (Foslie, 1900); Macquarie I. (Zaneveld & Sanford, 1980).

MATERIAL EXAMINED: Macquarie I.: collections MA-227, 381; Herbarium specimens: MELU, Microscope slides R/M 0623–0626, 0629.

DESCRIPTION: Crusts epilithic or epizoic, purplish-pink, coarse, 75–500 μm thick, thicker near conceptacles, generally flat with smooth to slightly convoluted surfaces but with prominent ridges at confluence of separate thalli, and composed of distinct tissues, with perithallium most extensive. Epithallium comprising 3–6 layers of isodiametric to twice elongate cells, with upper cells appearing to be periodically worn away (Fig. 76a); perithallium with upper layer of pyriform to more elongate (to 15 μm long) meristematic cells and numerous underlying layers of isodiametric (7–10 μm diameter) to ovoid cells *c.* twice as long as broad (to 13 μm long) arranged in simple vertical files often loosely associated with adjacent filaments and appearing delicate; hypothallium monostromatic, conforming to contour of substratum, composed of enlarged, angular, isodiametric to 1.5 times elongate cells that give rise to two vertical rows of smaller (perithallial) cells (Fig. 76c). Secondary pit connections absent; lateral fusions of two or more cells common throughout perithallium (Fig. 76b), infrequent in hypothallium, unknown in epithallium. Gametophytes monoecious, with reproductive structures borne in uniporate conceptacles flush with or slightly bulging above crust surfaces. Spermatangia produced by elongate unbranched spermatangial mother cells located only on conceptacle floors (Fig. 76e, f). Mature

*Sections taken from fragments of the type were kindly loaned to the author by Dr. R. L. Moe.

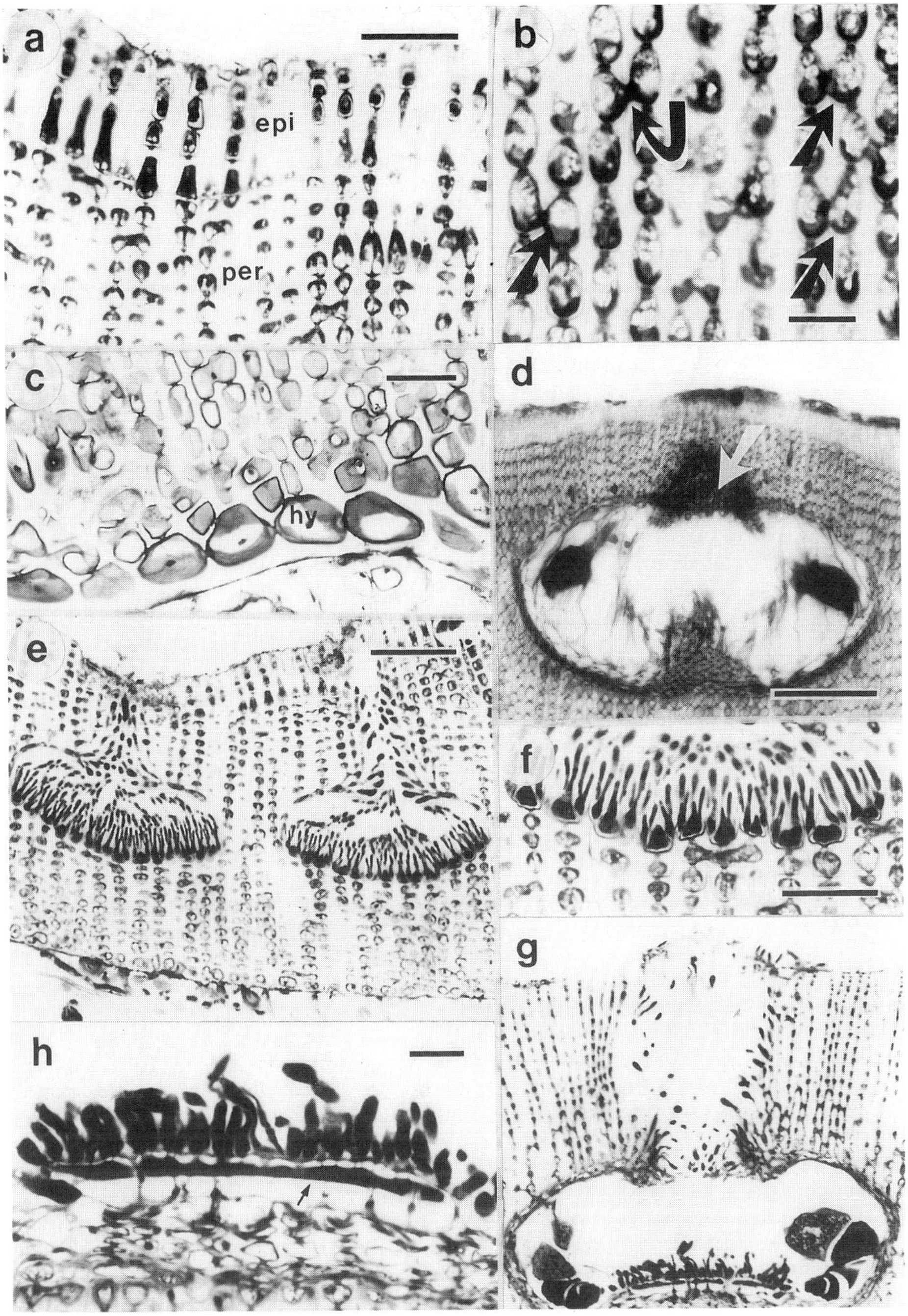

Fig. 76 *Pseudolithophyllum discoideum* (Foslie) Lemoine: MELU, Slides R/M 0623, 0624, 0626. **a.** Section through vegetative thallus showing chains of epithallial (epi) cells subtended by elongate meristematic cells and lower perithallial (per) cells. **b.** Cell fusions (arrows) in perithallium. **c.** Monostromatic hypothallium (hy). **d.** Eccentric section through tetrasporangial conceptacle showing edge of ostiole (arrow). **e.** Section showing two spermatangial conceptacles. **f.** Detail of spermatangial mother cells bearing spermatangia. **g.** Section through female conceptacle. **h.** Detail of fusion cell (arrow). Scales: a, c, f = 20 μm; b, h = 10 μm; d = 100 μm; e, g = 40 μm.

carposporophytes with central discoid fusion cell peripherally bearing carposporangia (Fig. 76g, h). Tetrasporangia zonately divided, oval, borne in uniporate conceptacles with mound of cells centrally positioned on conceptacle floor (Fig. 76d).

The above description is based on a few individuals that were growing on rocks and the shells of chitons and gastropods collected from the subtidal in 3–9 m depths. The distribution of this species around Macquarie I. is unknown.

NOTES: Mendoza (1976c) described and illustrated South American plants of *P. discoideum* with an extensive, compact perithallium to 1.5 mm thick composed of rounded and angular cells aligned in horizontal rows as Lemoine (1913) considered distinctive for *P. consociatum*. Macquarie plants have a similar habit and comparably structured hypothallium to the crusts illustrated by Mendoza (1976*c*, pl. 2, figs 4, 5), but lack the compact perithallium with uniform, horizontal cell layers as depicted by her (pl. 2, figs 1, 3).

Halymeniaceae Bory de Saint-Vincent, 1828: 158

Formerly known as either the Grateloupiaceae or the Cryptonemiaceae (see Guiry, 1978*a*), this family contains species that have life history patterns with an alternation of isomorphic somatic phases. Thalli are erect or prostrate, multiaxial, compressed or terete, and either broadly foliose or highly divided (Dixon, 1982). The medulla comprises a mesh of mostly periclinal, sometimes anticlinal, narrow, septate filaments. The cortex is compact or loose, often with a slippery or slimy surface texture, and consists of simple or branched anticlinal files of small rounded cells. Spermatangia usually are borne in superficial patches over thallus surfaces. Carpogonia terminate two-celled branches, whereas auxiliary cells are intercalary; both are produced in separate, specialized branch systems termed ampullae. Connecting filaments develop after fertilization and transfer zygote nuclei to auxiliary cells where gonimoblasts are initiated. Carposporophytes are embedded within female thalli, do not protrude above frond surfaces, and have most cells converted to carposporangia. Tetrasporangia are cruciately divided, embedded in cortical tissues, and sometimes localized in sori but usually scattered diffusely across blade surfaces (Kylin, 1956).

Chiang (1970) used structural differences between ampullar types to group genera and show stages of evolutionary development in the family, although Kraft (1977) has questioned the uniformity of these differences within genera and their importance in evolution of the group.

The Halymeniaceae is the second largest family in the order and includes some 18 genera (Kraft, 1981), of which nearly 85% occur in the Southern Hemisphere (Kylin, 1956). The family is poorly represented throughout the Subantarctic, however, and has been noted only recently for the first time from the Antarctic (Moe, pers. comm.). The southern tip of South America (Levring, 1960) and various islands south of New Zealand (Chapman & Parkinson, 1974) are the main localities where members of the Halymeniaceae, largely species of *Grateloupia* and *Pachymenia*, occur in the Subantarctic. The family is represented at Macquarie I. by one species of *Pachymenia*.

Pachymenia J. Agardh, 1876: 143

Lectotype: *P. carnosa* (J. Agardh) J. Agardh, 1876: 145.

Plants are erect, foliose, 15–90 cm tall, and have small discoid holdfasts that give rise to one or more thick, fleshy, cartilaginous to leathery fronds that are entire, laciniate, or irregularly lobed (Kylin, 1956). The medulla is composed of a dense mesh of slender, branched filaments; the cortex of anticlinal, branched files, with cells often having protuberances and sometimes producing rhizoidal outgrowths (Chiang, 1970). Gametophytes are dioecious, with spermatangia produced in sori; females are non-procarpic, with auxiliary cell and carpogonial branch ampullae each morphlogically distinct and embedded in cortical tissues. Mature carposporophytes are surrounded by compact, thin, filamentous pericaps. Tetrasporangia are cruciately divided and borne diffusely across blade surfaces.

Thalli of *Pachymenia, Aeodes, Epiphloea* and *Phyllymenia* are separated from others in the Halymeniaceae by their coarseness and foliose habits. Kylin (1956) distinguished each of the four genera by differences in blade shape, thickness, and the distribution of sporangia over frond surfaces. Chiang (1970), using degrees of branching in auxiliary cell ampullae, associated *Pachymenia* with *Aeodes*. although he considered the two distinct with respect to various vegetative and reproductive features.

The Southern Hemisphere genus *Pachymenia* has previously received up to 14 species, but only half of these are presently recognized, with *P. cornea* (Kützing) Chiang endemic to South Africa, *P. lusoria* (Greville) J. Agardh restricted to New Zealand, *P. carnosa* present at both South Africa and New Zealand, *P. apoda* J. Agardh, *P. prostrata* J. Agardh, and *P. stipitata* J. Agardh from Australia, and *P. cuticulosa* Howe from Peru (Chiang, 1970; Chapman & Parkinson, 1974). The genus has yet to be monographed and species are still distinguished on the seemingly variable characters of frond shape, thickness and degree of splitting. Using these features, the one *Pachymenia* species that occurs at Macquarie I. is closest to *P. stipitata* J. Agardh from South Australia. Because Macquarie plants also have similarities with several other species, however, it seems best to delay specific identification until *Pachymenia* species are circumscribed more clearly.

Pachymenia sp. cf. ***P. stipitata*** J. Agardh, 1890: 16

Fig. 77

DISTRIBUTION: Antarctic Peninsula (Moe, pers. comm.); Macquarie I.; southern coast of Australia from South Australia to Port Phillip Heads, Victoria (Lucus & Perrin, 1947).

MATERIAL EXAMINED: Macquarie I.: collections MA-46, 83, 137, 215; Herbarium specimens: MELU, R 0039, 0054, 0535, 0540, 3278, 3394–3397.
Pachymenia stipitata
Australia: Lectotype. 'Adelaide' (no collection data): LD, 22486 (female); Port Elliot, S. Australia, drift specimen (Kraft, 30 April 1971): MELU, K 3520.

DESCRIPTION: Plants to 90 cm long, reddish-brown (male) to dark purple (female—drying to nearly black) to purplish-brown (tetrasporophytes—lighter colour than female fronds and frequently mottled with light and dark patches), foliaceous, with terete to flattened stipes (to 1.5 cm long) grading through cuneate apophyses to ovate or orbicular, sometimes elongate and laciniate, rubbery fronds that become brittle to cartilaginous upon drying (Fig. 77a). Blades with velvety appearance when wet, 450–500 µm thick (Fig. 77b); cortex 75–110 µm thick, composed of anticlinal dichtomously branched filaments, with cells of adjacent filaments lacking secondary pit connections but fusing laterally (Fig. 77c, d); medulla 300–325 µm thick, composed of tightly interwoven mass of branched septate filaments. Spermatangia ovoid, borne on elongate spermatangial mother cells (Fig. 77c), produced superficially in indistinct sori over entire blade surfaces. Carpogonial branches two-celled, borne singly in carpogonial ampullae, with hypogynous cells additionally bearing one sterile lateral branch. (N.B. only one cell of this lateral branch is visible in Fig. 77e.) Carpogonial and auxiliary cell ampullae dissimilar, with the latter larger and more branched (up to three orders of alternately or pectinately branched filaments; Fig. 77f). Mature carposporophytes consisting mostly of carposporangia but with prominent sterile cell at base (not shown in Fig. 77g); pericarp composed of branched filaments laterally producing numerous, small, ovoid cells (Fig. 77h). Tetrasporangia cruciately to irregular-cruciately divided, embedded in cortex (Fig. 77i) and distributed diffusely over entire blade surfaces.

Plants occur on rocky surfaces from the lower intertidal, in pools and shallow (<2 m) surge channels, to depths of 7 m in protected coves. *Pachymenia* grows on vertical to horizontal surfaces that are shaded by overhead canopies of *Durvillaea* or exposed to direct light. The mat-forming green alga *Codium subantarcticum* often is associated with *P. 'stipitata'*, as is the foliose red alga *Iridaea cordata*. Although fronds of both red algae are rubbery and can be stretched somewhat when fresh, those of *Pachymenia* are distinctly thicker than *Iridaea* blades and also have a velvety appearance rather than the iridescent sheen characteristic of the latter.

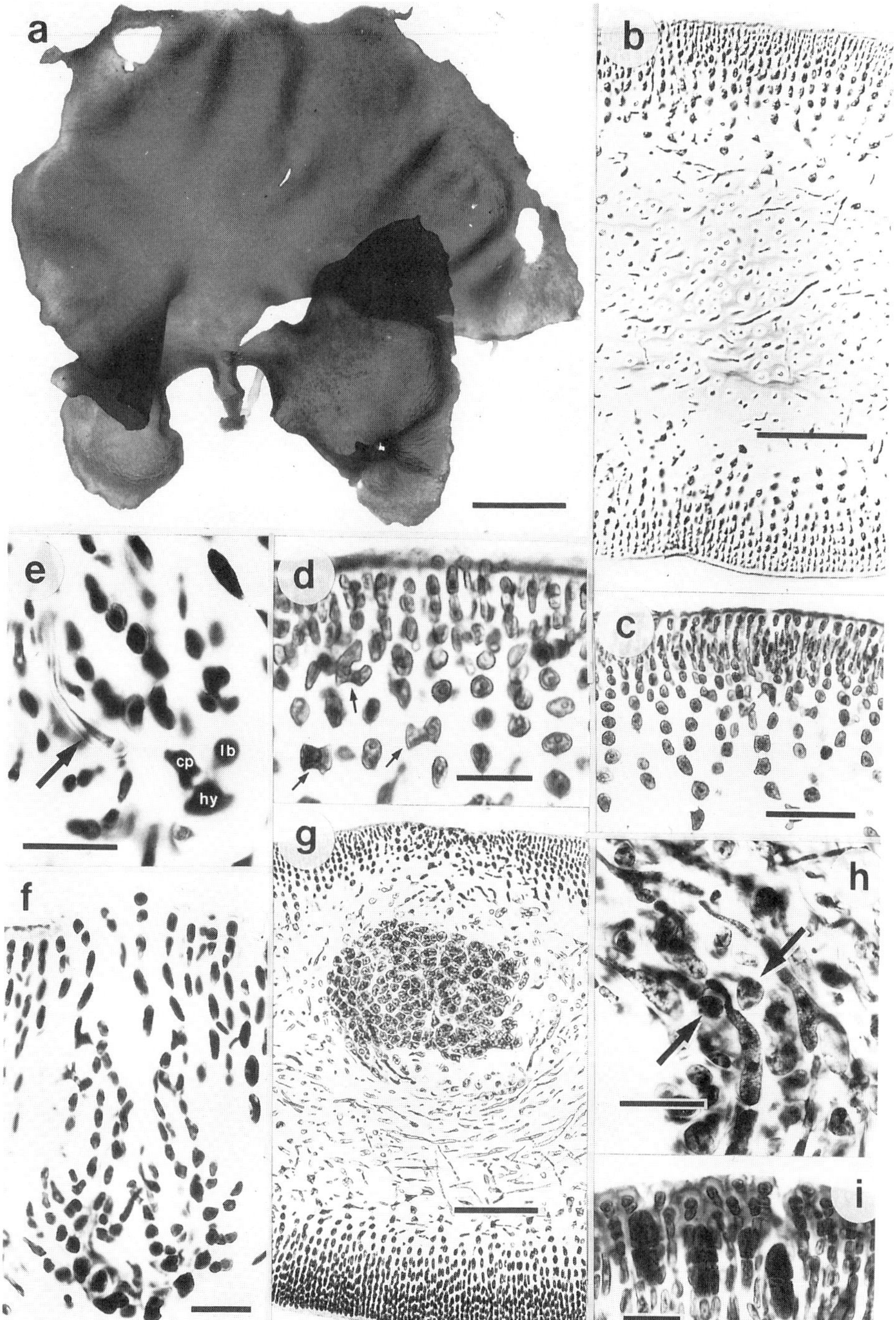
a
b
e
d
c
lb
cp
hy
g
f
h
i

NOTES: Most features considered by Chiang (1970) as taxonomically important, e.g., type of auxiliary cell ampullae, pericarp development, and production of lateral protuberances from cortical cells, have not been recorded previously for *P. stipitata*. Earlier descriptions of this South Australian alga (J. Agardh, 1890; Lucas & Perrin, 1947) have neglected detail on reproduction and referred largely to frond shape, colour and consistency. Based on present observations of blade morphology, mature carposporophyte structure, and pericarp development, *P. stipitata* from South Australia is similar to plants from Macquarie I. Plants from both areas have equally thick cortical tissues composed of branched, anticlinal files in which constituent cells frequently fuse laterally, although South Australian plants often have more dichotomies per cortical file and produce a thicker medulla (to 550 μm).

Hildenbrandiaceae Rosenvinge, 1917: 202

The Hildenbrandiaeae is characterized by asexual features of the cosmopolitan genus *Hildenbrandia*, although the family also is purported to include the New Zealand endemic genus *Apophloea* (Denizot, 1968; Chapman & Parkinson, 1974). The systematic position of the latter, however, remains uncertain (Hawkes, 1983). *Apophloea*, unlike the strictly encrusting *Hildenbrandia*, generally bears erect portions.

On the basis of ultrastructural studies, Pueschel & Cole (1982) recently proposed removal of the Hildenbrandiaceae from the Cryptonemiales to its own order, the Hildenbrandiales. Pit plugs that separate cells in *Hildenbrandia* have cap layers and are distinct from plugs of cryptonemialean algae that lack such layers. In the present survey, a more conservative scheme keeping the Hildenbrandiaceae in the Cryptonemiales is followed until life history patterns are known and this family can be related through Kylinian concepts (Kylin, 1956) to other red algal families.

Hildenbrandia represents the family throughout the Southern Ocean, with the circumpolar species *H. lecannellieri* Hariot common in the rocky intertidal at Macquarie I.

Hildenbrandia Nardo, 1834: 676

Type species: *H. prototypus* Nardo, 1834: 675.

Plants are encrusting, thick or thin, cartilaginous, tightly adherent to rocky surfaces, lacking rhizoids, and composed of compacted, cuboidal cells aligned in vertical rows. Tetrasporangia are zonately or irregularly divided, intermixed with paraphyses, and produced in conceptacles with apical pores. Sexual reproductive structures are unknown (Kylin, 1956).

DeCew & West (1977) recognized nine *Hildenbrandia* species from tropical to polar seas of both hemispheres, with most separated on differences in the sizes and shapes of conceptacles and tetrasporangia as well as sporangial division patterns (Denizot, 1968). Papenfuss (1964*a*) listed three species, *H. kerguelensis* (Askenasy) Chamberlain, *H. prototypus* and *H. lecannellieri* Hariot, from the Subantarctic. One, *H. lecannellieri*, also occurs in the Antarctic.

Hildenbrandia lecannellieri Hariot, 1887*a*: 72

Fig. 78

TYPE LOCALITY: Orange Bay, Fuegia.

TYPE: PC, unnumbered specimen.

Fig. 77 *Pachymenia sp.* cf. *P. stipitata* J. Agardh: MELU, R 0039, 0054, 0540, 3394. **a.** Habit of sterile thallus. **b.** Section through upper blade. **c.** Detail of male thallus showing superficial spermatangial layer. **d.** Cell fusions (arrows) between laterally adjacent cortical cells. **e.** Carpogonial branch ampulla showing carpogonium (cp), trichogyne (arrow), hypogynous cell (hy), and lateral branch cell (lb). **f.** Auxiliary cell ampulla. **g.** Oblique section through mature carposporophyte. (N.B. basal fusion cell missing from section.) **h.** Detail of involucral filaments showing branching and proliferation of small cells from filaments (arrows). **i.** Tetrasporangia. Scales: a = 2 cm; b, g = 100 μm; c = 50 μm; d–f, i = 20 μm; h = 30 μm.

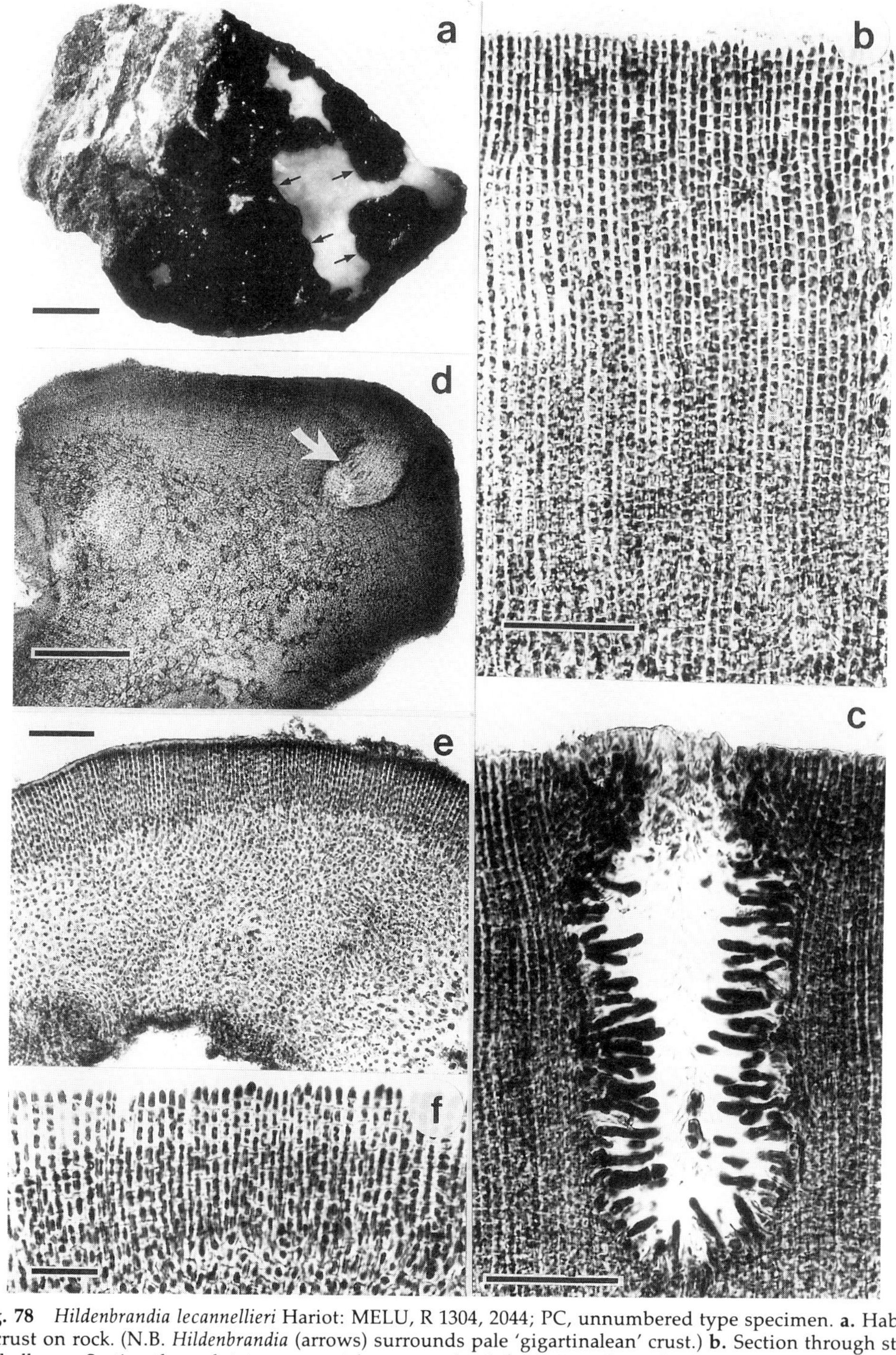

Fig. 78 *Hildenbrandia lecannellieri* Hariot: MELU, R 1304, 2044; PC, unnumbered type specimen. **a.** Habit of crust on rock. (N.B. *Hildenbrandia* (arrows) surrounds pale 'gigartinalean' crust.) **b.** Section through sterile thallus. **c.** Section through tetrasporangial conceptacle. **d–f.** Sections through rehydrated fragments of type material. **d.** Darkly pigmented structure of presumed fungal endophyte (arrow). **e.** Sterile thallus. **f.** Detail of cortical filaments. Scales: a = 5 mm; b, c, e = 50 μm; d = 200 μm; f = 25 μm.

DISTRIBUTION: Antarctic Peninsula (Gain, 1912; Skottsberg, 1953); Auckland Is (V. J. Chapman & Parkinson, 1974); Falkland Is (Kylin & Skottsberg, 1919; Cotton, 1915; Skottsberg, 1923); Fuegia (Hariot, 1887*a*; Askenasy, 1888); Macquarie I. (N.B. Papenfuss (1964*a*) additionally indicated distribution at Îles Kerguelen, but the original published record for this locality has not been located for the present report.)

MATERIAL EXAMINED:
1. Macquarie I.: collections MA-34, 45, 83, 111, 119, 134, 137, 227, 238, 278, 381; Herbarium specimens: MELU, R 0626, 1279, 1304, 1362, 1469, 1472, 1506, 1507, 1510–1515, 2044, 2050– 2053.
2. South America: Fragments from type specimen. Terre de Feu (1883): PC, unnumbered specimen.

DESCRIPTION: Crusts to 300 μm thick, but often thicker to at least 1 mm, reddish-brown to purple-black to black (Fig. 78a), rubbery to cartilaginous, with smooth slippery surfaces, and composed of closely packed vertical files of small (2–4 μm diameter) cuboidal cells (Fig. 78b, e, f). Sporangial conceptacles uniporate, oval (100 μm inner diameter × 140 μm inner height) to more elongate (70 μm inner diameter × 170 μm inner height), with tetrasporangia zonately divided, oblong (5 μm broad × 25 μm long), and intermixed with paraphyses (Fig. 78c). Gametophytes unknown.

Hildenbrandia grows most commonly in the mid- to lower-intertidal, but also occurs in pools and on shaded boulders in the upper intertidal and on rocky subtidal reefs to 9 m depths. Crusts adhere tightly and usually form extensive patches (>1 m diameter) that sometimes occur together with *Urospora penicilliformis* and *Pilayella littoralis* on shaded upper intertidal boulders, *Codium subantarcticum* along the walls of caldron-shaped pools in the mid-intertidal, *Acrosiphonia pacifica* and *Scytosiphon lomentaria* on unshaded rocky surfaces in the mid- to lower-intertidal, and various members of the Delesseriaceae in the subtidal.

NOTES: Hariot (1887*a*) described this species as encrusting and expansive, 5–8 mm thick, deep purple in colour, cartilaginous, with rough, uneven surfaces, and loosely adherent to rocks. His description of crust construction with closely packed vertical files of small, square cells could be applied to most species in the genus and hence is of little diagnostic value. Rather, he used morphological differences to distinguish his new species from *H. prototypus*, considered then to be the only member in the genus.

Plants from Macquarie I. are considerably thinner and less convoluted than those described by Hariot (1887*a*). Age might account for these differences, although Macquarie specimens of *Hildenbrandia* are reproductively mature. Skottsberg (1941*a*), when discussing Hariot's (1889) account of *H. prototypus* from Orange Bay, the type locality for *H. lecannellieri*, questioned whether or not this record might actually represent smooth young crusts of *H. lecannellieri*. Another possible explanation for the differences in crust thickness and shape, however, was suggested by examination of type material. Sections of this material revealed something resembling an endophytic ascomycete (Fig. 78d). Dark bodies, similar to those present in fungal-infected *Prasiola* (Fig. 21a, b), spot the surface of the otherwise yellowish-orange pieces of the type specimen. The surface convolutions of the crust could be caused by this endophyte, much as the saccate form of *Prasiola* apparently has resulted from infection by the ascomycete *Kohlmeyera*.

Kallymeniaceae W. R. Taylor, 1937: 248, 274

Members of this family have life history patterns with alternations of isomorphic somatic phases, whose thalli are multiaxial, generally free-living, and either broad and foliose or highly dissected into strap-like fronds. In the genus *Callocolax*, however, thalli are pulvinate and seemingly parasitic upon members of *Callophyllis*. Of the 18 constituent genera in the family (Wynne & Kraft, 1981), *Callophyllis* and *Kallymenia* contain the most species and also represent the two main types of thallus construction, from which the remaining genera appear as variations (Codomier, 1978). In the *Callophyllis*-type, blade interiors have a mixture of large, isodiametric cells and short-celled rhizoidal filaments, whereas those of the *Kallymenia*-type are filled with slender, septate filaments and occasional, specialized stellate cells (Kylin, 1956). Gametophytes are dioecious, with males bearing spermatangial sori and females producing either procarps or separate carpogonial and auxiliary cell branch systems (Dixon, 1982). Carpogonial branches are (2-)3-celled and produced singly (monocarpogonial) or in clusters (polycarpogonial) on supporting cells (Norris, 1957). Complex fusion cells are formed by fusion of the carpogonium with the supporting and adjacent cells. Zygote nuclei are transferred to auxiliary cells following fertilization. In the case of non-procarpic individuals, connecting filaments extend between carpogonia and auxiliary cells. Gonimoblasts develop inwardly, with mature carposporophytes embedded in gametophyte tissues, often bulging somewhat above frond surfaces, having one or more ostioles, and containing clusters of carposporangia separated by small amounts of sterile tissue (Kylin, 1956). Tetrasporangia are cruciately divided and scattered diffusely throughout the cortex (Dixon, 1982).

Norris (1957) used vegetative construction and features of the female reproductive apparatus to discuss evolution within the family. He hypothesized an evolutionary trend with primitive plants having a *Kallymenia*-type construction, separate carpogonial and auxiliary cell branch systems (i.e., non-procarpic), polycarpogonial condition, and a female reproductive system composed of more or less spherical cells.

The Kallymeniaceae is represented in tropical to polar seas of both hemispheres and has been monographed in three Southern Hemisphere localities (Australia—Womersley & Norris, 1971; New Zealand—V. J. Chapman & Parkinson, 1974; South Africa—Norris, 1964). Papenfuss (1964*a*) listed five genera in the family, *Callocolax*, *Callophyllis*, *Euthora*, *Kallymenia*, and *Nereoginko*, from the Antarctic and Subantarctic Zones, but Hooper & South (1974) considered *Euthora* to be congeneric with *Callophyllis*. This genus comprises two thirds of the 15 kallymeniaceous species listed from high-latitude Southern Ocean waters and also occurs with the parasite *Callocolax* at Macquarie I.

Callocolax Schmitz *ex* Batters, 1895: 318

Type species: *C. neglectus* Schmitz *ex* Batters, 1895: 316, 318.

Plants form wart-like cushions reputed to be parasitic upon *Callophyllis*. Because the anatomy of both host and 'parasite' is similar, however, it is difficult to distinguish connections between the two tissues where transfer of nutrients might confirm a parasitic relationship (Kylin, 1930). Whether *Callocolax* even is a distinct genus or simply an outgrowth from the host has been questioned by Chemin (1937*b*), who suggested that a bacterial infection could be responsible for the gall-like structures. The evidence supporting *Callocolax* as distinct from *Callophyllis*, however, is based on the presence of tetrasporic cushions on gametophytes and gametophytic cushions on tetrasporophytes.

Male gametophytes are unknown, but female thalli and tetrasporophytes have been reported from England (Irvine, 1983). Plants are procarpic, with each supporting cell bearing one three-celled carpogonial branch and one, rarely two, one-celled (subsidiary) sterile branch (Norris, 1957). Gonimoblasts develop outwardly, with mature cystocarps lacking filamentous pericarps and ostioles and remaining embedded in the gametophyte thallus but sometimes protruding slightly. Sporophytes produce cruciately divided tetrasporangia scattered throughout the cortex.

The genus has two species, with *C. fungiformis* Kylin (1925) larger than the type and known only from the west coast of North America.

Callocolax neglectus Schmitz *ex* Batters, 1895: 316, 318

Fig. 79e–g

Type locality: Dorset (Weymouth), England.

Type: BM. Isotypes: Algae Britannicae Rariores Exsiccatae No. 154. (N.B. lectotypified by Irvine (1983); specimens were not seen for the present survey.)

Southern Ocean distribution: Auckland Is (Laing, 1909); Îles Crozet (Levring, 1944); Macquarie I.; Tristan da Cunha (Baardseth, 1941).

MATERIAL EXAMINED: Macquarie I.: collection MA-46; Herbarium specimen: MELU, R 1164; Microscope slide R/M 0209.

DESCRIPTION: Plants pulvinate, rounded to irregularly lobed, 2–3 mm across, whitish to pale pink, and growing on *Callophyllis variegata* (Fig. 79e). Cortex 1–2 layers of small rounded cells; medulla comprising interwoven, interconnected (with secondary pit connections) rhizoidal filaments composed of simple or branched, elongate cells that penetrate host cortex and resemble medullary rhizoids of host (Fig. 79f, g). Reproductive structures not seen.

The above description is based on five plants that were growing on the lower parts of several individuals of *Callophyllis variegata* collected from ≤5 m depth in a generally calm cove. No information is available on the distribution or abundance of this 'parasite' around Macquarie I.

NOTES: Because reproductive structures have not been observed, it is unclear whether the cushions on *Callophyllis* at Macquarie I. are actually *Callocolax* or simply galls produced by bacterial infection of host tissues. On the basis of similarities with Batters (1895, figs 25, 28, 29) illustrations of *Callocalax* anatomy and morphology, however, Macquarie 'parasites' are tentatively placed in this genus.

Callophyllis Kützing 1843: 400

Lectotype: *C. variegata* (Bory de Saint-Vincent) Kützing, 1843: 401.

Plants are erect, flattened, membranous and soft to cartilaginous, foliose, entire to highly laciniate or irregularly branched in one plane, and internally constructed of pseudoparenchyma with a mixture of large, rounded cells and generally small-celled rhizoidal filaments. The cortex is composed of one to five layers of photosynthetic cells that decrease in size from the medulla outwards (Irvine, 1983). Female gametophytes are procarpic, with supporting cells bearing one or more three-celled carpogonial branches, and several (1-)2-celled sterile (subsidiary) branches (Womersley & Norris, 1971). Generally, supporting cells, basal cells of carpogonial branches, and subsidiary cells are variously lobed (Kylin, 1928). Following fertilization, supporting cells function as auxiliary cells and fuse with basal cells of subsidiary and carpogonial branches. Gonimoblasts develop directly from fusion cells, with most tissue eventually becoming carposporangia. Mature carposporophytes are embedded in gametophyte fronds, sometimes bulging slightly above blade surfaces, and are either scattered across the thallus or localized on marginal proliferations.

Callophyllis, with some 40 species, is the largest genus in the family and is present in temperate to colder waters of both hemispheres (Kraft, 1981). A poor understanding of the extensive variation in thallus habits has led to over-estimations of species numbers in areas such as the Pacific Coast of North America, where 22 species were reported for only 12 distinct taxa (Abbott & Norris, 1966). Papenfuss (1964*a*) listed 10 species from Subantarctic waters, with the type species, *C. variegata*, said to also occur along the coast of the Antarctic Peninsula. The genus has yet to be monographed from southern seas, however, with species numbers in this vicinity most likely yet to be reduced. The type species singly represents *Callophyllis* at Macquarie I.

Callophyllis variegata (Bory de Saint-Vincent) Kützing, 1843: 401

Fig. 79a–d

BASIONYM: *Halymenia variegata* Bory de Saint-Vincent, 1828: 179.

SYNONYMS: see Pujals, 1963.
Kallymenia multifida Reinsch 1888*b*: 146.
Callophyllis multifida (Reinsch) Kylin *in* Kylin & Skottsberg, 1919: 13.
Callophyllis corollata Baardseth, 1941: 61.

TYPE LOCALITY: Valparaiso, Chile.

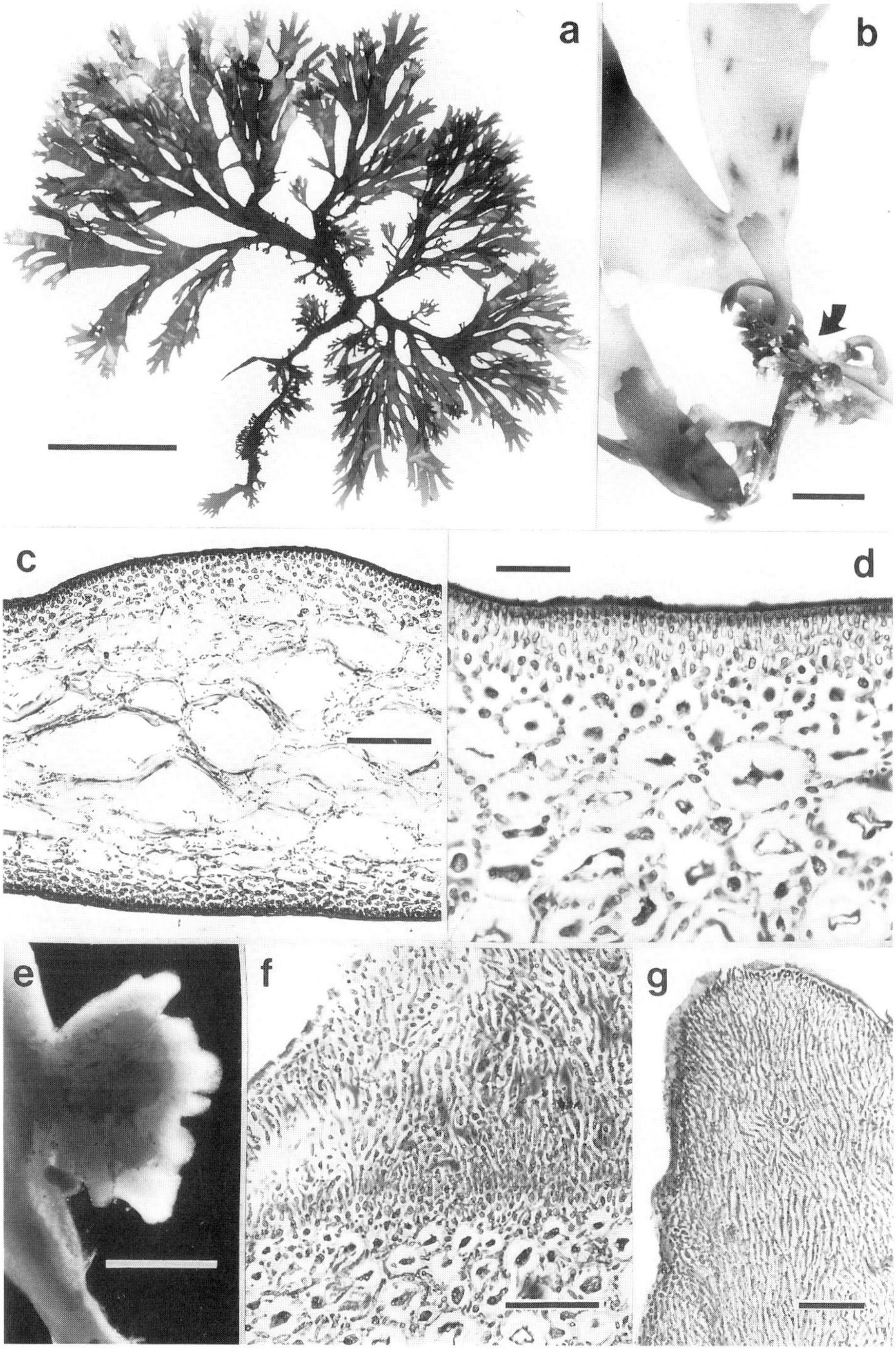

Fig. 79a–d *Callophyllis variegata* (Bory de Saint-Vincent) Kützing: MELU, R 1164, 1180. **e–g** *Callocolax neglectus* Schmitz: MELU, R 1164. **a.** Plant habit. **b.** Thallus base with haptera-like branches (arrow). **c.** T.S. mid-thallus. **d.** Detail of cortex. **e.** Plant habit. **f.** Section through 'parasite' (above) and host (below). **g.** Section through lobe of 'parasite'. Scales: a = 4 cm; b = 3 mm; c, f, g = 100 μm; d = 50 μm; e = 1 mm.

TYPE: Herb. Montagne, PC. (N.B. type in PC illustrated by V. J. Chapman & Parkinson (1974, pl. 77a); isotype (UC 435656; female) specimen, collected by Binder, illustrated by Norris (1957, pl. 35).)

DISTRIBUTION: Auckland Is (Laing, 1909); Campbell I. (Levring, 1945); Falkland Is (Cotton, 1915); Fuegia (Askenasy, 1888; Hariot, 1889); Îles Crozet (Levring, 1944); Îles Kerguelen (Dickie, 1879; Askenasy, 1888); Macquarie I.; New Zealand (V. J. Chapman & Parkinson, 1974); South Africa (Norris, 1964); South Georgia (Reinsch, 1890); South Orkney Is (Gepp & Gepp, 1905*c*); Tristan da Cunha (Baardseth, 1941). (N.B. Skottsberg (1906) reported this species from the Antarctic Peninsula, but subsequent floristic studies undertaken in this vicinity (e.g., Delépine, Lamb & Zimmermann, 1966; Lamb & Zimmermann, 1977) have not confirmed the presence of this alga.)

MATERIAL EXAMINED: Macquarie I.: collections MA-46, 83, 309, 381; Herbarium specimens: MELU, R 0189–0193, 1164, 1180, 1192–1194.
Callophyllis corollata Baardseth
Tristan da Cunha: Holotype (Baardseth): O, Baardseth no. 209 (female). (N.B. holotype has been illustrated by Baardseth (1941, fig. 29a).)
Kallymenia multifida Reinsch
South Georgia: Lectotype. 'Nordstrand der Landzünge' (Will, 3 July 1883): M, 67.83/36 (= Will, no. 81); Syntypes: M, 67.83/34, 67.83/35.
Callophyllis lambertii (Turner) J. Agardh, 1851: 300
Australia: Warrnambool, Victoria; drift behind the breakwater (Kraft & Kraft, 31 July 1977): MELU, K 6731 (two sheets).

DESCRIPTION: Plants erect, arising from hapteroid bases to 25 cm tall (Fig. 79b), yellowish-red to reddish-brown to blackish-purple, cartilaginous, highly laciniate, with branching in one plane, alternate to dichotomous to irregular, and often proliferous along lower margins (Fig. 79a). Blades to 425 μm thick; medulla with large, colourless, rounded cells encircled by small-celled rhizoidal filaments; cortex (1-)2-layered, composed of small, isodiametric to slightly elongate, pigmented cells (Fig. 79c, d). Reproductive structures not seen in Macquarie specimens.

Plants form clumps on rocks in the shallow subtidal to 6 m depths. Thalli are often partly covered by colonial bryozoans and frequented by the gastropod *Cantharidus coruscans*, which scrapes frond surfaces and probably accounts for the general absence of epiphytes. Blades sometimes have mottled appearances, with dark and light patches of unknown cause, and surfaces with scouring marks from snail radulae.

NOTES: *Callophyllis* species have usually been distinguished largely on differences in thallus habit. As advocated by Abbott & Norris (1966), however, species are better separated using a suite of characters such as degree and pattern of branching, shape of the ultimate branches, outline of margins, thallus colour, the presence of one or more carpogonial branches per suppporting cell, and the location of tetrasporangia on frond surfaces. Because reproductive structures were not developed in the specimens studied during the present survey, taxonomic judgements consequently have been founded upon comparative morphology of vegetative features. In studying some 80 specimens of *C. variegata* from Macquarie I., I observed extensive variation between individuals in the degree and pattern of branching. The range included forms comparable to the types of *C. variegata* illustrated by Norris (1957) and V. J. Chapman & Parkinson (1974) as well as the types of *Kallymenia multifida* Reinsch (1890) and *C. corollata* Baardseth (1941).

Kylin (*in* Kylin & Skottsberg, 1919), when moving *Kallymenia multifida* to *Callophyllis*, noted that Reinsch had accurately described the anatomy of this species with typical *Callophyllis* construction, but that he had incorrectly illustrated a kallymeniaceous-like blade anatomy (1890, pl. ii, fig. 5). On the basis of habit, Kylin also considered *C. multifida* to be closely related to, if not synonymous with the Australian entity *C. lambertii* (Turner) J. Agardh, but different from *C. variegata*. *Callophyllis lambertii* presently is considered to be distinct, however, because its thalli attach by discoid holdfasts and generally are coarser and more regularly branched.

When studying plants of *C. variegata* from South Africa, Norris (1964) observed considerable variation in thallus shape and external details of the cystocarp. Based on these observations, he considered *C. corollata* Baardseth (1941) from Tristan da Cunha likely to be a form of *C. variegata*. The present studies on thallus morphology further support Norris' observations and have led to synonymizing *C. corollata* with *C. variegata*.

Pugetia Kylin, 1925: 30

Type species: *P. fragilissima* Kylin, 1925: 30, 31.

Plants arise from discoid holdfasts and have short stipes subtending membranous fronds that are either orbicular and broadly foliose with irregular margins, or highly laciniate. Blades are composed of a layer of pigmented cells comprising one cortex and a medulla with pigmented filaments interspersed with two rows of large colourless cells. Female gametophytes produce separate auxiliary cell and carpogonial branch systems that join with connecting filmants.

Kylin (1925) considered *Pugetia* and *Callophyllis* to be closely related genera, each having a distinct anatomy. In the former, groups of assimilatory filaments are composed of nearly isodiametric cells running between large medullary cells, whereas in the latter, comparable filaments are composed of slightly more elongate cells that purportedly do not function in assimilation. Irvine (1983), however, reported plastids in the supposedly non-photosynthetic cells of the rhizoidal filaments of *Callophyllis*. The two genera can be distinguished more readily by differences in their female reproductive apparatuses, with species of *Pugetia* being non-procarpic and species of *Callophyllis* being procarpic (Norris, 1957).

Ten species of *Pugetia* are documented from temperate to colder waters of both hemispheres (Baardseth, 1941; Norris, 1957, 1964). *Pugetia kylinii* Baardseth (1941), supposedly endemic to Tristan da Cunha, and *P. delicatissima* Norris, known previously from New Zealand (Adams *et al.*, 1974) but also present at Macquarie I., represent the genus in the Subantarctic. Moe (pers. comm.) has recently identified two additional, as yet unpublished, species from the Antarctic Peninsula.

Pugetia delicatissima Norris, 1957: 273

Figs 80, 81

TYPE LOCALITY: Gore Bay, New Zealand.

TYPE: UC, 513609 (= Laing, no. 2443). (N.B. type illustrated by Norris (1957, pl. 31).)

DISTRIBUTION: Macquarie I.; New Zealand (Norris, 1957).

MATERIAL EXAMINED:

1. Macquarie I.: collections MA-119, 271, 302, 309; Herbarium specimens: MELU, R 1933, 1979, 2046, 2120, 2123, 2147, 2361.
2. Stewart I.: Harold Bay, in ≤5 m depths (Ricker & Torregrosa, 13 January 1979): Herb. Ricker, R 0402 (tetrasporophyte).

DESCRIPTION: Plants pale red to reddish-brown, arising from basal discs or several prostrate branches, to 27 cm tall, with terete stipes 1–5 mm long grading through flattened (sometimes with a slight midrib) cuneate apophyses to delicate membranous fronds to 130 µm thick and either broad and foliaceous or laciniate and pinnately to irregularly branched (Fig. 80a, b). Cortex monostromatic, with cells pigmented, 4–6 µm diameter; medulla to 100 µm thick, comprising 2–3 layers of large, isodiametric, colourless cells encircled by small-celled rhizoidal filaments

Fig. 80 *Pugetia delicatissima* Norris: MELU, R 1933, 2147, 2361. **a.** Habit of female gametophyte. **b.** Habit of tetrasporophyte. **c.** T.S. sterile thallus. **d, e.** Sections through female gametophyte. **d.** Supporting cell (sc) bearing two subsidiary cells and 3-celled carpogonial branch. **e.** Auxiliary cell (aux) with subsidiary cells. **f.** Squash preparation showing supporting cell (SC), three subsidiary cells (sub) and 3-celled carpogonial branch with trichogyne (arrow). First (1) and second (2) cells of branch out of focal plane. **g.** Section through mature cystocarp. **h.** Tetrasporangia (arrows). Scales: a = 4 cm; b = 5 cm; c, h = 50 µm; d, g = 30 µm; e, f = 20 µm.

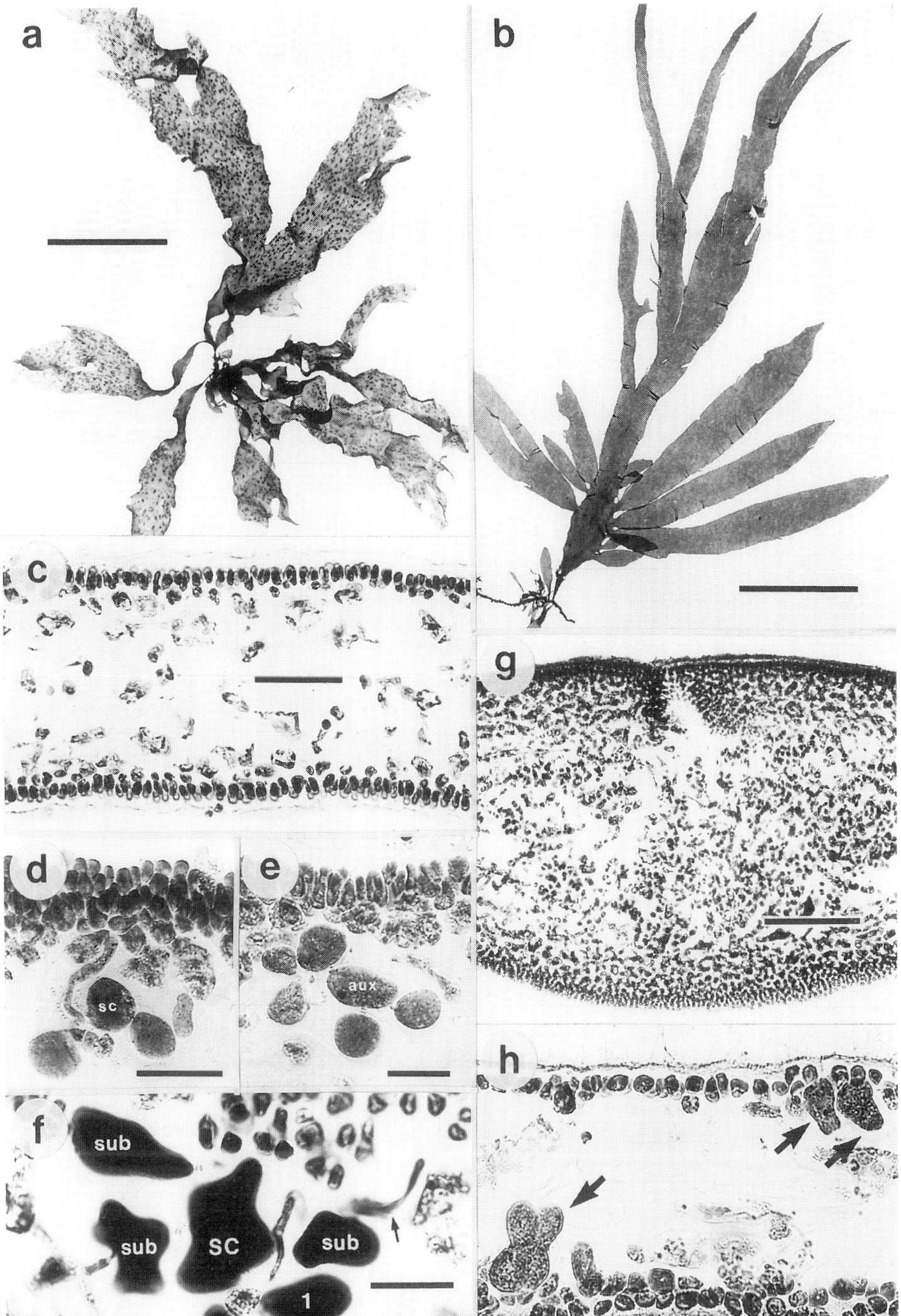
a
b
c
d
sc
e
aux
f
sub
sub
SC
sub
1
2
g
h

(Fig. 80c). Female gametophytes non-procarpic, with carpogonial branches and auxiliary cells borne between cortex and medulla, scattered across blades. Carpogonial branches three-celled, produced singly on supporting cells that also bear two or three (1-)2-celled subsidiary branches (Figs 80d, f, 81a); auxiliary cells bearing three (1-)2-celled subsidiary branches (Fig. 80e). Fusion cells (presumably post-fertilization products created by union of a carpogonial branch basal cell, supporting cell and associated subsidiary cells) giving rise to one or more non-septate connecting filaments (Fig. 81b). Cystocarps with one ostiole, embedded in fronds, visible macroscopically as dark splotches on fronds (Fig. 80a, g). Male gametophytes not seen. Tetrasporangia cruciately divided, oblong, 20–28 μm broad × 35–40 μm long, immersed between cortical and medullary tissues, and scattered diffusely across blades (Fig. 80h).

Some 20 plants were collected growing epiphytically on a few individuals of *Plocamium hookeri* and *Phyllophora appendiculata* in mid-intertidal pools and in the subtidal to depths of 14 m on the shaded floors of *Macrocystis* forests. *Pugetia delicatissima* appears to be rare at Macquarie I. but occurs from the northern to southern ends of the island.

NOTES: Macquarie specimens of *P. delicatissima* differ somewhat in habit from the New Zealand plants described by Norris (1957). The former tend to be larger, more laciniate, and sometimes have longer stipes with discoid attachments instead of a holdfast composed of prostrate filaments. Nevertheless, both groups of plants are considered to belong to the same species, because of reproductive similarities.

Gigartinales

The Gigartinales comprises more (28 total) families than any other order of red algae (Wynne & Kraft, 1981) and, together with the closely related Cryptonemiales, includes the most reproductively complex members of the Rhodophyta. Kylin (1932, 1956) separated constituent members of the two orders based on the positioning of auxiliary cells either in undifferentiated vegetative branches (Gigartinales) or in specialized accessory branches (Cryptonemiales), a feature which is difficult to see, let alone interpret.

Life history patterns include alternations of both isomorphic and heteromorphic somatic phases, with some species alternating between encrusting disc-like thalli and erect, uniaxial or

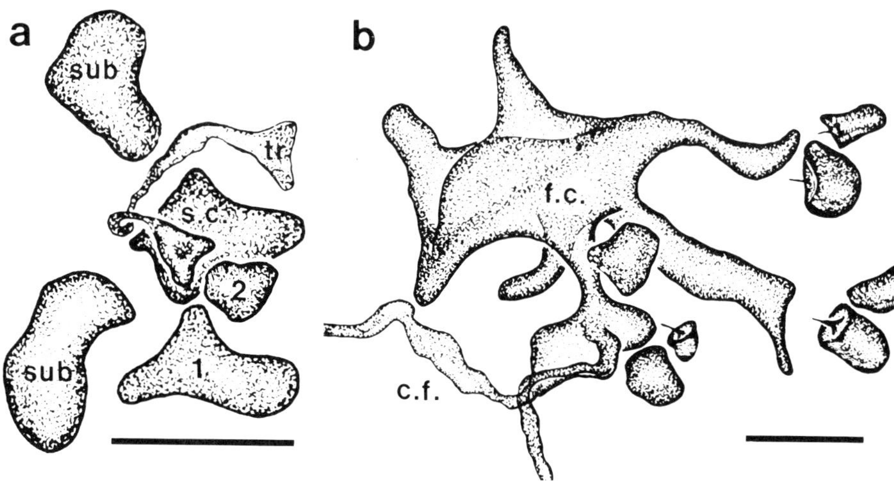

Fig. 81 *Pugetia delicatissima* Norris: MELU, R 2147, 2361. **a, b.** Camera lucida drawings. **a.** Carpogonial branch system with supporting cell (s.c.), 3-celled carpogonial branch terminated by a trichogyne (tr), and two subsidiary cells (sub). Carpogonial branch with elongate first cell (1), nearly spherical second cell (2) and pyramidal carpogonium. **b.** Fusion cell (f.c.) with connecting filament (c.f.). Scales: a, b = 20 μm.

multiaxial, pseudoparenchymatous, terete to foliose thalli composed of loosely to tightly compacted filaments (West & Hommersand, 1981).

Twelve families have been listed from southern high-latitude waters (Papenfuss, 1964*a*), of which four, the Gigartinaceae, Phyllophoraceae, Plocamiaceae and Cystocloniaceae (= Rhodophyllidaceae), totalling seven species, are represented at Macquarie I.

Gigartinaceae Bory de Saint-Vincent, 1828: 149

Kylin (1932, 1956) characterized members of this family as having multiaxial construction with loosely organized filamentous medullary tissues surrounded by small cortical cells, three-celled carpogonial branches borne on supporting cells that function as auxiliary cells following fertilization, inwardly developing gonimoblasts composed of a network of sterile filaments around masses of carposporangia, and cruciately divided tetrasporangia. Life history patterns within the family include alternations of both isomorphic or heteromorphic somatic phases. In the latter instance, the gametophyte is larger than the sporophyte (Dixon, 1982).

The Gigartinaceae is variously regarded as having from two to five genera (Kraft, 1981). Following the broadest concept, three of the genera, *Gigartina*, *Iridaea* and *Rhodoglossum*, occur in the Subantarctic. Southern Ocean records exist for the other two genera, *Chondrus* (Hariot, 1889) and *Petrocelis* (Holmes, 1912), but these probably reflect misidentifications. Only *Iridaea* occurs at Macquarie I.

Iridaea Bory de Saint-Vincent, 1826*c*: 15 (***nom. cons.***)

Type species: *I. cordata* (Turner) Bory de Saint-Vincent, 1826*c*: 16.

Kylin (1956) separated this genus from others in the family because its procarps are produced in unspecialized vegetative tissue (i.e., not in special fruiting branchlets), its gonimoblasts are surrounded by filamentous pericarps, and the tetrasporangia develop as accessory branches from medullary cells. The last criterion contrasts with development of sporangia from inner cortical cells, as supposedly is typical for the closely related genus *Rhodoglossum*. This feature usually is difficult to interpret and, consequently, is impractical for distinguishing *Iridaea* from *Rhodoglossum*, but both genera can still be recognized by other differences between their sporophytes. The cortex above mature sporangial sori in *Iridaea* is level with surrounding cortical tissues and the cells are similar to those from sterile tissues, whereas the cortex above soral patches in *Rhodoglossum* is sunken relative to the surrounding tissues and the cells are obviously meristematic and smaller than comparable cells in sterile parts (Leister, pers. comm.; pers. obs.).

Members of this genus have life history patterns with alternations of isomorphic somatic phases. Plants arise from discoid holdfasts and produce stipitate foliose fronds ranging from narrow to broad, and often are rubbery and iridescent when wet. Blade anatomy includes a filamentous medulla with narrow-diameter branched filaments and a cortex composed of small cells organized in anticlinal files. Cystocarps are immersed in the thallus, whereas spermatangia are produced superficially in sori.

Kylin (1956) estimated the existence of approximately 25 species of *Iridaea*, although present count would probably reveal considerably fewer species were the genus to be monographed using modern taxonomic concepts. Eight species are listed from southern high-latitude waters (Papenfuss, 1964*a*), and of these, *I. cordata* is the only one to extend into the Antarctic Zone. This alga, along with a previously undescribed species, represents the genus at Macquarie I.

Iridaea cordata (Turner) Bory de Saint-Vincent, 1826*c*: 16

Fig. 82

BASIONYM: *Fucus cordatus* Turner, 1809: pl. 116.

SYNONYMS: *Iridaea mawsonii* Lucus, 1919: 11 (Leister, pers. comm.).

Specimens of the following taxa have been studied by Leister (1977) and are believed to be

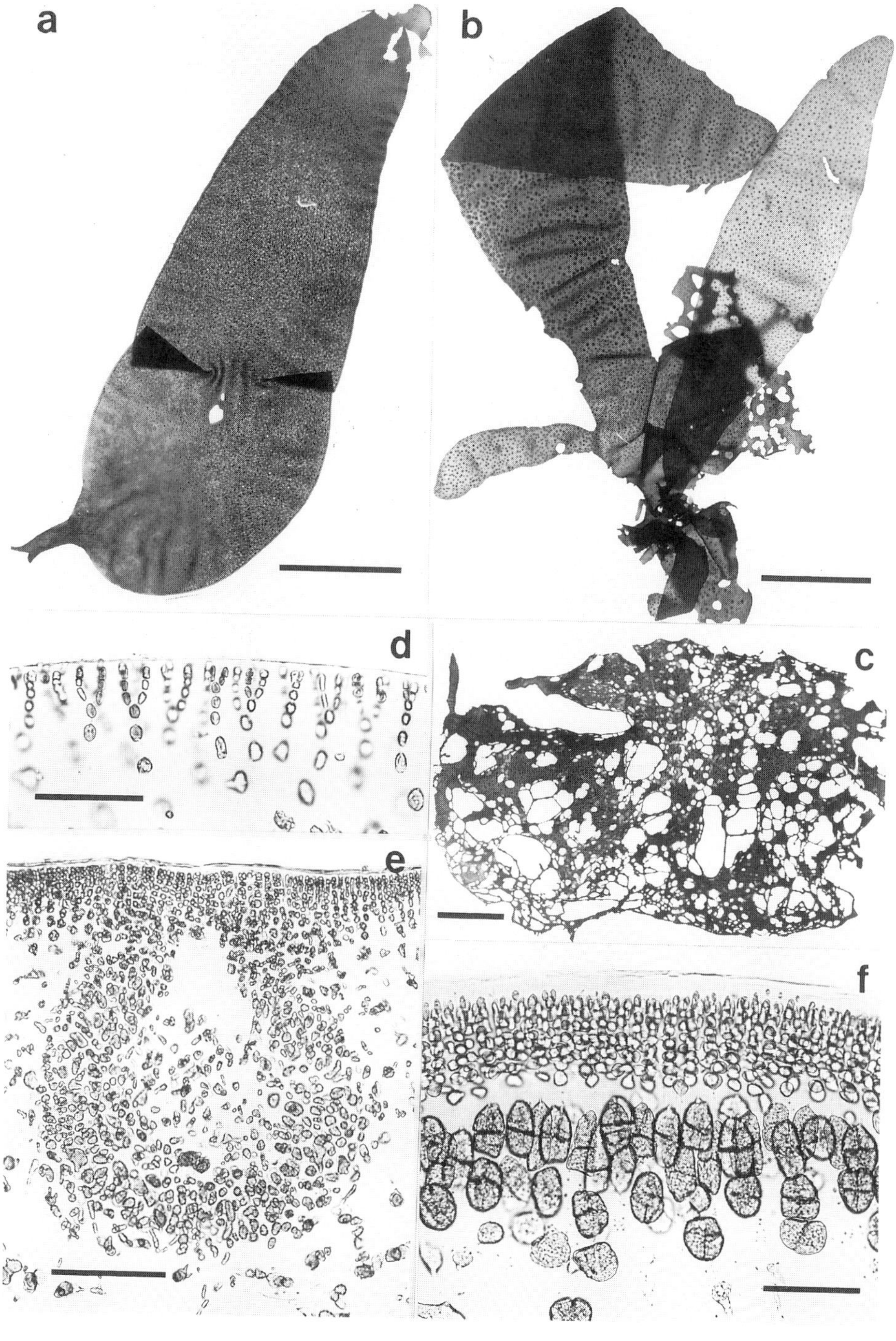

Fig. 82 *Iridaea cordata* (Turner) Bory de Saint-Vincent: MELU, R 0045, 3380–3382. **a.** Habit of sporophyte. **b.** Habit of female gametophyte. **c.** Habit of old female gametophyte. **d.** Detail of cortex. **e.** Section through young cystocarp with developing pericarp (Faserhülle). **f.** Section through tetrasporangial sorus. Scales: a–c = 5 cm; d, f = 50 μm; e = 100 μm.

conspecific with *Iridaea cordata:*
Iridaea micans Bory de Saint-Vincent, 1826*c*: 16.
Iridaea macrodonta Skottsberg, 1923: 8.
Iridophycus micans (Bory de Saint-Vincent) Setchell & Gardner, 1936: 470.
Rhodoglossum macrodontum (Skottsberg) Setchell & Gardner, 1936: 472.
Iridophycus caespitipes Setchell and Gardner, 1936: 470.
Iridaea caespitipes (Setchell & Gardner) Skottsberg 1941*a*: 82.
Iridaea (?) *proliferans* Skottsberg 1953: 557.
Rhodoglossum schotteri Delépine *in* Arnaud & Delépine 1964: 81.
Gigartina caespitipes (Setchell & Gardner) Kim, 1976: 39.
Gigartina cordata (Turner) Kim, 1976: 40.

Type locality: Argentina, Isla de los Estados.

Type: E (N.B. female specimen collected by Menzies, 26 January 1787 to 12 February 1787, chosen by Leister (1977) for lectotype.)

Distribution: Argentina, Antarctic Peninsula, Falkland Is, Îles Crozet and Kerguelen, South Georgia, South Shetland Is (Leister, 1977); Heard and Macquarie Is, Vestfold Hills, Antarctica.

Material examined:
1. Heard I.: Atlas Cove (Chittleborough, 30 August 1949): MELU, R 3718, 3719, 3721 (female), 3720 (juvenile); north of Wharf Point (transect no. 1), Atlas Cove, eulittoral rocks (Smith, 6 February 1983): MELU, R 3734 (juveniles).
2. Macquarie I.: collections MA-10, 14, 39, 46, 83, 111, 129, 137, 140, 215, 222, 227, 236, 244, 271, 295, 309, 381; Herbarium Specimens: MELU, R 0045, 0121, 0131, 1249–1251, 1702, 3380–3382.
3. Vestfold Hills, Antarctica: Confession Rock, on rocks in 5 m depths (Ricker & Tucker, 22 December 1981): MELU, R 2842 (female); Barratt I., on rocks in 7 m depths (Ricker & Beinssen, 3 February 1982): MELU, R 3252 (tetrasporic); Tidegauge Point, 1 m depth on boulders (Ricker & Butler, 14 December 1982): MELU, R 2736 (female).

Description: Plants to 35 cm tall, greenish-purple to brownish-orange to orange-red to burgundy in colour, cartilaginous with a rubbery elasticity, drying to either a satiny or dull sheen. Fronds erect, lanceolate to ovate to broad and foliose (Fig. 82a, b), becoming perforate with age (Fig. 82c), with cordate bases and flattened stipes 1–4 cm long arising from discoid attachments. Blade transections to 500 μm thick; medulla to 400 μm thick composed of loose mass of periclinal branched filaments; cortex to 50 μm thick, composed of anticlinal dichotomously branched files of 3–6 μm diameter cells (Fig. 82d). Cystocarps, with filamentous pericarps, prominent and distributed over blade surfaces (Fig. 82e). Sporophytes with rounded to oval sori distributed across blade surfaces, each sorus consisting of chains of cruciately divided tetrasporangia embedded beneath files of cortical cells (Fig. 82f). Male gametophytes not seen.

Iridaea cordata grows abundantly on rocks in lower intertidal pools to 10 m depths in the sublittoral. In the intertidal, *Iridaea* is commonly associated with the green algae *Acrosiphonia pacifica* and *Ulva rigida* and the brown alga *Adenocystis utricularis.* In the subtidal, it is sometimes a component of the red algal community growing on the floor of *Macrocystis* forests under canopies of *Desmarestia chordalis.* In this habitat, *I. cordata* occurs alongside *Delesseria lancifolia, Schizoseris condensata, S. dichotoma* and *Polysiphonia anisogona.* Blades frequently are infected by the green endophyte *Endophyton atroviridis,* which produces greenish-red splotches visible on thallus surfaces. Some irregularly shaped dark red patches were also present in the fronds, but these appear to be the result of bacterial infection.

Notes: At Macquarie I., *I. cordata, Cenacrum subsutum, Pachymenia sp.* (cf. *P. stipitata*) and *Palmaria decipiens* sometimes occur together in the subtidal. All of these taxa produce superficially similar large and foliose thalli, but fronds of *I. cordata* are distinctive in their elasticity and iridescence when wet. Other features such as blade anatomy and cystocarp morphology can also be used to distinguish these taxa.

Iridaea remuliformis R. W. Ricker sp. nov.

Fig. 83

HOLOTYPE (Fig. 83a): MELU, R 3742 (sporophyte); with isotypes in ADU, BM, Herb. Ricker, MELU, and UC.

TYPE LOCALITY: Hasselborough Bay, Macquarie I. (54°29′50″ S lat., 158°57′00″ E long.), on mid-littoral rocks (Ricker, 16 December 1977).

DISTRIBUTION: Macquarie I.

MATERIAL EXAMINED: Macquarie I.: collections MA-119, 120, 130, 278; Herbarium specimens: MELU, R 2027, 3360, 3361, 3742–3750.

DIAGNOSIS: Frondes stipitatae ex haptero discoideo, ad 3 cm altae, atropurpureae ad virescentes, foliaceae, remuliformes ad cochleariformes, plerumque simplices autem interdum ramificatione singulari, 475–500 µm crassitie, marginibus incrassatis, compositiae ex medulla filis angustis ramosis et cortice ordinibus anticlinialibus dichotomis cellularum 6–11 µm diametro. Sporangia cruciata, initio catenata, postea separata et inordinata, inter corticem medullamque inclusa, in soris ovalibus elevatis, interdum coalitis et agellos lineares super paginas laminae facientibus. Reproductio sexualis ignota.

The specific epithet refers to the paddle-shaped fronds of the newly described species.

DESCRIPTION: Fronds stipitate from discoid holdfasts, to 3 cm tall, dark-purple to greenish, foliaceous, paddle- to spoon-shaped, usually simple but sometimes once-branched, 475– 500 µm thick, with margins thickened (Fig. 83a, b). Medulla composed of branched narrow-diameter filaments (Fig. 83c); cortex with dichotomously branched anticlinal files of cells 6–11 µm diameter (Fig. 83d). Sporangia cruciately divided, initially catenate, subsequently becoming separated and disordered, embedded between cortex and medulla (Fig. 83e, f), forming raised oval-shaped sori that sometimes coalesce to form linear patches over blade surfaces. Sexual reproduction unknown.

Plants are gregarious and form turfs or clumps (Fig. 83a) on mid-to upper intertidal rocks. *Iridaea remuliformis* often grows alongside the turf-forming red alga *Bostrychia vaga* and frequently is associated at its base with the filamentous green alga *Rhizoclonium ambiguum.*

NOTES: This species is unlike any other previously reported from the Southern Ocean. It may be that *I. remuliformis* simply is a diminutive, ecological variant of *I. cordata,* which it resembles anatomically, but the two are recognized at present as distinct for several reasons. Fronds of the former characteristically have thickened margins that differ from the unthickened margins of the latter. The tetrasporangial sori also are larger and sometimes confluent in *I. remuliformis,* forming linear patches extending nearly the length of the fronds, whereas sori in *I. cordata* are usually discrete and oval in outline.

Phyllophoraceae Rabenhorst 1863: 281

The Phyllophoraceae forms a common component of most Antarctic and Subantarctic floras. The majority of phyllophoraceous algae collected from these regions, however, are usually sterile and cannot be distinguished vegetatively from members of the Gracilariaceae, Palmariaceae and Rhodymeniaceae. Species belonging to all four families exhibit terete to flattened multiaxial thalli which are dichotomously to irregularly to proliferously branched. Medullary tissues are compact, pseudoparenchymatous, and appear to be either cellular or filamentous. Cortical tissues are composed of small pigmented cells in anticlinal rows. Sexual material, although often scarce, is diagnostic for the Phyllophoraceae. The supporting cell of the carpogonial branch functions as the auxiliary cell. Gonimoblasts develop inward and result in mature cystocarps composed of irregular

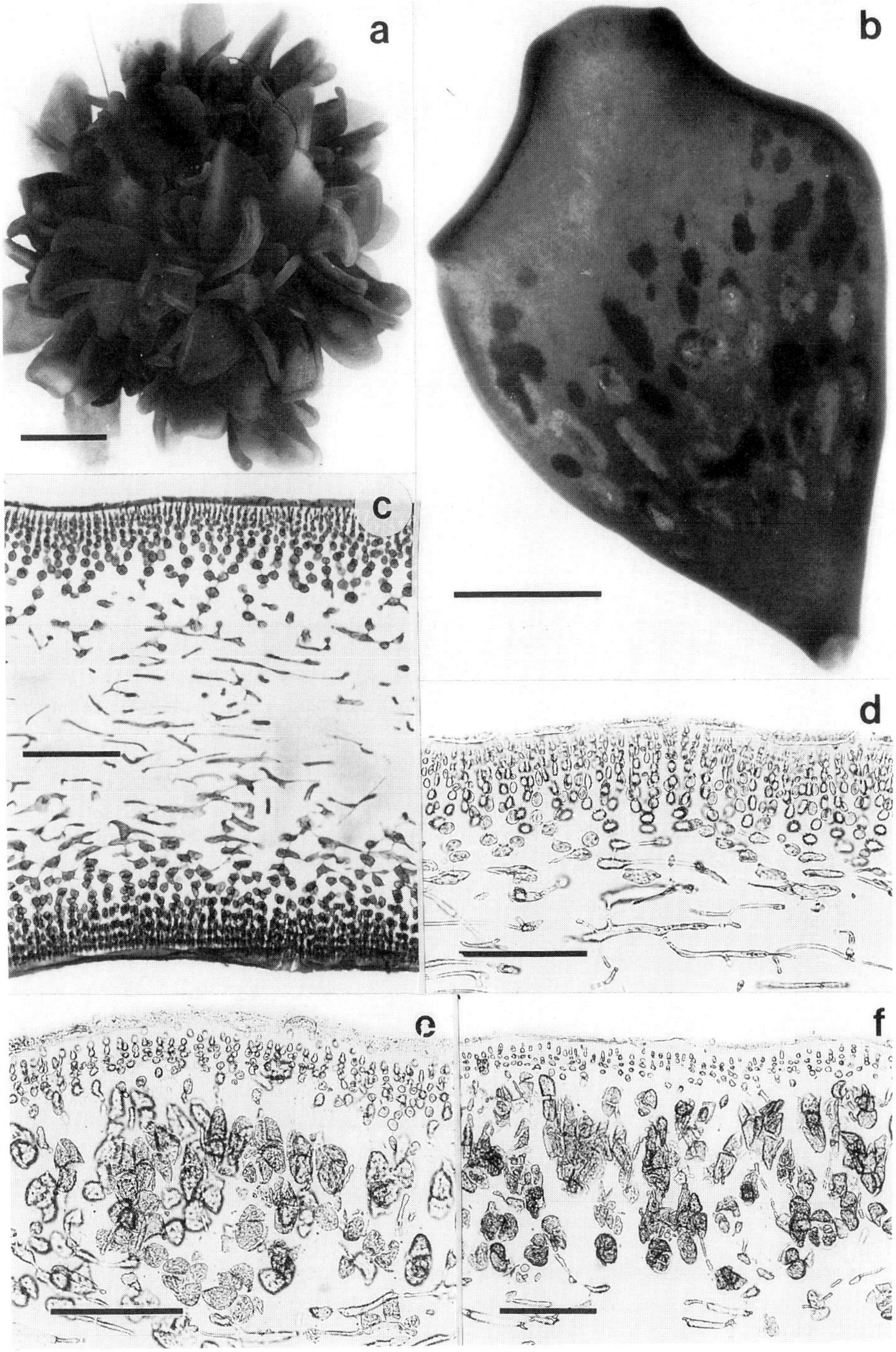

Fig. 83 *Iridaea remuliformis* R. W. Ricker: MELU, R 3360, 3361, 3742. **a.** Habit of holotype (R 3742) in water prior to drying as herbarium voucher; numerous individuals forming dense cluster of fronds. **b.** Habit of single frond with darkened tetrasporangial sori. **c.** Section through sterile thallus. **d.** Detail of cortex. **e, f.** Sections through tetrasporangial sori. Scales: a = 1 cm; b = 3 mm; c–f = 100 μm.

rows of carposporangia separated from one another by networks of sterile filaments. No specialized filamentous pericarps develop and mature cystocarps are either sunken in the thallus or protuberant. Sporophytes produce mostly cruciately divided tetrasporangia in nemathecia, although one report for *Ahnfeltia* (Chen, 1977) documents zonately divided tetrasporangia.

A variety of life history patterns, as reviewed by West & Hommersand (1981), have been reported for the 11 genera attributed to the family. Some genera, *Ozophora*, *Petroglossum*, *Schottera*, and *Stenogramme*, typically have alternations of isomorphic generations, whereas others, *Ahnfeltia*, *Gymnogongrus* and *Phyllophora*, have alternations of heteromorphic generations. Many species are imperfectly known through their gametophytes, probably because their sporophytes look vastly different and have not been associated with the gametophytes. Studies of *Gymnogongrus antarcticus* Skottsberg and *Phyllophora antarctica* A. & E. S. Gepp from the Vestfold Hills, Antarctica, both without known sporophytes, have revealed encrusting stages from the germination of carpospores (unpubl. pers. obs.). Although these crusts have yet to produce any reproductive structures, they presumably represent the sporophyte phases of both species.

Of the three phyllophoraceous genera, *Ahnfeltia*, *Phyllophora* and *Gymnogongrus*, occurring in the Southern Ocean, only the first two are known from Macquarie I.

Ahnfeltia Fries, 1835: 309

Lectotype: *A. plicata* (Hudson) Fries, 1835: 310.

One species of *Ahnfeltia*, *A. plicata*, is listed from the Antarctic and Subantarctic Zones (Papenfuss, 1964*a*), where it appears to have a circumpolar distribution. Kylin & Skottsberg (1919) reported this species from the Antarctic Peninsula, but subsequent studies in this vicinity (Lamb & Zimmermann, 1977; Moe, pers. comm.) have not confirmed this record. *Ahnfeltia* species have wiry, terete to flattened, dichotomously to irregularly branched, upright axes with numerous periclinal thick-walled medullary filaments surrounded by outwardly radiating small cortical cells in anticlinal rows. Reproduction is poorly understood in the genus, and sporophytes are unknown in most cases, although Chen (1977) reported zonately divided tetrasporangia for a Japanese specimen of *A. plicata*. Procarps, where known, are produced in nemathecia and are incompletely developed. Apparently these do not function in sexual reproduction, for the terminal cells of such procarps develop into monosporangia (Kylin, 1956). All species with erect gametophytes most probably alternate with crustose tetrasporophytes (West & Hommersand, 1981).

Excepting Kylin's & Skottsberg's (1919) unconfirmed Antarctic record, the Subantarctic appears to be the southernmost limit for this genus.

Ahnfeltia plicata (Hudson) Fries, 1835: 310

Fig. 84

Basionym: *Fucus plicatus* Hudson, 1762: 470.

Synonyms: see Papenfuss, 1964*a*.

Type locality: England.

Type: An undated Hudson specimen from an unspecified locality has been accepted as a provisional lectotype (Dixon & Irvine, 1977).

Distribution: Plants are distributed widely throughout the Northern and Southern Hemispheres, occurring in the Atlantic, Pacific and Southern Oceans. In the latter, plants are known from: Antarctic Peninsula, Falkland Is, Fuegia (Kylin & Skottsberg, 1919); Îles Crozet (Hemsley, 1884); Îles Kerguelen (Reinbold, 1907); Macquarie I.; South Georgia (Reinsch, 1890).

Material examined: Macquarie I.: collection MA-397; Herbarium specimens: MELU, R 0818, 1727; Hasselborough Bay, rock pool (Hallam, 9 February 1983): Monash University, HMI 028.

Description: Plants 6–7 cm tall, purple-black, loosely entangled to form springy clumps of wiry, terete, irregularly to radially to proliferously branched axes (Fig. 84a). Main and lateral branches equally thick (to 800 μm diameter), tapering minimally toward apices; medulla filamentous, 350–400 μm diameter, surrounded by a subcortical band (190 μm thick) of outwardly radiating cell rows and outer cortex several cells deep. In transection, medulla compact, composed of many small-diameter, thick-walled cells (Fig. 84b). Subcortical cells rectilinear, elongate, and thin-walled; outer cortical cells rectilinear and nearly isodiametric. Reproductive structures not seen.

Clumps of thalli lacking distinct holdfasts grow loosely attached to rocks in turbid waters of the shallow subtidal at the north end of the island. Fine sediment usually collects among the entangled branches and, together with worm tubes, cements *Ahnfeltia* branches into solid tufts that often are epiphytized by *Acrosiphonia, Adenocystis, Ectocarpus,* and *Geminocarpus*. The abundance and distribution of *A. plicata* around Macquarie I. is unknown.

Notes: *Ahnfeltia plicata* from Macquarie I. is similar to the European specimens studied by Rosenvinge (1931), but differs from the Japanese plant illustrated by Mikami (1965, fig. 4) which has smaller diameter branches, an indistinct subcortical band composed of rounded cells, and an outer cortex many cells deep. Macquarie plants also show distinct growth bands similar to those illustrated by Jönsson (1890) for European plants but not evident in Mikami's Japanese material.

Phyllophora Greville, 1830: 135 (*nom. cons.*)

Type species: *P. rubens sensu* Greville, 1830: lvi, 135 (= *P. crispa* (Hudson) Dixon 1964: 63).

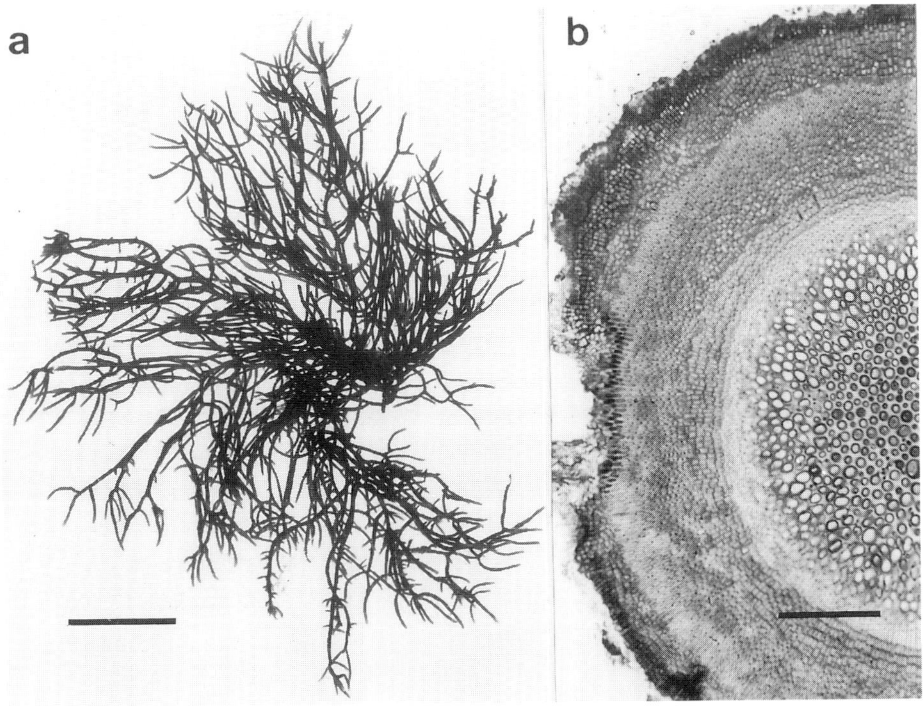

Fig. 84 *Ahnfeltia plicata* (Hudson) Fries: MELU, R 0818, 1727. **a.** Habit of sterile plant. **b.** T.S. vegetative branch. Scales: a = 2 cm; b = 100 μm.

Members of this genus are known by their generally flattened cartilaginous gametophytes. Life history patterns for most species are unknown, but apparently alternations of both heteromorphic (tetrasporoblastic—in which the diploid sporophyte develops directly from the carposporophyte) and isomorphic phases occur (West & Hommersand, 1981). Gametophytes usually arise from basal crusts and have stipitate fronds which are flattened or (less commonly) terete and either strap-like or more broadly expanded. Blade anatomy, comprising a compact cellular medulla and cortex composed of small cells in anticlinal rows, is uniform for most phyllophoraceous genera and is not useful for separating taxa below the family level. *Phyllophora* can be readily distinguished from other genera in the family, however, by its production of procarps in specialized fruiting branches.

Four Southern Ocean species of *Phyllophora* are listed by Papenfuss (1964*a*), but one, *P. abyssalis* Skottsberg, is most certainly a synonym of *P. antarctica* A. & E. S. Gepp, which is known only from the Antarctic (pers. obs.). *Phyllophora ahnfeltioides* Skottsberg occurs in both the Antarctic and Subantarctic Zones, whereas *P. appendiculata* Skottsberg, as presently conceived, is known only from the Subantarctic. It may be (as Skottsberg (1953) suggested) that the type of *Gymnogongrus turquetii* Hariot is conspecific with the type of *P. appendiculata*, in which case the name *turquetii* would have priority, and the known distribution of this entity would be extended into the Antarctic. If Hariot's (1907) description of 'pedicellate papillae' on *Gymnogongrus turquetii* actually refers to specialized cystocarpic branches (a feature considered to be diagnostic for *Phyllophora*), then the generic name *Phyllophora* would seemingly be better applied than that of *Gymnogongrus*. Re-evaluation of the taxonomy, nomenclature and biogeography of *P. appendiculata*, however, awaits critical examination of both Hariot and Skottsberg type specimens.

Phyllophora appendiculata Skottsberg ***in*** Kylin & Skottsberg, 1919: 10

Fig. 85

TYPE LOCALITY: South Georgia.

TYPE: Not designated; (N.B. syntype material from stations 17, 21, 30a in wet stack collection of Herb. Skottsberg, S).

DISTRIBUTION: Îles Kerguelen (Zonova, 1958); Macquarie I. (Law & Burstall, 1953); South Georgia (Kylin & Skottsberg, 1919; Skottsberg, 1923).

MATERIAL EXAMINED: Macquarie I.: collections MA-11, 34, 46, 83, 84, 118, 129, 137, 213, 222, 227, 236, 244, 271, 283, 284, 309, 381; Herbarium specimens: MELU, R 0095, 0589, 0618, 0624, 1184, 1187, 1216, 1289.

DESCRIPTION: Thalli to 30 cm tall, dark-red to red-brown to burgundy in colour, arising from extensive, epilithic dull-red crusts. Distinct main axes often lacking (Fig. 85c), with branching irregular from either blade faces or margins or both, often resulting in dense tangles of cartilaginous straps 3–10 mm wide to 17 cm long arising from upper ends of simple basal fronds (Fig. 85a). Outer cortex with pigmented cells in 2–3-celled anticlinal filaments (Fig. 85d); inner (sub)cortex composed of small, rounded, colourless cells grading inwards to zone of larger, unpigmented medullary cells (Fig. 85e). Carposporophytes in specialized branchlets (Fig. 85f) which, when young, resemble vegetative lateral initials. Mature cystocarps lacking ostioles and filamentous pericarps. Gonimoblasts embedded in medulla, composed mostly of carposporangia

---

Fig. 85 *Phyllophora appendiculata* Skottsberg: MELU, R 0095, 0618, 1216, 1289. **a.** Habit of sterile plant without base. **b.** Detail of 'a' showing *Actinococcus*-like galls on cartilaginous blades. **c.** Thallus habit with branching only from the margins. **d.** Detail of cortex. **e.** T.S. blade showing cortex, subcortex and medulla. **f.** Cystocarpic proliferation; bulge on branchlet (arrow) caused by mature gonimoblast. **g.** T.S. cystocarpic proliferation through mature gonimoblast. Scales: a = 3 cm; b = 1 cm; c = 5 cm; d = 25 μm; e = 100 μm; f = 2 mm; g = 200 μm.

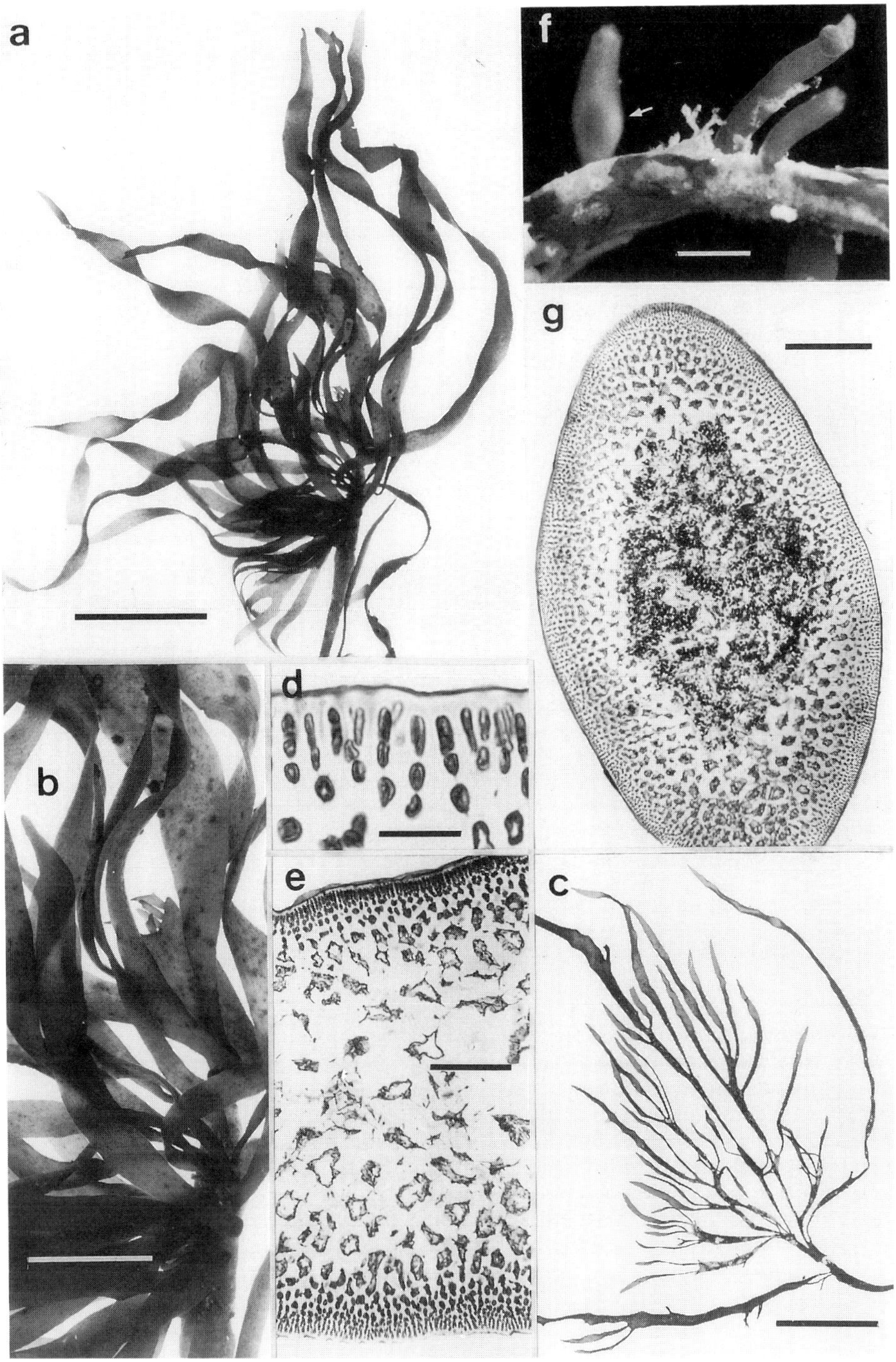
a
b
c
d
e
f
g

in groups separated by thin sterile filaments (Fig. 85g). Sporophytes and male gametophytes not seen.

Plants are gregarious and form clumps in shallow (<0.5 m) tidepools in the lower intertidal. A layer of sand often covers the tidepool floor and lower axes of *Phyllophora*, thereby obscuring details of thallus attachment to the underlying rock. Plants growing in 10 m depths atop rocks were not covered by sand and the erect axes clearly developed from extensive fleshy crusts. At its base, each crust has rounded colourless cells organized in assurgent files, with the upper cells gradually becoming 1–2 times elongate and densely pigmented toward the crust surface. Reproductive structures are unknown in crust tissues.

NOTES: *Phyllophora appendiculata* from Macquarie I. is anatomically similar to plants from the Antarctic Peninsula (Skottsberg, 1953) and comparable to Subantarctic specimens discussed and illustrated by Skottsberg (*in* Kylin & Skottsberg, 1919). Thalli from Macquarie I., however, grew more than twice as tall as those previously described from South Georgia and cystocarp morphology was more variable than indicated by Skottsberg. Cystocarps in Macquarie plants appear either as simple bulges in otherwise short terete branchlets (Fig. 85f) or as more elaborate swollen and knobby structures similar to those in Kylin's & Skottsberg's (1919) figures 5c–e. Cystocarpic branchlets, unlike vegetative lateral branch initials, cease growth following carposporophyte maturation.

Actinococcus-like galls have not been previously reported on *P. appendiculata*, but morphologically similar (bacteria-caused?) structures as present on Macquarie thalli (Fig. 85b) are known to develop on other species of *Phyllophora*.

Plocamiaceae Kützing, 1843: 442, 449

This family consists of two genera, with *Plocamiocolax* monotypic, known only from the Pacific coast of North America, and parasitic upon certain species of the more diverse cosmopolitan genus *Plocamium*. The latter was originally placed in the Rhodymeniales, but later moved by Bliding (1928) to the Gigartinales because of its uniaxial construction, the presence of zonately divided tetrasporangia, and the development of carposporophytes directly from supporting cells that function as auxiliary cells. The tetrasporangial division pattern and procarp organization are typically gigartinalean, but the vegetative construction of *Plocamium* and its production of stichidia are atypical for this order (Kraft, 1981).

Plocamium Lamouroux, 1813: 137 (***nom. cons.***)

Type species: *P. vulgare* Lamouroux, 1813 (= *P. cartilagineum* (Linnaeus) Dixon 1967: 58; see South & Adams (1979) re. synonymy).

Plants arise from a basal system of entangled, terete, prostrate branches and have distinct apical cells giving rise to erect, compressed thalli, often with axial filaments visible, that branch sympodially. Branching is pinnate, with series of 2–5 laterals arising alternately along main axes. Gametophytes and sporophytes are isomorphic. Spermatangia are developed in sori over frond surfaces. Carpogonial branches are three-celled, with the supporting cells functioning as auxiliary cells. The zygote nucleus is transferred to the auxiliary cell via a fusion of this cell with the carpogonium and the gonimoblast initial develops toward the thallus surface (Dixon, 1982). Mature carposporophytes are enclosed in cellular pericarps lacking distinct openings. Tetrasporangia are zonately divided and immersed in the cortex of branched stichidia.

This genus comprises some 40 species (Kraft, 1981), with six to seven occurring in the Southern Ocean (Papenfuss, 1964*a*; South & Adams, 1979). One of these, *P. cartilagineum*, is broadly distributed in both Northern and Southern Hemispheres and occurs in the Cool Temperate (Womersley, 1971*b*), Subantarctic (Kylin & Skottsberg, 1919), and Antarctic (Lamb & Zimmermann, 1977) Zones. *Plocamium hookeri* and *P. secundatum* grow at Macquarie I. and apparently are restricted to, but spread throughout the Subantarctic*, whereas the remaining species are known from smaller geographic areas [New Zealand—*P. leptophyllum* Kützing & *P. angustum* (J. Agardh) Hooker f. & Harvey (South & Adams, 1979); Îles Crozet—*P. minutum* Levring (1944); Fuegia—*P. telfairiae* Harvey (source for record presently unknown)]. South (1979*b*) considered the *Plocamium* species present in New Zealand Subantarctic islands to be largely derived from New Zealand, instead of imports (via the West Wind Drift) from more distant lands.

Plocamium hookeri Harvey *in* Hooker f. & Harvey, 1845: 257

Fig. 86

TYPE LOCALITY: Îles Kerguelen.

TYPE: presumably in TCD (Koster, 1969), but specimens were not seen.

DISTRIBUTION: Heard I. (Dickie, 1876*d*); Îles Kerguelen (Dickie, 1876*c*, 1876*e*, 1879); Macquarie I. (Laing, 1909); South Georgia (Reinsch, 1890); South Orkney Is (Gepp & Gepp, 1905*a*).

MATERIAL EXAMINED:
1. Heard I.: Atlas Cove (7 May 1953): MELU, R 3731 (female).
2. Macquarie I.: collections MA-11, 14, 34, 39, 113, 114, 116, 117, 222, 227, 298, 300, 302, 303, 306, 309, 377; Herbarium specimens: MELU, R 0009, 0795, 0846, 1373, 1454–1458, 1721.
3. South Georgia: Station 21, Cumberland Bay, Moraine fjord (Skottsberg, 23 May 1902): UPS, unnumbered specimen (tetrasporophyte).
4. South Orkney Is: Bay A (Scotia Bay), 9–10 fathoms (Brown, May 1903; July 1903): E, unnumbered specimens.

DESCRIPTION: Plants to 12 cm tall, rose to reddish-orange to reddish-brown, bushy, arising from base of entangled terete branches (Fig. 86a–c). Main axes distinct, flattened, 1–2 mm broad, bearing series of 2–3 laterals on alternate sides (Fig. 86e, f), with branches curving acropetally and either determinate or indeterminate. Determinate branches leaf-like, with entire or slightly dentate margins (Fig. 86d); indeterminate branches near thallus apices with adaxial pectinate series of 2–3(–4) branchlets. Cystocarps prominent, rounded, borne adaxially in axils of lateral branches or laterally along main axes (Fig. 86i). Sporangia zonately divided, borne in irregularly lobed to much-branched stichidia (Fig. 86g, h). Male gametophytes unknown.

Plants commonly grow on coralline-encrusted or bare rocky surfaces on the shaded floors of *Macrocystis* forests in the subtidal from 3–10 m depths. In some instances, the basal portions of *P. hookeri* are covered by sponges and colonial tunicates. Other taxa frequently associated with this species include *Delisea pulchra* and members of the Delesseriaceae.

NOTES: At Macquarie I., both sterile plants and sporophytes greatly outnumbered recognizable gametophytes. Immature thalli were easily identified by their enlarged leafy determinate laterals. In some of the older individuals, however, the size of determinate blades is less pronounced (although still enlarged by comparison with other *Plocamium* species), which previously has caused some taxonomic confusion and misidentifications. Kenny's & Haysom's (1962) report of *P. cartilagineum* from Macquarie I. actually refers to mature specimens of *P. hookeri*. These two species, however, can be easily distinguished by differences in the shapes of their stichidia if sporphytes are available.

* Skottsberg (1906) reported both of these species from the Antarctic Peninsula, but neither's presence has been confirmed in subsequent floristic studies for this locality (e.g., Lamb & Zimmermann, 1977).

Fig. 86 *Plocamium hookeri* Harvey; MELU, R 0009, 1373, 1454, 1456, 1721. **a.** Habit of juvenile thallus. **b.** Habit of tetrasporophyte. **c.** Habit of female gametophyte. **d.** Detail of upper thallus. **e, f.** Branching along main axes. **g, h.** Tetrasporophylls. **i.** Cystocarp (arrow) on adaxial surface of determinate lateral. Scales: a, c = 2 cm; b = 4 cm; d = 1 mm; e, f, i = 2 mm; g, h = 500 μm.

Plocamium secundatum (Kützing) Kützing, 1866: 15

Fig. 87

BASIONYM: *P. coccineum* var. *secundatum* Kützing, 1849: 883.

SYNONYMS: see Papenfuss, 1964*a*.

TYPE LOCALITY: Fuegia.

TYPE: Herb. Kützing (L). (N.B. plants illustrated by Kützing (1866, pl. 42).)

DISTRIBUTION: Falkland Is (Cotton, 1915); Fuegia (Hariot, 1889); Îles Kerguelen (Levring, 1944); Macquarie I. (Zinova, 1958); South Georgia (Kylin & Skottsberg, 1919); South Orkney Is (Holmes, 1905, 1912).

MATERIAL EXAMINED:

1. Fuegia: Type specimen. Cape Horn, N.W. Bay, Hermite I. (J. D. Hooker): L, unnumbered sheet.
2. Macquarie I.: collections MA-34, 39, 46, 56, 83, 84, 94, 98, 113, 114, 116, 117, 119, 213, 215, 222, 227, 244, 283, 295, 298, 300, 302, 303, 364, 377, 378; Herbarium specimens: MELU, R 0030, 1229–1240, 1382–1385.
3. South Georgia: Station 21, Cumberland Bay, Moraine fjord (Skottsberg, 23 May 1902?): UPS, unnumbered specimen (sterile).
4. South Orkney Is: Scotia Bay (Brown, 1903): E, 123/80 17 (fragments).

DESCRIPTION: Plants to 13 cm tall, reddish-brown to burgundy, bushy, arising from base of entangled, flattened to terete branches (Fig. 87a). Main axes distinct, flattened, 2–3 mm broad, often with pronounced zigzag, and in mid-thallus bearing series of 3–4 laterals on alternate sides (Fig. 87c), with branches curving acropetally and either determinate or indeterminate. Determinate branches lanceolate or narrow and elongate, usually with smooth margins, tapering gradually toward apices, sometimes with one edge curved and the other straight. Indeterminate branches near apices with up to six branchlets in pectinate series (Fig. 87b). Cystocarps prominent, rounded, borne singly or in series of four or more along margins of main axes and lateral branches (Fig. 87f). Stichidia oppositely branched, adaxial on lateral branches, containing zonately divided tetrasporangia (Fig. 87d, e). Male gametophytes unknown.

NOTES: Plants from Macquarie I. are similar in habit to specimens from Hermite I. (Cape Horn) illustrated by Kützing (1866, pl. 42) but differ in producing fewer laterals along the adaxial margins of indeterminate branches in the upper thallus. Kützing illustrated pectinate series with up to 13 laterals on a side, in contrast to six for Macquarie thalli.

Plocamium secundatum and *P. hookeri* sometimes occur together in the same habitat, but there is no difficulty distinguishing the two, as the former lacks leafy determinate laterals and produces indeterminate branches with longer series of pectinate branchlets.

Cystocloniaceae Kützing, 1843: 390, 404

Commonly known as the Rhodophyllidaceae (see Guiry, 1978*a*), this family consists of ten genera (Min-Thein & Womersley, 1976; Hansen, 1980) totalling some 35 species. All are characterized as having uniaxial thalli, isomorphic gametophytes and sporophytes, procarps with auxiliary cells derived from the supporting cells, primarily inwardly developed gonimoblast initials, mature carposporophytes that lack filamentous pericarps and have either a central fusion cell or mass of small nutritive cells surrounded by outwardly radiating chains of carposporangia, and zonately divided tetrasporangia.

Fig. 87 *Plocamium secundatum* (Kützing) Kützing: MELU, R 0030, 1240. **a.** Habit of sporophyte. **b.** Detail of upper thallus. **c.** Branching along main axis. **d.** Immature tetrasporophylls on adaxial side of indeterminate lateral. Compound stichidia (arrow) rare. **e.** Mature stichidium. **f.** Cystocarps (arrows) developed adaxially and abaxially. Scales: a = 3 cm; b, f = 1 mm; c = 4 mm; d = 500 μm; e = 200 μm.

The majority of genera and species in this family occur in the Southern Hemisphere, with most represented along the southern coastline of Australia (Min-Thein & Womersley, 1976). Four genera, *Acanthococcus*, *Calliblepharis*, *Cystoclonium*, and *Rhodophyllis*, have been listed from the Subantarctic (Papenfuss, 1964*a*), but none from the Antarctic. *Acanthococcus* is monotypic and apparently restricted to the Subantarctic. In contrast, *Rhodophyllis* has more species and is known from the Subantarctic as well as from temperate to colder waters of both hemispheres. The family is represented at Macquarie I. by one species of *Rhodophyllis*.

Rhodophyllis Kützing, 1847: 23 (*nom. cons.*)

Type species: *R. bifida* (Lamouroux) Kützing, 1847: 23 (= *R. divaricata* (Stackhouse) Papenfuss 1950: 190).

Members of this genus are flattened and foliose, arise from discoid attachments, and have uniaxial blades with distinct apical cells positioned among marginal cells. Fronds are irregularly to alternately to pinnately branched and composed of a filamentous medulla surrounded by 1–2(–4) layers of cortical cells. The cortex consists of larger inner cells that are surrounded (on the thallus surface) by rosettes of smaller outer cells. Carpogonial branches are three-celled; gonimoblasts have chains of carposporangia surrounding either a large fusion cell or a central mass of small nutritive cells (Kylin, 1932). Spermatangia are distributed over blade surfaces (Min-Thein & Womersley, 1976). Sporangia are zonately divided, produced by cortical cells, and embedded in fronds but not localized in sori.

Rhodophyllis had *c.* 16 species, most of which occur in the Southern Hemisphere (Hansen, 1980), with Australia (Min-Thein & Womersley, 1976) and New Zealand (Cotton, 1908; V. J. Chapman, 1979) together accounting for seven. Four species, *R. acanthocarpa* (Harvey) J. Agardh, *R. angustifrons* Hooker f. & Harvey, *R. divaricata* (Stackhouse) Papenfuss and *R. reptans* (Suhr) Papenfuss, are reported from the Subantarctic, with none extending into the Antarctic, and all except *R. angustifrons* are listed from the Auckland Is (Papenfuss, 1964*a*). The occurence of *R. acanthocarpa* at Macquarie I. represents the southernmost range of the genus.

Rhodophyllis acanthocarpa (Harvey) J. Agardh, 1876: 364

Fig. 88

BASIONYM: *Callophyllis acanthocarpa* Harvey, 1855: 251.

SYNONYMS: see V. J. Chapman, 1979. *Rhodophyllis laingii* Cotton, 1908: 98

TYPE LOCALITY: New Zealand; unspecified between Middle Is., East Coast, or Cape Cooper.

SYNTYPES: TCD, unnumbered specimens. (N.B. holotype not designated.)

DISTRIBUTION: Auckland I. (Laing, 1909); Macquarie I.; New Zealand (Cotton, 1908); Stewart I. (Adams *et al.*, 1974).

MATERIAL EXAMINED:

1. Macquarie I.: collections MA-34, 46, 54, 56, 83, 94, 95, 98, 117, 227, 248, 271, 283, 298, 300, 302, 303, 309, 364, 377, 378; Herbarium specimens: MELU, R 0141–0148, 0612, 0613, 0846, 1335–1343, 2028.
2. New Zealand: East coast (Lyall, April 1849): TCD, unnumbered specimen; (Colenso, undated): TCD, unnumbered specimen.

DESCRIPTION: Plants to 10 cm tall, rose to reddish-brown to blackish-red, foliose and occasionally stipitate, with cuneate bases, rounded to acute apices, and discoid holdfasts or proliferously branched terete basal axes with multiple attachment pads. Fronds membranous, usually crisp, 3–25 mm broad, dichotomously to palmately to pinnately to irregularly branched, often laciniate

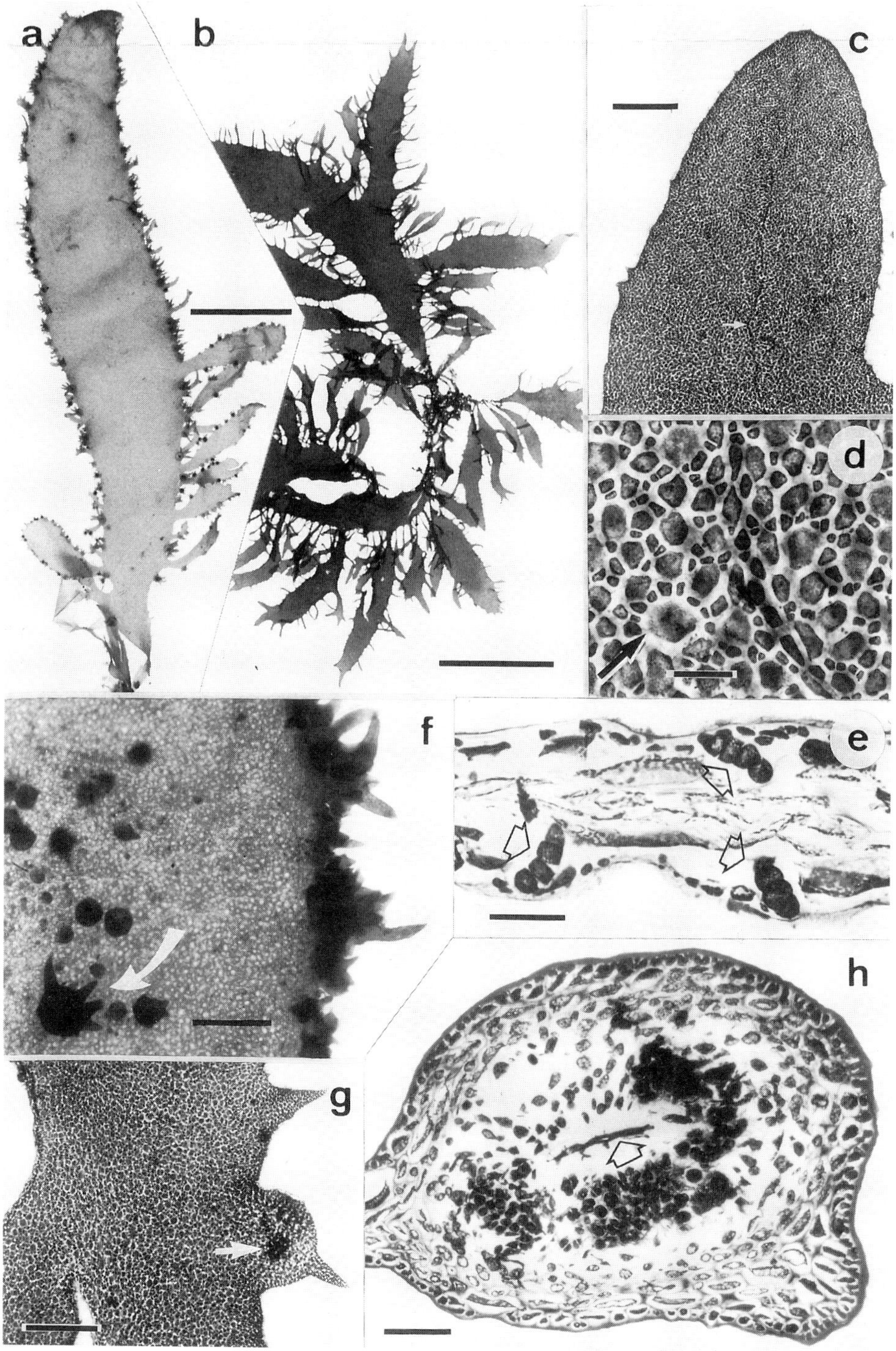
a
b
c
d
e
f
g
h

and with marginal proliferations (Fig. 88a, b), uniaxial with axial filament sometimes visible from blade surface (Fig. 88c), with distinctive cortex composed of small pigmented cells outermost and forming rosettes around larger unpigmented inner (subcortical) cells (Fig. 88d). Medulla slight, with one or more filaments running between more prominent layers of large subcortical cells. Cystocarps mostly marginal but occasionally on blade faces, covered with spinose projections (Fig. 88f, g). Gonimoblasts with large, centrally positioned fusion cells surrounded by outwardly radiating chains of carposporangia (Fig. 88h). Tetrasporangia zonately divided, immersed in thallus, distributed diffusely over blade surfaces (Fig. 88e). Male gametophytes unknown.

Plants grow on rocks and various algae (e.g., *Callophyllis variegata, Delisea pulchra, Plocamium hookeri*) on the shaded floors of *Macrocystis* forests to 20 m depths and commonly occur alongside the red alga *Cladodonta lyallii* and brown alga *Microzonia velutina. Pterothamnion plumula* sometimes grows epiphytically on the bases of *R. acanthocarpa.*

NOTES: Kylin (1932) described gonimoblasts of *R. acanthocarpa* with several centrally positioned relatively large cells, but the present studies have confirmed the observations of Min-Thein & Womersley (1976), who studied New Zealand specimens and noted a single large fusion cell in the centre of each gonimoblast.

Plants from Macquarie I. agree anatomically and morphologically with illustrations by De Toni & Forti (1922) of New Zealand specimens. On the basis of some 100 individuals studied during the present survey, it appears that sporophytic thalli generally are branched to a greater degree than female gametophytes. Thallus morphology (of either reproductive phase) is extremely variable and includes forms such as represented by the type specimen of *R. laingii* Cotton (V. J. Chapman, 1979, pl. 143) and possibly that of *R. acanthocarpa* var. *lacerata* (Harvey) V. J. Chapman (1979, pl. 142). Chapman distinguished this variety from the type variety by the acute frond apices and cartilaginous filiform stipes of the former—features that are also developed to various degrees in Macquarie thalli.

Rhodymeniales

Members of this order are multiaxial and pseudoparenchymatous, have life history patterns with alternations of isomorphic somatic phases, and produce either encrusting or erect thalli that are terete, compressed, or foliose and either solid or hollow. Plants are procarpic, each supporting cell bearing a (3–)4-celled carpogonial branch and a two-celled branch with a terminal auxiliary cell. Following fertilization, the diploid zygote nucleus is transferred to the auxiliary cell by either direct fusion of the carpogonium with the auxiliary cell or through a connecting cell. Gonimoblasts develop outwards, produce masses of carposporangia subtended by one large fusion cell, and are encased in well developed ostiolate pericarps that protrude above thallus surfaces. Spermatangia are produced superficially from assimilatory cells and often occur in sori. Tetrasporangia, being either terminal or intercalary, are cruciately or tetrahedrally divided and often localized in sori.

Bliding (1928), Kylin (1931), Sparling (1957) and Lee (1978) reviewed the majority of the 38 genera that now constitute this order. Lee (1978) recognized three families, the Champiaceae, Lomentariaceae and Rhodymeniaceae, separated mostly on vegetative characteristics, the carposporophytes being extremely uniform throughout the order. Kraft (1981) estimated 265 species for the order, with roughly 75% belonging to the Rhodymeniaceae, which is the only family present at Macquarie I.

Fig. 88 *Rhodophyllis acanthocarpa* (Harvey) J. Agardh: MELU, R 0148, 0612, 0613, 2028. **a.** Habit of female gametophyte with marginal cystocarps. **b.** Habit of sporophyte. **c.** Surface view of blade apex showing axial filament (arrow). **d.** Detail of cortical cell rosettes surrounding underlying subcortical cells (arrow). **e.** Section through blade with zonate tetrasporangia (hollow arrows). **f.** Surface view of spiny cystocarps on blade margin and face (arrow). **g.** Surface view of immature cystocarp with single spine and central core of gonimoblast tissue (arrow). **h.** Collapsed fusion cell (hollow arrow) and surrounding carposporangia in sectional view of mature cystocarp. Scales: a = 1 cm; b = 2 cm; c, g = 300 μm; d, e = 50 μm; f = 1 mm; h = 100 μm.

Rhodymeniaceae Harvey, 1849*a*: 75, 120

The Rhodymeniaceae is seemingly well represented throughout southern high-latitude waters by the three genera *Cenacrum, Epymenia,* and *Rhodymenia,* with the last most widely distributed and comprising the largest number of species. *Rhodymenia* is also the only genus in the family reported from the Antarctic, although this distribution is based upon a single record from Queen Mary Coast (Antarctica).* Subsequent studies in Eastern Antarctica (Moe, pers. comm.; pers. obs.), however, have failed to confirm *Rhodymenia*'s occurrence in the Antarctic.

The genus *Leptosarca*†, which occurs abundantly throughout Antarctic and Subantarctic waters, was placed by Kylin & Skottsberg (1919) in the Rhodymeniaceae. For reasons presented in the discussion of *Palmaria,* however, members of this genus, as well as some species previously attributed to *Rhodymenia* (*R.* (?) *antarctica* Skottsberg, *R. georgica* Reinsch, *R. palmatiformis* Skottsberg), have been moved to the Palmariales in the present survey.

The family is represented at Macquarie I. by *Cenacrum* and *Rhdymenia.*

Cenacrum R. W. Ricker & G. T. Kraft, 1979: 435

Type species: *C. subsutum* R. W. Ricker & G. T. Kraft, 1979: 435.

Plants are multiaxial and erect, have life history patterns with alternations of isomorphic somatic phases, and consist of foliose thalli subtended by terete stipes arising from discoid holdfasts. Blade anatomy is distinctive, with young portions having medullary cavities surrounded by filaments. Older regions of blades lack medullary cavities and are filled with a mixture of filaments and large rounded cells. Cortical tissues consist of 1–2 inner layers of colourless oblong cells and an outer region with (2-)3-celled anticlinal filaments composed of small pigmented cells. Plants are procarpic, with supporting cells bearing (3-)4-celled carpogonial branches and two-celled branches with terminal auxiliary cells. Following fertilization, diploid zygote nuclei are transferred to auxiliary cells by connecting cells and gonimoblasts develop toward thallus surfaces. Ostiolate cystocarps are prominent, lack a *tela arachnoidea,* and have mature carposporophytes with 2–3 dense lobes of carposporangia subtended by a short stalk produced by the fusion of auxiliary and auxiliary mother cells. Spermatangia are produced superficially in continuous sori on both blade faces; tetrasporangia are cruciately divided and produced in raised sori (= nemathecia). The genus is monotypic and known from islands in the New Zealand Subantarctic and from Macquarie I.

Cenacrum subsutum R. W. Ricker & G. T. Kraft 1979: 435

Fig. 89

Holotype: MELU, R 0752 (= M23999; female); (Ricker & Kraft, 1979, fig. 1); with isotypes in ADU, BM, Herb. Ricker, and UC.

Type locality: Garden Bay, Macquarie I. (54°29′56″ S lat., 158°57′30″ E long.); growing on shaded vertical rock walls in 11–15 m depths (Ricker, 1 February 1978).

Distribution: Auckland, Macquarie, and Snares Is (Ricker & Kraft, 1979).

Material Examined:

1. Auckland Is: subtidal (Johnson, 19 February 1973); WELT, A 8351.

*The distribution of *Rhodymenia* in Antarctic waters is indicated by Papenfuss (1964*a*), but the original report upon which this record is based has not been located for the present survey.

†Kylin (1956) incorrectly considered this genus to be congeneric with *Leptosomia,* whose type, *L. cliftoni* (Harvey) J. Agardh from Western Australia, is generically distinct (Kraft, pers. comm.).

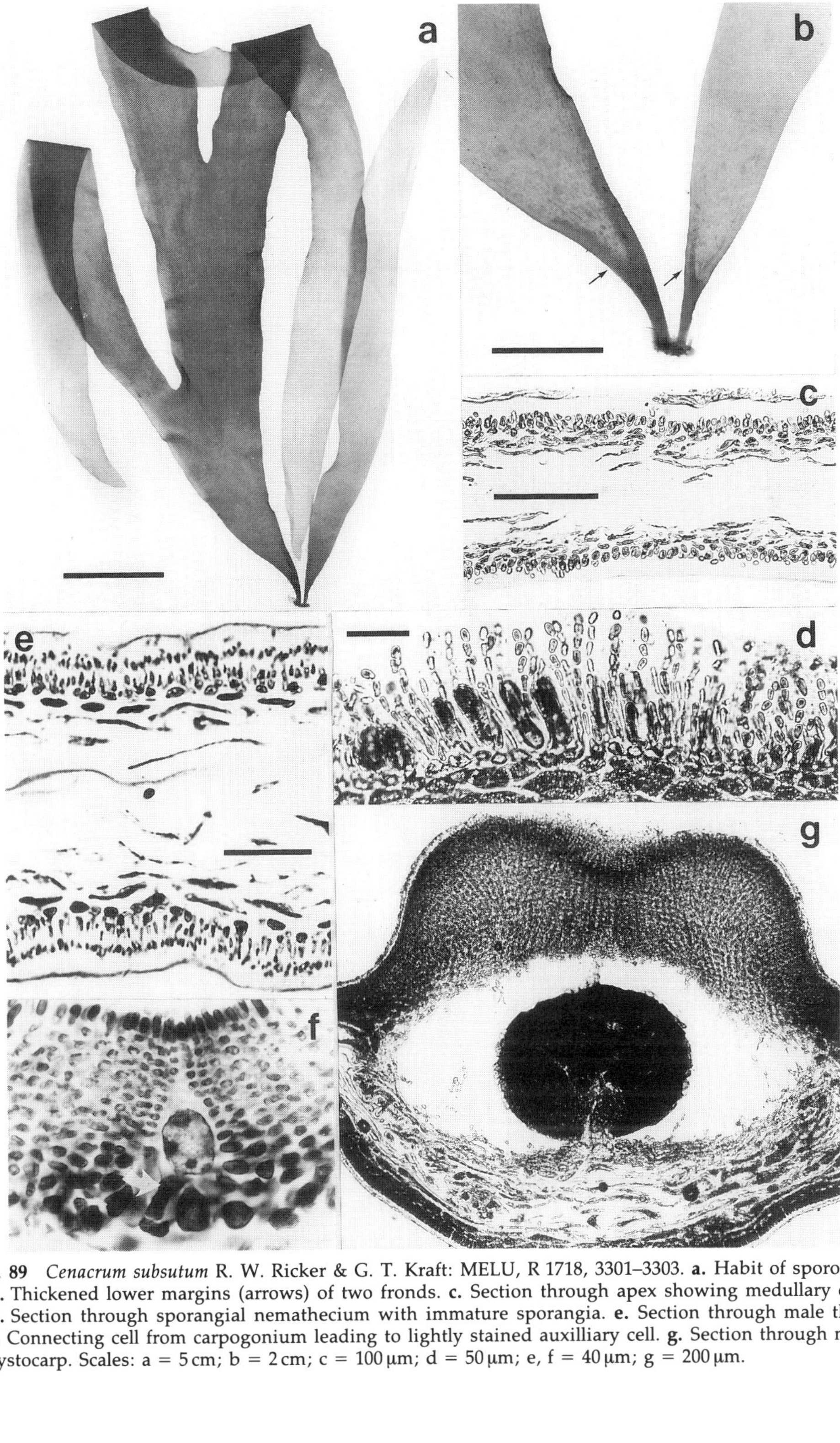

Fig. 89 *Cenacrum subsutum* R. W. Ricker & G. T. Kraft: MELU, R 1718, 3301–3303. **a.** Habit of sporophyte. **b.** Thickened lower margins (arrows) of two fronds. **c.** Section through apex showing medullary cavity. **d.** Section through sporangial nemathecium with immature sporangia. **e.** Section through male thallus. **f.** Connecting cell from carpogonium leading to lightly stained auxilliary cell. **g.** Section through mature cystocarp. Scales: a = 5 cm; b = 2 cm; c = 100 µm; d = 50 µm; e, f = 40 µm; g = 200 µm.

2. Macquarie I.: collections MA-14, 34, 39, 46, 54, 83, 94, 95, 98, 113, 129, 137, 140, 143, 215, 222, 227, 237, 244, 248, 271, 283, 295, 300, 302, 303, 309, 364, 377, 378; Herbarium specimens: MELU, R 0751–0775, 1718, 3301–3303.

3. Snares Is: 1–3 m deep (Horning, 20 January 1975): CHR 248540.

DESCRIPTION: Plants to 85 cm long, solitary or gregarious, arising from discoid holdfasts, pale red (juvenile & male thalli) to deep burgundy (mature female gametophytes) in colour, soft distally and somewhat turgid basally when living (becoming flaccid after formalin preservation), with slippery surfaces (Fig. 89a). Thalli with 5–15 mm long stipes, apophyses with noticeably thickened margins (Fig. 89b), and foliose fronds simple to once-forked to highly dissected, sometimes becoming proliferous along margins of older or damaged plants. Blade interior near apices with filaments surrounding medullary cavities (Fig. 89c), becoming solid below. Medullary filaments sometimes bearing densely staining (with aniline blue) 'gland' cells. Plants dioecious; spermatangial mother cells elongate, producing spermatangia in continuous patches covering most of frond surface (Fig. 89e). Procarps surrounded by thickened cortex; connecting cells forming after fertilization for transfer of zygote nuclei to auxiliary cells (Fig. 89f). Cystocarps ostiolate, embedded in thallus but bulging above surrounding vegetative blade surface, with carposporophyte mostly composed of carposporangia (Fig. 89g). Tetrasporangia cruciately divided, oblong, to 56–70 μm long by 22–30 μm broad, produced in nemathecia composed of cortical files 8–10 cells long (Fig. 89d).

Cenacrum is abundant in sheltered habitats around Macquarie I., less so in rougher wave-beaten areas. This species extends from the sublittoral fringe immediately below the *Durvillaea* zone, where it grows along with *Ballia callitricha* on shaded rocks encrusted with coralline algae, to depths of 20 m or more, where it is associated with *Plocamium hookeri* and various members of the Delesseriaceae. The largest numbers of plants occur on rocky surfaces exposed to full light in shallow (<5 m) wave-sheltered coves or immediately below the *Durvillaea* band on shaded rocky walls of unprotected intertidal platforms along the relatively calm east coast of the island.

NOTES: Plants of this species, although sometimes occurring together with superficially similar foliose red algae such as *Iridaea cordata* and *Palmaria decipiens*, are readily recognized by their conspicuously thickened basal margins.

Rhodymenia Greville, 1830: xlviii

Type species: *R. palmetta* (Lamouroux) Greville, 1830: xlviii [= *R. pseudopalmata* (Lamouroux) Silva, 1942: 265].

Plants are either erect or prostrate, sessile or stipitate, to 1 m long, narrow and compressed to broadly foliose, simple or dichotomously to irregularly branched or lobed, and usually cartilaginous. Medullary tissues lack rhizoids and consist of large, axially elongated cells that appear parenchymatous in blade transections. Cortical tissues are usually with 2–3 layers of small pigmented cells. Gametophytes are dioecious, although most species are known only through sterile plants or sometimes sporophytes. Tetrasporangia are cruciately divided and formed from assimilatory filaments in sori.

Rhodymenia is a large genus with species distributed in tropical to subpolar waters of both hemispheres (Guiry, 1977). Some 40 (Kylin, 1956) to 54 (Dawson, 1941) species have been recognized, although Sparling (1957) has suggested that these numbers are probably exaggerated owing to poorly defined species concepts. Dawson (1941) divided the 54 species into five sections (Pertusae, Palmatae, Palmettae, Clinophora, Dendrymeniae) based on criteria such as: sessile vs. stipitate fronds; blade shape, size, and texture; branching pattern; and tetrasporangia shapes and disposition on thallus surfaces. The last criterion is considered to be extremely important taxonomically, yet tetrasporangia are unknown in nearly half of the described species, and cystocarps unknown in an even larger percentage (Dawson, 1941). Most species are described using only sterile plants.

Nine species have been listed from the Southern Ocean (Papenfuss, 1964*a*), but three of these have been moved during the present survey to *Palmaria*. *Rhodymenia* is represented at Macquarie I. by one species.

Rhodymenia subantarctica R. W. Ricker *nom. nov.*

Fig. 90

REPLACED NAME: *Rhodymenia cuneifolia* (Hooker f. & Harvey) W. R. Taylor 1939: 147. (N.B. this name is a later homonym of *R. cuneifolia* Okamura (1934). In the absence of any (available) synonym, a new name is necessary.)

BASIONYM: *Phyllophora cuneifolia* Hooker f. & Harvey, 1845: 260
Non *Rhodymenia cuneifolia* Okamura, 1934: 16, pl. vii [= *Gracilaria cuneifolia* (Okamura) I. K. Lee & Kurogi (1977: 177)].

SYNONYMS: see Papenfuss (1964*a*) under *Rhodymenia cuneifolia*.

TYPE LOCALITY: St. Salvador Bay, East Falkland Is.

LECTOTYPE: K *in* BM; selected by J. H. Price (1 March 1985) from available syntypes; J. D. Hooker annotation: 'Found on the beach St. Salvador Bay East Falkland I. April 1842'; A. Gepp annotation: 'Type' (Price, pers. comm.).

DISTRIBUTION: Falkland Is (Hooker f. & Harvey, 1845); Fuegia (Hariot, 1889); Îles Kerguelen (Dickie, 1879); Juan Fernandez Is (Levring, 1941); Macquarie I. (N.B. Based upon Laing's (1909) uncertain record of *Rhodymenia corallina*, Papenfuss (1964*a*) listed the Auckland Is as a possible locality for this species. Zinova (1985) indicated distribution at Queen Mary Coast (Antarctica), but the specimens upon which this record is based have not been studied and may belong to the superficially similar *Phyllophora antarctica*, which is common in Eastern Antarctic waters.

MATERIAL EXAMINED: Macquarie I.: collections MA-98, 99, 302; Herbarium specimens: MELU, R 0055–0057, 1867, 1894–1899.

DESCRIPTION: Plants to 12 cm tall, reddish-brown to rose-red, with discoid holdfasts, stipes to 2 cm long, and dichotomously, palmately, or irregularly branched, strap-like fronds with rounded, truncate or apiculate apices (Fig. 90a, b). Blades crisp, cartilaginous, to 130 μm thick, composed of two layers of large, rounded, colourless medullary cells, a single layer of inner cortical cells, and a single layer of small, roughly isodiametric, outer cortical cells (Fig. 90d, e). Cortical tissues of stipes much thicker (to 75 μm), with subcortex comprising (5-)7-celled anticlinal files, outer cortex (3-)4-layered, with periclinally elongated oval cells (Fig. 90c). Reproductive structures unknown.

Rhodymenia subantarctica was collected from shaded rock walls in 5 m depths in the wave-sheltered waters of Aerial Cove and on shaded boulders along the floor of *Macrocystis* forests in 8–9 m depths. Plants usually were embedded, at least at their bases, in sponge and sometimes were associated with the red algae *Cladodonta lyallii* and *Rhodophyllis acanthocarpa*. Coralline algae, bryozoans and tube worms commonly encrusted *Rhodymenia* fronds.

NOTES: *Rhodymenia* is often confused with *Palmaria* and *Phyllophora*, all three having species with similar vegetative construction. Sporophytic thalli of *Rhodymenia* are readily separated from those of *Palmaria* by using differences in tetrasporangial structures (Guiry, 1975, 1978*b*), whereas sterile thalli of Rhodymenia can be distinguished from those of *Phyllophora* by differences in cortical cell arrangements (Guiry, 1977). Near the tips of *Rhodymenia* fronds, the cortical cells are irregular in size and loosely arranged over the subcortical cells, allowing the latter to be seen from the blade surfaces. In comparable portions of *Phyllophora* fronds, the cells are uniform in size and closely packed, thereby blocking the view of the subcortical cells from blade surfaces. Based on personal observations of Southern Ocean taxa, this distinction holds true for *R. subantarctica* (as known

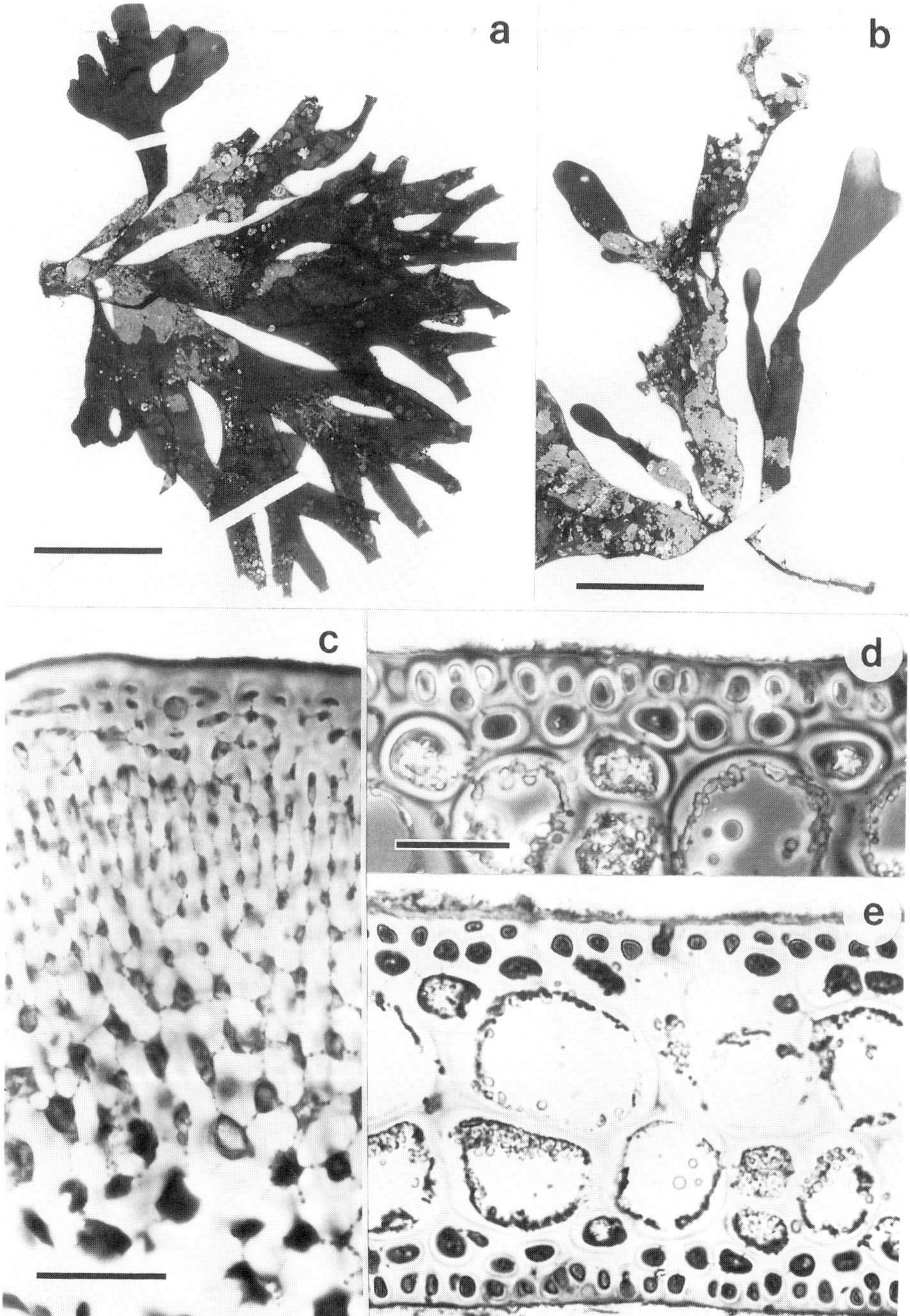

Fig. 90 *Rhodymenia subantarctica* R. W. Ricker: MELU, R 0055, 0056. **a, b.** Habits of sterile thalli. (N.B. Plant 'a' lacking stipe and with apices truncated by grazing invertebrates.) **c.** T.S. through stipe. **d.** Detail of cortex in section through frond. **e.** Section through frond. Scales: a, b = 2 cm; c = 30 μm; d, e = 25 μm.

from Macquarie I.) and two species of *Phyllophora* (*P. appendiculata*—from Macquarie I.; *P. antarctica*—from the Vestfold Hills, East Antarctica). At Macquarie I., sterile plants of *Rhodymenia*, *Palmaria* and *Phyllophora* can also be easily recognized by their distinctive habits.

Palmariales

The Palmariales was established by Guiry & Irvine (Guiry, 1978*b*) to accommodate the single family Palmariaceae, which previously had been placed in the Rhodymeniales. In contrast to the triphasic life history patterns in rhodymenialean algae, diphasic life history patterns, which lack carposporophytes, occur in the Palmariales (Irvine & Guiry *in* Irvine, 1983). Pueschel & Cole (1982) have supplied ultrastructural evidence further supporting the order.

Two especially distinctive features of palmarialean algae are the presence of dimorphic gametophytes in species that are dioecious and successive production of tetrasporangia from generative 'stalk cells' (Kraft, 1981). Gametophytes are either monoecious (Irvine & Guiry *in* Irvine, 1983) or dioecious. In the latter instance, male plants are macroscopic and morphologically similar to sporophytic thalli, whereas female plants are smaller (often microscopic) and produce sessile carpogonia (van der Meer & Todd, 1980). Tetrasporophytes develop directly from zygotes contained within the carpogonia, the female gametophytes then becoming overgrown and incorporated into the bases of the developing tetrasporophytes. Sporangia are cruciately divided, embedded in the cortex, either localized in sori or scattered across thallus surfaces, and are subtended by sporangial mother cells that appear as short stalks and are capable of repeatedly producing tetrasporangia (Guiry, 1974, 1978*b*).

Palmariaceae Guiry, 1974: 522

Members of this family are vegetatively similar to certain taxa in the Rhodymeniales, Gracilariaceae and Phyllophoraceae. Thalli are multiaxial, pseudoparenchymatous, free-living or parasitic, solid or hollow, and either encrusting, saccate, or foliose. Medullary cells are large, colourless, loosely coherent, nearly isodiametric, and arranged in 1–3 layers. The cortex is either (1-)2-layered or composed of anticlinal filaments with small pigmented cells.

The Palmariaceae includes six genera [*Coriophyllum* (Pueschel & Cole, 1982); *Devaleraea* (Guiry, 1982); *Halosaccion* and *Palmaria* (Guiry, 1974, 1975); *Neohalosacciocolax* (Lee & Kurogi, 1978); *Rhodophysema* (DeCew & West, 1982)]; although Irvine & Guiry (*in* Irvine, 1983) have provisionally added a seventh (*Halosacciocolax*) partly on the strength of ultrastructural studies by Pueschel & Cole (1982). All seven genera occur in boreal waters, but only one, *Palmaria* (= *Leptosarca*), extends into the Southern Hemisphere, where it is widespread in circumpolar seas and abundant at Macquarie I.

Palmaria Stackhouse, 1801: xxxii

Lectotype: *P. palmata* (Linnaeus) O. Kuntze, 1891: 909 (see Ruprecht (1850: 76) re. lectotypification).

Thalli arise from discoid holdfasts and have stipitate flattened fronds that are dichotomously, palmately, or irregularly branched and often proliferous along margins. Plants are dioecious, with male gametophytes and tetrasporophytes. Female gametophytes are diminutive and bearing erect blades composed of several to many cells clustered into irregularly shaped masses that produce single-celled sessile carpogonia. Male gametophytes develop sori in which the outer cortical cells bear pairs of spermatangial mother cells that elongate anticlinally and give rise terminally or obliquely to successive crops of spermatangia (Guiry, 1974). Tetrasporangia are cruciately divided, subtended by generative stalk cells, and distributed in cortical tissues diffusely over blade surfaces.

Palmaria seems to have been considered to be a genus with primarily boreal distribution. Recently, it has been reported to contain two species, the type (*P. palmata*) and another species transferred from *Rhodymenia*, *P. stenogona* (Perestenko) Perestenko (1980). Gepp & Gepp (1905*a*,

1905*b*) described and illustrated the genus *Leptosarca* based on sterile plants from the Southern Hemisphere. The anatomy of this genus is identical to that of *Palmaria*, and the present observations of stalked tetrasporangia in various members of this southern genus have led to synonymizing it with *Palmaria*. The genus is represented at Macquarie I. by two species, both based on Reinsch (1888*b*) specimens.

Palmaria decipiens (Reinsch) R. W. Ricker comb. nov.

Figs 91, 92

BASIONYM: *Rhodymenia decipiens* Reinsch, 1888*b*: 148.

SYNONYMS:
Rhodymenia palmata sensu Reinsch, 1890: 379.
Rhodymenia palmetta var. *multiloba* Reinsch, 1890: 379.
Leptosarca dumontioides A. & E. S. Gepp, 1905*a*: 108.
Leptosarca simplex A. & E. S. Gepp, 1905*a*: 108.
Gracilaria simplex (A. & E. S. Gepp) A. & E. S. Gepp, 1905*c*: 195.
Gracilaria dumontioides A. & E. S. Gepp, 1907: 10.
Leptosarca decipiens (Reinsch) Skottsberg *in* Kylin & Skottsberg, 1919: 26.
Leptosarca antarctica Skottsberg *in* Kylin & Skottsberg, 1919: 26.
Leptosarca alcicornis Skottsberg *in* Kylin & Skottsberg, 1919: 28.
Rhodymenia (?) *antarctica* Skottsberg, 1953: 561.
Leptosomia alcicornis (Skottsberg) Kylin, 1956: 336.
Leptosomia antarctica (Skottsberg) Kylin, 1956: 336.
Leptosomia decipiens (Reinsch) Kylin, 1956: 336.
Leptosomia simplex (A. & E. S. Gepp) Kylin, 1956: 336.

TYPE LOCALITY: South Georgia.

LECTOTYPE: M, 67.83/66.

DISTRIBUTION: Adelie Coast (Lucas, 1919); Antarctic Peninsula (Skottsberg, 1906; Gain, 1912); Enderby Land, MacRobertson Coast, Wilkes Land (Skottsberg, 1953); Heard and Macquarie Is, Queen Mary Land; Îles Kerguelen (Levring, 1944); South Georgia (Reinsch, 1888*b*, 1890; Kylin & Skottsberg, 1919); South Orkney Is (Gepp & Gepp, 1905*a*); Victoria Land (Gepp & Gepp, 1907).

MATERIAL EXAMINED:

1. Heard I.: Atlas Cove, on buoy (Chittleborough, 15 December 1949): MELU, R 3757–3759.
2. Macquarie I.: collections MA-34, 46, 111, 137, 264, 271; Herbarium specimens: MELU, R 0126, 1617, 1700, 1744–1749; Garden Cove, on rock shelf below lowest tide (Taylor, 21 March 1951): MELU, R 3751 (= Taylor no. 537) (tetrasporophyte); Buckles' Bay, reef just below low tide mark (Haysom, 13 January 1950): MELU, 3752–3754, 3756.
3. Mawson Coast (Antarctica): Kista Strait, dredge 350–400 m depth (Krummel, 2 March 1982): Herb. Ricker, R 3379.
4. Rauer Is (Queen Mary Land, Antarctica): Unnamed island directly west of the southern tip of Varyag I., 9 m depth (Ricker & Tucker, 23 December 1981): MELU, R 2888.
5. Vestfold Hills (Queen Mary Land, Antarctica): Unnamed island between Sørsdal Glacier and Zolotov and Kazak Is, at mouth of Crooked Fjord, on rocks in 14 m depth (Ricker & Butler, 15 January 1982): MELU, R 3238; Southern point at mouth of Heidemann Bay, 5 m depth (Ricker, Butler, & Tucker, 14 December 1981): MELU, R 2728.

Leptosarca alcicornis
South Georgia: Syntype. Por harbour, Cumberland Bay (Skottsberg, 22 May 1902): UPS, 165.
Leptosarca antarctica
Graham Land (Antarctic Peninsula): Syntypes. St. 4, Louis Philippe harbour, Cape Roquemaurel (63°35′ S lat., 58°35′ W long.), littoral (Skottsberg, 14 January 1902): UPS, two unnumbered specimens.

Fig. 91 *Palmaria decipiens* (Reinsch) R. W. Ricker: MELU, R 0126, 1700, 2888, 3238. **a.** Habit of plants growing with *Durvillaea antarctica* in lower intertidal. **b.** Habit of plant with simple frond. **c.** Habit of plant with branched frond. **d.** Habit of (Antarctic) plant with proliferations from blade faces. **e.** Habit of (Antarctic) plant with laterals produced on blade margins and faces. Scales: b, c, d = 5 cm; e = 3 cm.

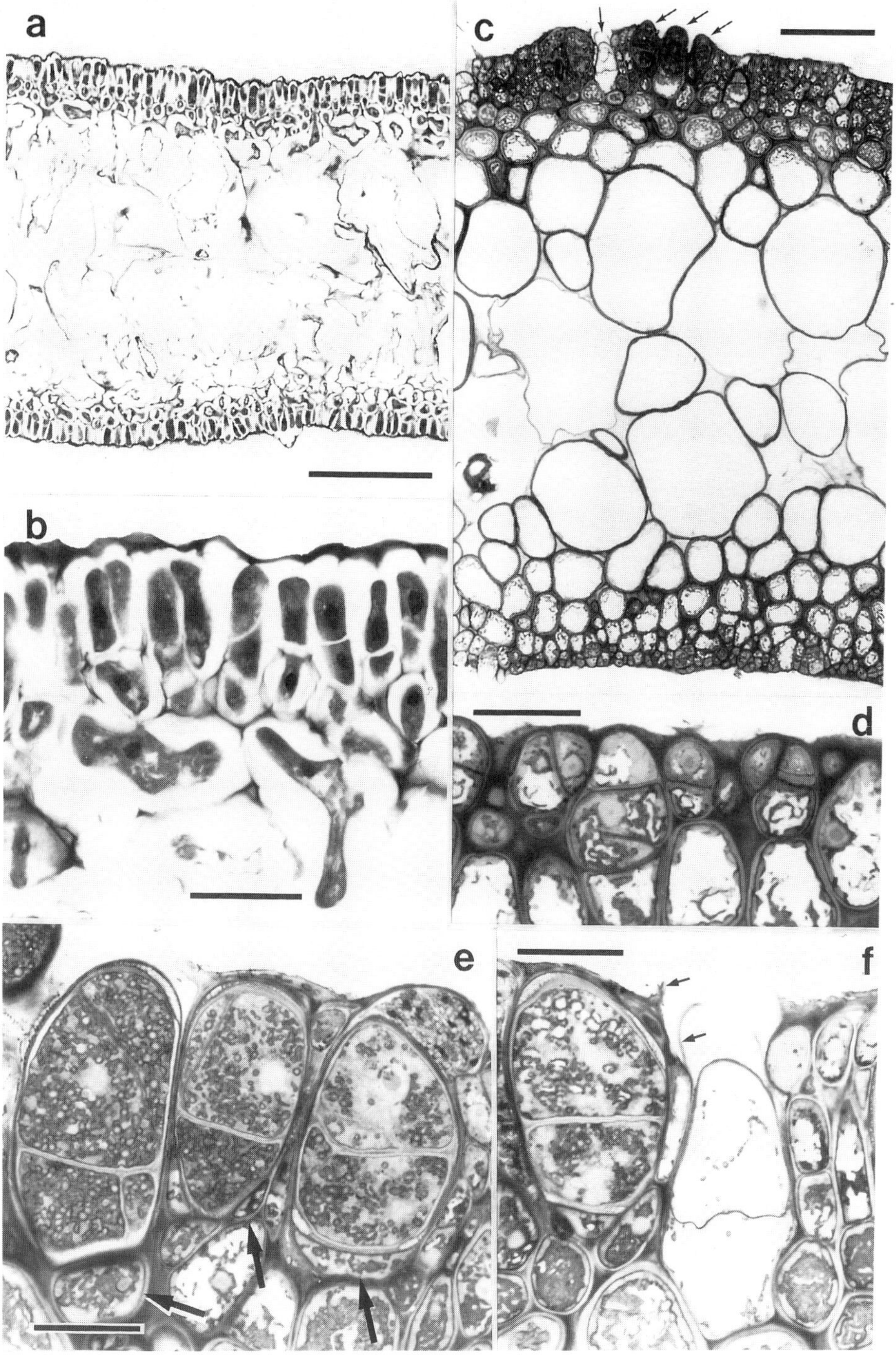

Fig. 92 *Palmaria decipiens* (Reinsch) R. W. Ricker: MELU, R 2728, 3379. **a.** Section through young sporophyte frond. **b.** Detail of cortex from 'a'. **c.** Section through mature sporophyte frond showing tetrasporangia (arrows). **d.** Detail of cortex from 'c'. **e.** Tetrasporangia within stalk cells (arrows). **f.** Proliferation of tetrasporangium within old sporangial walls (arrows). Scales: a, c = 100 µm; b, d–f = 20 µm.

Leptosarca simplex
Graham Land: St. 5, Paulet I., 100–150 m depth (Skottsberg, 15 January 1902): UPS, unnumbered specimen.
Rhodymenia decipiens
South Georgia: Lectotype. (Will, 1883): M, 67.83/66.
Rhodymenia palmata sensu Reinsch
South Georgia: (Will, February 1883): M, 67.83/85, 67.83/86.
Rhodymenia palmetta var. *multiloba*
South Georgia: Lectotype. 'Strand unterhalb der Station, durch Sturm ausgeworfen' (Will, February 1883): M 67.83/84 (= Will, no. 64).

DESCRIPTION: Tetrasporophytes and sterile thalli 35–45(–60) cm tall, reddish-brown to brownish-purple to dull purple, arising from discoid holdfasts, with terete stipes grading through flattened cuneate apophyses to linear and strap-like or broad and foliose fronds that have slippery glossy surfaces and are simple to dichotomously to palmately to irregularly branched or sometimes proliferous from thallus margins or blade (Fig. 91a–e). Fronds to 300 μm thick, medulla comprising 2–3 layers of large nearly isodiametric cells either grading outwards to smaller, rounded, colourless cells (Fig. 92a, c) or abruptly abutting (1-)3-layered cortex composed of rounded to anticlinally elongate, pigmented cells (Fig. 92b, d). Tetrasporangia cruciately divided, 25–30 μm broad × 50–55 μm long, successively produced by stalk cells within walls of discharged sporangia, interspersed between anticlinal files comprising 3–5 slightly elongate cortical cells (Fig. 92e, f). Gametophytes not observed in the field.

Tetraspores released from cultures of Antarctic plants germinated into either discs that gave rise to erect fronds (presumably males thalli, but reproductive structures never developed) or microscopic, spherical to irregularly shaped cell-masses 50–75 μm diameter (female gametophytes) producing slightly bulging sessile carpogonia with colourless, slender and terete, simple or branched trichogynes 250–550 μm long. Trichogynes began appearing shortly after the tetraspores had settled and divided a few times, with more than one usually borne on each young female gametophyte. This is similar to female gametophyte development in *Palmaria palmata* (van der Meer & Todd, 1980).

Plants usually form extensive mats covering surf-exposed rocky faces of reef platforms in the lower intertidal immediately above or slightly overlapping the upper limit of the *Durvillaea* zone. Within this belt, *P. decipiens* and *P. georgica* both occur, although the former is more abundant and the latter mostly restricted to crevices. A barren rocky strip often delimits the upper edge of the mat, but clumps of *Adenocystis* and *Chaetangium* sometimes border it. Kenny & Haysom (1962) and Simpson (1976) both referred to the *Palmaria* belt as an 'upper red zone' dominated by *Rhodymenia sp.*, having considered *P. decipiens* and *P. georgica* as one polymorphic entity.

NOTES: Gain (1912, pl. vi) illustrated some of the extreme morphological variation that occurs in *P. decipiens*. Its fronds range from simple and strap-like (comparable to the illustration of *Leptosarca simplex* by Gepp & Gepp (1905*a*) to palmately to irregularly lobed and sometimes proliferous (equivalent to the illustrations and description of *Rhodymenia* (?) antarctica by Skottsberg (1953)). Of the two new species, *Leptosarca alcicornis* Kylin & Skottsberg and *Leptosarca antarctica* Skottsberg, and one new combination, *Leptosarca decipiens* (Reinsch) Skottsberg, proposed by Kylin & Skottsberg (1919), Skottsberg (*in* Kylin & Skottsberg, 1919) considered *L. decipiens* closest to the type, *L. simplex*. He distinguished each of the four species on variable characteristics such as blade colour (*L. simplex*—carmine to aniline-red), small cortical cell size (*L. decipiens*), highly compacted internal tissues coupled with distinctive breadth to length blade ratios (*L. antarctica*), and high degree of frond splitting (*L. alcicornis*). After studying over 200 specimens, however, I have observed a range of intermediate conditions in all of these characters that blurs such distinctions.

Palmaria decipiens is morphologically similar to the type species, *P. palmata*, but probably should be considered as a separate entity until comparative studies can be undertaken. One distinctive feature of southern representatives of the genus is the relatively rapid softening of blades following fixation in 4% formalin-seawater. Although this feature has not been described for *P. palmata*, the tendency to soften is taxonomically useful for distinguishing *Palmaria* species from other Southern Ocean foliose red algae, which generally remain firm after formalin preservation. This change in thallus texture is probably responsible for the poor rehydration of tissues following drying. Internal anatomy of fronds is best studied using wet material, for medullary cells collapse upon drying and usually resist rehydrating treatment (Skottsberg *in* Kylin & Skottsberg, 1919).

Palmaria georgica (Reinsch) R. W. Ricker comb. nov.

Fig. 93

BASIONYM: *Rhodymenia georgica* Reinsch, 1888*b*: 147.

SYNONYMS: *Rhodymenia palmatiformis* Skottsberg *in* Kylin & Skottsberg, 1919: 21 (including all forms described in Kylin & Skottsberg (1919) but considered by Skottsberg (1941*a*) to be synonymous with this species).

TYPE LOCALITY: South Georgia.

LECTOTYPE: M, 67.83/81

DISTRIBUTION: Campbell I. (Levring, 1945); Falkland Is (Kylin & Skottsberg, 1919); Fuegia (Levring, 1960); Heard I.; Îles Kerguelen (Zinova, 1973); Macquarie I. (Law & Burstall, 1956; Zinova, 1958, 1963, 1964); Queen Mary Coast (Zinova, 1958); South Georgia (Reinsch, 1888*b*; Kylin & Skottsberg, 1919). (N.B. Papenfuss (1964*a*) additionally indicated distribution at Victoria Land, but the original published record for this locality has not been located for the present report.)

MATERIAL EXAMINED:

1. Heard I.: North of Wharf Pt., Atlas Cove, or rocky surfaces in the lower eulittoral (Smith, 6 February 1983): MELU, R 3637–3641.

2. Macquarie I.: collections MA-119, 120, 129, 130, 131, 133, 213, 278; Herbarium specimens: MELU, R 1871–1882, 1893; Buckles' Bay, clefts and channels in intertidal reef (Haysom, 13 January 1950): MELU, R 3760–3762; Garden Bay, intertidal (Kenny, 8 February 1949): MELU, R 3763.

Rhodymenia georgica

South Georgia: Lectotype. 'Klippen am Nordufer der Landzunge, bis zur Grenze des Niedrigwassers' (Will, 3 July 1883): M, 67.83/81 (= Will no. 66).

Rhodymenia palmatiformis

South Georgia: Pinguinbay (Mosthaff, 3 July 1883): S, unnumbered specimen (det. Skottsberg, 1940).

Rhodymenia palmettoides (Bory de Saint-Vincent) var. *austrogeogica* Skottsberg

South Georgia: Station 14, Cumberland Bay, Jason Harbour (Skottsberg, 23 April 1902): UPS, unnumbered specimen.

DESCRIPTION: Plants to 6 cm tall, purple-brown to reddish-brown, with terete stipes grading through flattened cuneate apophyses to strap-like, simple or dichotomously to irregularly branched fronds with rounded to acute apices (Fig. 93a, b). Blades 240–260 μm thick (Fig. 93c); medulla comprising 1–2 layers of large, colourless, nearly isodiametric cells; cortex monostromatic in upper portions, otherwise distromatic, becoming thicker and noticeably filamentous with anticlinal files 3–4 cells long in sporophytic fronds. Outer cortical cells paired, usually anticlinally elongate, borne on rounded subcortical cells, sometimes producing unicellular hairs (Fig. 93d).

Fig. 93 *Palmaria georgica* (Reinsch) R. W. Ricker: MELU, R 1893. **a, b.** Plant habits. **c.** Section through mid-thallus showing organization of pigmented cortical and subcortical cells and colourless medullary cells. **d.** Cortical hairs. **e, f, g.** Tetrasporangia in various stages of maturity. Note stalk cells (arrows) in 'e'. **h.** Section through cortex showing immature tetrasporangia (arrows). Scales: a, b = 1 cm; c = 50 μm; d–g = 20 μm; h = 25 μm.

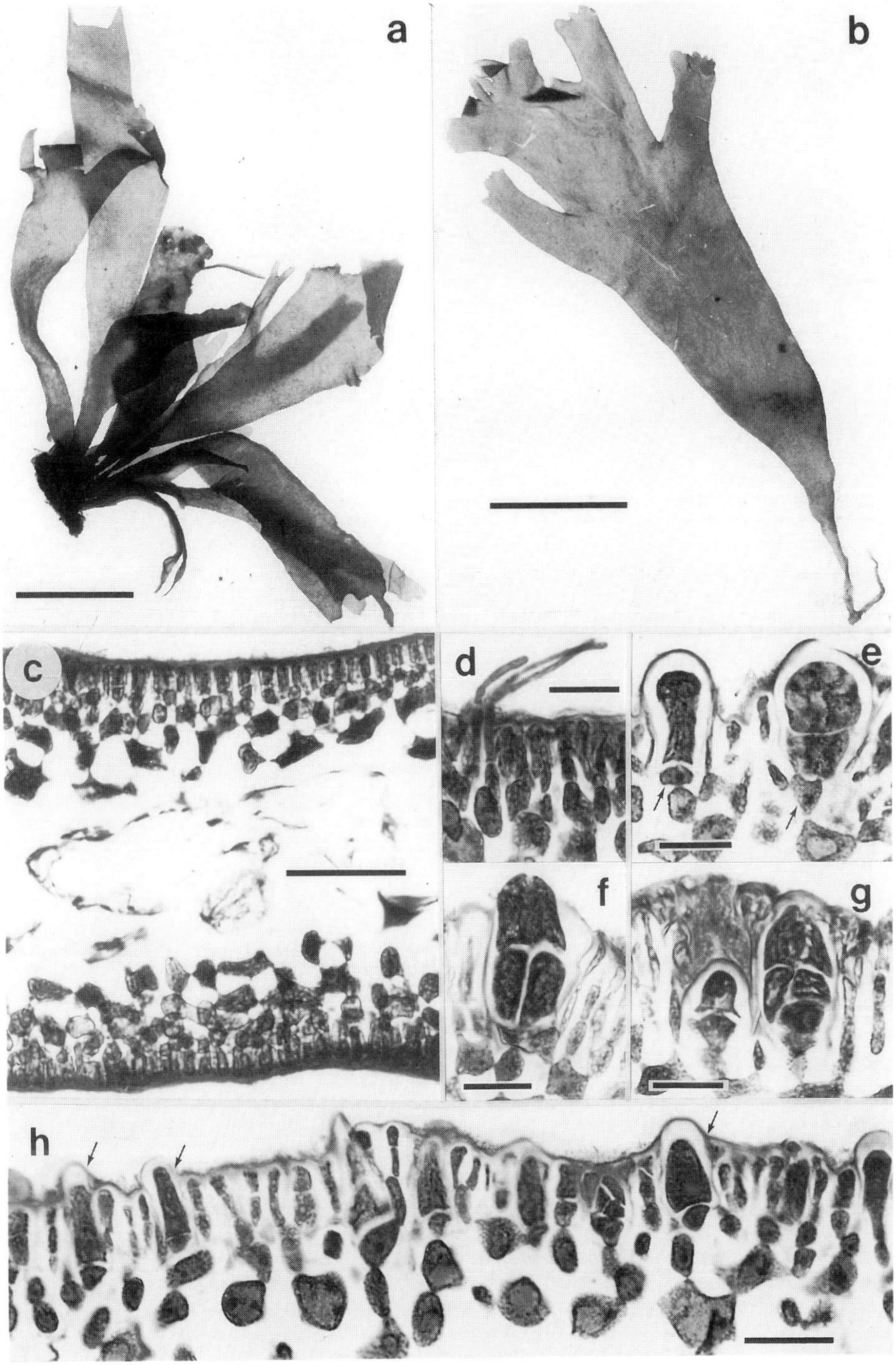
a
b
c
d
e
f
g
h

Tetrasporangia cruciately divided, oval, 26 µm broad × 46 µm long, borne on stalk cells, scattered diffusely throughout cortex (Fig. 93e–h). Gametophytes unknown.

Plants are gregarious, with numerous fronds arising from a single discoid base (Fig. 93a), and line clefts and channels of rocky platforms in the lower intertidal immediately above the upper limit of *Durvillaea*. *Rhizoclonium ambiguum* and *Iridaea remuliformis* often grow intermixed with this species. *P. decipiens*, although growing adjacent to *P. georgica*, occupies seemingly more exposed rocky surfaces that do not retain water as well as the creviced rock habitats dominated by *P. georgica*.

NOTES: *Rhodymenia palmetta* var. *multiloba* Reinsch has been synonymized with *P. decipiens* but is intermediate in size between *P. georgica* and larger individuals of *P. decipiens*. The degree to which environmental conditions determine blade morphologies is unknown and it is possible that thallus variation is even greater than presently conceived. It may be that *P. georgica* is merely a diminutive form or an ecotype of *P. decipiens*.

Although thallus morphologies of the two species do overlap (cf. Figs 91d, 93b), both taxa are recognized in the present report because *P. georgica* is more gregarious, bushier, forms tufts composed of many fronds, produces cortical hairs, and occurs in separate habitats from *P. decipiens*.

Ceramiales

The Ceramiales contains approximately half the genera and a third of the species in the Rhodophyta, and in recent times has been the focus of intensive taxonomic research (Kraft, 1981). In spite of its broad diversity of forms, reproduction in the group is extremely uniform, with gametophytes forming procarps in which auxiliary cells develop after fertilization (Kylin, 1956). A few exceptions to this have been observed, however, in which supporting cells directly bear, or are suspected to bear, gonimoblast initials instead of auxiliary cells (Brauner, 1979; Gordon-Mills & Kraft, 1981). Most taxa, therefore, are distinguished by vegetative features.

The families Ceramiaceae, Delesseriaceae, Dasyaceae and Rhodomelaceae constitute the order, with the four purported to total over 150 species in high-latitude southern seas (Papenfuss, 1964*a*). All four families occur at Macquarie I., the first two comprising the largest number of species.

Ceramiaceae Dumortier, 1822: 71, 100

The Ceramiaceae is considered to be the most primitive family in the Ceramiales (Parsons, 1975). Its members often are delicate and generally consist of monopodially branched monosiphonous filaments that lack pericentral cells (*sensu* Min-Thein & Womersley, 1976) but sometimes become thickened by cortication that results in a pseudoparenchyma. Consolidated pericarps generally are not produced (Kraft, 1981; Dixon, 1982). Over 100 genera are included in the family (Wynne & Kraft, 1981), some of which have been studied critically in various publications dealing with either selected tribes (e.g., Cruanieae, Spyridieae and Ceramieae—Hommersand, 1963; Crouanieae—Wollaston, 1968; Wrangelieae, Sphondylothamnieae and Spermothamnieae—Gordon, 1972; Antarcticothamnieae and Ptiloteae—Moe & Silva, 1979) or regional floras (e.g., Argentina—Mendoza, 1969*a*; California—Abbott & Hollenberg, 1976; Japan—Itono, 1977).

Genera are distinguished partly on the location and characteristics of reproductive structures, but more on vegetative features such as thallus habit, branching patterns, degree of cortication, presence or absence and structure of gland cells, and the number of nuclei per cell.

At least 13 genera of Ceramiaceae* occur in high-latitude southern waters (pers. obs.), with an additional five (*Aglaothamnion, Carpoblepharis, Halurus, Microcladia, Pleonosporium*) not confirmed through personal observations, but also listed from this vicinity (Papenfuss, 1964*a*). The Ceramiaceae is represented at Macquarie I. by 11 species from 8 genera. All, save a newly described species of *Antithamnionella*, have been previously reported from various Subantarctic localities.

Antithamnionella Lyle, 1922: 347

Lectotype: *A. sarniensis* Lyle, 1922: 348.

Lyle (1922) distinguished *Antithamnionella* from the similar-appearing *Antithamnion* by differences in the seemingly variable patterns of branching and tetrasporangial division. Kylin (1956) and Wollaston (1968, 1971*a*, 1971*b*) have further supported autonomy of the genus by supplying additional diagnostic features. *Antithamnionella*, as circumscribed by Wollaston (1968), has erect uncorticated thalli with distinct central axes which are monosiphonous and bear variable numbers of whorl-branches (= 'whorl-branchlet' or 'pinnae' of Wollaston, 1968) from each axial cell. Attachment is generally by creeping filamentous bases, but some species apparently lack prostrate axes. Indeterminate branches, where present, replace whorl-branches. Gland cells are sessile and usually singly produced on the adaxial surface of the second to fifth cell of whorl-branches, although *A. glandifera* Wollaston (1968) bears multiple gland cells on both the abaxial and adaxial surfaces of whorl-branches. Spermatangial mother cells are laterally initiated on short, adaxial branches (= 'pinnules' of Wollaston, 1968) along whorl-branches. Carpogonial branches are 4-celled and borne singly on the basal cells of reduced whorl-branches near indeterminate branch apices, with fertile axial cells producing only one whorl-branch. Tetrasporangia are ovoid to sub-spherical, cruciately, tetrahedrally, or obliquely divided, and sessile or stipitate.

Antithamnionella species are distinguished largely on differences in habit, branching uniformity, the presence or absence of gland cells, and the positioning of tetrasporangia. Using such criteria, 14 species have been described previously from tropical to cold temperate waters of both hemispheres (Lyle, 1922; Baardseth, 1941; Wollaston, 1968; Abbott & Hollenberg, 1976; Abbott, 1979; Itono, 1977), but only three of these [*A. australis* Baardseth; 1941 [*A. seriata* Baardseth, 1941; and *A. ternifolia* (Hooker f. & Harvey) Lyle, 1922] occur in the Subantarctic. The genus is represented at Macquarie I. by *A. ternifolia* and a newly described species.

Antithamnionella alternans R. W. Ricker sp. nov.

Fig. 94

Holotype: (Fig. 94c): MELU, R 3767 (= Microscope slide no. R/M 0420 (male), with isotype in Herb. Ricker.

Type locality: Anchor Rock, Macquarie I. (54°29′15″ S lat., 158°56′ 50″ E long.), growing at a depth of 13 m in association with *Heterosiphonia berkeleyi* (Ricker, 8 December 1977).

Distribution: Macquarie I.

Material examined: Macquarie I.: collections MA-94, 95; Herbarium specimens: MELU, R 1785, 3312; Microscope slide: MELU, R/M 0420.

Diagnosis: Thalli pallide saturatiusve rosei; axibus erectis 500–1000 µm altitudine, orientibus ex filis prostratis quae rhizoidea, quodque pulvillo terminali discoideo pro suffictione, ferunt.

*Including: *Antarcticothamnion* [probably including *Spongoclonium orthocladum* A. & E. S. Gepp (pers. obs.)], *Antithamnionella, Ballia, Callithamnion, Ceramium, Dasyptilon* [probably including *Antithamnion ptilota* (Hooker f. & Harvey) Gibson (pers. obs.)], *Euptilota, Falklandiella, Georgiella, Griffithsia, Medeiothamnion, Plumariopsis,* and *Pterothamnion.*

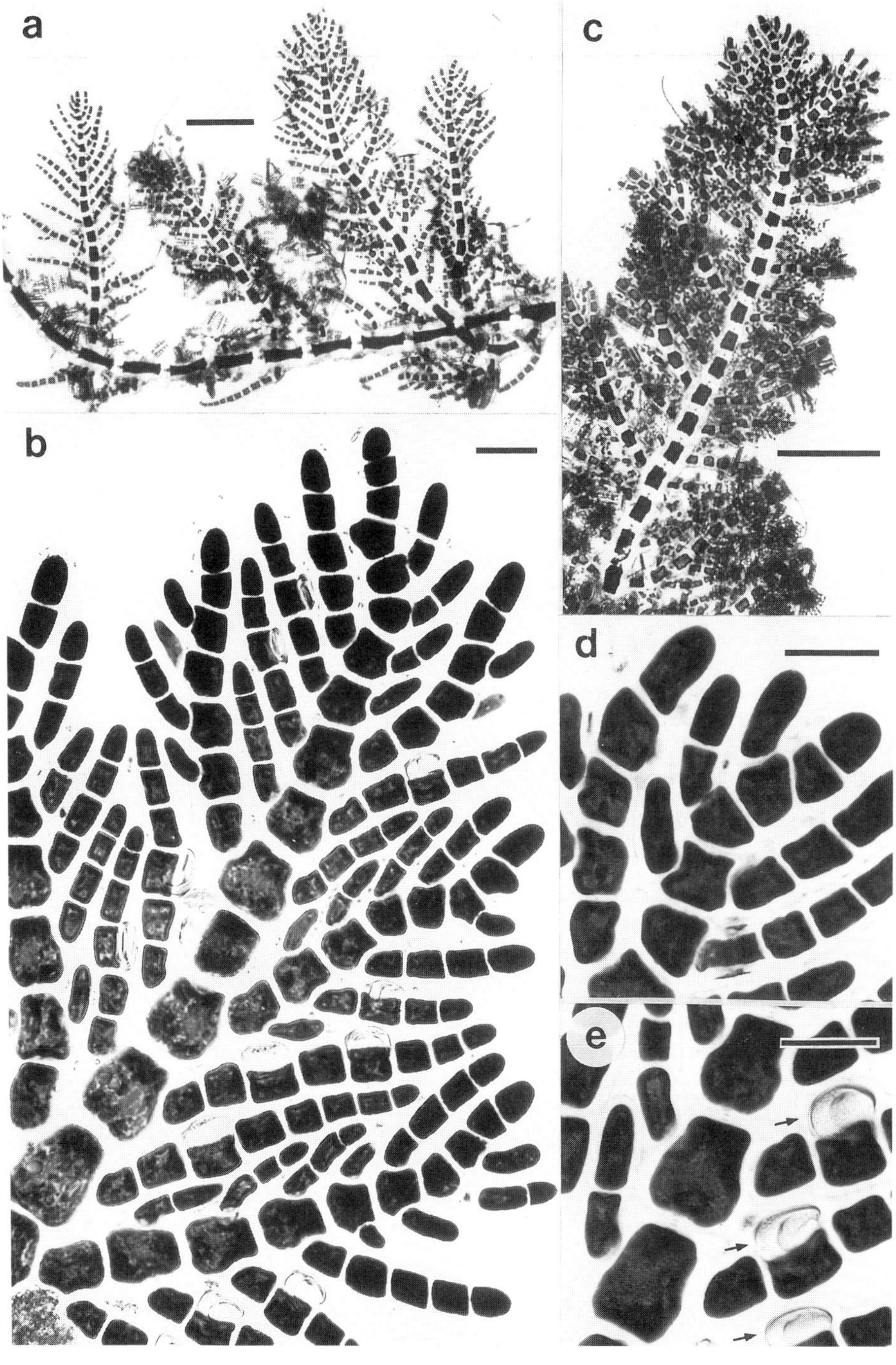

Fig. 94 *Antithamnionella alternans* R. W. Ricker: MELU, R 1785, 3312, 3767. **a.** Habit of isotype (R 3767) showing prostrate and erect axes. **b.** Detail of upright axis. **c.** Portion of holotype (R 3767—male gametophyte). **d.** Apical portion of main axis. **e.** Gland cells (arrows). Scales: a, c = 100 µm; b, d, e = 20 µm.

Ramificatio axium erectorum disticha; determinatis verticillorum ramis simplicibus, rectis, cellulis apicalibus transverse dividentibus, cellulis basalibus alias aequantibus, plerumque paribus oppositis, interdum mucronatis autem frequentius obtusis; indeterminatis alternis, ramo determinato unico in omni altero pari generatim permutatis, horum partibus superioribus valde flexuosis, cellulis apicalibus versus latera alterna oblique dividentibus. Glandulae infrequentes, ovales, ramis determinatis insidentes, quaeque paginam cellulae unicae adaxialem tegens. Maternicellulae spermatangiales verticillatae terna quaternave, laterale ex cellulis axialibus ramulorum brevium dense fasciculatorum qui in ramis determinatis adaxialiter vel interdum abaxialiter exoriuntur. Gametophyta feminea tetrasporophytaque ignota.

Antithamnionella alternans is named for the regular alternation of indeterminate lateral branches along opposite sides of main axes.

DESCRIPTION: Thalli pale to deep rose-coloured, with erect axes 500–1000 μm tall arising from prostrate filaments bearing rhizoids with terminal discoid attachment pads (Fig. 94a). Branching opposite-distichous in one plane. Whorl-branches (determinate) usually paired, simple, straight, with basal cells *c.* equal in size to other whorl-branch cells. Apical cells of whorl-branches transversely dividing, sometimes mucronate but usually obtuse. Indeterminate laterals generally arising alternately from successive cells of main axes and replacing one (determinate) member of each whorl-branch pair (Fig. 94b). Upper portions of indeterminate branches strongly flexuous, with apical cell divisions oblique in alternate directions (Fig. 94b, d). Gland cells infrequent, ovoid, and covering adaxial surfaces of single cells on determinate laterals (Fig. 94e). Spermatangial mother cells produced laterally in whorls of 3–4 from axial cells of short branches developed in dense clusters on adaxial, sometimes abaxial, surfaces of determinate laterals (Fig. 94c). Female gametophytes and tetrasporophytes unknown.

Only a few plants were collected, all of which were inconspicuous in the field and either epizoic on hydroid tubes or associated with the red algae *Heterosiphonia berkeleyi*, *Polysiphonia anisogona*, and an unidentified member of the Acrochaetiaceae. Because of its infrequent occurrence in collections and single known locality (Anchor Rock), *A. alternans* is presumed to be rare at Macquarie I.

NOTES: Most species of *Antithamnionella* generally have variable branching patterns and numbers of whorl-branches, in contrast to the regular arrangement (either decussate or distichous) and number of laterals in species of *Antithamnion*. In this regard, *A. alternans* differs from other *Antithamnionella* species and strongly resembles members of *Antithamnion*. The latter, however, differ from *A. alternans* in having whorl-branches with basal cells smaller than adjacent cells of the branch. Additionally, gland cells in *Antithamnion* overlap several cells of specialized short branches that develop on whorl-branches (Wollaston, 1971*b*), rather than overlying single cells of whorl-branch axes as typically occurs in *Antithamnionella*. Spermatangial mother cells in *Antithamnion* are borne in terminal whorls and not laterally as in Antithamnionella (Wollaston, 1968).

Antithamnionella alternans is much smaller than other species in the genus. This difference in size, combined with a distichous branching pattern and uniform alternation of indeterminate laterals to opposite sides along main axes, separates it from *A. ternifolia*, the only other member of the genus at Macquarie I.

Antithamnionella ternifolia (Hooker f. & Harvey) Lyle, 1922: 350

Fig. 95

BASIONYM: *Callithamnion ternifolium* Hooker f. & Harvey, 1845: 272.

SYNONYMS: see Pujals, 1963.

TYPE LOCALITY: St. Martin's Cove, Cape Horn.

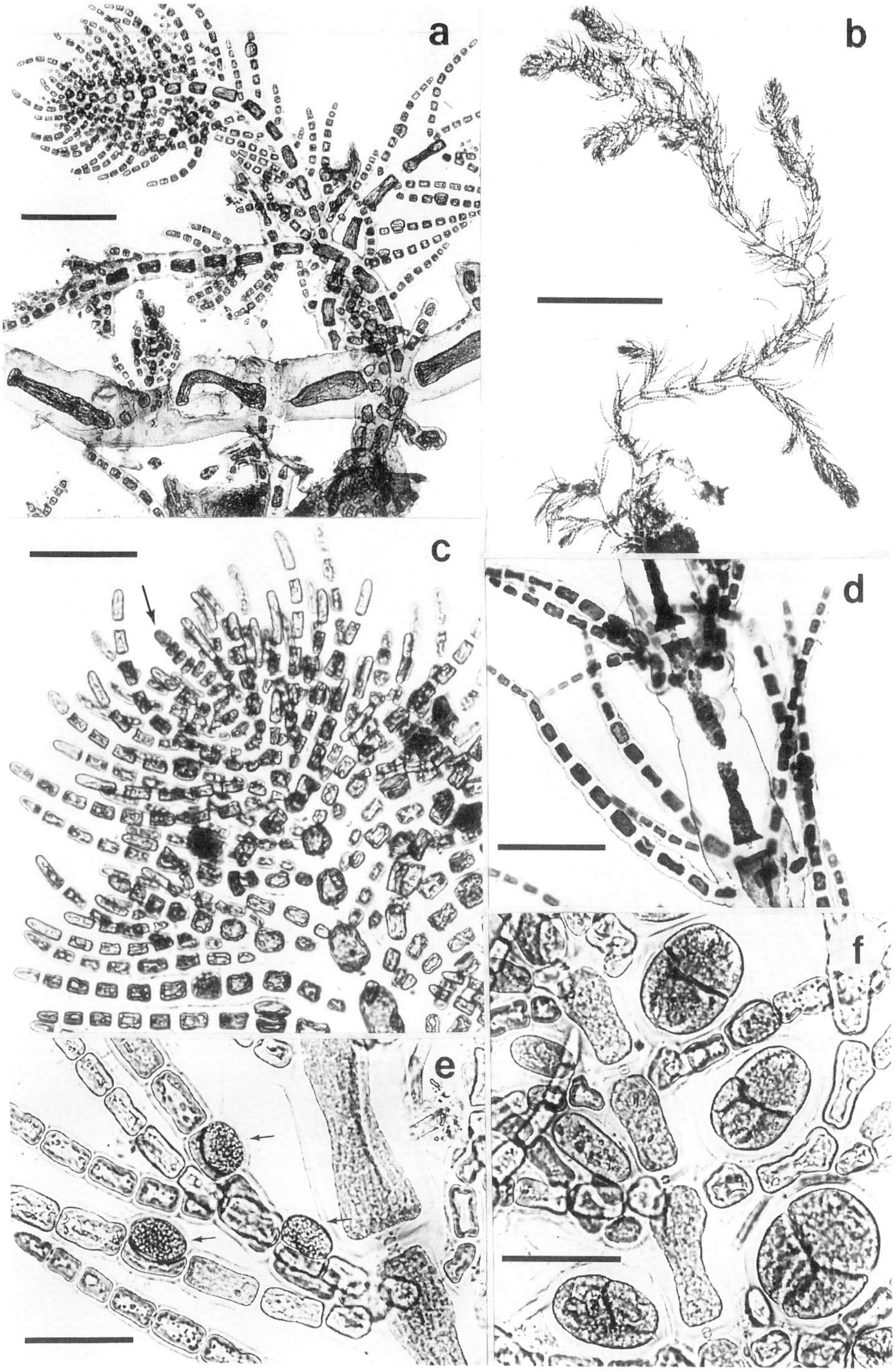
a
b
c
d
e
f

LECTOTYPE: BM; specimen designated by M. H. Hommersand (November 1978) as 'Holotype?', later selected by J. H. Price for lectotypification. Herbarium sheet with J. D. Hooker annotation: 'St. Martin's Cove, Cape Horn (in deep water) . . .'; (Price, pers. comm.).

DISTRIBUTION: Argentina (Mendoza, 1969*a*); Fuegia (Hooker f. & Harvey, 1845); Macquarie I.; New Zealand (Laing, 1905).

MATERIAL EXAMINED: Macquarie I.: collections MA-94, 98, 364; Herbarium specimens: MELU, R 0496, 3311; Microscope slides R/M 0119, 0415.

DESCRIPTION: Thalli epiphytic, with erect axes to 8 mm tall, pink to rose-coloured, arising from prostrate filaments bearing attachment rhizoids (Fig. 95a, b). Branching radial, with (determinate) whorl-branches usually tetrastichous (Fig. 95d), except near indeterminate branch apices, where 1–3 whorl-branches arise from axial cells and overtop apices of bearing axes (Fig. 95c). Whorl-branches simple, borne on upper ends of axial cells, gradually tapering from bases to rounded apices, and with cells generally of similar size throughout. Indeterminate branches irregularly produced and replacing one determinate branch of whorl; apical cells of indeterminate axes dividing slightly obliquely causing upper portions of branches to be weakly flexuous. Gland cells ovoid, solitary, and overlying single cells on adaxial surfaces (usually 2–6 cells from bases) of whorl-branches (Fig. 95e). Tetrasporangia tetrahedrally divided, ovoid to subspherical, *c.* 40 μm broad × 50 μm long, and sessile on adaxial surfaces of basal cells of whorl-branches (Fig. 95f). Sexual reproductive structures not observed.

Antithamnionella ternifolia was collected infrequently and only from the subtidal in 5–20 m depths at the north end of the island. Plants usually grew prostrate across the flabellate blades of *Microzonia velutina* and less often across encrusting coralline algae or on the bases of *Cladodonta lyallii* and *Myriogramme livida*.

NOTES: Macquarie and South American plants are morphologically identical, although the latter are reported to be larger (to 2–4 cm long) and are also known by their female gametophytes (Mendoza, 1969*a*).

Ballia Harvey, 1840: 191

Type species: *B. brunonia* Harvey, 1840: 191 (= *B. callitricha* (C. Agardh) Kützing, 1843: 293).

Ballia is defined mostly by vegetative features, with details concerning sexual reproduction (known for only two Australian species) suggesting its placement in the Tribe Antithamnieae (Wollaston, 1974). Wollaston (1968) characterizes the genus as having densely corticated filamentous thalli that are stiff, erect, and distichously or tristichously branched. Apical cells are prominent, pit connections are distinctive with caps on either side, gland cells are absent, and holdfasts are composed of matted rhizoids. Cortical rhizoids develop from the basal cells of whorl-branches, beginning near indeterminate branch apices, and descending basipetally along main axes and short branches.

Thirteen species of *Ballia* have been described (*Index Nominum Algarum*), with one reported from a freshwater habitat in Malaysia (Kumano, 1978). Of the six species documented from southern Australia, Wollaston (1968) suggests that two, *B. hirsuta* and *B. scoparia*, probably should be excluded from the genus. More recently, Pujals (1981) erected a new genus in the Acrochaetiaceae based on Montagne's (1839) *Conferva oxyclada* (*Camontagnea oxyclada* (Montagne) Pujals), with *B. scoparia* considered to be a later synonym. The genus is represented in high-latitude southern waters by *B. callitricha* (Papenfuss, 1964*a*).

Fig. 95 *Antithamnionella ternifolia* (Hooker f. & Harvey) Lyle: MELU, R 0496, 3311; Slides R/M 0119, 0415. **a.** Habit showing erect and prostrate axes. **b.** Habit of erect portion. **c.** Apex (arrow) of indeterminate axis. **d.** Tetrastichous whorl-branches on main axis. **e.** Gland cells (arrows). **f.** Tetrasporangia. Scales: a, d = 100 μm; b = 1 mm; c = 50 μm; e, f = 40 μm.

Ballia callitricha (C. Agardh) Kützing, 1843: 293

Fig. 96

BASIONYM: *Sphacelaria callitricha* C. Agardh, 1824: 166.

SYNONYMS: see Lamb & Zimmermann, 1977.

TYPE LOCALITY: Falkland Is.

TYPE: Herb. Agardh (LD), no. 19357 (illustrated by Lamb & Zimmermann, 1977: 205, Fig. 75).

DISTRIBUTION: Plants are distributed throughout the Antarctic (Skottsberg, 1906; Zinova, 1966; Neushul, 1968; Lamb & Zimmermann, 1977), Subantarctic (Kylin & Skottsberg, 1919; Laing, 1909), and Cold Temperate (Argentina—Mendoza, 1969*a*; Australia—Wollaston, 1968; New Zealand—J. Agardh, 1878; Laing, 1901, 1905) Zones. Levring (1945) previously recorded this species from Macquarie I.

MATERIAL EXAMINED:

1. Heard I.: Southwest Bay (Chittleborough, 25 April 1949): MELU, R 3730 (= Chittleborough Catalogue no. 98, Specimen no. 11).
2. Macquarie I.: collections MA-10, 11, 34, 39, 46, 83, 98, 99, 111, 113, 222, 244, 248, 255, 271, 283, 295, 302, 303, 309, 364, 367; Herbarium specimens: MELU, R 0031, 0872, 1159, 1167, 1171; UC, R 1206.
3. South Georgia: 'Nordstrand der Landzunge' (Will, 3 July 1883): M, 67.83/3.
4. Vestfold Hills, Antarctica: small turf at base of rock wall in 5 m depth (Ricker & Dick, 7 December 1981); Herbarium specimen: MELU, R 2600.

DESCRIPTION: Plants arising from conical basal masses of tightly interwoven rhizoids, forming compact tufts to 5 cm tall or loose bushes to 25 cm tall (Fig. 96a). Thalli rose-coloured to purplish-brown, monosiphonous, and appearing feathery with prominent central axes composed of uninucleate cells distichously bearing oppositely paired, simple or branched whorl-branches, one of which sometimes being replaced by an indeterminate branch. Indeterminate lateral branches morphologically similar to main axes, developing in the same plane as determinate whorl-branches, and often producing flabellate branching habit; numerous 'fans' usually overlapping one another. Upper portions of indeterminate branches ecorticate, with simple mucronate whorl-branches borne along tapered upper halves of axial cells (Fig. 96b, c). Nearly mature whorl-branches in mid-thallus having rounded apices and appearing feathery by bearing branchlets pinnately or bipinnately; fully mature whorl-branches sometimes spicate, with whorl-branch rachis elongating and not producing additional branchlets from distal axial cells. Lower axes corticated, composed of elongate (200–225 μm long × 55–85 μm diameter) parallel-walled cells often lacking whorl-branches. Cortication dense along lower axes, produced by basipetal rhizoids and short arcuate branches arising from the basal cells of whorl-branches (Fig. 96b). Reproductive structures not seen.

Ballia is abundant around the entire island and grows from the intertidal to at least 20 m depths. It sometimes occurs in mid-littoral pools, but usually underlies *Durvillaea* fronds and forms dense tufts on rocky surfaces along the sublittoral fringe. In the subtidal, plants are longer, bush-like, and grow on unshaded boulders and rock walls on the floors of kelp forests. In the Vestfold Hills, Antarctica, plants were absent from the intertidal but common in the sublittoral, where they often formed compact tufts beneath the coarse fronds of the brown alga *Himantothallus* or grew as short turfs 2–3 cm tall on boulders unshaded by larger algae (pers. obs.).

Various seaweeds such as *Mesophyllum patena*, *Rhodochorton concrescens*, and *Sphacelaria bornetii* sometimes grow epiphytically on *B. callitricha*, and the marine ascomycete *Spathulospora antarctica* Kohlmeyer (1973) commonly forms dark bodies (ascocarps) inside and on cells of this same host.

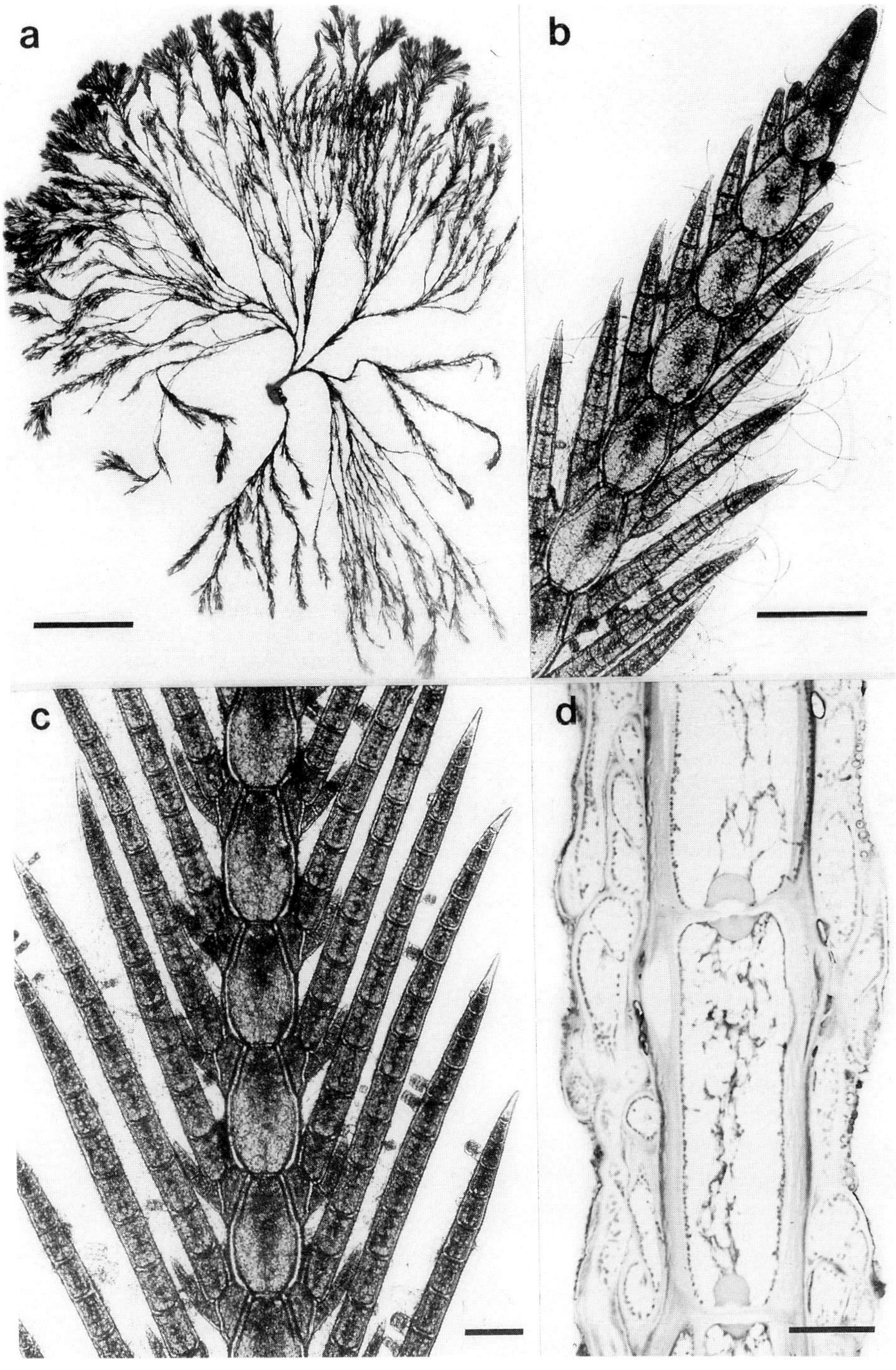

Fig. 96 *Ballia callitricha* (C. Agardh) Kützing: MELU, R 0872, 1206. **a.** Plant habit. **b.** Upper branch. **c.** Mid-thallus showing mucronate whorl-branches. **d.** L.S. lower axis showing cortical rhizoids. Scales: a = 5 cm; b = 200 μm; c = 100 μm; d = 50 μm.

NOTES: Antarctic and Macquarie plants agree closely with Mendoza's (1969*a*) and Wollaston's (1968) accounts on the vegetative structure of *B. callitricha*. Wollaston, however, described axial cells with convex upper and concave lower end walls, whereas those in the present study were usually flat. Neither of the above authors reported gametophytes, but both described cruciately divided tetrasporangia borne on specialized branches produced from the basal cells of whorl-branches.

The presence of *Spathulospora* on *B. callitricha* has been described previously (Kohlmeyer, 1973; Kohlmeyer & Kohlmeyer, 1975), as has the growth of *Mesophyllum* (as *Polyporolithon patena*) on this same species (Levring, 1945).

Callithamnion Lyngbye, 1819: 123

Lectotype: *C. corymbosum* (Smith) Lynbye, 1819: 125.

Callithamnion species have corticated or uncorticated, uniseriate axes arising from rhizoidal bases. Kylin (1956) distinguished *Callithamnion* from other genera in the Ceramiaceae by its monosiphonous construction with multinucleate cells, the production of single laterals from each axial cell, tetrahedral division pattern for tetrasporangia, the development of carpogonial branches from pericentral cells arising laterally near the apices of indeterminate axes, and the presence of gonimoblasts with rounded lobes of carposporangia. Abbott & Hollenberg (1976) presented additional distinguishing criteria including an alternate branching pattern along main axes and major laterals (indeterminate branches), the development of solitary or paired tetrasporangia that usually are sessile and borne on the adaxial surfaces of determinate laterals, the production of spermatangia in flat colourless tufts on the adaxial surfaces of determinate laterals, and the production of carpogonial branches from only one of the two pericentral cells on a fertile axial cell. After fertilization, both pericentral cells develop auxiliary cells that give rise to gonimoblast filaments.

Dixon & Price (1981) reviewed the genus as it occurs in the British Isles and documented considerable variation in some features previously reported as diagnostic for the genus or used to characterize species. They expanded the concept of *Callithamnion* to include plants with uninucleate and multiuncleate cells, alternate, spiral, distichous, or subdichotomous branching, and several types of asexual sporangia (e.g., bisporangia, parasporangia, tetrasporangia, or a combination of these structures). Although thallus colour, plastid shape, presence or absence of unicellular hairs, and the degree of cortication vary greatly depending upon plant age and environmental conditions, Dixon & Price use some of these features, along with thallus habit, nuclear condition, and cell sizes, to distinguish species.

Kylin (1956) credited the genus with approximately 50 species from tropical to sub-polar seas in both hemispheres. Baardseth (1941) described two species from Tristan da Cunha and Papenfuss (1964*a*) listed another eight from more southern Subantarctic waters, with none present in the Antarctic. The genus is represented at Macquarie I. by *Callithamnion gracile* Hooker f. & Harvey.

Callithamnion gracile Hooker f. & Harvey *in* Harvey & Hooker f., 1845: 191

Fig. 97

TYPE LOCALITY: Campbell I.

LECTOTYPE: BM, no. 1311 (6).

DISTRIBUTION: Campbell I. (Hooker f. & Harvey, 1845); Macquarie I. (Zinova, 1958).

Fig. 97 *Callithamnion gracile* Hooker f. & Harvey: MELU, R 0583, 1932, 1983. **a.** Habit of tetrasporophyte epiphytic on *Lophurella hookeriana*. **b.** Portion of upper thallus stained with acetocarmine to show multinucleate (arrows) cells. **c.** Upper branch bearing tetrasporangia. **d.** Detail of tetrasporangia with rounded tetraspores. **e.** Habit of spermatangia-bearing axes. **f.** Detail of spermatangial clusters. **g.** Female branch with fertile pericentral cells (arrows). **h.** Carposporophyte with angular gonimolobes. Scales: a = 4 mm; b, d, f = 50 μm; c, e, h = 100 μm; g = 25 μm.

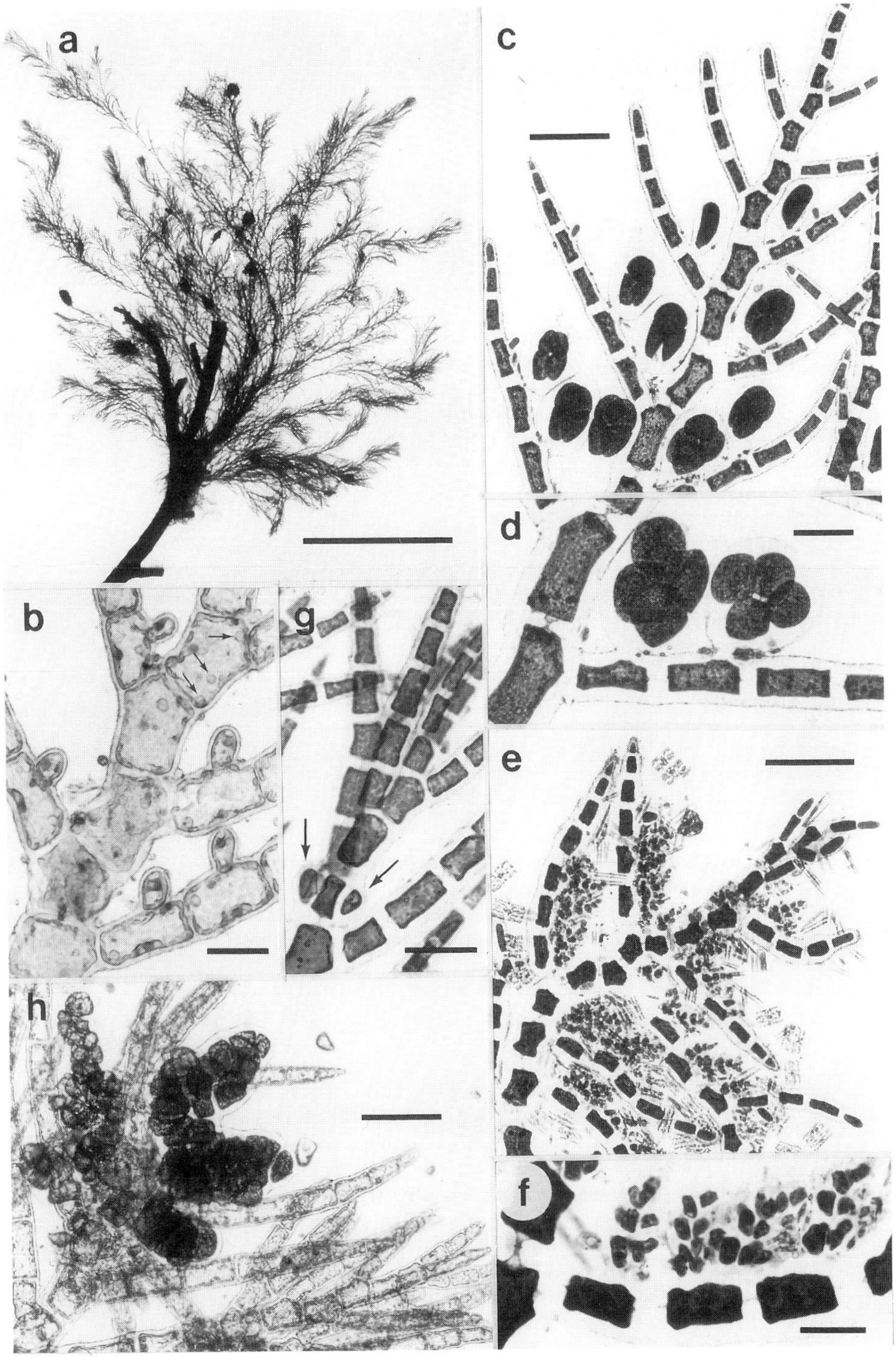
a
b
c
d
e
f
g
h

MATERIAL EXAMINED:

1. Campbell I.: Lectotype. (J. D. Hooker, 1840). (N.B. the sheet has been signed by Hooker but not designated specifically as a type; it contains a clump of tangled filaments wrapped around a hydroid.)
2. Macquarie I.: collections MA-244, 248, 271, 302: Herbarium specimens: MELU, R 0583, 1932, 1983.

DESCRIPTION: Thalli erect, to 3 cm tall, rose-coloured to dark-red to purplish-red, monosiphonous, uncorticated, branched, composed of multinucleate cells, and arising from entangled rhizoidal bases (Fig. 97a, b). Branching alternate, with simple determinate laterals singly produced from axial cells of indeterminate axes. Gametophytes dioecious; spermatangia borne in adaxial clusters on lateral branches (Fig. 97e, f). Fertile pericentral cells paired, borne along indeterminate axes (Fig. 97g); gonimoblast development retarding growth of fertile axes; mature carposporophytes naked and subterminal on stunted indeterminate axes. Carposporangial masses elongated (180–200 μm), usually angular, and each with gelatinous envelope (Fig. 97h). Tetrasporangia tetrahedrally to sub-cruciately divided, oval to subspherical, *c.* 90 μm broad × 120 μm long, and sessile on adaxial surfaces of determinate laterals (Fig. 97c, d).

Callithamnion gracile appears to be rare at Macquarie I. Some 20 plants, growing either as epiphytes on *Lophurella hookeriana* or on rocks adjacent to the bases of *Ballia callitricha*, were collected at 10–20 m depths along the floors of *Macrocystis* forests.

NOTES: Feldmann-Mazoyer (1940) erected the genus *Aglaothamnion* for *Callithamnion*-like plants that otherwise differ in having uninucleate cells and gonimoblasts with oblique cross walls and elongate, angular lobes of carposporangia. *Callithamnion gracile* has *Aglaothamnion*-like gonimoblasts with elongate and sometimes angular carposporangial masses, but differs from this genus in having multinucleate cells. This combination of features from both genera supports Dixon's & Price's (1981) conclusion that *Callithamnion* should be conceived broadly to include species that were previously placed in *Aglaothamnion*, and that the latter name should be considered as a later synonym of the former.

Ceramium **Roth, 1797: 146 (*nom. cons.*)**

Type species: *C. virgatum* Roth, 1797: 148 (= *C.rubrum* (Hudson) C. Agardh, 1811: 17. see Silva (1952) re. synonymy).

Members of this genus generally are rhizoidally attached and have terete, filamentous, branched thalli that are at least partially corticated and composed of uninucleate cells with numerous discoid to fusiform plastids. Cortication is initiated at nodes, with 3–10 periaxial (= pericentral) cells usually giving rise to files of small cortical cells. Some species are distinctive in having cortical bands separated by uncorticated internodal regions, whereas other species are corticated throughout and lack banding (Womersley, 1978). Branching is pseudodichotomous, alternate, or irregular, with apical portions often appearing forcipate. Gametophytes are dioecious, with spermatangia produced in sori that eventually cover cortical bands; procarps develop on first-formed periaxial cells, which function as supporting cells, and at maturity have either one or two four-celled carpogonial branches, a sterile cell group, and, following fertilization, a single auxiliary cell (Kylin, 1923; Hommersand, 1963, fig. 24j). Mature gonimoblasts are often subterminal and have 1–3 rounded gonimolobes surrounded by a gelatinous envelop that is subtended by short, spicate, involucral branches (Abbott & Hollenberg, 1976). Tetrasporangia, usually tetrahedrally but sometimes cruciately divided, arise from periaxial cells or inner cortical cells and are either exposed or embedded in the cortex.

Sexual reproduction is extremely uniform throughout the genus and of little taxonomic value at infrageneric levels. Species are separated largely on differences in cortication patterns, with periaxial cells producing cortical files either acropetally, basipetally, or in both directions; cortical cells divide either synchronously (resulting in cortical bands with straight margins) or irregularly (Womersley, 1978). Filament dimensions and branching patterns are also considered to be taxonomically useful, but the presence or absence of variably produced gland cells and unicellular hairs less so.

Ceramium is the largest genus in the family and has some 70 species reported from tropical to polar seas in both hemispheres (Kraft, 1981). Papenfuss (1964*a*) listed 13 species from the Subantarctic, with one of these, *C. involutum* Kützing, also recorded as washed ashore along MacRobertson Coast in the Antarctic (Skottsberg, 1953). The genus is represented at Macquarie I. by *C. strictum* Kützing.

Ceramium strictum (Kützing) Harvey, 1849: 163*

Fig. 98

BASIONYM: *Gongroceras strictum* Kützing, 1841: 735.

TYPE LOCALITY: German Ocean (= North Sea).

TYPE: presumably in L (Koster, 1969), but specimens were not seen.

SOUTHERN OCEAN DISTRIBUTION: Argentina (Mendoza, 1969*a*); Falkland Is (J. Agardh, 1851; Cotton, 1915); Macquarie I.

MATERIAL EXAMINED: Macquarie I.: collections MA-54, 56, 244, 283, 284, 300, 302, 303, 309; Herbarium specimens: MELU, R 1357–1359, 1669, 1816, 3293–3296.

DESCRIPTION: Thalli erect, to 2 cm tall, dark red to purplish-red, monosiphonous, branched, composed of axial cells that are small and rounded near apices, oblate (to 250 µm diameter) in mid-thallus, more elongate and cylindrical toward rhizoidal base; cortical bands prominent at nodes, to twice as broad as high in mid-thallus, and derived from five periaxial cells (Fig. 98a). Branching appearing sub-dichotomous to dichotomous, with uppermost furcations either straight or slightly forcipate (Fig. 98b, g). Initial cortication acropetal, with two cells developing from each periaxial cell and giving rise to (2-)3-celled, branched files; cortical cell divisions synchronous and resulting in straight margins along upper edges of cortical bands (Fig. 98c). Basipetal cortication secondary, with shorter files usually produced by synchronous cell divisions, but sometimes seemingly asynchoronous and resulting in slightly uneven lower margins. Spermatangia in sori covering nodal cortication (Fig. 98f); gonimoblasts subapical, composed mostly of carposporangia in 2–3 rounded gonimolobes (to 360 µm diameter) surrouned by gelatinous envelopes and subtended by 2–5 short spicate involucral branches (Fig. 98e). Tetrasporangia tetrahedrally divided, oval to subspherical, and fully embedded to partially emergent in nodal cortication (Fig. 98d).

Plants usually grow as epiphytes partly embedded in mats of *Codium subantarcticum*, but sometimes occur on the bases of *Hymenena decumbens*. Both hosts thrive in shaded and unshaded sites in the subtidal between 6–20 m depths. Aside from *Ballia callitricha*, *C. strictum* is probably the most common member of the Ceramiaceae at Macquarie I.

NOTES: Species of *Ceramium* are often described and illustrated poorly, either unidentified or incorrectly named, and not understood in terms of morphological plasticity. Additionally, many taxa have been presumed (without substantial justification) to be cosmopolitan, with European types commonly being reported from remote localities. As a result, species names and distributions frequently are dubious.

Ceramium (Gongroceras) strictum was described initially from the North Sea, illustrated first by Harvey (1851, pl. 334), and reported later from the Southern Ocean (J. Agardh, 1851; Cotton, 1915; Mendoza, 1969*a*). Similarities between Northern and Southern Hemisphere plants, however, are questionable. Individuals of *Ceramium* at Macquarie I. are strictly epiphytic and shorter than the

* Newton (1931) and Parke & Dixon (1964, 1968, 1976) credited Harvey with this species, Papenfuss (1964*a*) credited Greville ex Harvey, and J. Agardh (1851) and Mendoza (1969*a*) attibuted it to Greville & Harvey. Harvey (1849, 1851), however, based *C. strictum* on Kützing's (1841) *Gongroceras strictum* and did not indicate any connection with Greville.

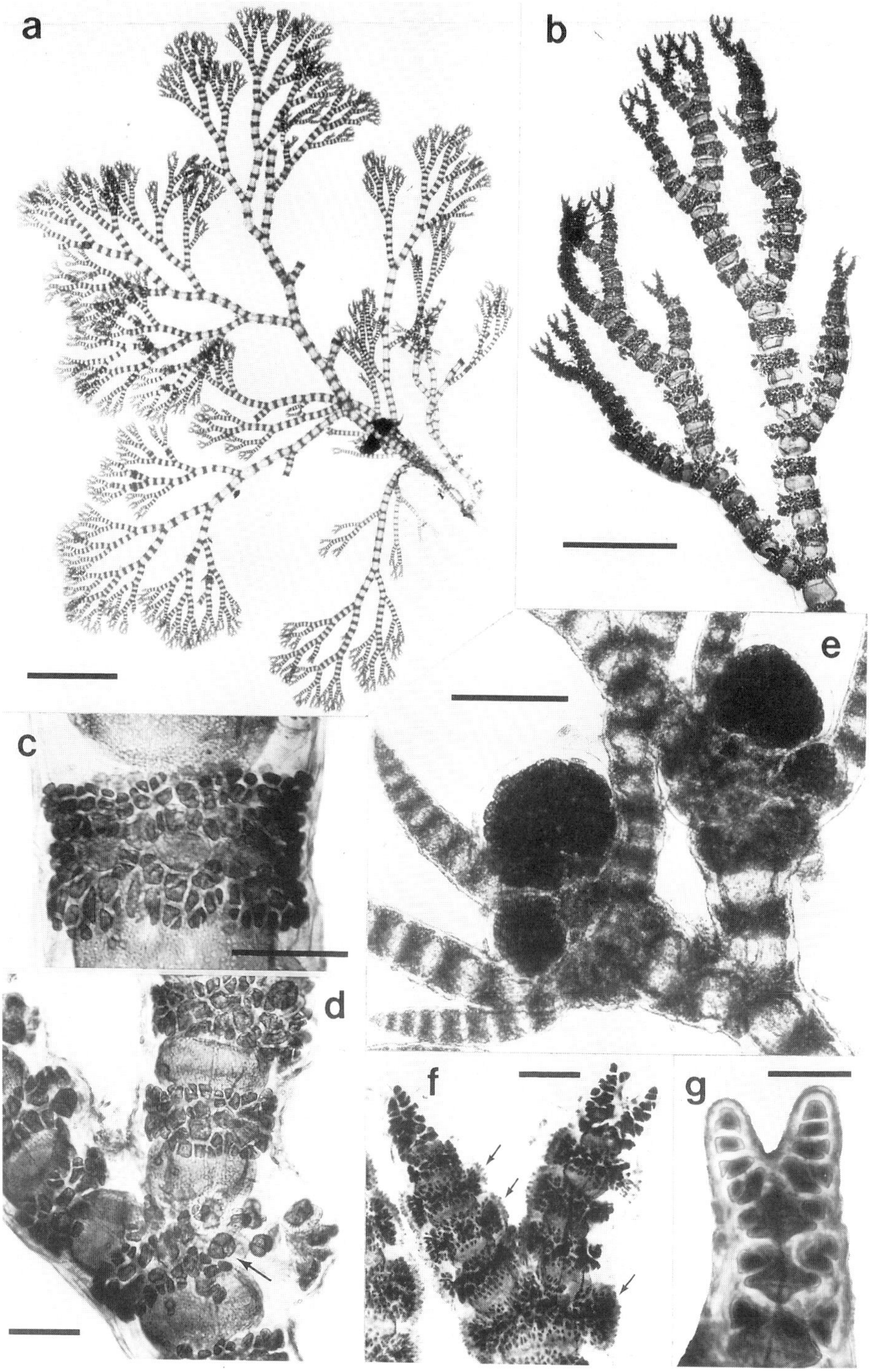
a
b
c
d
e
f
g

epizoic *C. strictum* documented by Harvey (1851) from England. The more critical feature of cortication pattern has been poorly described in plants from northern populations (Harvey, 1849, 1851; Newton, 1931) and consequently cannot be compared with the patterns in Southern Hemisphere plants.

In spite of habit differences, Macquarie plants are provisionally allied with the European species *C. strictum* because of similarities to the South American plants that Mendoza (1969*a*, figs 43–45) accredited to this species. Macquarie specimens are also similar in some respects to Mendoza'a (1969*a*, figs 39–42) illustrations of *C. diaphanum* (Lightfoot) Roth, although the latter species produces more extensive nodal cortication and has strongly forcipate upper branches (= 'apices' of Mendoza), in contrast to the straight upper branches considered to be distinctive for *C. strictum*. Macquarie thalli, having both forcipate and straight distal branches, are somewhat intermediate between the two species as conceived by Mendoza and support the designation '*diaphanum-strictum* species complex' that Rueness (1978) applied to European plants. Should these two species be united, the older name *diaphanum* would have priority, and the known distribution for this species would additionally include the Auckland Is, Fuegia, Îles Kerguelen and South Georgia.

Dasyptilon G. Feldmann, 1950: 308

Type species: *D. pellucidum* (Harvey) G. Feldmann, 1950: 307.

Kylin (1956) and Papenfuss (1958) considered *Dasyptilon* and *Falklandiella* to be congeneric, probably because the two are similar in some vegetative features (e.g., branching patterns and growth of indeterminate axes with obliquely dividing apical cells overtopped by surrounding determinate branches). Moe & Silva (1979), however, distinguished the genera by differences in sexual and asexual reproductive structures. *Dasyptilon* produces procarps directly on basal cells of lateral branches, whereas *Falklandiella* has supporting cells cut off from fertile axial cells in a plane perpendicular to that of lateral branches. Tetrasporangia of *Dasyptilon* are cruciately divided and positioned laterally along the adaxial surfaces of determinate branches, similar to those produced by members of the Tribe Antithamnieae (Yoshida, 1974; Itono, 1977). The tetrasporangia of *Falklandiella* are tetrahedrally divided and usually terminal (sometimes lateral—pers. obs.) on short branches, as is typical for members of the Tribe Ptiloteae (Moe & Silva, 1979). Based on the present observations, cortication patterns can also be used to separate species of the two genera. *Dasyptilon* produces cortical rhizoids that are basipetal, initially parallel to the indeterminate branch axis, either simple or unilaterally to pinnately branched, and composed of elongate cells. In contrast, *Falklandiella* (similar to *Plumariopsis eatoni*) has short-celled cortical rhizoids that also grow downward, but are initially directed inward toward the midline of the indeterminate branch axis.

Dasyptilon has been described by Moe & Silva (1979) as monotypic, but in the present survey it is expanded to accommodate two female gametophytes of uncertain affinity, which uniformly produced horizontally oriented carpogonial branches, a feature considered to be characteristic of *Falklandiella* and several other southern hemisphere members of Ptiloteae (Moe & Silva, 1979) as distinct from the vertically oriented ones of the type species, *D. pellucidum*. For reasons presented below, these two specimens are not formally described, but are provisionally aligned with *Dasyptilon*. The genus is represented at Macquarie I. by this unnamed species and *D. pellucidum*.

Fig. 98 *Ceramium strictum* (Kützing) Harvey: MELU, R 1357, 3294, 3296. **a.** Thallus habit showing banded axes. **b.** Branch habit (tetrasporophyte). **c.** Detail of cortical band. **d.** Detail of swollen cortical band bearing tetrasporangia (arrow). **e.** Habits of two carposporophytes. **f.** Portion of male gametophyte bearing spermatangial patches (arrows). **g.** Detail of vegetative branch apex. Scales: a = 2 mm; b = 500 μm; c = 100 μm; d = 70 μm; e = 200 μm; f = 50 μm; g = 25 μm.

Dasyptilon pellucidum (Harvey) G. Feldmann, 1950: 307

Figs 99, 105c

BASIONYM: *Ptilota pellucida* Harvey (*in* J. D. Hooker), 1855: 257.

SYNONYMS:
Plumaria pellucida (Harvey) Schmitz, 1896: 7.
Euptilota pellucida Laing, 1905: 399.
Plumariopsis pellucida (Harvey) De Toni, 1924: 489.
Falklandiella pellucida (Harvey) Kylin, 1956: 391.

TYPE LOCALITY: Stewart I., New Zealand.

LECTOTYPE: BM, largest plant on unnumbered herbarium sheet (specimen with annotation by Hommersand (1978) designating it as best syntype for lectotypification).

DISTRIBUTION: Macquarie I.; New Zealand (Laing, 1905); Stewart I. (Adams *et al.*, 1974).

MATERIAL EXAMINED:
1. Macquarie I.: collections MA-39, 94, 227, 248, 309, 364, 366, 378, 397; Herbarium specimens: MELU, R 0703, 1413, 1415, 1416.
2. Stewart I.: Evening Cove, tetrasporophyte epiphytic on *Laingia* in 3–5 m depth (Kraft, 9 November 1972): MELU, K4369 (Microscope slide R/M 0633); Lectotype and syntypes. East side of South I., N. Z., on the beach (Lyall, January 1851): BM, unnumbered specimen.
Callithamnion ptilota Hooker f. & Harvey, 1845: 272
Crozet Is: Holotype. On *Macrocystis* (J. D. Hooker, April 1840): BM, no. 635.

DESCRIPTION: Thalli to 3 cm tall, pink to red, branched in one plane, corticated, and attached by rhizoids (Fig. 99a). Main axes indeterminate, slightly to markedly flexuous above, with apical cells obliquely dividing and overtopped by surrounding determinate branches (Fig. 99b). Branching distichous, initially alternate (sometimes with alternating unilateral series of branches), but becoming opposite 4–10 cells behind apices of indeterminate axes; lateral branches usually developing first from high sides of axial cells (Fig. 99b) and remaining larger than second-formed laterals. Most first-formed and all second-formed laterals determinate; indeterminate branches sometimes developing in place of first-formed determinate branches and arising alternately, rendering thallus subdichotomous. Cortication rhizoidal, basipetal, beginning 12–25 cells behind branch apices, and arising from two cortical initials produced on abaxial surfaces of each basal cell of lateral branches; rhizoids initially simple and aligned parallel to main axes, becoming unilaterally, alternately, or oppositely branched while spreading across axes, and composed of cells 2–10 times as long as broad (Fig. 99c). Gametophytes dioecious; spermatangial mother cells catenate and laterally producing spermatangia in whorls of four; spermatangial branches usually single, rarely paired, and produced on adaxial surfaces at distal ends of whorl-branch cells (Fig. 99e). Procarps replacing lateral branches; supporting cells borne in plane of branching and eventually producing on abaxial surfaces 4-celled carpogonial branches that curve upwards around supporting cells and orient vertically (i.e., parallel to fertile axis; Figs 99f, 105c). Gonimoblast development retarding growth of fertile (indeterminate) axis. Mature carposporophytes with rounded lobes of carposporangia subtended by prominent stalk cells; involucre weakly developed, with involucral filaments initiated before fertilization and arising from 1–4 cells subtending fertile axial cells (Fig. 99g). Tetrasporangia sequentially divided (mostly decus-

Fig. 99 *Dasyptilon pellucidum* (Harvey) G. Feldman: MELU, R 0703, 1413, 1415. **a.** Habit of sterile thallus with epiphytic foraminifera (arrows). **b.** Apex of indeterminate branch. **c.** Cortication of main axis c. 1 cm behind apex. **d.** Portion of tetrasporophyte with adaxial sessile tetrasporangia. **e.** Detail of spermatangial branches. **f.** Upper portion of indeterminate axis. Carpogonial branch developed in plane of lateral branches. **g.** Mature carposporophyte. Scales: a = 2 mm; b, e, f = 50 µm; c = 100 µm; d, g = 200 µm.

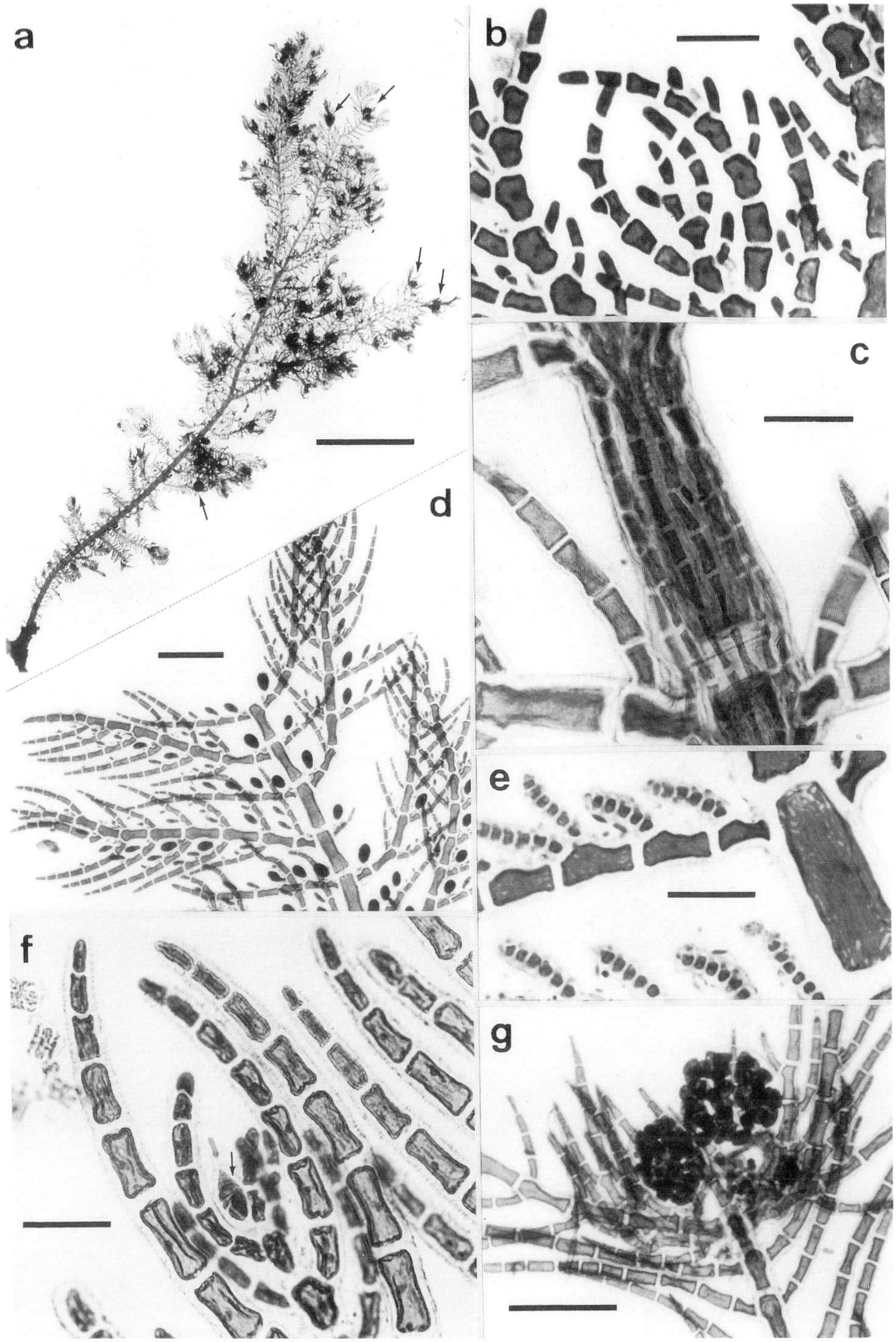
a
b
c
d
e
f
g

sate), oval, sessile, solitary, and on adaxial surfaces at distal ends of whorl-branch cells (Fig. 99d).

Dasyptilon pellucidum grows embedded in sponges or entangled with algae covering rocky surfaces between 8–20 m depths. These subtidal habitats generally are calm and shaded by either *Desmarestia* fronds or overhead canopies of *Macrocystis*. Other species associated with *D. pellucidum* include *Antithamnionella alternans, Heterosiphonia berkeleyi, Pterothamnion spp.* and *Rhodophyllis acanthocarpa*. In one instance, *D. pellucidum* was epiphytic on *Pterothamnion simile*.

NOTES: Macquarie plants compare well with specimens examined from Stewart I. and agree in most respects with the descriptions and illustrations of Feldmann (1950) and Erskine (1955). Erskine (1955, figs 1a, 2b) illustrated six vertically aligned carpogonial branches on one fertile axis, with single carpogonial branches produced alternately on opposite sides of the main axis, and borne on the abaxial surfaces of basal cells of lateral branches that have arisen from successive fertile axial cells near apices of indeterminate axes. In Macquarie plants, upper portions of fertile axes usually bear only 1–2 procarps, which develop in place of lateral branches. The supporting cell is equivalent to a highly reduced (single-celled) lateral branch.

Callithamnion ptilota Hooker f. & Harvey (1845) may be identical to *D. pellucidum*, as both species have similar apical development and branching patterns. Examinations of the remaining material of the type specimen of *C. ptilota*, which is stored in the British Museum (Natural History), revealed no reproductive structures, however, leaving the taxonomy unresolved. Should the two taxa be conspecific, the name *ptilota* would have priority, and the known distribution of *Dasyptilon* would be expanded throughout the Subantarctic to include Îles Crozet (Hooker f. & Harvey, 1845; Levring, 1944), Îles Kerguelen (Dickie, 1879), and South Georgia (Kylin & Skottsberg, 1919). Kylin & Skottsberg (1919) have given the only record of cystocarps in *C. ptilota*. Based on their illustration (1919, fig. 37b) and the present observations of reproductive structures in *D. pellucidum*, cystocarp position and the degree of involucre development are comparable for both species.

Zinova's (1958) Macquarie I. record of *Antithamnion ptilota* (= *C. ptilota*) probably refers to *D. pellucidum*.

Dasyptilon sp.

Figs 100, 105a

DISTRIBUTION: Macquarie I.

MATERIAL EXAMINED: Macquarie I.: collections MA-366, 378; MELU: Microscope slides R/M 0503, 0526.

ADDITIONAL SPECIES EXAMINED:
Falklandiella harveyi (Hooker f.) Kylin 1956: 391
Falkland Is: Oestliche Falklands-Inseln; Ueberzogen von Grammatophora marina (leg. ?): HBG, no. 174 (as *Euptilota harveyi* Kg.).
Georgiella confluens (Reinsch) Kylin, 1956: 391
South Georgia: Lectotype. 'Nordstrand der Landzunge, Pinguin Bay' (leg. ?, 3 Juli 1883): M, 67.83/54 (tetrasporophyte) (N.B. specimen identified by Reinsch and probably used for his habit illustration (1890, taf. III, fig. 5) of the species); Syntype: HBG, unnumbered specimen (determined by Reinbold as *Ptilota harveyi* and later (1918) annotated by Kylin as *Euptilota confluens*).

DESCRIPTION: Thalli similarly constructed to *Dasyptilon pellucidum*, with branches lax and feathery (Fig. 100a). Main axes mostly straight, developing from obliquely dividing apical cells overtopped by surrounding lateral branches (Fig. 100c, d). Branching initially alternate or unilateral, soon becoming opposite; members of branch pairs unequal in size, with first developed branches remaining larger (Fig. 100b). Cortication rhizoidal, with two initials arising on abaxial surfaces of each lateral branch basal cell and growing basipetally and parallel to main axis (Fig. 100e); rhizoids soon branching (unilaterally, alternately, or oppositely) and covering axial cells (Fig. 100f, g). Procarps replacing lateral branches; supporting cells produced in plane of branch-

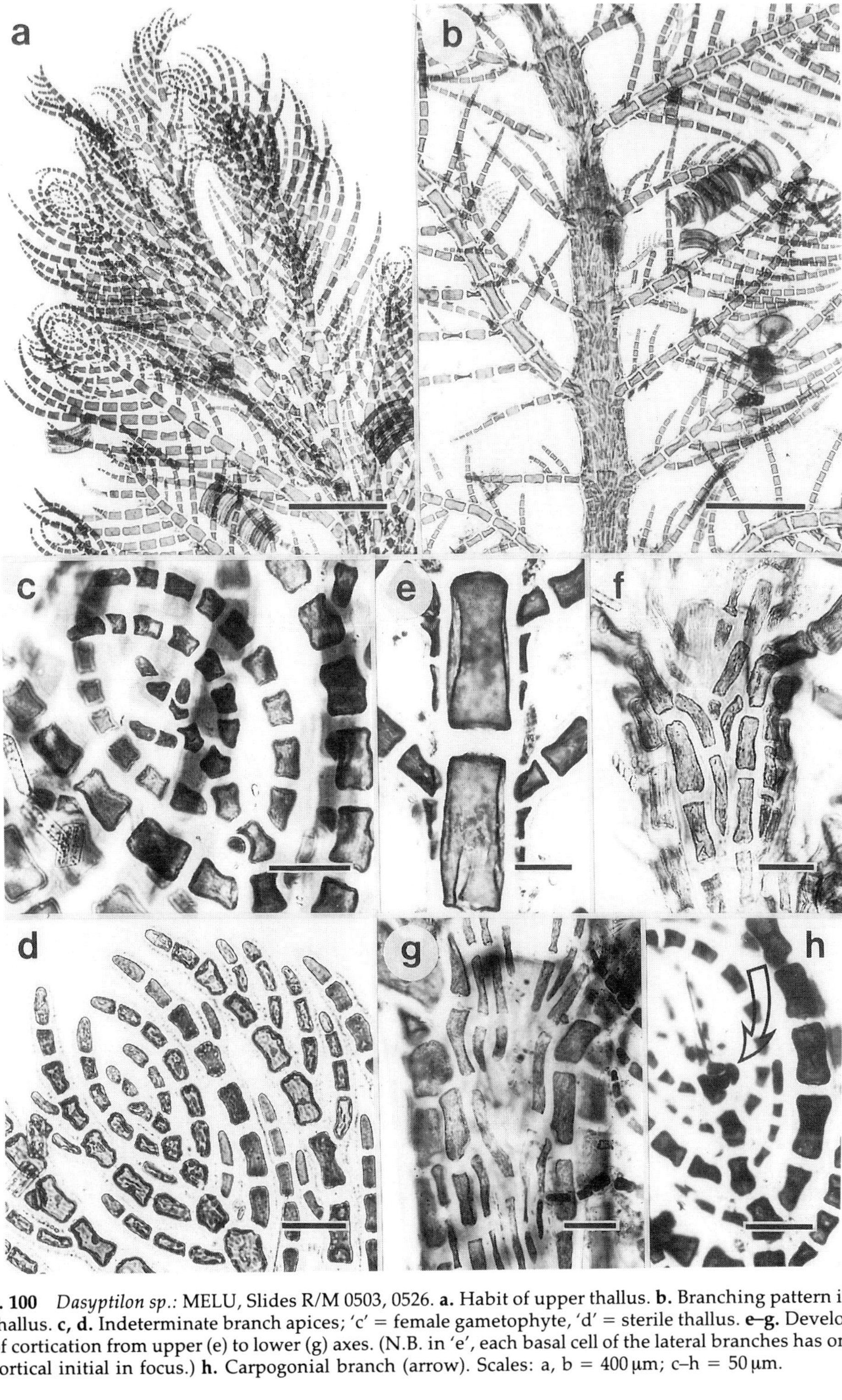

Fig. 100 *Dasyptilon sp.*: MELU, Slides R/M 0503, 0526. **a.** Habit of upper thallus. **b.** Branching pattern in mid-thallus. **c, d.** Indeterminate branch apices; 'c' = female gametophyte, 'd' = sterile thallus. **e–g.** Development of cortication from upper (e) to lower (g) axes. (N.B. in 'e', each basal cell of the lateral branches has only one cortical initial in focus.) **h.** Carpogonial branch (arrow). Scales: a, b = 400 μm; c–h = 50 μm.

ing and bearing 4-celled horizontally oriented carpogonial branches that wrap around fertile axial cells and curve acropetally (Figs 100h, 105a). Male gametophytes and tetrasporophytes unknown.

The above description is based on two plants collected from two subtidal localities at the northern end of the island. In both instances, the undescribed species was growing intermixed with *D. pellucidum* and embedded in a sponge matrix. The sponge covered boulder surfaces and vertical rock walls on the floor of *Macrocystis* forests between 11–14 m depths.

NOTES: Four genera of the Ptiloteae that occur in the Subantarctic have strong superficial similarities. Each, however, has a unique combination of features by which it can be recognized (Table 3). The two plants described above have some characteristics common to all four genera, but are most similar to *Dasyptilon*. The type species, *D. pellucidum* and the unnamed plants have similar habits (generally more delicate than the other three genera), apical cells of indeterminate branches obliquely divided and overtopped by surrounding determinate branches, and comparable patterns of cortication. Concerning the last feature, two cortical initials are produced on the abaxial surfaces of each basal cell of lateral branches. This is unlike the pattern in *Falklandiella*, in which two initials develop on the flanks and one from the abaxial surface of each basal cell of lateral branches, and also unlike the patterns described by Moe & Silva (1983) for *Georgiella* (one initial) and *Plumariopsis* (three initials). Cortical initials in *Dasyptilon* and the plants in question also develop parallel to main axes, unlike development in the other three genera (see Table 3).

Vegetative differences appear to be minor between the type species and the unknown species from Macquarie I. For example, the upper portions of indeterminate axes in *D. pellucidum* are usually strongly flexuous, whereas those in the unidentified species are mostly straight. Some seemingly important reproductive features, however, distinguish these two taxa. Carpogonial branch orientation, a feature considered important in the taxonomy of several genera (Moe & Silva, 1979, 1983), is not uniform, the carpogonial branches of *D. pellucidum* being vertically oriented and not wrapping around fertile axial cells and those of the unidentified species wrapping horizontally around fertile axial cells before curving acropetally. Carpogonial branches of *Falklandiella* also wrap horizontally, not around fertile axial cells, but rather around supporting cells that are cut off fertile axial cells in a plane perpendicular to that of the lateral branches (Moe & Silva, 1979).

Although carpogonial branch orientation in the unidentified species superficially resembles that in *Falklandiella*, which might suggest placement of the unknown plants in this genus, overall similarities to *D. pellucidum* (especially regarding the origins of supporting cells and the details of cortication) indicate a stronger relationship to *Dasyptilon*. It might be that *Dasyptilon* has only the one species that produces both vertically and horizontally oriented carpogonial branches. If so, the taxonomic importance of carpogonial branch orientation for separating species, let alone genera, would be devalued. *Dasyptilon* and *Falklandiella*, nevertheless, could still be distinguished by various differences in vegetative and reproductive features (see Table 3).

Medeiothamnion Pujals, 1970: 290

Type species: *M. sanctacrucense* Pujals, 1970: 291 (= *M. flaccidum* (Hooker f. & Harvey) Brauner, 1979: 339).

Pujals (1970) erected the genus *Medeiothamnion* to accommodate plants with a unique combination of characters, some of which also occur in *Spermothamnion*, *Ptilothamnion* (both in the Tribe Spermothamnieae) and *Sphondylothamnion* (Tribe Sphondylothamnieae). She described the thalli as arising from rhizoidal bases and having erect, uncorticated, branched axes composed of multinucleate cells with more or less discoid plastids. Gordon (1972) expanded the genus to include plants with prostrate and erect axes, either of which can predominate, that sometimes are corticated by rhizoids produced from basal cells of whorl-branches. Axial cells of main axes each bear 2–4(–5) whorl-branches, one of which may be replaced with an indeterminate branch. Branching of indeterminate axes ranges from irregular and sparse to unilateral to opposite or alternate-distichous. Spermatangial branches and tetrasporangia are positioned adaxially on ultimate branches; three pericentral cells are produced on fertile subapical cells, with one pericentral supporting a four-celled carpogonial branch.

Table 3. Comparison of genera in the Tribe Ptiloteae

	Dasyptilon[1,2,3,5]	*Falklandiella*[3,5]	*Georgiella*[4,5]	*Plumariopsis*[4,5]
Indeterminate branch apex	Obliquely divided Overtopped	Obliquely divided Overtopped	Obliquely divided Not overtopped	Obliquely divided Overtopped/Not overtopped
Upper portion of indeterminate branch	Strongly flexuous	Weakly flexuous	Nearly straight	Straight/Weakly flexuous
Cortication	Rhizoidal	Subrhizoidal	Parenchymatous	Subrhizoidal/ Parenchymatous
(initial direction)	Vertical/ Downward	Oblique/ Downward	Upward/Across/ Downward	Oblique/ Downward
Cortical initials	Two	Three	One	Three
Spermatangia	Branchlets	?	Superficial sori	Marginal filaments
Supporting cells	Lateral	Transverse	Transverse	Lateral
Sterile cells	Absent	Absent	Absent	Absent
Carpogonial branch	Vertical/ Horizontal	Horizontal	Horizontal	Vertical
Gonimolobe initials	Clavate	?	Spheroidal	Clavate
Involucral branches	Oppositely branched/ Uncorticated	?	Alternately branched/ Uncorticated	Oppositely branched/ Corticated
Tetrasporangia	Cruciate	Tetrahedral	Tetrahedral	Cruciate
Tetrasporangium position	Lateral & sessile on whorl-branches	Terminal & lateral on short branchlets	Terminal & lateral on short branchlets	Terminal & lateral on short branchlets

[1]Erskine, 1955; [2]Feldmann, 1950; [3]Moe & Silva, 1979; [4]Moe & Silva, 1983; [5] Pers. obs. of type and unidentified species from Macquarie I.

Certain aspects of the female gametophyte and postfertilization development that supposedly typify the genus have been variously described. Procarps have been said to produce two (Pujals, 1970), one (Gordon, 1972), or no (Brauner, 1979) auxiliary cells. In the last instance, in which no auxiliary cell is specially produced, the supporting cell functions as the auxiliary cell. Concerning an involucre, Pujals (1970) reported none or 1–2 involucral filaments developing from hypogenous cells (those subtending fertile subapical cells), whereas Gordon (1972) characterized the genus as having only one involucral filament borne on the distal (= 'adaxial' of Gordon) end of each hypogenous cell. Later, however, she described a new species, *M. norrisii*, with hypogenous cells bearing three involucral filaments (Gordon-Mills, 1977). Brauner (1979) reduced the taxonomic value of both the number and position of involucral filaments by finding great variation of these features within the type species.

Medeiothamnion species are separated largely on differences in vegetative features, such as prostrate vs. erect habits, branching pattern, the number, orientation, and organization of whorl-branches, axial cell sizes, the presence or absence of cortication, and the production of either sessile or stalked tetrasporangia (Gordon, 1972). Using these criteria, six species have been recognized. All are restricted to the Southern Hemisphere, with three [*M. halurum* (Harvey) Gordon, *M. protensum* (Harvey) Gordon, *M. repens* Gordon (1972)] from Australia, two [*M. lyallii* (Harvey) Gordon (1972), *M. norrisii* Gordon-Mills (1977)] from New Zealand, and the type from South America and at least two Subantarctic islands. Although absent from the Antarctic, the genus is represented in the New Zealand Subantarctic by both New Zealand species, and around Fuegia, the Falkland Is., and Macquarie I. by *M. flaccidum*.

Medeiothamnion flaccidum (Hooker f. & Harvey) Brauner, 1979: 339

Fig. 101a–c, f–h

BASIONYM: *Callithamnion flaccidum* Hooker f. & Harvey, 1845: 273.

SYNONYMS: see Brauner, 1979.

TYPE LOCALITY: Fuegia.

LECTOTYPE: Herb. Hookerianum (BM), unnumbered specimens. (N.B. lectotypes of *C. flaccidum* var. *oppositifolium* (= *Medeiothamnion flaccidum*) and *C. flaccidum* var. *alternifolium* selected and illustrated by Brauner (1979, figs 20, 21).)

DISTRIBUTION: Argentina, Chile (Brauner, 1979); Falkland Is (Cotton, 1915); Fuegia (Hooker f. & Harvey, 1845); Macquarie I.

MATERIAL EXAMINED:
Medeiothamnion flaccidum
1. Macquarie I.: collections MA-34, 39, 54, 83, 98, 248, 302, 303, 309, 364, 377, 378; Herbarium specimens: MELU, R 1422–1426.
2. Cape Horn: Lectotype selected by Brauner (9 August 1976). St. Martin's Cove, in deep water (leg. Hooker): BM, unnumbered specimen.
3. Chile: Southwest of Isla May (49°52.5′ S lat., 75°14.4′ W long.), Station 72-21-7: Brauner slide 1016; Puerto Eden (50°11.5′ S lat., 74°47.5′ W long.), Station 72-27-14: Brauner slide 1026; Puerto Alert (47°57.5′ S lat., 74°40.5′ W long.), Station 72-31-52: Brauner slide 1033; South of Isla San Pedro (47°44.1′ S lat., 74°53.1′ W long.), Station 1046: Brauner slide 1046; Bahia San Andreas (46°36.6′ S lat., 75°28.5′ W long.), Station 72-38-3: Brauner slide 1052.

ADDITIONAL SPECIES EXAMINED:
Medeiothamnion lyallii (Harvey) Gordon, 1972: 59
New Zealand: Shag Point, Palmerston, tetrasporophytes from east coast of South I. (Parsons, 10 February 1981): CHR 368026; Ruapuke Foveaux Straits (Lyall, January 1851): unnumbered specimens in Herb. Hookerianum, BM. (N.B. plants have been designated by Gordon (25 June 1970) as type material.)

DESCRIPTION: Thalli to 35 mm tall, rose to reddish-brown, delicate, uncorticated, branched, with prostrate and erect axes arising from rhizoidal bases (Fig. 101a); indeterminate axes composed of tubular axial cells 2–2.5 mm long × 100–300 μm diameter, pectinately bearing whorl-branches near indeterminate branch apices (Fig. 101b), becoming oppositely branched below. Some cells of lateral branches producing attachment rhizoids (Fig. 101c). Indeterminate branches sometimes developing in place of whorl-branches, with branching sparse and irregular to alternate along main axes. Tetrasporangia tetrahedrally divided, oval to spherical, 90 μm broad × 110 μm long, sessile, and forming clusters of up to three tetrasporangia on adaxial surfaces of whorl-branch cells (Fig. 101f). Gametophytes unknown from Macquarie I.

Plants grow between 5–20 m depths and are distributed in subtidal habitats along all sides of Macquarie I. Thalli usually lack distinct points of attachment and are entangled loosely among a variety of algae or partially embedded in a sponge matrix. In one instance, however, *M. flaccidum* was rhizoidally attached to the base of *Rhodophyllis acanthocarpa*.

Fig. 101a–c, f–h *Medeiothamnion flaccidum* (Hooker f. & Harvey) Brauner: MELU, 1422, 1426; Brauner Slide 1016. **d, e** *Medeiothamnion lyallii* (Harvey) Gordon: CHR 368026. **a.** Tetrasporophyte habit. **b.** Upper portion of tetrasporophyte indeterminate axis with unilateral branching. **c.** Lateral branch bearing rhizoids. **d.** Upper portion of indeterminate axis from *M. lyallii* with opposite branching. **e–g.** Tetrasporangia; **e.** *M. lyallii* from New Zealand, **f.** *M. flaccidum* from Macquarie I., **g.** *M. flaccidum* from Chile. **h.** Spermatangiophores of *M. flaccidum* from Chile. Scales: a = 5 mm; b, d–g = 100 μm; c = 250 μm; h = 50 μm.

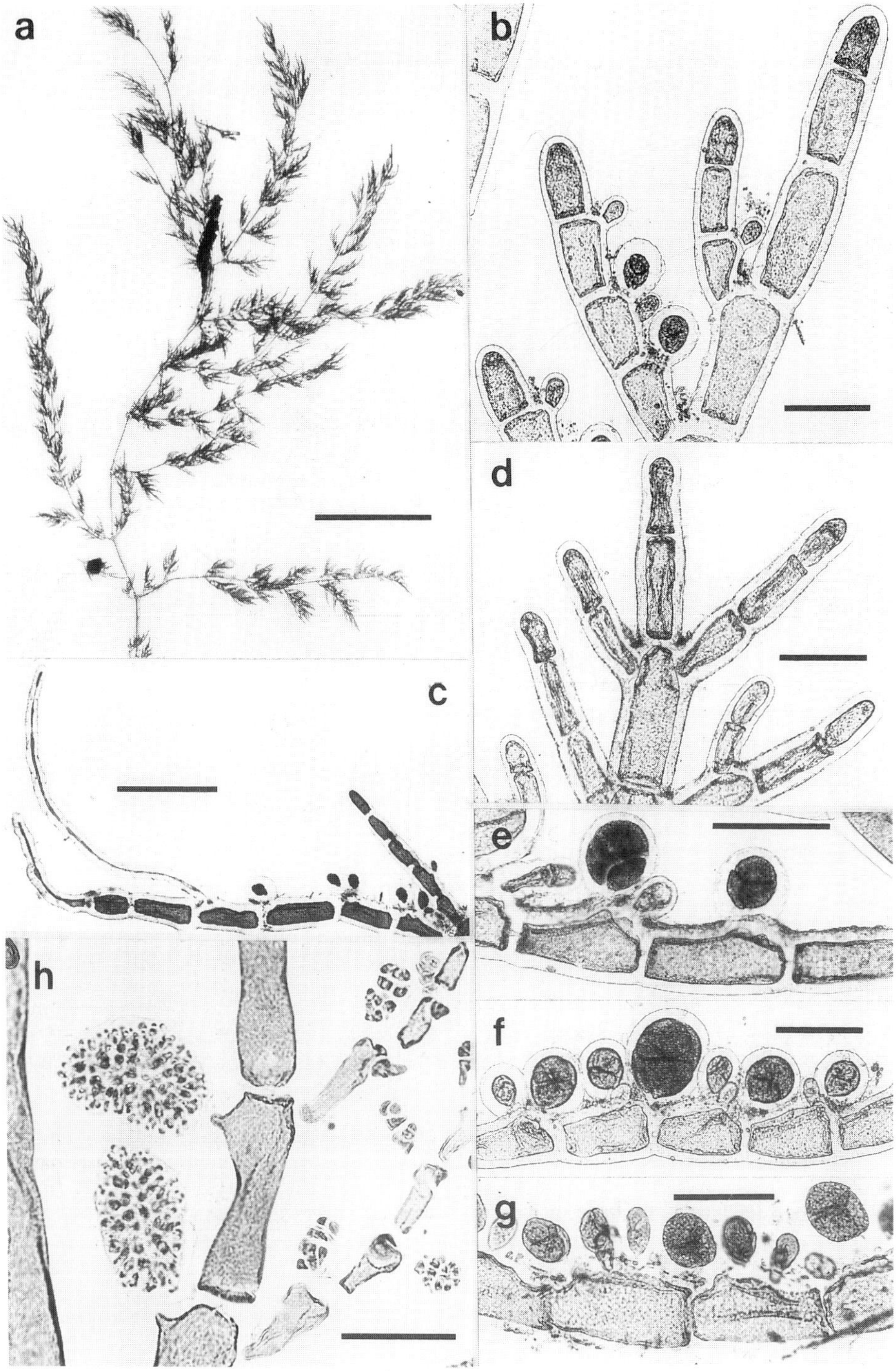
a
b
c
d
e
f
g
h

NOTES: *Medeiothamnion* at Macquarie I. is known chiefly by sterile plants, with some mature tetrasporophytes collected from sponges at 15 m depth during late summer. No gametophytes were found, although in some 50 collections of South American plants studied by Brauner (pers. comm.), nearly 50% were female or tetrasporangial, 30% male, and only 20% sterile. Brauner (1979, figs 3–11) has thoroughly described and illustrated the critical features of postfertilization development in the type species, and an illustration of spermatangial branches from his South American material is included here (Fig. 101h). Spermatangial branches are sessile and develop irregularly or alternately along indeterminate and determinate axes.

Macquarie and South American plants are similar excepting a few seemingly minor differences. The apical cells of Macquarie thalli are somewhat isodiametric and 50–65 μm diameter, whereas those of South American thalli are more elongate and only 35–50 μm diameter. Also, a maximum of three tetrasporangia are borne per cell on the ultimate branches in Macquarie plants (Fig. 101f), whereas 2–5(–7) tetrasporangia often develop on the cells of ultimate and penultimate branches in South American plants (Fig. 101g).

Medeiothamnion flaccidum and the New Zealand *M. lyallii* are similar in habit and disposition of tetrasporangia. Both species may be closely related, although some differences distinguish the two (Gordon, 1972; pers. obs.). In Macquarie specimens of *M. flaccidum*, initial branching near the apices of indeterminate axes is pectinate until several laterals have been produced, at which time branching becomes opposite (Fig. 101b). The few specimens of *M. lyallii* that were studied during the present survey were oppositely branched near the apices of indeterminate branches and lacked pectinate branching (Fig. 101d). When describing South American specimens of *M. flaccidum*, however, Pujals (1970, fig. 4a) illustrated a similar pattern of branching to that presently described for *M. lyallii*, thus suggesting some variation in this feature. Tetrasporangia in both species are sessile, although usually several per cell in the former (Fig. 101f) and singly produced on cells of the latter (Fig. 101e). *M. lyallii*, however, can produce 2–3 tetrasporangia per cell. Gordon (1972) distinguished the two species largely by differences in cortication and branching patterns with *M. lyallii* corticated by rhizoids and bearing up to three whorl-branches per axial cell of indeterminate axes. In contrast, *M. flaccidum* is uncorticated and bears a maximum of only two whorl-branches.

Plumariopsis De Toni, 1903: 1385

Type species: *P. eatoni* (Dickie) De Toni, 1903: 1385.

Plumariopsis and *Georgiella*, both members of the Tribe Ptiloteae, are endemic to the Southern Ocean and have branched, somewhat feathery thalli in which all branches lie in one plane. Moe & Silva (1983, table 1) recently distinguished the two genera by differences in various vegetative and reproductive features. *Plumariopsis*, as characterized by the type species, *P. eatoni* (Dickie) De Toni, and a recently described species from the Antarctic Peninsula, *P. peninsularis* Moe & Silva (1983), has indeterminate axes developing from obliquely dividing apical cells and determinate axes developing from transversely dividing apical cells. Cortication is extensive on indeterminate axes and is subrhizoidal (*P. eatoni*) or pseudoparenchymatous (*P. peninsularis*). Three cortical initials develop on the abaxial sides and flanks of the basal cells of lateral branches and branchlets. Gametophytes are dioecious, with spermatangial branches and procarps replacing vegetative laterals. Supporting cells lack sterile cell groups and each bears one vertically aligned (i.e., parallel to the fertile axis) four-celled carpogonial branch and one auxiliary cell. Gonimoblast initials are clavate, and mature carposporophytes are surrounded by oppositely branched, corticated involucral branches. Tetrasporangia are cruciately divided.

Of the two species known for the genus, *P. eatoni* is restricted to the Subantarctic Zone, and *P. peninsularis* to the Antarctic Zone (including the South Shetland Is).

Plumariopsis eatoni (Dickie) De Toni, 1903: 1385

Figs 102, 105b

BASIONYM: *Ptilota eatoni* Dickie, 1876*a*: 51.

TYPE LOCALITY: Swain's Bay, Îles Kerguelen.

LECTOTYPE: BM. (N.B. a tetrasporophyte was illustrated by Dickie (1879) and later lectotypified and illustrated by Moe & Silva (1983, fig. 2).)

DISTRIBUTION: Heard and Macquarie Is (Moe & Silva, 1983); Îles Kerguelen (Dickie, 1876*a*); South Georgia (Kylin & Skottsberg, 1919).

MATERIAL EXAMINED:

1. Heard I.: Atlas Cove (Chittleborough, 30 August 1949): MELU, R 2416 (tetrasporophyte); (Chittleborough, 11 September 1949): MELU, R 2417 (female); Southeast of Rogers Head, on side of large boulder (Smith, 7 February 1983): MELU, R 3642 (male), 3643–3649.
2. Îles Kerguelen: Lectotype and paratypes. Swain's Bay (Eaton, 16, 28, 30 January 1875): BM, four unnumbered sheets representing ten clumps of plants.
3. Macquarie I.: Buckles Bay, drift specimen (Merilees, 25 June 1967): MELU, R 2418.

DESCRIPTION: Thalli to 9 cm tall, dark red-purple, with one to several coarse, densely corticated, branched main axes arising from discoid bases (Fig. 102a). Branching opposite, in one plane, usually with members of pair unequal in size. Lateral branches of pair mostly determinate but sometimes with one branch being indeterminate. Apical cells of indeterminate axes obliquely dividing, overtopped by surrounding determinate branches; division planes alternating from side to side resulting in apical derivatives having short and long sides (Fig. 102b, c). First-developed lateral branches arising from high sides of cells and generally remaining larger than second-formed laterals produced from low sides of cells; all first-formed and most second-formed laterals determinate and having transversely dividing apical cells (Fig. 102d). Second-formed laterals sometimes indeterminate. Cortication subrhizoidal, with cortical initials produced on basal cells of lateral branches and branchlets; short-celled rhizoids developing inwards toward mid-line of indeterminate axis, forming 'chevron' pattern (Fig. 102e). Gametophytes dioecious; spermatangia densely clustered around cells of pinnate branchlets of determinate lateral branches (Fig. 102g). Supporting cells borne on indeterminate axes in plane of branching and replacing lateral branches, each bearing one (acropetal) vertical 4-celled carpogonial branch and not producing sterile cell groups (Figs 102f, 105b). Tetrasporangia sequentially divided (usually decussate), oval (*c*. 60 μm broad × 80 μm long) to subspherical, and terminal on short uncorticated determinate lateral branches (Fig. 102h).

This species is presumed to be very rare around Macquarie I., as it is known by only one plant that was collected in drift near the ANARE station. It is more common at Heard I., however, where plants grow epiphytically on various red algae [e.g., *Callophyllis fastigiata* (J. Agardh) J. Agardh and *Microrhinus carnosus* (Reinsch) Skottsberg] attached to large boulders in the mid- to lower intertidal.

NOTES: The above description is based on wet and dried material from Heard I. and one dried herbarium specimen from Macquarie I. The present findings agree with Askenasy's (1888, pl. ix, fig. 8) illustration of the carpogonial branch position and orientation, and although lacking detail on postfertilization development, agree in most respects with the comprehensive studies of Moe & Silva (1983) on type specimens from Îles Kerguelen. One difference between the two accounts, however, concerns the shape of the indeterminate branches. The upper portions of these axes were described by Moe & Silva (1983) as straight, but have been seen in the present account to be slightly flexuous.

At Macquarie I., *P. eatoni* might be confused with *Dasyptilon pellucidum*, *D. sp.*, or *Pterothamnion simile*, all of which produce feathery fronds branched in only one plane. *Plumariopsis*,

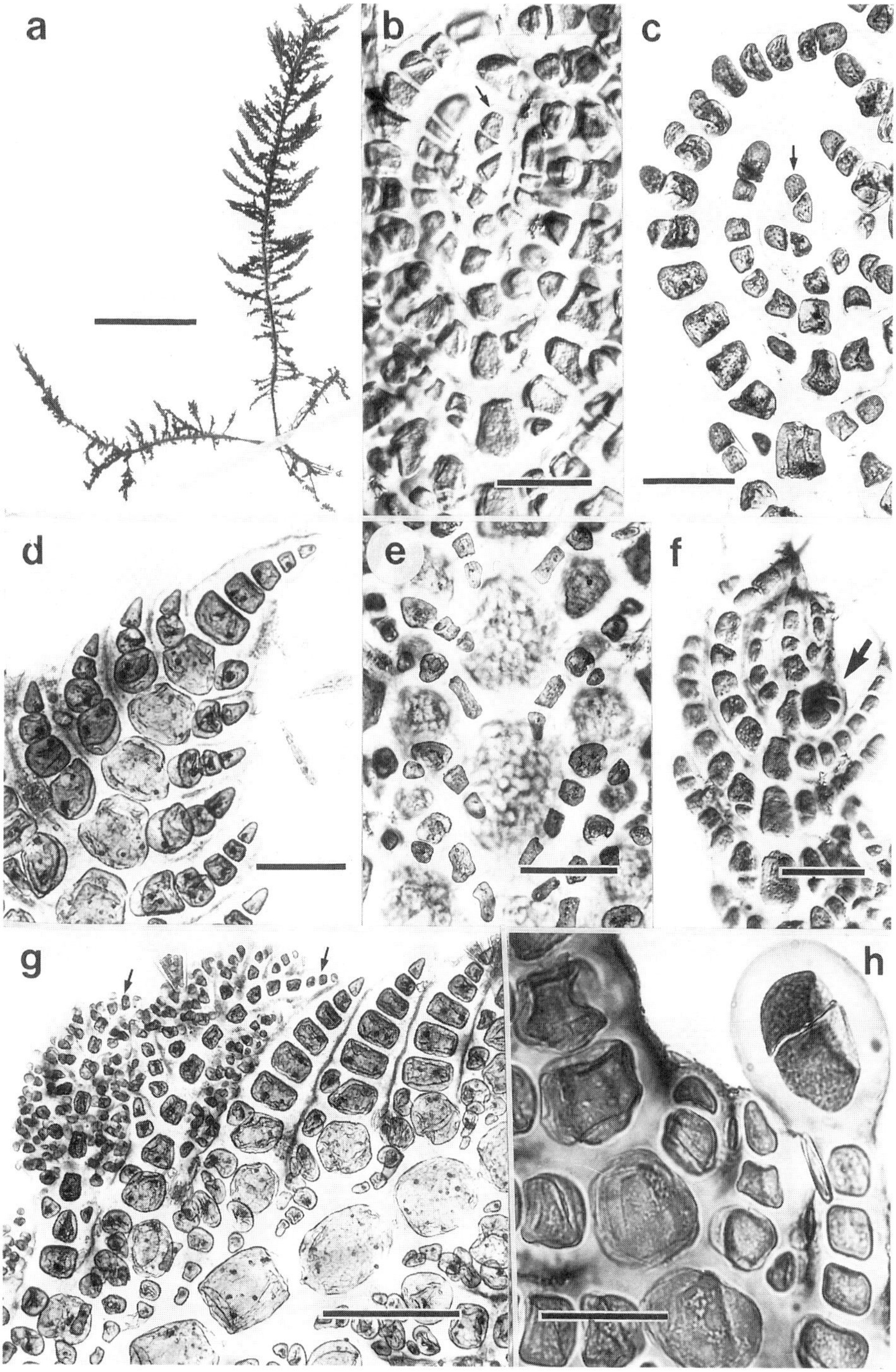
a
b
c
d
e
f
g
h

however, has considerably coarser thalli with a distinctive 'chevron' cortication pattern that does not occur in these species.

Pterothamnion Nägeli *in* Nägeli & Cramer, 1855: 66

Lectotype: *P. plumula* (Ellis) Nägeli *in* Nägeli & Cramer, 1855: 66.

Nägeli (*in* Nägeli & Cramer, 1855) originally included distichously branched plants in *Pterothamnion*, but later (1861) incorporated thalli with more than two whorl-branches per axial cell. The genus was subsequently split into several genera that now belong to the Antithamnieae (Wollaston, 1979). It has been redefined recently by Moe & Silva (1980) and is probably related most closely to *Platythamnion* J. Agardh (1892), which produces a uniform arrangement of whorl-branches on each axial cell of main axes, including two short (determinate) branches oriented perpendicular to two longer (indeterminate) branches. All indeterminate axes are restricted to one plane. *Pterothamnion* also has all indeterminate axes lying in a single plane and has a comparable, although less uniform, whorl-branch pattern. Axial cells of main axes produce two (major) whorl-branches in the main plane of branching and 1–2 facultatively produced, usually less developed transverse (minor) whorl-branches. Indeterminate axes usually replace one member of a pair of major whorl-branches and produce an alternate pattern of branching. Occasionally, however, indeterminate branches develop from the basal cells of whorl-branches. Gland cells are produced by non-specialized cells of whorl-branches or branchlets and are solitary or arranged in series of 3–4. Spermatangial branches replace either whorl-branches or branchlets on whorl-branches. Individual procarps arise on the basal cells of ordinary major whorl-branches, with up to four per fertile axis. Gonimoblast development sometimes suppresses further growth of indeterminate fertile axes, and mature carposporophytes can be naked [e.g., *P. antarcticum* (Kylin) Moe & Silva, 1980] or surrounded by weakly developed involucres composed of whorl-branches produced by axial cells above and below the fertile segment [e.g., *P. plumula* (Ellis) Nägeli (Kylin, 1923)]. Tetrasporangia are sessile or stipitate, solitary or clustered, and arise in place of branchlets on the adaxial surfaces of whorl-branches.

Three species have been attributed previously to *Pterothamnion* (Wollaston, 1979; Moe & Silva, 1980), with only the type occurring in the Northern Hemisphere. A fourth species, known from the Juan Fernandez Is, is transferred in the present survey to this genus. Based on similarities of branching habit, disposition of tetrasporangia and carpogonial branches, carposporophyte structure, and spermatangial branch structure between *P. simile* and Levring's illustrations (1941, fig. 18) of *Antithamnion minutissimum* Levring (1941, p. 646), the latter should be known as *P. minutissimum* (Levring) comb. nov.

All four species of *Pterothamnion* occur in the Southern Hemisphere, but only two, *P. plumula* and *P. simile*, grow in the vicinity of Macquarie I.

Pterothamnion plumula (Ellis) Nägeli *in* Nägeli & Cramer, 1855: 55

Figs 103, 105e

Basionym: *Conferva plumula* Ellis, 1768: 425.

Synonyms: see De Toni, 1903.

Type locality: England.

Type: considered to be lost (Dixon, 1960).

Fig. 102 *Plumariopsis eatoni* (Dickie) De Toni: MELU, R 2416–2418, 3642. **a.** Plant habit. **b, c.** Indeterminate branch apices (arrows). **d.** Upper portion determinate branch. **e.** Short-celled rhizoids forming 'chevron' cortication pattern. **f.** Subapical position of carpogonial branch (arrow). **g.** Spermatangial branches (arrows). (N.B. 'c–g' are from plants from Heard I.) **h.** Tetrasporangial mother cell after first division. Scales: a = 2 cm; b–f, h = 50 µm; g = 100 µm.

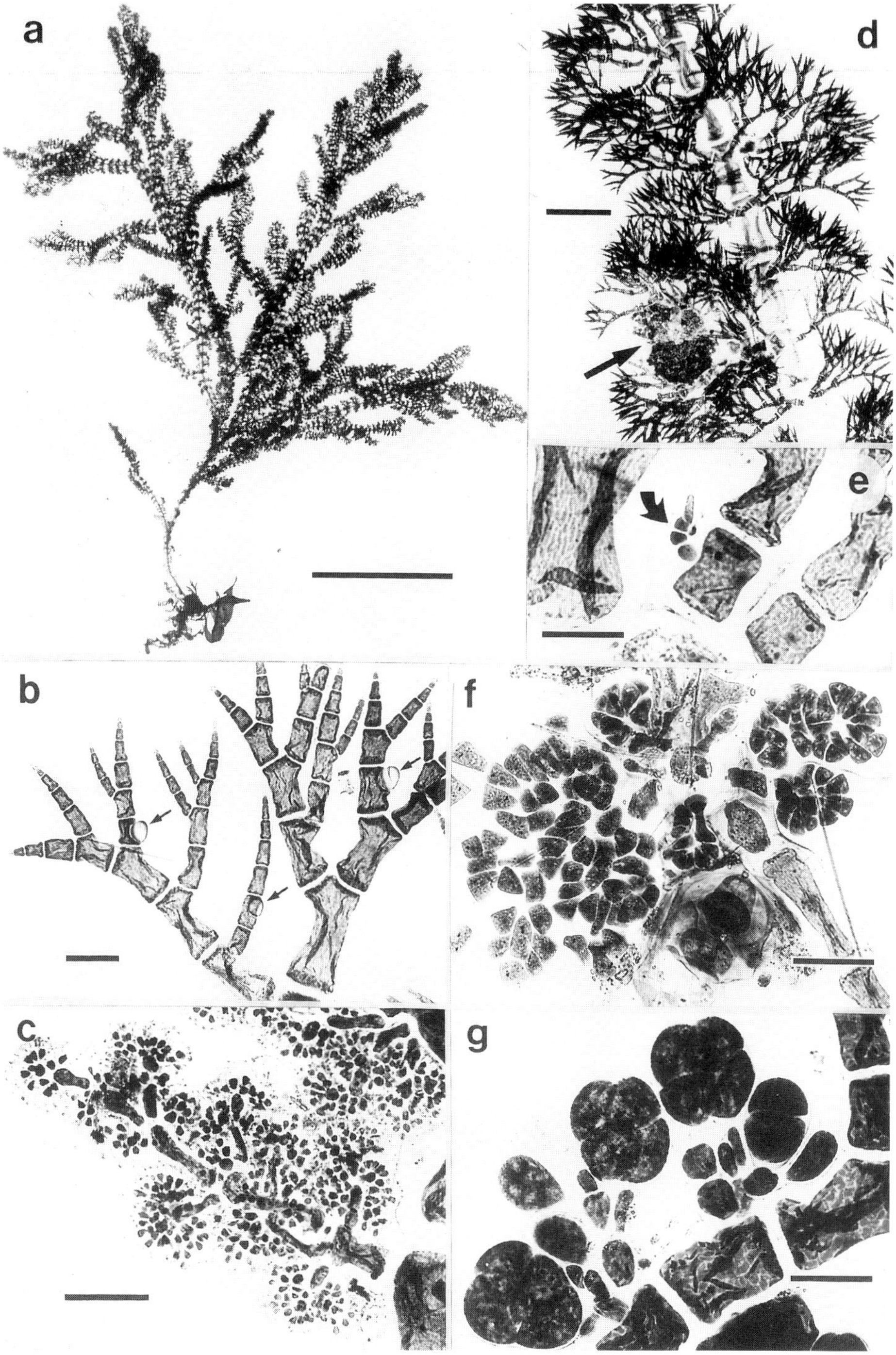

Fig. 103 *Pterothamnion plumula* (Ellis) Nägeli: MELU, R 1695, 3279–3281. **a.** Thallus habit. **b.** Gland cells (arrows). **c.** Spermatangiophores. **d.** Portion of female gametophyte with carposporophyte (arrow). **e.** Nonfunctional carpogonial branch (arrow). **f.** Detail of gonimoblast. **g.** Tetrasporangia. Scales: a = 1 cm; b, c, f = 50 μm; d = 300 μm; e, g = 30 μm.

Southern Ocean distribution: Australia (Wollaston, 1968); Fuegia (J. D. Hooker, 1847); Macquarie I.; New Zealand (J. Agardh, 1878—as *Callithamnion plumula*; Laing, 1901, 1905). (N.B. this species is widely distributed in the Northern and Southern Hemispheres, being best known in the former from both Pacific (Okamura, 1922) and Atlantic (Rueness & Rueness, 1975) Oceans.)

Material examined: Macquarie I.: collections MA-39, 94, 111, 303, 364, 377, 378; Herbarium specimens: MELU, R 1695, 1711, 3279–3281.

Description: Thalli to 45 mm tall, rose-coloured, uncorticated, branched, and attached by rhizoids (Fig. 103a). Main axes indeterminate, linear throughout, with each axial cell usually bearing tetrastichous (rarely only two or three) whorls of determinate branches. Two major whorl-branches in prominent plane of branching, with minor whorl-branches perpendicular to this, generally less developed than major whorl-branches, and when paired, with members often subequal. Whorl-branches branched along adaxial surfaces, with first-order branchlets often diverging in several planes; branchlets mucronate, unilaterally branched 1–3 times, with second- and third-order branchlets produced on either abaxial or adaxial surfaces. Indeterminate branches infrequent, arising in one plane, replacing one major whorl-branch of pair, and generally alternating from side to side along main axes. Gland cells solitary and produced on adaxial surfaces of whorl-branches or on abaxial or adaxial surfaces of branchlets (Fig. 103b). Gametophytes dioecious; spermatangia borne in whorls around spermatangiophores developed oppositely on distichously branched spermatangial branches produced in place of major whorl-branches (Fig. 103c). Procarps 1–2 per fertile axis; carpogonial branches 4-celled, initiated mostly on abaxial sides of basal cells from major whorl-branches, and developing around as well as (acropetally) along whorl branches (Figs 103e, 105e). Carposporophytes composed of dense carposporangial masses subtended by stalk cells, lacking specialized involucres but generally surrounded by adjacent whorl-branches (Fig. 103f). Gonimoblast development not retarding growth of indeterminate fertile axes (Fig. 103d). Tetrasporangia sequentially divided (mostly decussate but sometimes irregularly divided), subspherical, *c.* 40 μm diameter, borne in dendroid clusters on adaxial surfaces of whorl-branches; tetraspores mostly rounded (*c.* 20 μm diameter) in mature tetrasporangia (Fig. 103g).

Pterothamnion plumula grows in shaded habitats along all sides of Macquarie I. and extends from the sublittoral fringe to 20 m depths. It sometimes is epiphytic (along with *P. simile*) on larger red algae such as *Delisea pulchra*, but more commonly is embedded in a sponge matrix without apparent connection to other plants. An extensive covering of diatoms and sponge usually obscures detail of the thallus.

Notes: At least three of the four species of *Pterothamnion* (*P. minutissimum* was not studied) exhibit considerable variation in branching, which sometimes leads to similarities in their general morphologies. *P. plumula*, however, is usually bushier than the others and appears to be most densely branched. It is also the only member of the genus to grow in both Northern and Southern Hemispheres.

Okamura (1922, vol. iv, pl. 189) illustrated Japanese specimens of *P. plumula* that are similar to *P. simile* in habit and sessile (rarely stipitate) disposition of tetrasporangia, but differ from this species in the production of solitary gland cells, which are comparable to those produced by *P. plumula* from Macquarie I. The amount of plasticity in the various vegetative and reproductive features used to distinguish the four species in the genus is not yet understood, although some work suggests that considerable variation is possible. For example, branching differences between two interfertile varieties of *P. plumula* in the Atlantic Ocean have been explained by a single gene change (Rueness & Rueness, 1975). In another instance, morphological variation of thalli belonging to the same species has been attributed to differences in environmental conditions where the plants occur (Rosenvinge, 1923–24). In the present survey, *P. plumula* (which has linear main axes, solitary gland cells, uniform production of tetrastichous whorl-branches from each axial cell of indeterminate axes, and pinnately branched spermatangial branches) and *P. simile* (which has strongly flexuous upper portions of main branches, gland cells in series of 3–4, 2–3 whorl-branches per axial cell, and unilaterally branched spermatangial branchlets) are considered to be distinct species.

Pterothamnion simile (Hooker f. & Harvey) Nägeli, 1861: 376

Figs 104, 105d

BASIONYM: *Callithamnion simile* Hooker f. & Harvey, 1845: 271.

SYNONYMS: see Wollaston, 1979.

TYPE LOCALITY: Christmas Harbour, Kerguelen's Land.

LECTOTYPE: Herb. Hookerianum (BM), no. 248. (Wollaston, 1979, Fig. 1).

DISTRIBUTION: Enderby Land, Antarctica (Skottsberg, 1953); Îles Kerguelen (Hooker f. & Harvey, 1845); Macquarie I.; South Georgia (Kylin & Skottsberg, 1919).

MATERIAL EXAMINED:

1. Îles Kerguelen: Type specimens. Christmas Harbour, on the beach (J. D. Hooker, July 1840): nos. 248, 682; Valdivia Expedition (plant determined by Reinbold as *Antithamnion simile*); M, 67.83/2.
2. Macquarie I.: collections MA-34, 39, 94, 95, 227, 248, 364, 366, 377, 378; Herbarium specimens: MELU, R 0701, 0702.
3. South Georgia: no collection data (det. Reinbold): M, 67.83/1.

ADDITIONAL SPECIES EXAMINED:
Pterothamnion antarcticum (Kylin) Moe & Silva, 1980: 12
Antarctic Peninsula: Gamma I. (Isla Observatorio), north side, 150 m offshore form Gallows Point, growing on an ascidian in 29 m depth. (N.B. Lamb's Antarctic algal collections (FH) lack specimens of this species; Lamb & Zimmermann (1977) based their illustration (fig. 74) of *Antithamnion antarcticum* Kylin upon a 35 mm micrograph transparency (A-875, filed on herbarium sheet no. 259).)

DESCRIPTION: Thalli to 3 cm tall, pale-red to dark-red, uncorticated, branched, and arising from rhizoidal bases (Fig. 104a). Main axes indeterminate, strongly flexuous in upper portions, with each axial cell bearing two major whorl-branches; minor transverse whorl-branches singly produced, infrequent, often on axial cells bearing indeterminate branches (Fig. 104b). Whorl-branches simple or pectinate along adaxial surfaces, with branchlets simple or unilaterally branched; apices of branchlets mucronate or rounded. Indeterminate branches arising in one plane, replacing one major whorl-branch of pair (generally every 3–6 axial cells apart), alternating from side to side along main axes. Gland cells either solitary or in series of 2–4 on adaxial surfaces of whorl-branch and branchlet cells (Fig. 104c). Gametophytes dioecious; spermatangia borne in whorls around spermatangiophores developed unilaterally on branchlets (Fig. 104e, f). Procarps to four per fertile axis; carpogonial branches 4-celled, initiated on abaxial sides of basal cells from major whorl-branches, and developing around as well as (acropetally) along whorl-branches (Fig. 105d). Gonimoblast development not retarding growth of indeterminate fertile axes. Mature carposporophytes with dense rounded lobes of carposporangia encased in gelatinous envelopes surrounded by simple or sparsely branched involucral branches (Fig. 104g). Tetrasporangia sequentially divided (mostly decussate), oval, *c.* 45 μm broad × 65 μm long, sessile to stipitate, mostly solitary, sometimes clustered, and terminal or lateral on whorl-branches and branchlets (Fig. 104d).

Fig. 104 *Pterothamnion simile* (Hooker f. & Harvey) Nägeli: MELU, R 0701, 0702. **a.** Habit of plant with epiphytic foraminifera (arrows). **b.** Upper portions of indeterminate axes showing unilateral and opposite development of determinate branches. **c.** Darkly staining gland cells. **d.** Tetrasporangia. **e.** Portion of male gametophyte. **f.** Detail of spermatangial clusters. **g.** Mature carposporophyte at apex of lateral branch. Scales: a = 2.5 mm; b, f, g = 100 μm; c, d = 50 μm; e = 250 μm.

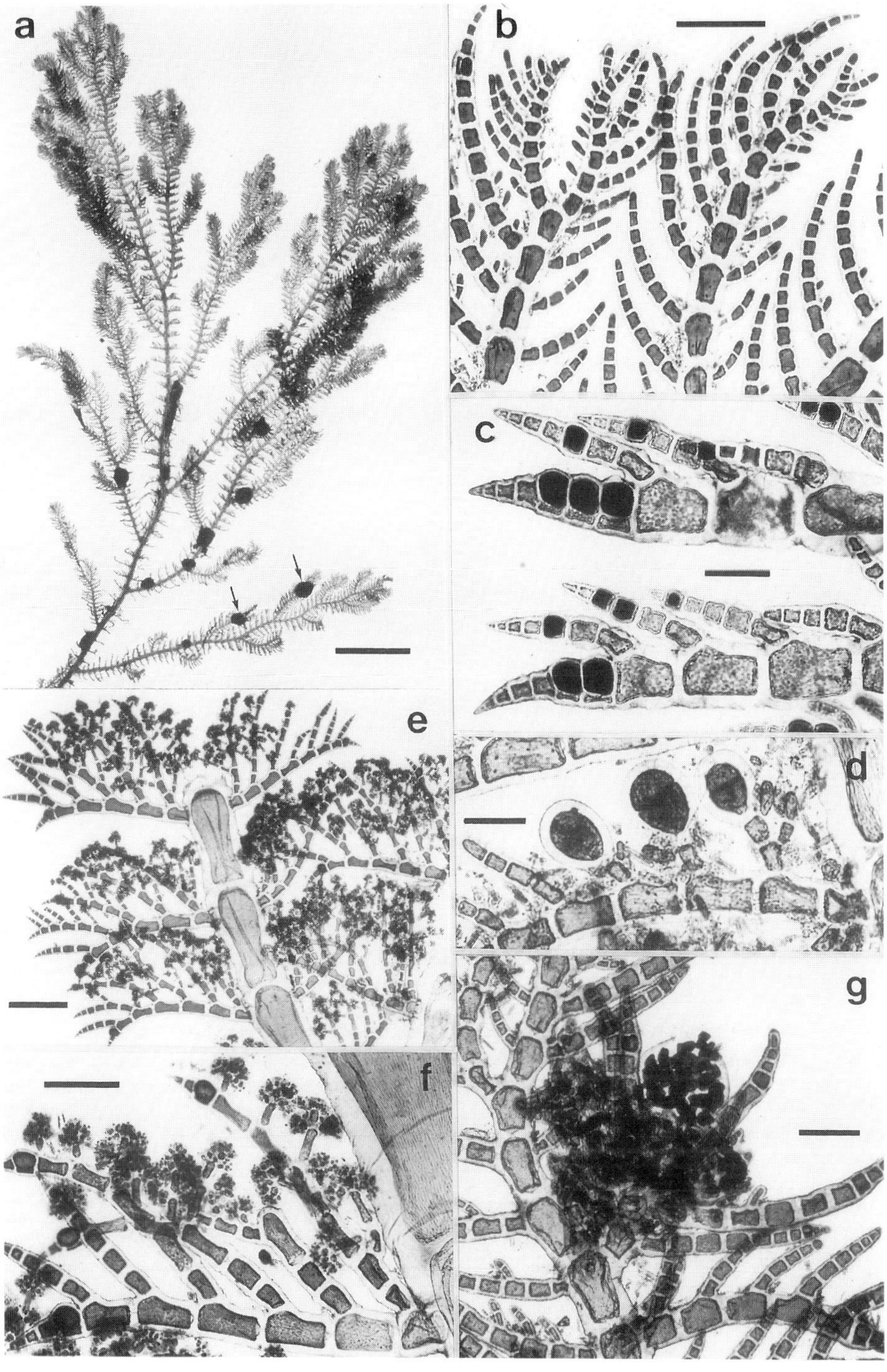
a
b
c
d
e
f
g

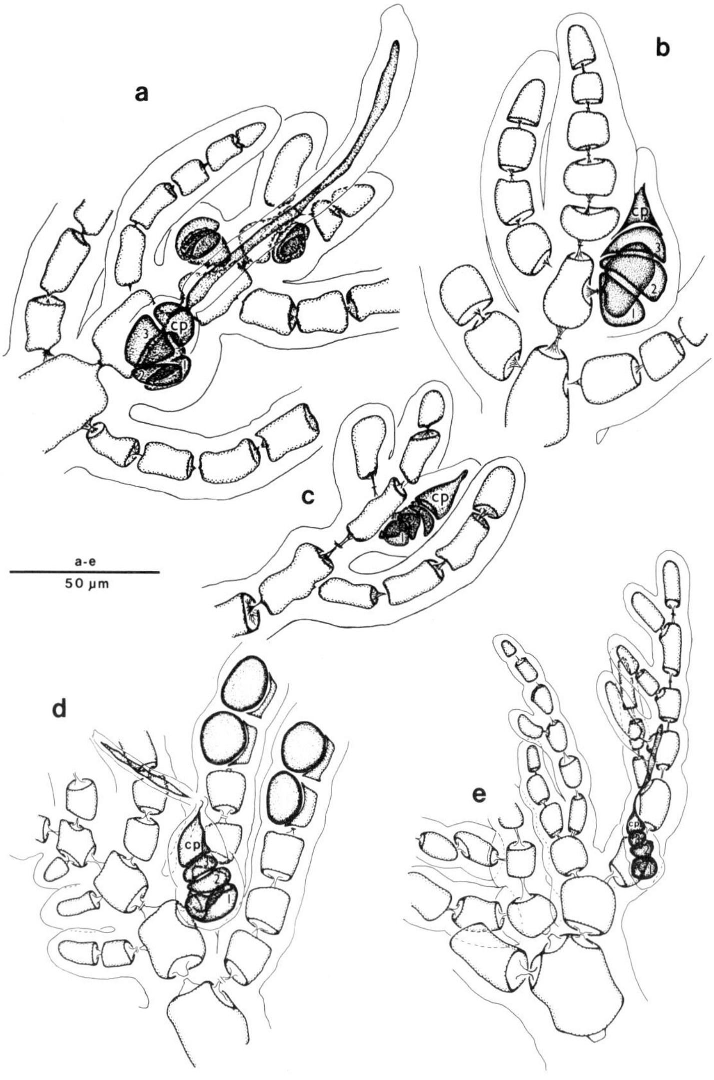

Fig. 105a *Dasyptilon sp.*: MELU, Slide R/M 0503. **b** *Plumariopsis eatoni* (Dickie) De Toni: MELU, R 2417. **c** *Dasyptilon pellucidum* (Harvey) G. Feldmann: MELU, R 0703. **d** *Pterothamnion simile* (Hooker f. & Harvey) Nägeli: MELU, R 0701. **e** *Pterothamnion plumula* (Ellis) Nägeli: MELU, R 3279. **a.** Supporting cells borne in plane of branching; immature and mature carpogonial branches (shaded) curving (horizontally) around fertile axial cells of indeterminate axis. **b, c.** Supporting cells borne in plane of branching; carpogonial branches (vertically) aligned parallel to indeterminate branch axes. **d, e.** Carpogonial branches borne abaxially on and curving around basal cells of lateral branches. Scales: a–e = 50 µm.

Pterothamnion simile is one of the more common delicate members of the Ceramiaceae at Macquarie I. It grows abundantly between 8–20 m depths, often occurring on the sides of boulders along the floors of *Macrocystis* forests or on rocky walls bordering narrow chasms. Plants are usually embedded in sponges or attached by multicellular rhizoids (produced from cells on lateral branches and lower parts of main axes) to various algae. Other species associated with *P. simile* include *P. plumula, Antithamnionella alternans* and *Dasyptilon pellucidum*.

NOTES: The present observations on *P. simile* from Macquarie I. largely agree with Askenasy's (1888, pl. viii) fine illustrations of vegetative and reproductive features of this species, but additionally have revealed considerable variation in thallus consistency (from lax to rigid fronds) and branching pattern (from extremely flattened branches bearing only two major whorl-branches to bushier fronds with minor whorl-branches consistently produced and well developed). The different morphologies observed encompass the earlier descriptions and illustrations of this species from various Subantarctic and Antarctic localities (Dickie, 1879; Askenasy, 1888; Kylin & Skottsberg, 1919; Skottsberg, 1953; Mendoza, 1969a), as well as overlap the thallus habits of two other species in the genus.

Recent work by Wollaston (1979) supports Skottsberg's (1953) earlier belief that this species is closely related to the Juan Fernandez alga *Pterothamnion minutissimum*; the present observations on spermatangial branches in *P. simile* further confirm a close relationship between the two taxa. Lamb & Zimmermann (1977) illustrated a branching pattern in *P. antarcticum* resembling one of the branching habits of *P. simile,* thereby suggesting a close relationship between these species also. *Pterothamnion antarcticum,* however, is generally slightly more robust, with whorl-branches branched to a greater extent and somewhat intermediate between *P. simile* and *P. plumula*. These two species, however, can be distinguished by differences in the disposition of their gland cells as well as by various other features.

The relationships between *P. antarcticum, P. minutissimum* and *P. simile* remain uncertain and need to be resolved by combining culture studies with more exacting field observations on the ecological preferences of the various morphological forms of each species.

Dasyaceae Kützing, 1843: 413, 414

Members of the Dasyaceae, which can superficially resemble some plants in the Rhodomelaceae, have terete to compressed uniaxial thalli with polysiphonous portions that sometimes bear pigmented monosiphonous laterals (= trichoblasts). Unlike the taxa in the Rhodomelaceae and other families in the Ceramiales, members of the Dasyaceae have sympodial growth, in which the apical cell of the main axis is continually displaced by a lateral cell that becomes the new apex (Rosenberg, 1933; Kylin, 1956). The original apical cell eventually terminates a lateral branch. Parsons (1975) detailed this distinctive type of growth and used the term 'pseudolateral' for such side branches terminated by apical cells displaced from main axes. Members of the Dasyaceae are also characterized by several reproductive features. These include postfertilization development, in which fusion cells initially comprise the auxiliary cell and axial cell of a fertile axis, and the production of tetrasporangial stichidia that lack trichoblasts and have whorls of tetrahedrally divided tetrasporangia borne on each stichidial segment (Parsons, 1975.).

The Dasyaceae is the smallest family in the Ceramiales, with only 12 genera and some 100 species (Kraft, 1981), but is well represented in tropical to cold temperate seas of both hemispheres. *Dasya* and *Heterosiphonia* together account for nearly 80% of all dasyaceous species (Kylin, 1956), although only three species of the latter, along with one species of *Colacodasya*, are listed from high-latitude southern waters (Papenfuss, 1964*a*). The family is represented at Macquarie I. by both *Colacodasya* and *Heterosiphonia*.

Colacodasya Schmitz ***in*** Schmitz & Falkenberg, 1897: 473

Type species: *C. inconspicua* (Reinsch) Schmitz *in* Schmitz & Falkenberg, 1897: 474.

Plants are parasitic on *Heterosiphonia*, weakly dorsiventral and form basal cushions anchored firmily by rhizoids penetrating host tissues. Erect axes arise from the basal cushions and are mostly terete, similarly constructed to host branches, and often pseudodichotomously branched. Gametophytes are dioecious, with males producing spermatangial stichidia and females bearing 4-celled carpogonial branches. Carposporophytes, producing short chains of carposporangia, are encased in well developed pericarps. Cystocarps are spherical to egg-shaped, sessile, and borne at the bases of pseudolaterals. Tetrasporangia are tetrahedrally divided and borne in stichidia, with tiers of 4–6 sporangia per fertile segment.

Four species have been described for this genus, but two, *Colacodasya sinicola* Setchell & Gardner (1924*a*) and *C. verrucaeformis* Setchell & McFadden *in* McFadden (1911), have been moved to the rhodomelaceous genus *Jantinella* Kylin (1941). Of the two remaining species, *C. californica* Hollenberg (1970), is known only from California, while the type, *C. inconspicua*, is restricted to the Southern Hemisphere but distributed throughout the Subantarctic.

Colacodasya inconspicua (Reinsch) Schmitz *in* Schmitz & Falkenberg, 1897: 474

Fig. 106

BASIONYM: *Polysiphonia inconspicua* Reinsch 1888*b*: 146.

SYNONYMS: see Papenfuss, 1964*a*.

TYPE LOCALITY: South Georgia.

TYPE: Possibly in M, but specimens were not found with bulk of Reinsch Herbarium.

DISTRIBUTION: Argentina (Mendoza, 1970*a*); Campbell I. (Levring, 1945); Falkland Is (Kylin & Skottsberg, 1919); Fuegia (pers. obs.); Îles Kerguelen (Levring, 1944); Macquarie I.; South Georgia (Reinsch, 1888*b*, 1890).

MATERIAL EXAMINED:

1. Fuegia: Magellan. *Heterosiphonia berkeleyi* Montagne bearing *Colacodasya* (D'Urville): PC, unnumbered specimen.
2. Macquarie I.: collections MA-83, 95, 237, 302, 304; Herbarium specimens: MELU, R 0542, 0594.

DESCRIPTION: Thalli 'parasitic', forming subspherical (to 3 mm diameter) clusters of branches on *Heterosiphonia* (Fig. 106a). Branches polysiphonous, radiating outwards from a basal cushion anchored to the host by penetrating rhizoids (Fig. 106h). Growth sympodial, with pseudolateral formation obvious at branch apices (Fig. 106b). Gametophytes dioecious, with female thalli *c.* one fifth to one third the size of either male plants or tetrasporophytes. Spermatangia covering all except uppermost portions of subdichotomously forked branches (Fig. 106f); procarps subapical (Fig. 106c, d), with 4-celled carpogonial branches (Fig. 106e). Tetrasporangia tetrahedrally divided, borne in branched stichidia, with whorls of 4–6 sporangia produced per axial segment (Fig. 106g).

Although *Heterosiphonia berkeleyi* is common at Macquarie I., few plants were infected with *C. inconspicua*. The 'parasite' is known from only five collections made between 7–15 m depths along both east and west coasts at the northern and southern ends of the island. The collections suggest an equally broad distribution, although not abundance, for host and 'parasite' around the island.

Fig. 106 *Colacodasya inconspicua* (Reinsch) Schmitz: MELU, R 0542, 0594. **a.** Habit of tetrasporophyte on *Heterosiphonia berkeleyi*. **b.** Apex showing old apical cell (large arrow) and new apical cell (small arrow). **c.** Apex of female gametophyte with procarp (arrow). **d.** Upper branch of female gametophyte with procarps (arrows). **e.** Squash preparation showing carpogonial branch. Cell '3' below the carpogonium (cp) is not in focus. **f.** Upper branch of male gametophyte. **g.** Tetrasporangial stichidia. **h.** T.S. host at site of infection showing deep penetration of parasite rhizoids. Scales: a = 1 mm; b = 30 μm; c, g = 50 μm; d, f, h = 100 μm; e = 20 μm.

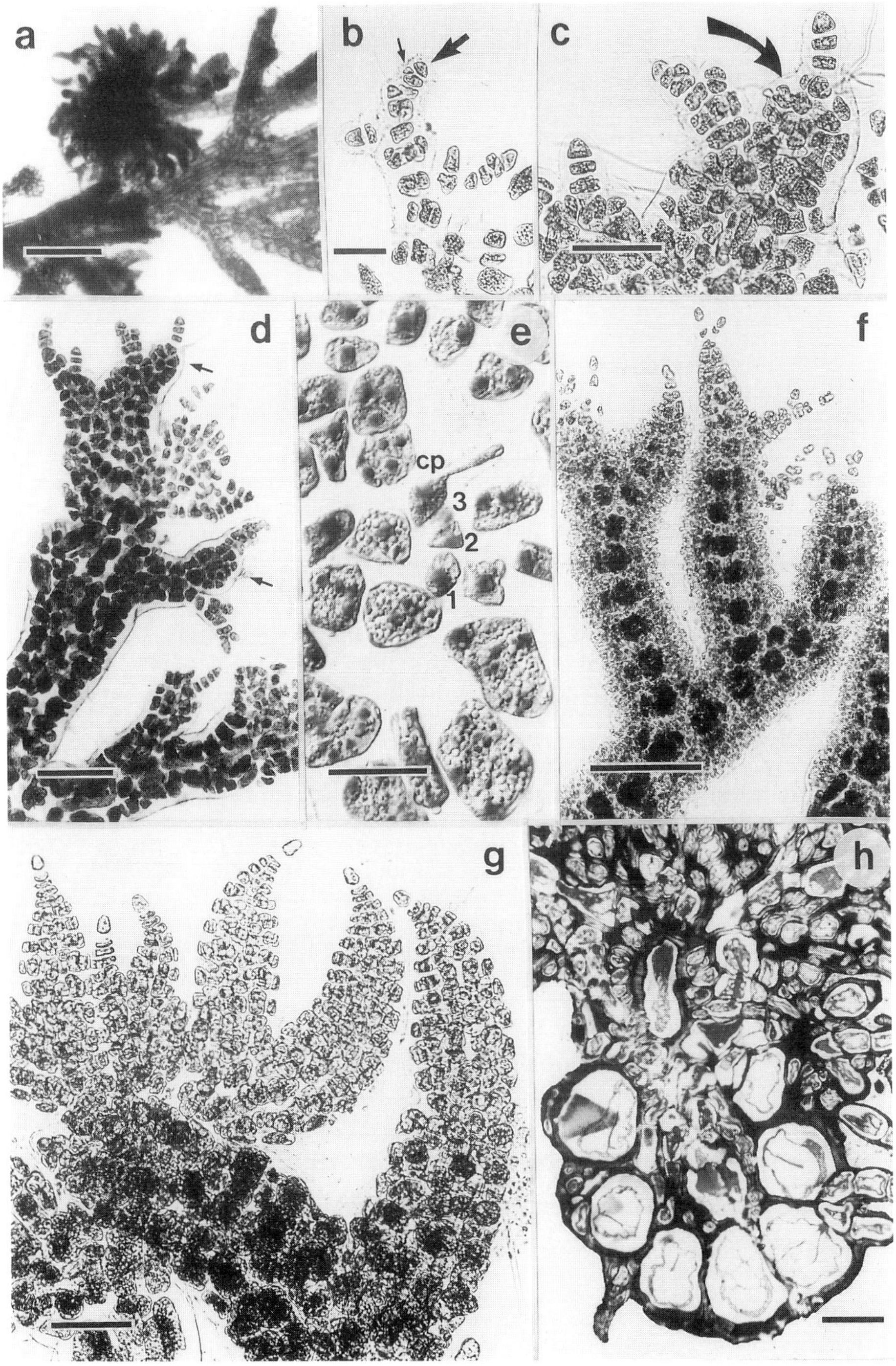
a
b
c
d
e
cp
3
2
1
f
g
h

Most plants were reproductive, with the majority being tetrasporophytes. Of the gametophytes, males were more common than females.

NOTES: The present report of female thalli remaining smaller than male thalli is the first record of slightly dimorphic gametophytes in this genus. This observation, however, is based on only one female gametophyte, and such differences in thallus size might not be consistent among a larger number of individuals. In other respects, Macquarie plants are comparable to those from Kerguelen (Levring, 1944) and South America (Mendoza, 1970a), although South American plants were abundant in the mid- to lower-intertidal as opposed to being strictly subtidal.

Colacodasya inconspicua is the only Southern Hemisphere 'parasite' in the Dasyaceae. It is structurally very similar to its host, but the two can usually be distinguished readily by differences in size, with the 'parasite' being much smaller. Tetrasporangial stichidia are also distinctive, with those of *Colacodasya* branched (Mendoza, 1970a, pl. 1, figs 3, 4) and those of *Heterosiphonia* simple (Falkenberg, 1901).

Heterosiphonia Montagne 1842*b*: 4 (*nom. cons.*)

Type species: *H. berkeleyi* Montagne 1842*b*: 5.

Heterosiphonia was so named because its thalli consist of siphons (pericentral cells) with various diameters (Montagne, 1842*b*). Additional features of the genus, as summarized by Parsons (1975), include discoid holdfasts, either erect or prostrate habit and main axes that are corticate or ecorticate, usually slightly compressed, dorsiventrally organized, composed of 4–12 pericentral cells, and distichously or subdichotomously branched in one plane, with pseudolaterals separated by internodes of 2–9 axial segments. Pigmented, simple or branched, monosiphonous laterals are present and persistent on some polysiphonous axes. Spermatangia are produced on stichidia; procarps, with 4-celled carpogonial branches and two groups of sterile cells, are borne on one of the pericentral cells subtending subdichotomies of pseudolaterals. Mature gonimoblasts have basal fusion cells, carposporangia in chains or terminating sterile filaments, and well developed pericarps with single ostioles. Tetrasporangia are tetrahedrally divided, produced in stichidia, with most stichidial segments bearing whorls of 4–6 sporangia, and covered at maturity by four cells.

Reproductive features in the genus are mostly uniform and not used to delineate infrageneric groups, although Falkenberg (1901) separated some species partly by the production of tetrasporangial stichidia with either monosiphonous or polysiphonous bases. Parsons (1975) used differences in thallus habit and size, diameters of main axes, the presence or absence of cortication, the number of pericentral cells, and the number of segments (= the number of axial cells) between nodes where branches arise.

Heterosiphonia includes some 40 species and occurs in both hemispheres (Kylin, 1956). It is absent from the Antarctic but credited with three species in the Subantarctic (Papenfuss, 1964*a*), of which the type, *H. berkeleyi*, is most widespread and present at Macquarie I.

Heterosiphonia berkeleyi Montagne 1842*b*: 5

Fig. 107

SYNONYMS: *Heterosiphonia merenia* Falkenberg, 1901: 653 (see Papenfuss, 1964*a*).

TYPE LOCALITY: Fuegia.

TYPE: PC, unnumbered specimens.

Fig. 107 *Heterosiphonia berkeleyi* Montagne: MELU, R 1714, 3304, 3305. **a.** Habit of sterile thallus associated with *Myriogramme*. **b.** Branch apices with mature (m) and developing (d) pseudolaterals and new indeterminate axis (arrow). **c.** T.S. mature axis. **d.** Tetrasporangial stichidium. **e.** Mature cystocarp. **f.** Branch apex of male gametophyte with immature spermatangial branches (arrows). **g.** Mature spermatangial branches. Scales: a = 2 cm; b, c, f, g = 100 μm; d = 200 μm; e = 500 μm.

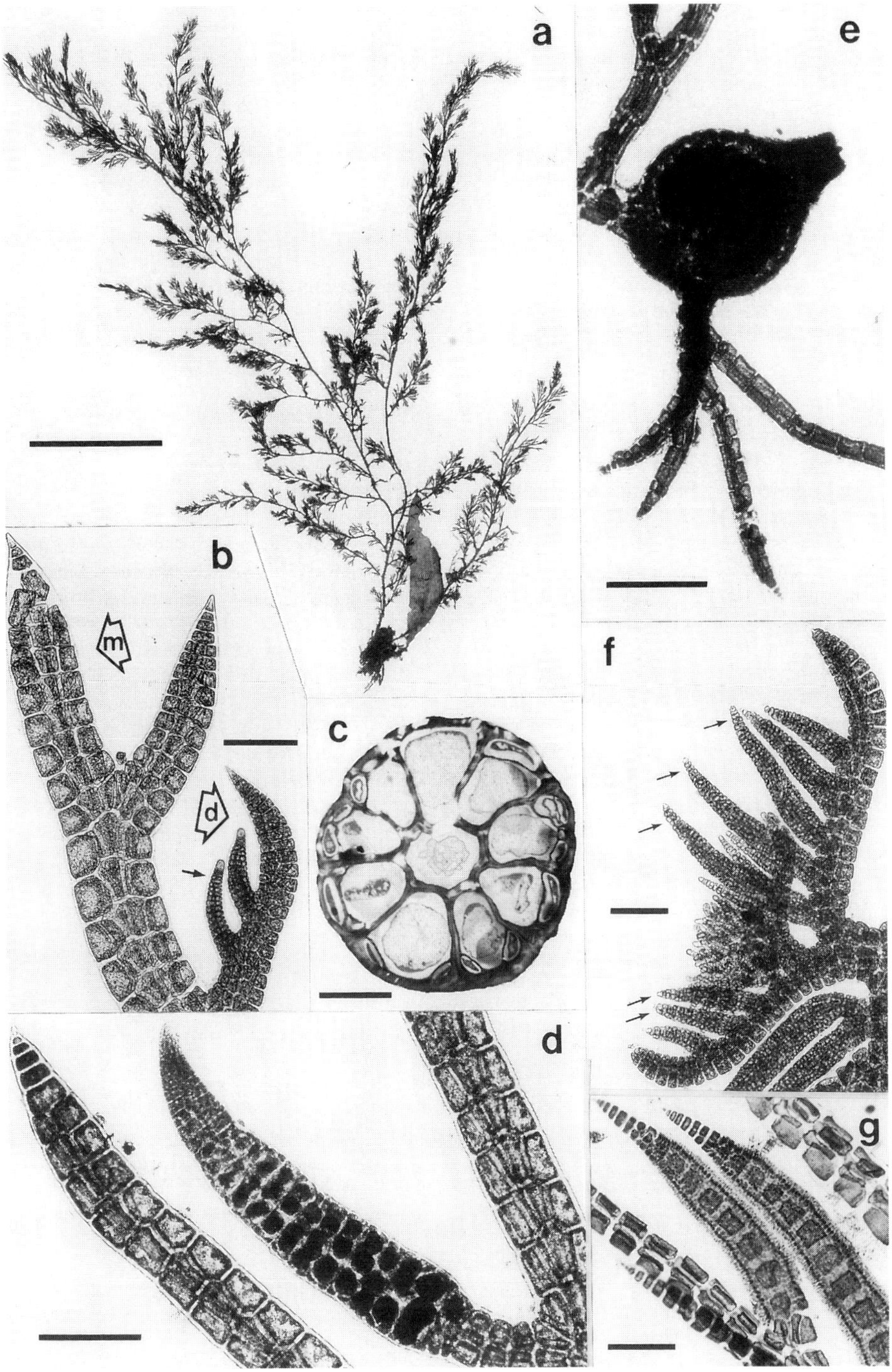
a
b
m
d
c
d
e
f
g

DISTRIBUTION: Argentina (Pujals, 1963); Auckland Is (Laing, 1909); Campbell I. (Levring, 1945); Chile (Levring, 1960); Falkland Is, South Georgia (Skottsberg, 1923); Fuegia (Montagne, 1842*b*, 1845); Îles Crozet and Kerguelen (Levring, 1944—as *H. merenia*); Marion I. (Dickie, 1876*b*); Macquarie I. (N.B. Papenfuss (1964*a*) additionally indicated distribution at the Prince Edward Is., but the original published record for this locality has not been located for the present report.)

MATERIAL EXAMINED:

1. Auckland Is: no collection data available M, 67.83/62 (male).
2. Fuegia: Type specimens. Magellan (Berkeley): PC, unnumbered specimen; Magellan (D'Urville): PC, unnumbered specimen bearing *Colacodasya*.
3. Îles Kerguelen: no collection data available; specimens intermixed with *Ceramium*; det. Reinbold: M, 67.83/61 (female).
4. Macquarie I.: collections MA-14, 39, 46, 54, 83, 95, 98, 113, 222, 227, 237, 271, 283, 300, 302, 303, 306, 309, 364, 377, 378; Herbarium specimens: MELU, R 0542, 1444, 1714, 1717, 1934, 2034, 2035, 3304, 3305; Buckles Bay, washed ashore (Haysom, 30 January 1950): MELU, R 3732 (= Haysom no. MI/49/×31).

DESCRIPTION: Thalli erect, to 11 cm tall, rose-coloured to reddish-brown to purplish-red, with distinct branched main axes arising from discoid holdfasts (Fig. 107a). Axes lax, terete to slightly flattened, alternately to subdichotomously branched in one plane, composed of monosiphonous and polysiphonous (with 7–10 pericentral cells) portions, to 400 μm diameter in mid-thallus (Fig. 107c), and with apices of indeterminate axes overtopped by surrounding (determinate) pseudolaterals (Fig. 107b, f). Monosiphonous upper portions of indeterminate and determinate branches pigmented, simple, and to 10 cells long. Lower portions corticated slightly by short rhizoidal cells produced only at nodes. Spermatangia covering stichidial surfaces except at monosiphonous upper portions (Fig. 107g); mature cystocarps with well-developed, ostiolate, pyriform pericarps surrounding dense carposporangial masses (Fig. 107e). Tetrasporangia tetrahedrally divided, with whorls of six sporangia per tier in unbranched stichidia (Fig. 107d). Tetrasporangial stichidia with polysiphonous bases.

Plants are common but not abundant in various shaded and unshaded localities between 3–14 m depths around Macquarie I. *Heterosiphonia* usually grows epiphytically on other red algae, most often on *Phyllophora appendiculata*, but sometimes occurs on rocks or bryozoan colonies. Several members of the Delesseriaceae (e.g., *Hymenena laciniata*, *Myriogramme livida* and *Schizoseris dichotoma*) and the encrusting coralline *Melobesia membranacea* frequently grow adjacent to this species.

NOTES: Falkenberg (1901) described *H. berkeleyi* as 12–15 cm tall, with axes uncorticated throughout, constructed of 4–10 pericentral cells (two of which are characteristically flattened and larger than the others), and producing unbranched tetrasporangial stichidia. Older plants from Macquarie I. become weakly corticated along lower axes but otherwise closely match Falkenberg's description. The differences in siphon (pericentral cell) diameters in Macquarie plants are apparent close to the branch apices but less obvious in mid- to lower-thallus transections (Fig. 107c).

In the same publication, Falkenberg (1901) erected the new species *H. merenia*, based on Reinsch's (1890) *Merenia microcladioides*, which supposedly has thalli with partially corticated lower axes. Macquarie plants are similarly corticated and in other aspects agree completely with various descriptions of *H. merenia* (Falkenberg, 1901; Levring, 1944; Mendoza, 1970*a*; Parsons, 1975). Askenasy (1888, pl. xi) illustrated plants, which he identified as *H. berkeleyi*, with lightly corticated main axes and other vegetative and reproductive features similar to those in Macquarie thalli. In addition to matching Falkenberg's (1901) descriptions of both species and agreeing with Askenasy's concept of the type species, Macquarie plants are similar to type material of *H. berkeleyi*. Because the fragments of the type species studied during the present survey represented only upper portions of the thallus, it was not possible to check for nodal cortication along lower axes in the type. The numerous similarities in habit, branching pattern, apical development, lengths of monosiphonous portions, main axis diameters, and numbers of pericentral cells, however, support the present belief that the taxa are identical, with *H. merenia* being a later synonym of *H. berkeleyi*.

Delesseriaceae Bory de Saint-Vincent, 1828: 181

Members of this family are mostly free-living, although four genera are parasitic. Thalli are usually delicate and either monopodially or sympodially constructed with pseudoparenchyma, with fronds generally flattened and foliaceous, although sometimes terete or pulvinate, and either simple or branched, perforate, lattice-like, or entire. Sporophytes and gametophytes are morphologically similar. Reproductive development generally is uniform throughout the family and not taxonomically useful at infra-familial levels, although features such as the number of carpogonial branches and sterile cells per procarp, the location of reproductive organs, and the construction of mature gonimoblasts are distinctive for some taxa. Mostly, however, constituent genera are distinguished vegetatively, with closely related taxa often separated by single characters. These include thallus habit, frond thickness and shape, degree and pattern of innervation, apical morphology (i.e., with either a single apical cell or a row of marginal cells), cell division patterns (i.e., with intercalary divisions occurring in either primary, secondary, or tertiary cell rows), and presence or absence of rhizoidal cortication (Kylin, 1956; Wynne, 1983).

The Delesseriaceae often forms a prominent component of marine floras, especially in temperate to colder waters of both hemispheres. Extensive research in the family through monographs (Kylin, 1924; Wagner, 1954), regional studies (Aleutian Is—Wynne, 1970; Argentina—Mendoza, 1969*b*, 1969*c*, 1974; Australia—Womersley & Shepley, 1982; New Zealand—Kylin, 1929; Northern Pacific—Zinova, 1972, 1976; Kerguelen & Macquarie Is—Zinova, 1963; South Sandwich Is— Wynne, 1982*a*) and descriptions of new taxa (e.g., Mendoza, 1975, 1976*b*; Zinova, 1981) has led to the present count of nearly 100 genera totalling some 300 species (Kraft, 1981). Papenfuss (1964*a*) listed 56 species from the Southern Ocean, although (my) present estimations, based on various synonymies, place the number somewhere between 40 and 45. At Macquarie I., the family is represented by seven genera and nine species, many of which are prominent members of subtidal communities.

Cladodonta Skottsberg, 1923: 36

Type species: *C. lyallii* (Hooker f. & Harvey) Skottsberg, 1923: 38.

The monotypic genus *Cladodonta* was described first from the Kerguelen and Falkland Is but has been collected since from other Subantarctic localities. Its thalli are superficially similar to those of four other Southern Ocean species, *Anisocladella serratodentata, Erythroglossum bipinnatifida, Phycodrys austrogeorgica,* and *P. quercifolia,* in having foliose fronds, single transversely dividing apical cells, intercalary divisions in primary cell rows, no rhizoidal cortication, and scattered procarps (Kylin, 1956). *Cladodonta* is distinct, however, in having polystromatic blades with prominent midribs and paired lateral veins. Additionally, blade margins are usually thickened and produce numerous marginal teeth that sometimes give rise to lateral blades (Hooker f. & Harvey, 1845).

Cladodonta lyallii (Hooker f. & Harvey) Skottsberg, 1923: 38

Fig. 108

Basionym: *Delesseria lyallii* Hooker f. & Harvey, 1845: 252.

Synonyms: see Papenfuss, 1964*a*.

Type locality: Not indicated (Falkland or Kerguelen Is).

Holotype: not designated; syntypes in BM, M, and MEL.

Distribution: Argentina (Mendoza, 1969*b*); Falkland Is (Hooker f. & Harvey, 1845); Fuegia (Skottsberg, 1923); Îles Kerguelen (Hooker f. & Harvey, 1845; Levring, 1944); Macquarie I.; Marion I. (Dickie, 1876*b*); Tristan da Cunha (Baardseth, 1941). (N.B. Papenfuss (1964*a*) addition-

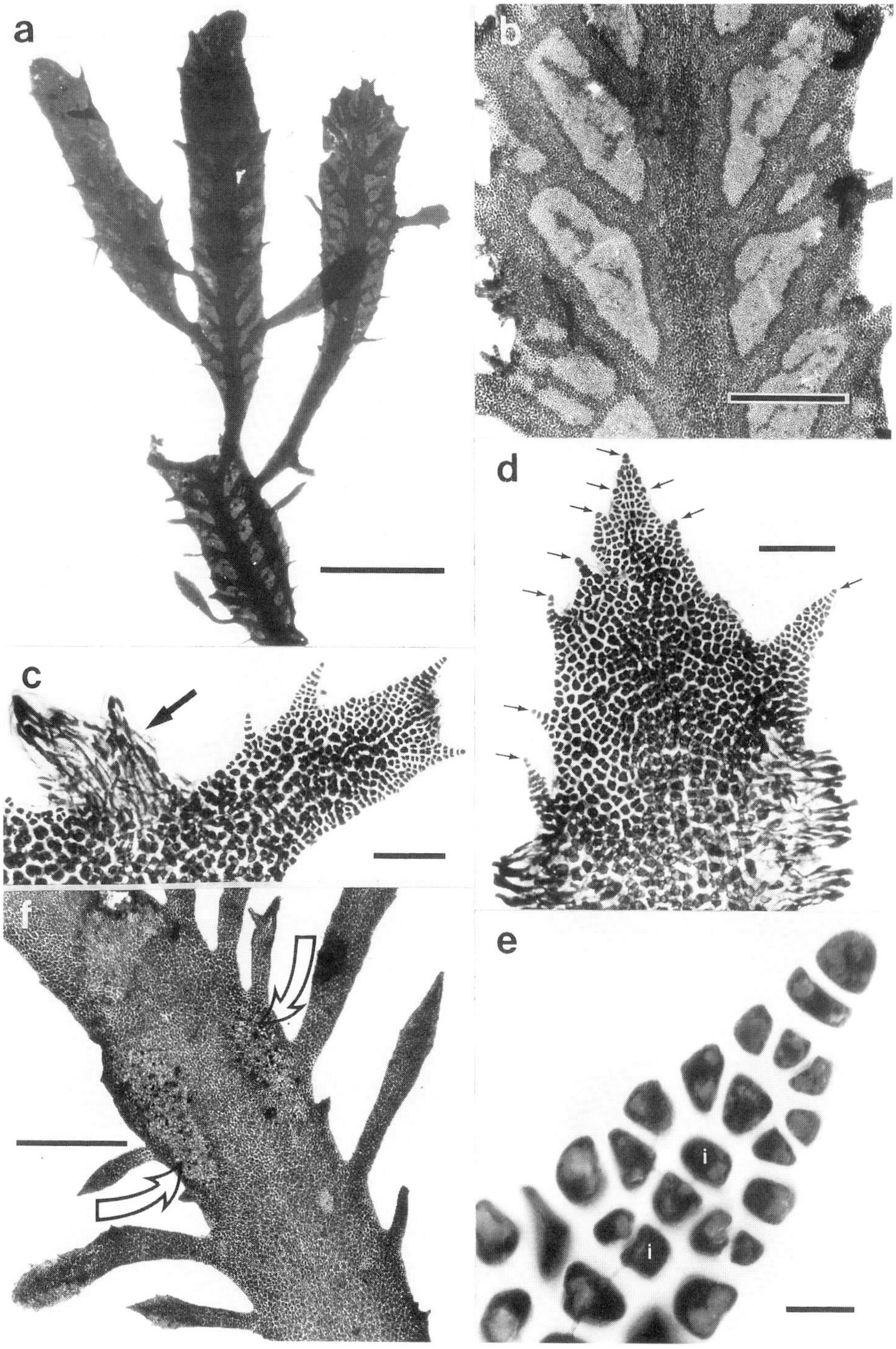
a
b
c
d
e
f
i
i

ally indicated distributions at the Prince Edward Is, but the original published records for these localities have not been located for the present report.)

MATERIAL EXAMINED:

1. Falkland Is: Hohenacker, Algae marin. exs. no. 249 (Oestliche Falklands-Inseln): M, 67.83/21.
2. Iles Kerguelen: (plant identified by Reinbold): M, 67.83/22 (tetrasporophyte); Syntype (J. D. Hooker): M, 67.83/23 (tetrasporophyte); Syntype (J. D. Hooker): MEL, 573342.
3. Macquarie I.: collections MA-98, 303; Herbarium specimens: MELU, R 1536, 1766–1782.
4. South America: Isthmus Bay (Paessler, 1908): M, 67.83/20 (tetrasporophyte).

DESCRIPTION: Thalli prostrate, to 3 cm long, reddish-orange, and with fronds foliose, polystromatic throughout, and having pronounced midribs with paired lateral veins (Fig. 108a, b). Blade margins thickened, dentate, giving rise to lateral fronds resembling primary blades, and bearing patches of rhizoids for attachment (Fig. 108c). Growth from prominent, transversely dividing, dome-shaped apical cells; intercalary divisions present in primary cell rows (Fig. 108d, e). Tetrasporangia tetrahedrally divided, localized in sori between veins (Fig. 108f). Sexual reproductive structures not seen.

Plants were collected only twice and on both occasions from shaded subtidal localities. In one collection from 5 m depth at the northern end of the island, plants formed patches on the underside of a rocky overhang partially covered by sponges and tunicates. The second collection was from the southern end of the island, where plants grew on boulders in 20 m depth on the floor of a *Macrocystis* forest.

NOTES: Macquarie specimens of *Cladodonta* are similar to the type species in the critical features of blade development, anatomy, and morphology, but differ in their smaller size, prostrate habit, and rhizoidal attachment. Hooker f. & Harvey (1845) described plants as erect, arising from discoid holdfasts, and with primary blades to 23 cm (nine inches) long. All subsequent descriptions of this species, excepting the present record of prostrate individuals, also refer to erect thalli (Skottsberg, 1923; Baardseth, 1941; Mendoza, 1969*b*). In spite of these differences, the overall similarities to *C. lyallii* suggest tentative placement of Macquarie plants in this species until information is available regarding environmental influences on its thallus morphology.

Certain vegetative and reproductive features are variously expressed in plants from different localities, the same locality, or, in some instances, single specimens. The marginal thickening noted first by Hooker f. & Harvey (1845), illutrated later by Mendoza (1969*b*), and recorded here in Macquarie thalli, is sometimes present (Falkland plant—M, 67.83/21 and Macquarie thalli), absent (South American plant—M, 67. 83/20), or developed only in the lower thallus and absent from the upper thallus (Kerguelen thalli—M, 67.83/22, 67.83/23). Rhizoids for attachment were present along thallus margins only on plants from Macquarie I. and one individual from the Falkland Is. Tetrasporangial sori are produced on either principal blade faces (Macquarie thalli), lateral proliferations (South American thalli), or both main blade faces and lateral proliferations (Kerguelen thalli).

Gametophytes were not observed during the present survey but were described and illustrated earlier by Mendoza (1969*b*, pls. 4–7). According to her, spermatangia are localized in sori and scattered across blade surfaces, and carposporophytes bear chains of carposporangia encased in bulbous cystocarps positioned between veins.

Fig. 108 *Cladodonta lyallii* (Hooker f. & Harvey) Skottsberg: MELU, R 1536, 1782. **a.** Habit of upper frond. **b.** Detail of blade venation. **c.** Tuft of rhizoids on blade margin. **d.** Upper blade with multiple apical cells (arrows). **e.** Detail of indeterminate apex showing primary cell row with cells (i) produced by intercalary divisions. **f.** Surface view of tetrasporangial sori (arrows). Scales: a = 3 mm; b, f = 1 mm; c, d = 100 µm; e = 10 µm.

Delesseria Lamouroux, 1813: 122 (*nom. cons.*)

Type species: *D. sanguinea* (Hudson) Lamouroux, 1813: 124.

Members of this genus generally are delicate and have flattened foliose fronds that are monostromatic except along thickened midribs and lateral veins. Thalli usually produce proliferations along costae. *Delesseria* is distinguished mostly by vegetative features, including monopodial construction, growth from single transversely dividing apical cells, presence of intercalary cell divisions in only second and higher order cell rows, failure of tertiary initials to reach thallus margins, and development of rhizoids amongst larger nearly isodiametric cells of midribs (Kylin, 1924, 1956; Wynne, 1983).

Procarps of *Delesseria* are borne on primary cell rows, have one carpogonial branch per supporting cell and, following fertilization, result in mature cystocarps occurring only on midribs of fertile blades or bladelets. Spermatangia and tetrasporangia are borne in sori that generally develop to either side of midribs (Kylin, 1956).

Kylin (1956) estimated 15 species for the genus, most of which are known only from the Southern Ocean, although some also occur in temperate to colder waters of the Northern Hemisphere. Of the eleven species listed by Papenfuss (1964*a*) from Antarctic and Subantarctic waters, two were recently synonymized (Wynne, 1982*a*: *D. salicifolia* = *D. stephanocarpa*), and another two (*D. belayevii, D. uschakovii*) are considered in the present survey to be synonyms of the morphologically variable *D. lancifolia* (Hooker f.) J. Agardh.

The genus is represented at Macquarie I. by *D. lancifolia*, which occurs throughout the Subantarctic and is also known from the Antarctic Peninsula.

Delesseria lancifolia (Hooker f.) J. Agardh, 1872: 59

Figs 109a, b, d, e, 110a–c, e–i

Basionym: *Delesseria sanguinea* var. *lancifolia* Hooker f. 1847: 470.

Synonyms:

Delesseria lancifolia var. *minor* Laing, 1896: 450.
Paraglossum lancifolium (Hooker f.) J. Agardh, 1898: 217.
D. belayevii A. Zinova 1963: 53.
D. uschakovii A. Zinova 1963: 62.

Type locality: Fuegia.

Type: LD, 31749.

Distribution: Antarctic Peninsula (Lamb & Zimmermann, 1977); Falkland Is (Cotton, 1915); Fuegia (J. D. Hooker, 1847; Kylin & Skottsberg, 1919); Îles Kerguelen (Zinova, 1963); Macquarie I. (Laing, 1896; Zinova, 1963); New Zealand (Kylin, 1929).

Material examined:

1. Antarctic Peninsula: Melchoir Archipelago, Gamma I., N. side 150 m off Gallows Pt., depth 29 m, on large tunicate on the vertical rock face (Lamb, 25 January 1965): FH, A-855a.
2. Fuegia: St. Martin's Cove (O. Smith, 1842): LD, 31749.
3. Macquarie I.: collections MA-10, 11, 34, 39, 94, 95, 111, 129, 137, 140, 222, 227, 244, 271, 278, 303, 364, 377; Herbarium specimens: MELU, R 0102, 0785, 0888, 1254, 1720, 1730, 1734–1736, 3738.

Additional species examined:

Delesseria salicifolia Reinsch 1888*b*, 149

South Georgia: Lectotype. 'Strand unterhalb der Station durch Sturm ausgeworfen' (Will, February 1883): M, 67.83/49 [N.B. this sterile specimen was illustrated by Reinsch (1890, pl. 4, fig. 1)]; Syntypes: M, 67.83/48 (four separate sheets with sterile fragments, two of which are mixed

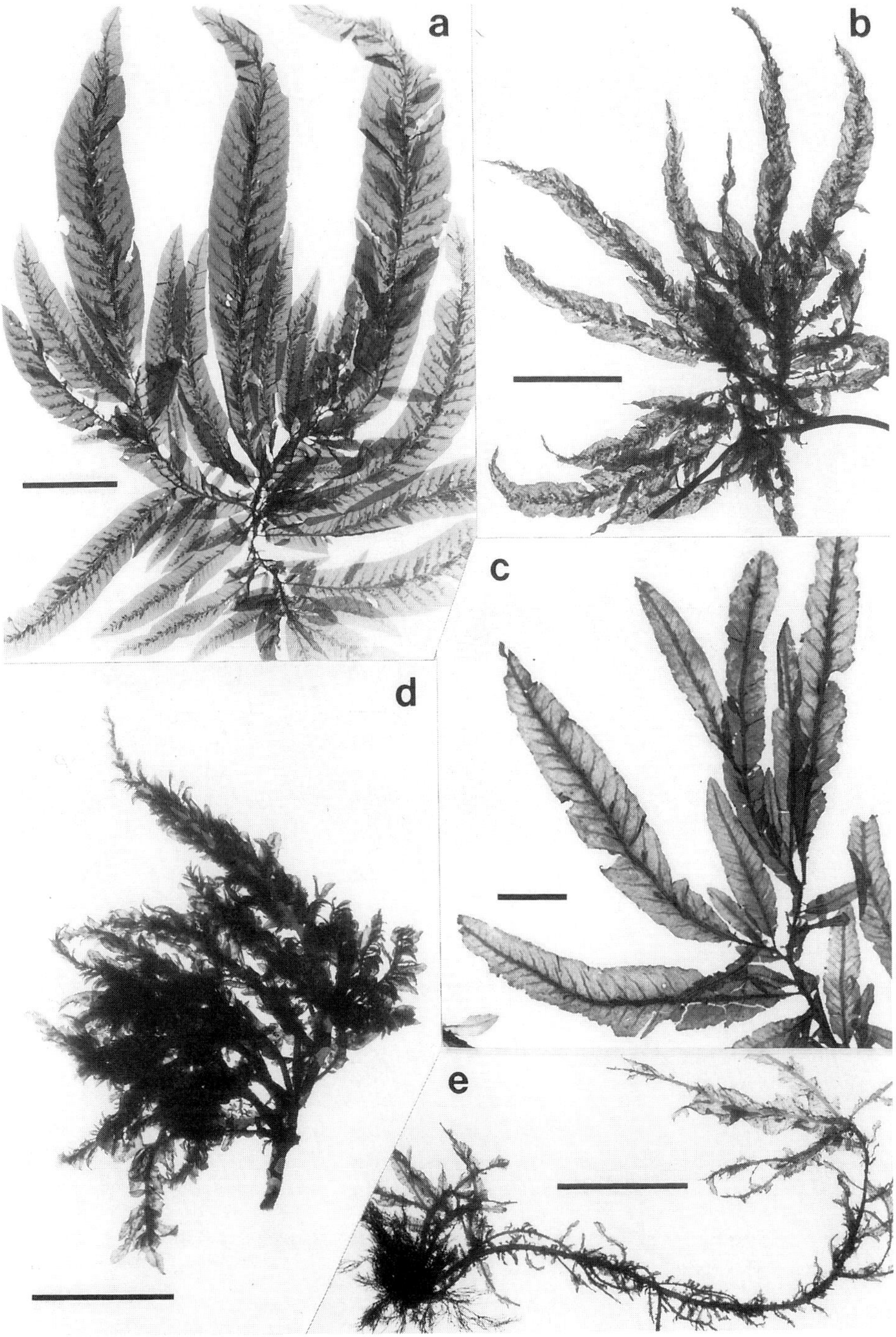

Fig. 109a, b, d, e *Delesseria lancifolia* (Hooker f.) J. Agardh: MELU, R 0102, 1720, 1730, 1736. **c** *Delesseria salicifolia* Reinsch: M, 67.83/49. **a.** Habit of epilithic plant. **b.** Habit of plant epiphytic on *Desmarestia chordalis*. **c.** Portion of lectotype of *Delesseria salicifolia*. **d.** Habit of coarse thallus. **e.** Habit of eroded thallus. Scales: a, b, e = 5 cm; c = 1 cm; d = 4 cm.

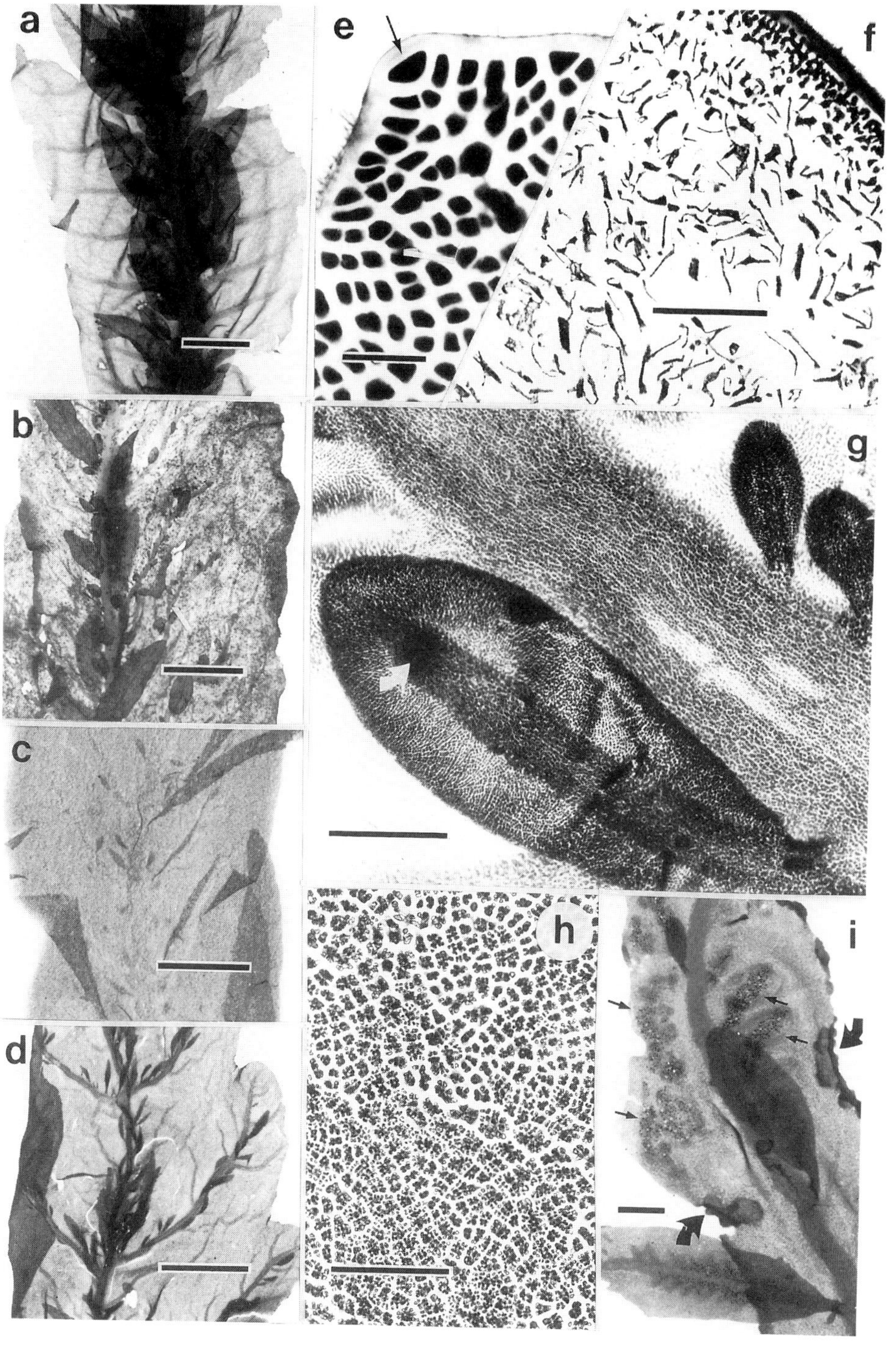
a
b
c
d
e
f
g
h
i

with *Pterothamnion simile* and *Ballia callitricha*), 67.83/50 (= Will no. 64), 67.83/51 (tetrasporophyte).

DESCRIPTION: Thalli erect, to 50 cm tall, rose-coloured to deep red, branched, and with fronds foliose, stipitate, mostly monostromatic, delicate, generally entire and lanceolate to 10 cm broad (Fig. 109a, b), although sometimes highly eroded with narrow (<2 cm) fronds and coarse stipes (Fig. 109e). Branching from percurrent midribs, with branches usually arising alternately or oppositely but sometimes irregularly or proliferously (Fig. 109d). Growth from transversely dividing apical cells (Fig. 110e). Microscopic veins developed to various degrees, mostly appearing as extensions of larger veins; midribs coarse, thickened partly with internal rhizoids (Fig. 110f), and bearing macroscopic lateral veins—usually simple and straight in young thalli, irregularly branched, sinuous, and sometimes anastomosing in older thalli (Fig. 110a–c). Bladelets (having midribs bearing or lacking lateral veins) borne singly or in clusters along midribs and lateral veins of primary blades. Spermatangia in irregularly shaped sori located between veins and margins of major blades (Fig. 110h); soral patches tristromatic, with spermatangial mother cells rounded to anticlinally elongate and dividing obliquely at distal ends producing clusters of rounded to elongate spermatangia. Procarps borne on midribs, usually 1–2 per bladelet (Fig. 110g), although sometimes singly produced near apices of main blades; mature cystocarps conical, ostiolate,with tapered elongated necks. Tetrasporangia tetrahedrally divided, aggregated in irregularly shaped sori located between veins and frond margins (Fig. 110i); sori mostly lining midribs of small bladelets but also forming extensive patches on primary blades.

Plants are common in intertidal and subtidal habitats along most stretches of Macquarie's coastline. In the mid- to lower-intertidal, *D. lancifolia* usually is either epilithic or epiphytic on encrusting coralline algae and sometimes is epiphytized by species such as *Acrosiphonia pacifica*, *Ectocarpus constanciae* and *Schizoseris dichotoma*. Intertidal plants, rarely exceeding 20 cm in length, generally have profusely branched thalli with coarse, terete to flattened stipes and fronds that are narrow and lanceolate or wiry and greatly eroded. Subtidal plants, which grow either on rocks or as epiphytes on fleshy algae (usually on *Desmarestia chordalis*) at 3–20 m depths, generally have broad delicate fronds that are simple or branched and up to 50 cm in length. Most of the larger plants were epilithic and formed a part of the red algal understory that proliferates along the floors of *Macrocystis* forests. In this habitat, *D. lancifolia* frequently grows adjacent to *Schizoseris dichotoma* and *Polysiphonia anisogona* and sometimes is epiphytized by *Pterothamnion plumula*.

NOTES: *Delesseria* from Macquarie I. forms a morphologically diverse group, some individuals matching Skottsberg's (1923) account of *D. epiglossum* J. Agardh from the Falkland Is, and others agreeing with the type descriptions of *D. lancifolia* J. Agardh (1872), *D. belayevii* Zinova (1963), and *D. uschakovii* Zinova (1963). The two extremes in the range of morphologies would be the coarse intertidal plants with narrow blades (which resemble *D. epiglossum* and *D. uschakovii*) and the delicate subtidal plants with broad blades (which resemble *D. lancifolia* and *D. belayevii*). Study of over 100 specimens has revealed numerous intermediate morphologies that apparently reflect degrees of frond erosion and prevent clear delineation of more than one species. For this reason, *D. lancifolia* is interpreted broadly and considered to be the only member of the genus at Macquarie I. *Delesseria epiglossum* is either closely related to or identical with *D. lancifolia*, but because type material of the former has not been studied during the present survey, the two are maintained as separate species.

Delesseria salicifolia Reinsch (Fig. 109c) and *D. lancifolia* have similar habits, comparable patterns of venation (Fig. 110a–d), and identical disposition of bladelets. The two species appear to be closely related, possibly even conspecific, but tentatively are distinguished largely by

---

Fig. 110a–c, e–i *Delesseria lancifolia* (Hooker f.) J. Agardh: MELU, R 1734, 1735, 3738; LD, 31749. **d.** *Delesseria salicifolia* Reinsch: M, 67.83/51. **a–d.** Patterns of blade venation. **a.** Young frond. **b.** Old frond. **c.** Type specimen of *D. lancifolia*. **d.** Portion of syntype of *D. salicifolia*. **e.** Blade apex (arrow). **f.** T.S. main axis. **g.** Female gametophyte with developing gonimoblast (arrow). **h.** Spermatangial patches covering blade surface of male gametophyte. **i.** Frond with tetrasporangial sori (straight arrows) and bacterial infections (curved arrows). Scales: a, c, d = 5 mm; b = 1 cm; e = 25 μm; f, h = 200 μm; g = 500 μm; i = 1 mm.

differences in cystocarp morphology. Pericarps of *D. salicifolia* have spinose projections, whereas those of *D. lancifolia* are smooth (Wynne, 1982*a*). The production of tetrasporangia only on specialized sporophylls with serrated margins has been purported to be another feature diagnostic for *D. salicifolia* (Kylin & Skottsberg, 1919; Wynne, 1982*a*). Syntype material of this species has confirmed the presence of such sporophylls (vs. those with smooth margins in *D. lancifolia*), but also shows development of tetrasporangia on main blade surfaces as occurs in *D. lancifolia*.

Hymenena Greville, 1830: xlviii

Type species: *H. fissa* Greville, 1830: xlviii (= *H. venosa* (Linnaeus) Krauss, 1846: 209. see Kylin (1924) re. synonymy).

Members of this genus have flattened and foliose fronds developed from marginal meristems, are monostromatic above and polystromatic below, and are dichotomously branched. Microscopic veins occur in all species, and macroscopic veins sometimes also develop. Plants are either erect and seemingly stipitate, this caused by erosion of blades in the lower thallus and persistence of only the midrib, or prostrate and estipitate. Thickened midribs lack rhizoids and are composed of large-celled pseudoparenchyma. Procarps are distributed over frond surfaces, and mature gonimoblasts produce terminal carposporangia. Spermatangia and tetrasporangia are produced in sori that are scattered across frond surfaces.

Kylin (1956) estimated 20 species for the genus, of which most occur in temperate to colder seas in both hemispheres (Kylin, 1924; Abbott & Hollenberg, 1976). *Hymenena* is represented especially well in the Southern Hemisphere by the type species from South Africa (Greville, 1830), a total of nine species from Australia and New Zealand (Kylin, 1924), and some five species listed by Papenfuss (1964*a*) from the Subantarctic. The genus is unknown in the Antarctic but is represented at Macquarie I. by two species which can be distinguished easily by their prostrate (*H. decumbens*) and erect (*H. laciniata*) habits.

Hymenena decumbens Levring, 1941: 650

Fig. 111

Type locality: Juan Fernandez Is.

Type: Göteborg? (N.B. not in LD.)

Distribution: Juan Fernandez Is (Levring, 1941); Macquarie I.

Material examined: Macquarie I.: collections MA-98, 303, 306; Herbarium specimens: MELU, R 0498, 0839, 1911.

Description: Thalli prostrate, to 2 cm long, rose-coloured to dark red, with fronds foliose and rhizoidally attached along the ventral surfaces of blades (Fig. 111a, b). Rhizoids solitary, multicellular, colourless, with terminal attachment discs (Fig. 111d, f). Blades flabellate to irregularly lobed, monostromatic above, thicker below, with apices rounded and bordered by zones of meristematic cells (Fig. 111c). Most veins microscopic, distinct, and persisting almost to blade margins (Fig. 111e), although some in lower thallus macroscopic and having indistinct margins (Fig. 111g). Reproductive structures not seen.

Plants are seemingly uncommon at Macquarie I., although are present at both northern and southern ends of the island. *Hymenena decumbens* was collected from shaded habitats beneath ledges in relatively shallow (5 m depth) waters and on rocky surfaces to 18 m depths along the floors of *Macrocystis* forests where *Codium subantarcticum*, *Plocamium hookeri*, *P. secundatum*, and *H. laciniata* are common. In the latter environment, *H. decumbens* and *Microzonia velutina* usually grow intermixed, although the former sometimes occurs by itself and covers coralline encrusted rocks shaded by *Desmarestia willii*.

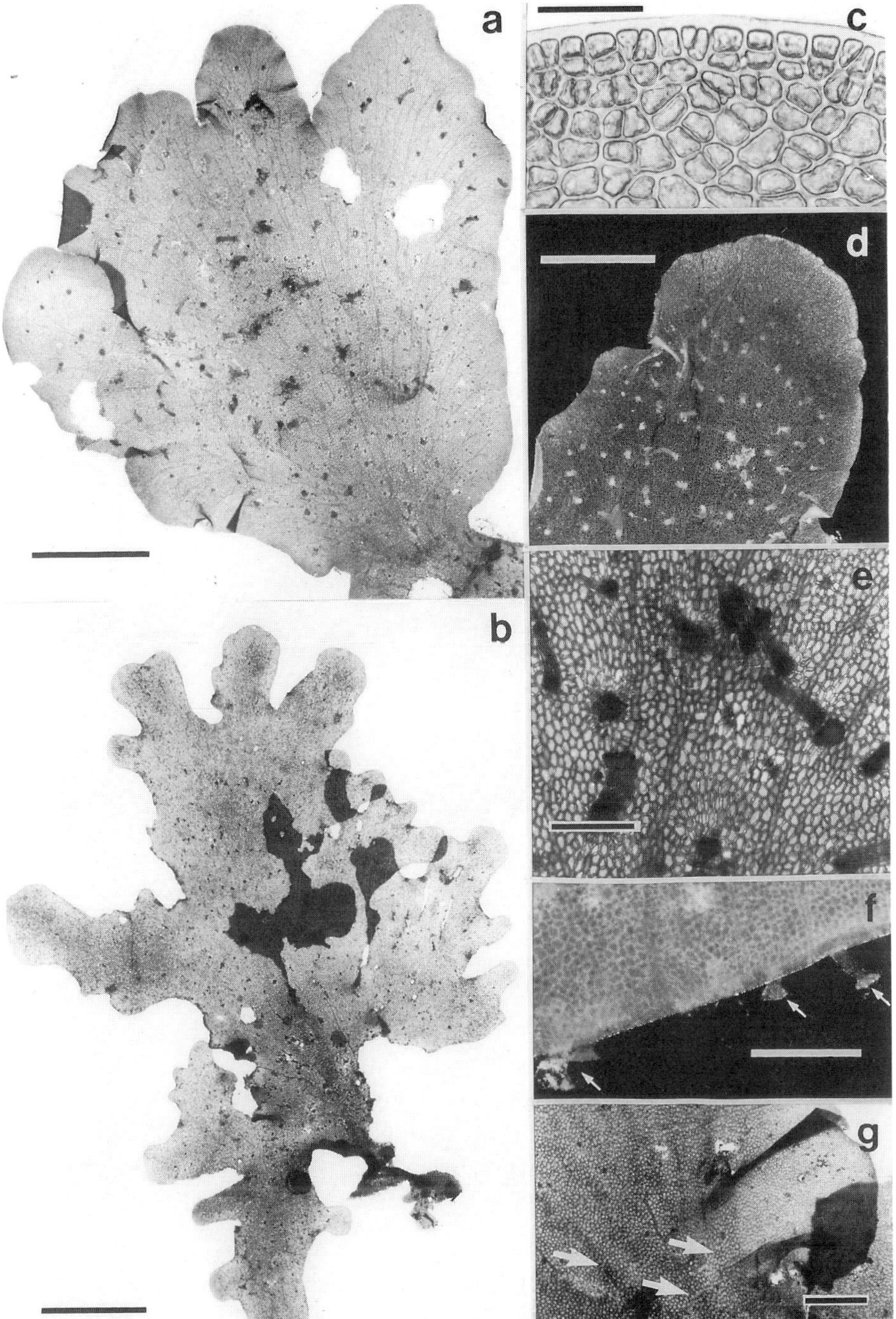

Fig. 111 *Hymenena decumbens* Levring: MELU, R 0498, 0839. **a, b.** Habits of sterile thalli. **c.** Thallus margin. **d.** Upper thallus. **e.** Mid-thallus showing prominent venation. **f.** Ventral surface of frond with attachment rhizoids (arrows). **g.** Lower thallus with coarse veins (arrows) lacking distinct margins. Scales: a, d = 2 mm; b = 3 mm; c = 40 μm; e, f = 500 μm; g = 1 mm.

NOTES: Macquarie thalli usually are lobed and morphologically similar to the plants that Levring (1941, fig 21b–d, f–h) illustrated from the Juan Fernandez Is, although sometimes fronds are broader and mostly undivided. Tetrasporophytes and female gametophytes are known only from Levring's (1941) original description of the species and male plants remain unknown.

Hymenena laciniata (Hooker f. & Harvey) Kylin, 1924: 82

Fig. 112

BASIONYM: *Nitophyllum laciniatum* Hooker f. & Harvey, 1845: 256.

SYNONYMS: see Papenfuss, 1964*a*.

TYPE LOCALITY: Not designated in the original account. Lectotype from the Falkland Is.

LECTOTYPE: K *in* BM; specimen with 'fruits' selected by J. H. Price (28 February 1985). Herbarium sheet with annotation: 'Cast up on the beach Berkeley [sic!] Sound.—June 1842'. (N.B. specimen mounted on same herbarium sheet containing three young plants 'without fruit'—Price, pers. comm.)

DISTRIBUTION: Falkland Is, Fuegia (Hooker f. & Harvey, 1845); Îles Kerguelen (Levring, 1944). (N.B. Papenfuss (1964*a*) indicated distribution at Macquarie I., but the original published record for this locality has not been located for the present report.)

MATERIAL EXAMINED: Macquarie I.: collections MA-39, 54, 83, 84, 98, 140, 244, 248, 283, 302, 303, 309, 364, 378; Herbarium specimens: MELU, R 0152, 1345–1356, 1712, 3391.

DESCRIPTION: Thalli erect, to 25 cm tall, rose-coloured to reddish-brown, and with fronds foliose, mostly monostromatic, sometimes soft and delicate but usually slightly crisp, regularly branched to highly laciniate, stipitate, and arising from hapteroid bases comprising irregularly branched flattened axes (Fig. 112a, b). Blades initially flabellate, later strap-like and alternately to sub-dichotomously branched, with branches 8–15 mm broad and having smooth or crenulate (Fig. 112e) margins and rounded apices bordered by zones of meristematic cells (Fig. 112c). Fronds with veins; most veins microscopic (Fig. 112d) except in lower thallus where they generally become coarse and prominent, with one often forming a polystromatic midrib. Midrib persisting following erosion of monostromatic lower portions of frond resulting in mature thalli with elongate stipes. Tetrasporangia tetrahedrally divided, borne in rounded to elliptical sori produced on main blade surfaces in upper thallus (Fig. 112f). Gametophytes not seen.

Plants are abundant on rocky surfaces from immediately below the sublittoral fringe to at least 18 m depths around Macquarie I. Overstories of *Desmarestia chordalis* and *D. willii*, as well as canopies of *Macrocystis* fronds, usually shade underlying clumps of *H. laciniata*. The lower branches of *H. laciniata* are generally partially covered by tube worms and bryozoan colonies and less often by various epiphytic algae such as members of the Ectocarpales and the red algae *Rhodophyllis acanthocarpa* and *Myriogramme livida*.

Most specimens collected were sterile, although tetrasporophytes were present in a few collections.

NOTES: *Hymenena laciniata* from Macquarie I. is morphologically similar to the South American specimen illustrated by Mendoza (1969*b*, pl. 11). Her description and illustrations of tetrasporphytes and male and female gametophytes constitute the most complete account for this species.

Myriogramme Kylin, 1924: 55

Type species: *M. livida* (Hooker f. & Harvey) Kylin, 1924: 58.

Members of this genus are erect and have fronds that are foliose, stipitate and either undivided,

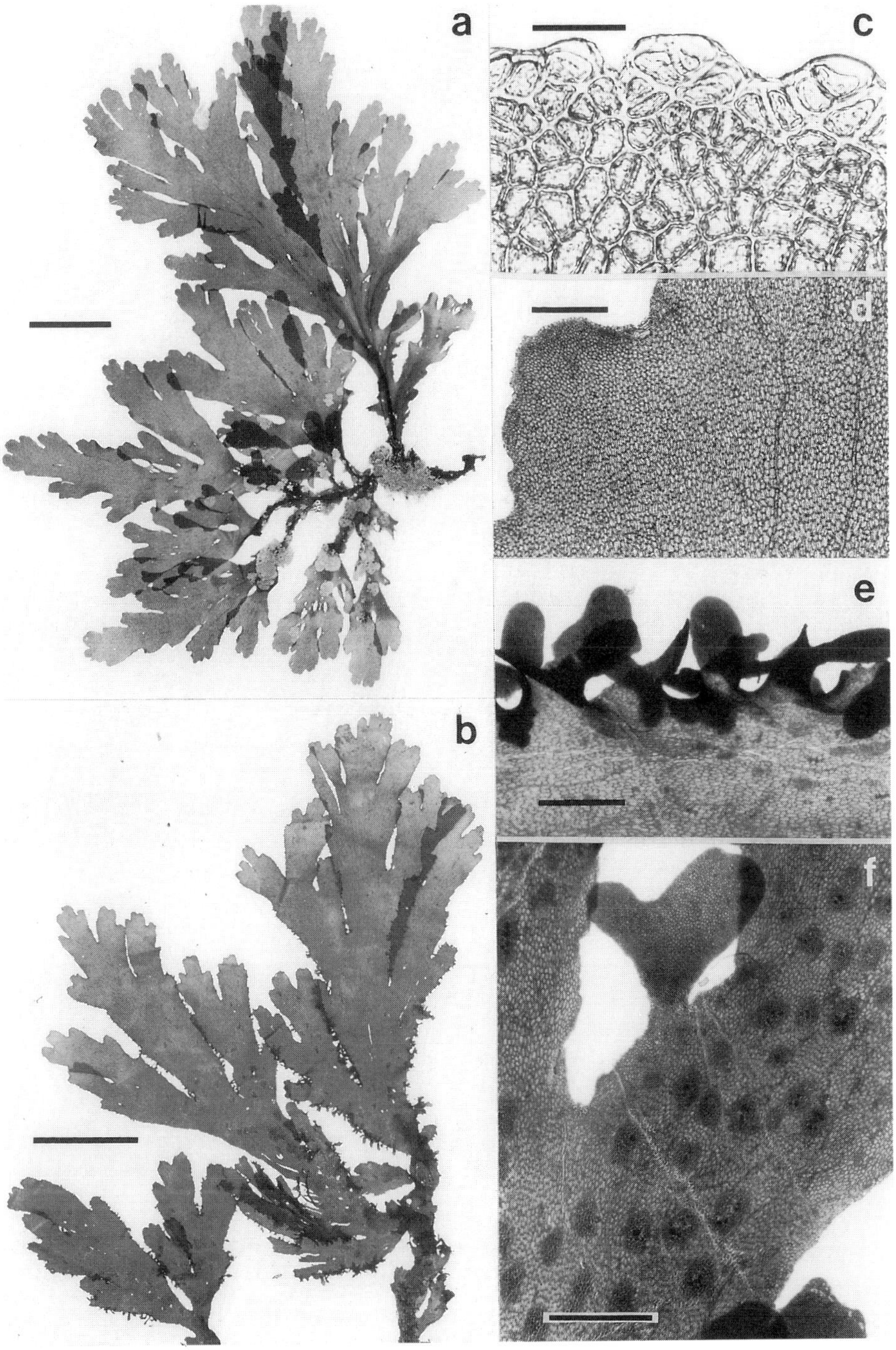

Fig. 112 *Hymenena laciniata* (Hooker f. & Harvey) Kylin: MELU, R 0152, 1712, 3391. **a, b.** Habits of sterile thalli. **c.** Thallus margin. **d.** Mid-thallus showing venation. **e.** Detail of ruffled thallus margin. **f.** Blade surface with tetrasporangial sori. Scales: a, b = 2 cm; c = 40 μm; d = 500 μm e, f = 1 mm.

irregularly to regularly lobed, or highly laciniate. Growth is from a marginal zone of meristematic cells. Blades usually are monostromatic above and polystromatic below and sometimes have coarse veins that dichotomize once in the lower thallus. Microscopic veins are absent. Macroscopic veins, which often are present, lack internal rhizoids and generally do not extend beyond the apophyses. Procarps, spermatangial sori, and tetrasporangial sori are distributed across blade surfaces. Mature gonimoblasts have basal fusion cells with outwardly radiating lobes that give rise to chains of carposporangia.

Myriogramme is similar to, and might be confused with, several other foliaceous genera in the Delesseriaceae. Most species now placed in *Myriogramme* were originally members of *Nitophyllum*. Species of the latter, however, are distinguished by having gonimoblasts with terminal instead of catenate carposporangia (Kylin, 1956). *Platyclinia* and *Schizoseris* also occur in the Southern Ocean and are closely related to *Myriogramme*. Fronds of *Platyclinia* are polystromatic throughout but otherwise morphologically identical to those of *Myriogramme* in which some species are also predominantly polystromatic (Wynne, 1983). Differences in medullary construction, however, separate the two genera. In regions away from midribs, stipes, and veins, *Platyclinia* has a central layer of large cells surrounded by cells that gradually become smaller toward frond surfaces, whereas *Myriogramme* has medullary cells mostly of equal size. Fronds of *Schizoseris* have predominant, thickened, branched, macroscopic veins; in older thalli, these veins persist after much of the monostromatic tissue has been eroded, eventually creating a highly branched thallus habit (pers. obs.). Fronds of *Myriogramme* also become eroded and highly laciniate, but blade erosion in this genus does not result in highly branched thalli because thickened macroscopic veins are lacking or developed only at blade bases.

Kylin (1924) initially included 18 species in *Myriogramme*, but later (1956) estimated a total of 25 species from various temperate to colder seas in both hemispheres. Papenfuss (1964*a*) listed ten species from the Southern Ocean, with only two, *M. mangini* (Gain) Skottsberg and *M. smithii* (Hooker f. & Harvey) Kylin, occurring in Antarctic waters. Of the remaining eight species, two, *M. multinervis* (Hooker f. & Harvey) Kylin and *M. macquariensis* A. Zinova, are considered in the present survey to be members of *Schizoseris*. The genus is represented at Macquarie I. by the type species, *M. livida*.

Myriogramme livida (Hooker f. & Harvey) Kylin, 1924: 58

Figs 113, 114, 115

BASIONYM: *Nitophyllum lividum* Hooker f. & Harvey, 1845: 253.

SYNONYMS: see Papenfuss, 1964*a*.

TYPE LOCALITY: Falkland Is.

LECTOTYPE: K *in* BM; female specimen (basis for Hooker's illustration in *Flora Antarctica* vol. 1(2), pl. clxxix) selected from syntypes by J. H. Price (28 February 1985). Herbarium sheet with annotation: 'Port William East Falkland I. July 1842. Captn Cr-[ozier]'. (Price, pers. comm.) (N.B. lectotype is uppermost specimen of three plants mounted on same herbarium sheet.)

DISTRIBUTION: Falkland Is, Fuegia (Hooker f. & Harvey, 1845); Îles Kerguelen (Zinova, 1963); Macquarie I.

MATERIAL EXAMINED:

1. Falkland Is: Station 34, Berkeley (?) I., Point Louis, 16 m depth on sand (Skottsberg, 19 July 1902): UPS, 6 (tetrasporophyte).

2. Macquarie I.: collections MA-11, 14, 34, 39, 46, 98, 117, 118, 140, 143, 227, 244, 271, 295, 303, 309, 364, 377, 378; Herbarium specimens: MELU, R 0110, 0177, 0878, 0881, 0882, 1713, 1860, 2019, 3473, 3477, 3484, 3485, 3736.

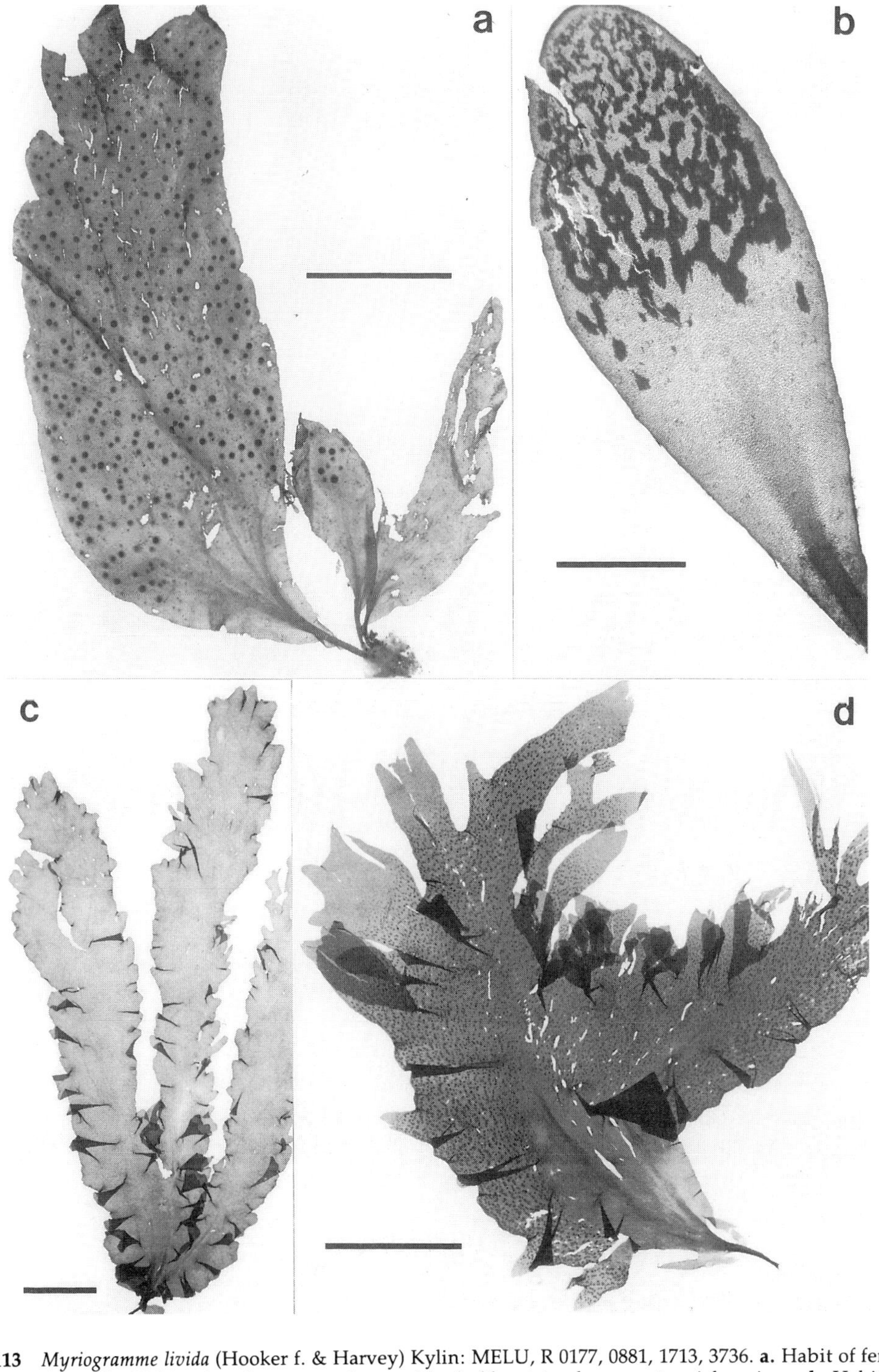

Fig. 113 *Myriogramme livida* (Hooker f. & Harvey) Kylin: MELU, R 0177, 0881, 1713, 3736. **a.** Habit of female gametophyte. **b.** Habit of male gametophyte with darkly stained spermatangial sori. **c, d.** Habits of tetrasporophytes. Scales: a = 2 cm; b, c = 3 cm; d = 4 cm.

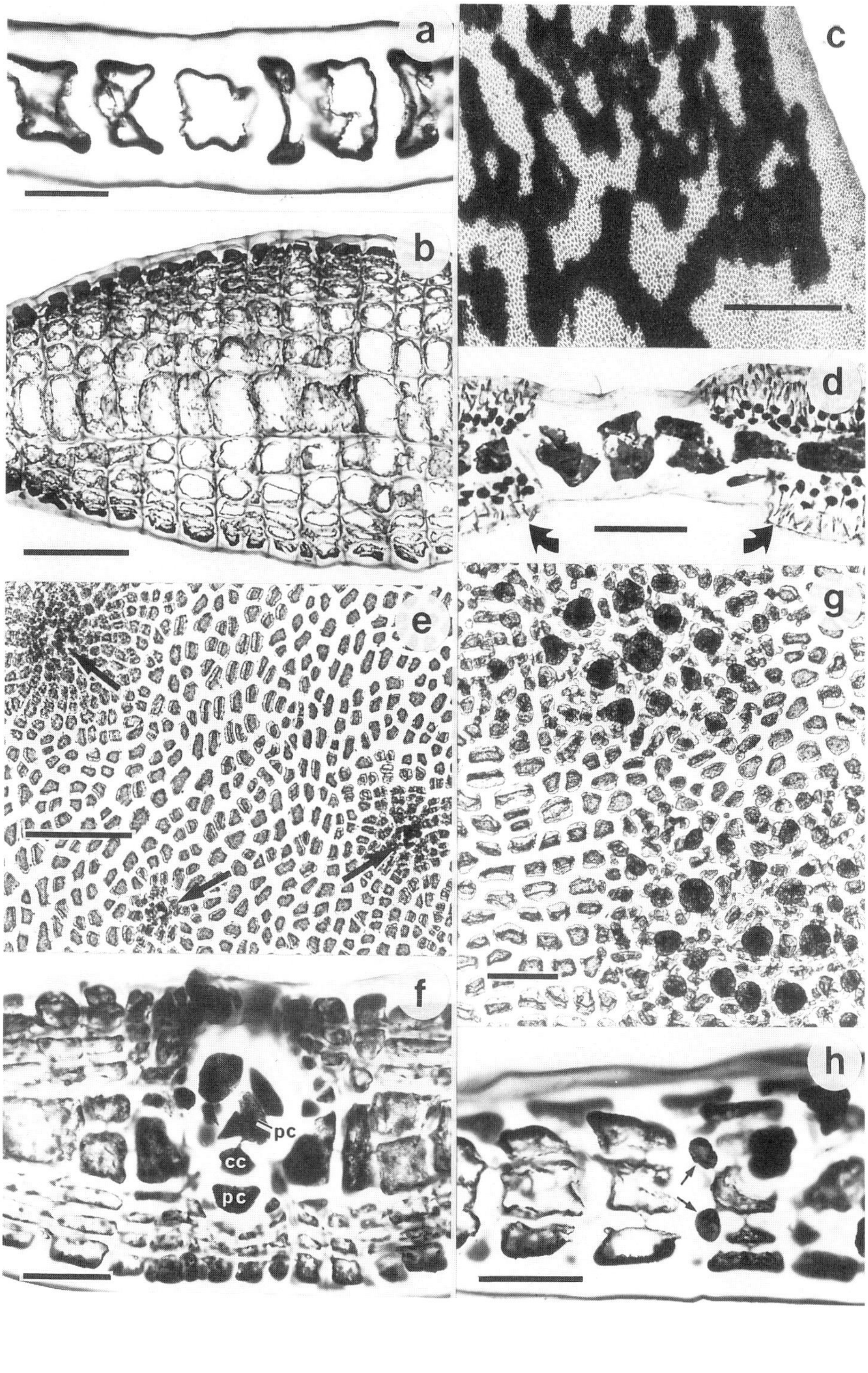
a
b
c
d
e
f
g
h
pc
cc
pc

DESCRIPTION: Thalli (Fig. 113a–d) erect, to 40 cm tall, pink to reddish-brown to bluish-purple, and with fronds foliose, stipitate, mostly monostromatic (Fig. 114a), and either broad and undivided or highly laciniate and irregularly lobed. Thallus margins either smooth (Fig. 113a, b) or ruffled (Fig. 113c), entire or somewhat dentate with rounded teeth. Growth from numerous marginal initials. Microscopic veins absent; macroscopic veins simple or once-branched, sometimes absent but usually present at least in apophyses, occasionally persisting through lower third of frond; veins composed of central layer of large cells surrounded by layers of smaller cells and lacking internal rhizoids (Fig. 114b). Spermatangia in irregularly shaped sori developed on both surfaces of blades in upper thallus (Figs 113b, 114c, d). Procarps scattered across fronds (Fig. 114e), borne on central cells of mostly tristromatic portions of thallus; central cell producing two anticlinal pericentral cells, one of which functioning as supporting cell and bearing one 4-celled carpogonial branch, two 1-(2-)celled sterile groups, and (following presumptive fertilization) one auxiliary cell (Fig. 115a–d); gonimoblasts developing from rounded initials (Figs 114f, 115c). Cystocarps ostiolate, embedded in thallus but protuberant, maturing simultaneously across entire frond; mature gonimoblasts with massive fusion cells bearing chains of carposporangia (Fig. 115e). Tetrasporangia tetrahedrally divided and aggregated in small, rounded to oblong sori scattered across blade surfaces (Fig. 114g). Blades thickened (often tristomatic) near sporangial sori, with most cells of central layer bearing four tetrasporangial mother cells (Fig. 114h).

Plants are abundant in shaded habitats at 4–18 m depths around Macquarie I., most commonly colonizing rock walls and small boulders, although sometimes growing epiphytically on plants such as *Ballia callitricha* and *Phyllophora appendiculata*. *M. livida* is usually a major component of the red algal understory beneath fronds of *Desmarestia chordalis*, *D. willii* and *Macrocystis pyrifera*. It often grows adjacent to *Delesseria lancifolia*, *Schizoseris dichotoma*, *Plocamium hookeri*, *P. secundatum* and *Heterosiphonia berkeleyi* and sometimes has one or more ceramiaceous algae (e.g., *Dasyptilon pellucidum*, *Pterothamnion plumula* and *P. simile*) at its base. Additionally, stipes and apophyses of *M. livida* are occasionally infected with *Polycoryne radiata*.

Tetrasporophytes and female gametophytes are most common, with male gametophytes relatively infrequent.

NOTES: Macquarie thalli match the original description of *M. livida* (= *Nitophyllum lividum* Hooker f. & Harvey, 1845) as well as a more recent description of this species from Argentina (Mendoza, 1969*c*). Additionally, observations made during the present survey have revealed extensive variation in thallus size, colour and shape, with some plants also resembling J. D. Hooker's & Harvey's (1845) description of *M.* [*Nitophyllum*] *crozieri* (Hooker f. & Harvey) Kylin (1924) and Mendoza's illustration (1969*c*, pl. 3) of a tetrasporophyte from Argentina. Mendoza (1969*c*) separated these two species largely on blade morphology and the presence of either simple (*M. crozieri*) or bifurcate (*M. livida*) basal veins, features which presently appear to be too variable for species distinctions. The boundaries between both taxa need to be more clearly defined.

Polycoryne Skottsberg *in* Kylin & Skottsberg, 1919: 36

Type species: *P. radiata* Skottsberg *in* Kylin & Skottsberg, 1919: 36.

Fig. 114 *Myriogramme livida* (Hooker f. & Harvey) Kylin: MELU, R 0882, 3484, 3485, 3736. **a.** Section through upper blade near margin. **b.** T.S. stipe. **c.** Blade surface of male gametophyte showing detail of spermatangial sori stained with aniline blue. **d.** Section through sterile and reproductive (arrows) portions of male gametophyte. **e.** Blade surface of female gametophyte showing procarps (arrows). **f.** Section through female gametophyte thallus showing young gonimoblast developing from pericentral cell (pc) borne on central cell (cc) (details in Fig. 115c). **g.** Blade surface showing tetrasporangial sori. **h.** Section through tetrasporangial sorus showing immature tetrasporangia (arrows). Scales: a, d, f, h = 50 µm; b, e = 200 µm; c = 1 mm; g = 100 µm.

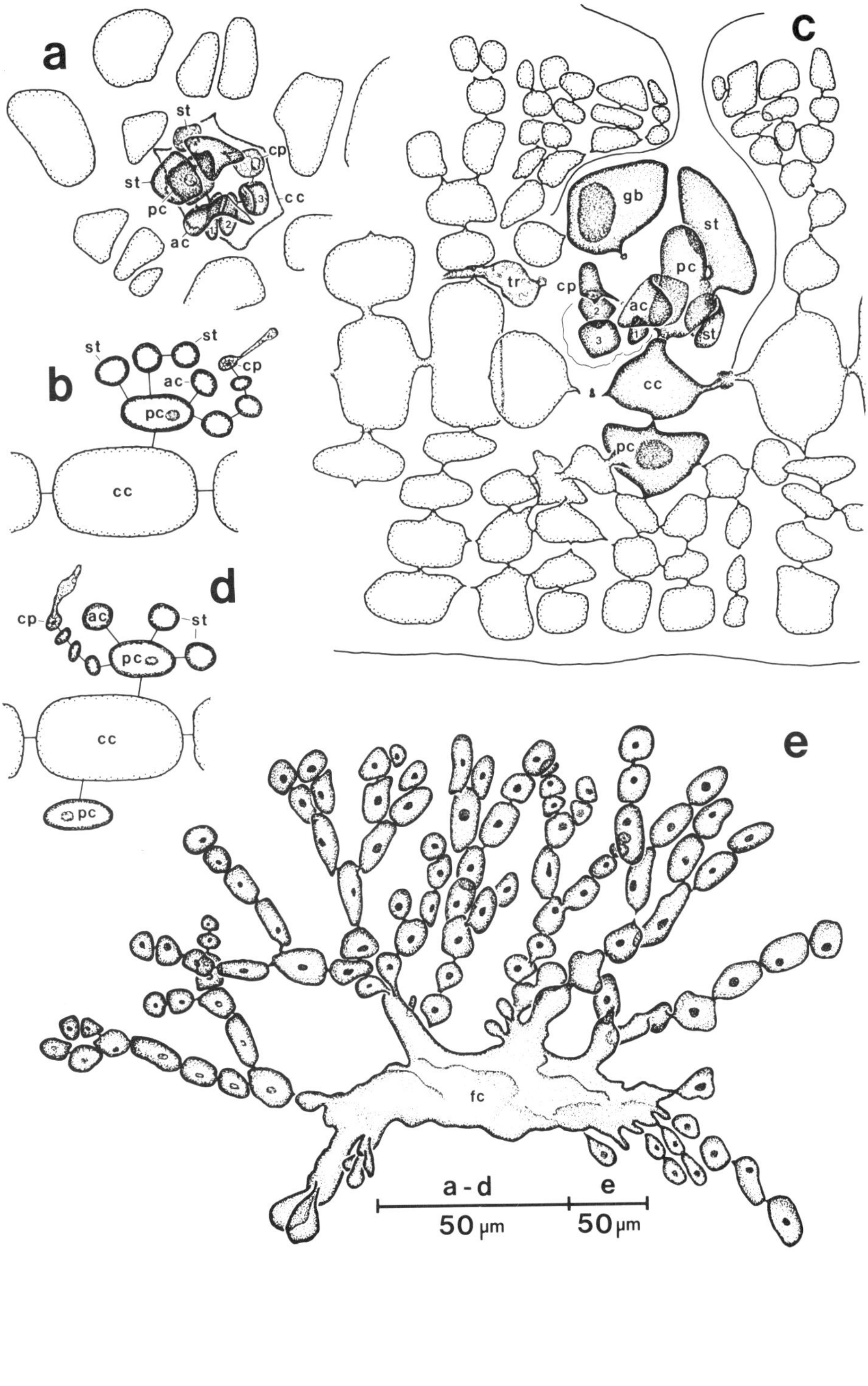
a
st
cp
st
pc
cc
ac
b
st
st
ac
cp
pc
cc
c
gb
st
pc
tr
cp
ac
st
cc
pc
d
cp
ac
st
pc
cc
pc
e
fc
a - d
e
50 μm
50 μm

Goff (1982) listed four adelphoparasite genera in the Delesseriaceae, of which only *Asterocolax** and *Gonimophyllum* (Abbott & Hollenberg, 1976) occur in the Northern Hemisphere. Of the two strictly Southern Hemisphere genera, *Polycoryne* and *Gonimocolax*, the former is most like *Asterocolax* in its production of basal cushions with outwardly radiating acicular branches. *Gonimocolax* and *Gonimophyllum*, in contrast, produce tufts of small foliose fronds (Kylin, 1956). *Polycoryne* and *Asterocolax* are morphologically similar, produce tetrasporangia and spermatangia over thallus surfaces, and develop gonimoblasts with chains of carposporangia. They have been separated, however, by the obliquely dividing apical cells of the former, and transversely dividing apical cells of the latter (Feldmann & Feldmann, 1951; Kylin, 1956). Although the present studies on *Polycoryne* largely agree with Kylin's (1924, fig. 48a–d) illustrations of the type species showing obliquely dividing apical cells, occasional exceptions exist. A few branches appear to have transversely divided apical cells, an observation that obscures the distinction between *Polycoryne* and *Asterocolax*.

The Subantarctic genus *Polycoryne* comprises two species which Zinova (1963) distinguished by various vegetative features. She erected *P. compacta* for plants from Îles Kerguelen that lack uniseriate filamentous portions, thicken immediately behind apices, and have compact thalli with short conical protuberances. The present observations on *P. radiata* from Macquarie I. indicate overlap in some of these features and cloud Zinova's distinctions of the two species. The precise relationship between these two taxa awaits critical study of type material of *P. compacta*, which was not available during the present survey.

Polycoryne radiata Skottsberg ***in*** Kylin & Skottsberg, 1919: 36

Figs 116, 117

Type locality: South Georgia.

Type: presumably in S, but specimens were not seen.

Distribution: Macquarie I. (Zinova, 1963); South Georgia (Kylin & Skottsberg, 1919).

Material examined: Macquarie I.: collections MA-34, 39, 118, 227, 237, 271, 377; Herbarium specimens: MELU, R 0704, 0861, 1811–1815, 1821–1828.

Description: Thalli presumed to be parasitic, 1–3 mm diameter, whitish, with knobby to spike-like, terete to slightly flattened, uniaxial branches arising from pseudoparenchymatous basal cushions attached by endophytic rhizoids penetrating delesseriaceous host tissues (Fig. 116a–c, e, f). Juvenile thalli with uniseriate branches (Fig. 117a); mature plants generally lacking uniseriate portions, with cortication beginning immediately behind apices (Fig. 116d). Apical cells mostly obliquely dividing, with cell divisions alternating in opposite directions and producing zigzag pattern of central axis; transversely dividing apical cells rare (Fig. 117b). Spermatangia covering majority of thallus except near branch apices; spermatangial mother cells

* Feldmann & Feldmann (1951) initially credited this genus with a newly described species and one transferred from *Polycoryne*, *P. denticulata* Tokida (1934). They later (1958) incorporated another *Polycoryne* species, *P. gardneri* Setchell (1923), into *Asterocolax*. Abbott & Hollenberg (1976) considered *P. phycodricola* Dawson (1944) to be synonymous with *A. gardneri* (Setchell) Feldmann & Feldmann.

Fig. 115 *Myriogramme livida* (Hooker f. & Harvey) Kylin: MELU, R 3484. **a, c, e.** Camera lucida drawings. **a.** Procarp as seen from blade surface. **b.** Schematic representation of procarp in blade section. (N.B. interpretation based on 'a' with only one of two anticlinal pericentral cells illustrated.) **c.** Section through frond showing fertilized procarp and developing gonimoblast. **d.** Schematic representation of procarp in blade section. (N.B. interpretation based on 'c'.) **e.** Mature carposporophyte removed from cystocarp. ac = auxiliary cell; cc = central cell; cp = carpogonium; 1, 2, 3 = carpogonial branch cells; fc = fusion cell; gb = gonimoblast initial; pc = pericentral cell; st = sterile cell; tr = trichogyne. Scales: a–e = 50 μm; (N.B. scales for b and d are approximate.)

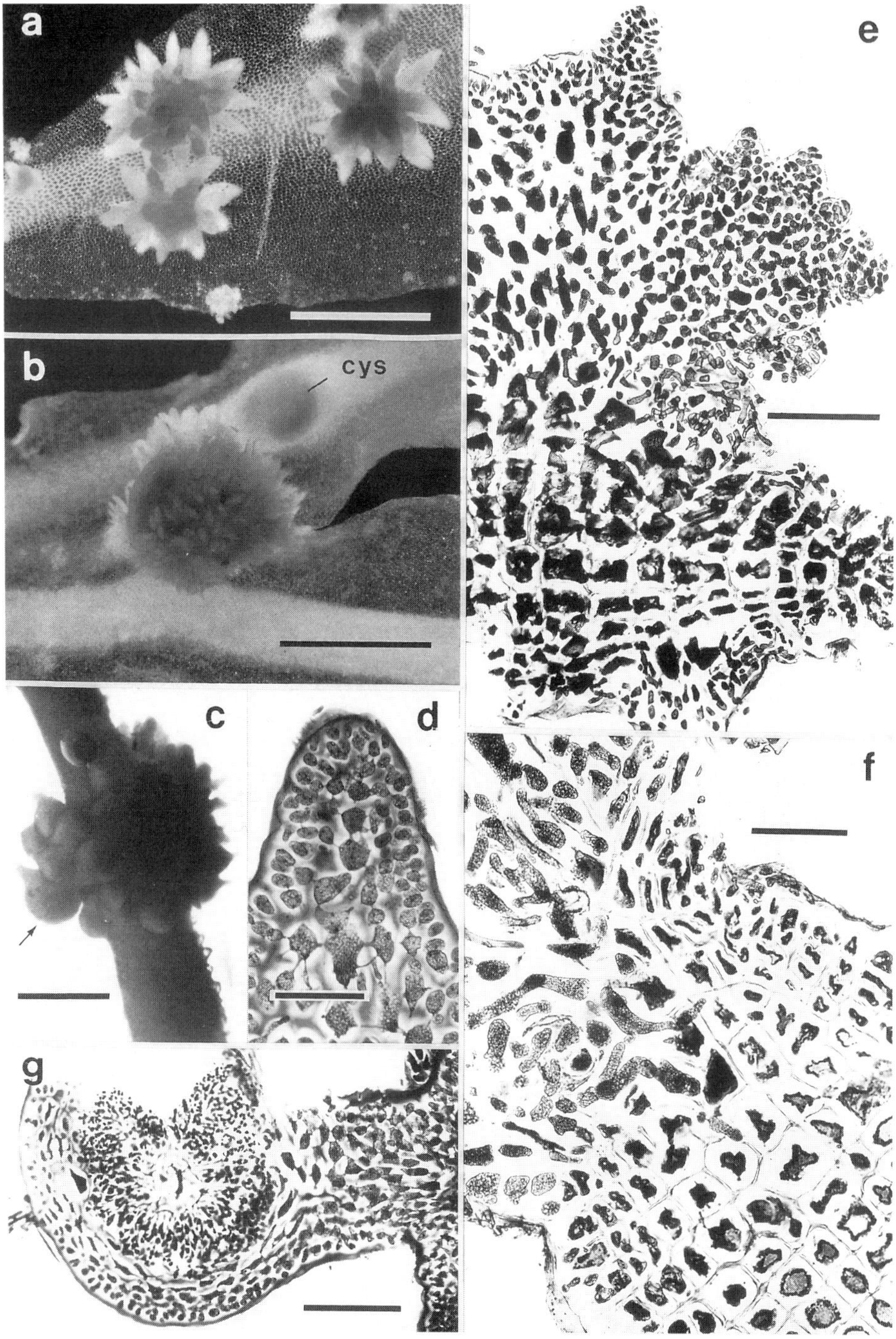

Fig. 116 *Polycoryne radiata* Skottsberg: MELU, R 0704, 0861. **a.** Habit of tetrasporophyte on *Myriogramme livida.* **b.** Habit of male gametophyte on *Schizoseris dichotoma.* cys = cystocarp. **c.** Habit of female gametophyte on *M. livida.* **d.** Branch apex. **e.** Section through parasite and host *(Myriogramme).* **f.** Detail of granular (starch-filled) cytoplasm of parasite rhizoids penetrating host thallus. **g.** Section through cystocarp. Scales: a, b = 2 mm; c = 1 mm; d, f = 100 μm; e, g = 200 μm.

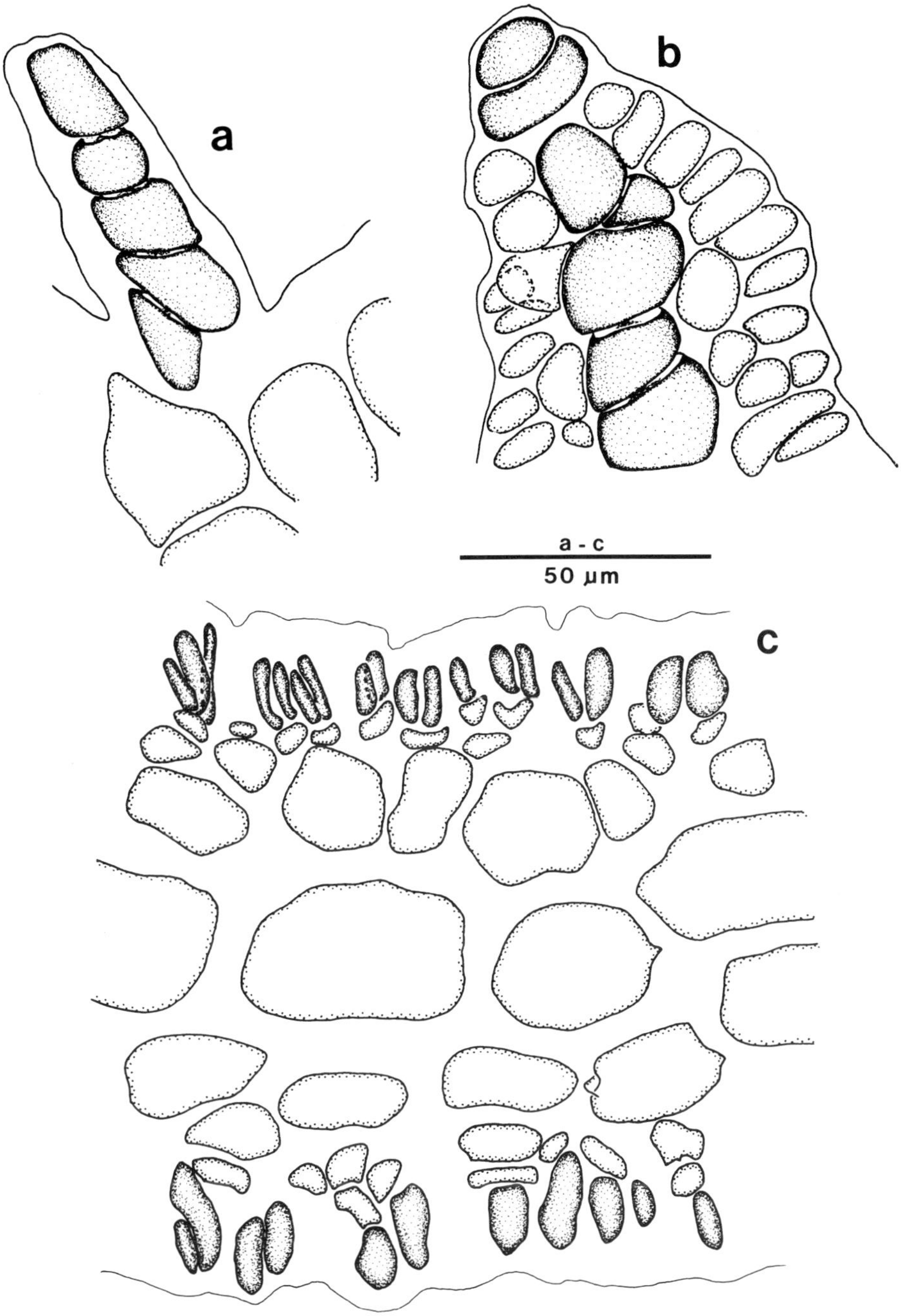

Fig. 117 *Polycoryne radiata* Skottsberg: MELU, R 0704. **a–c.** Camera lucida drawings. **a.** Sporeling development on *Myriogramme livida*. **b.** Apical portion of spermatangial frond. **c.** L.S. terete branch of male gametophyte. Scales: a–c = 50 μm.

anticlinally elongated (Fig. 117c). Cystocarps singly produced and sub-apical on branches, with well developed ostiolate pericarps; mature gonimoblasts with central fusion cells and outwardly radiating fascicles of branched sterile filaments ending in chains of carposporangia (Fig. 116c, g). Tetrasporangia tetrahedrally divided, spherical (38 μm diameter) to oblong (25 μm broad × 55 μm long), sunken in cortex, and distributed across frond surfaces.

Plants grow on the stipes and apophyses of *Myriogramme livida* and *Schizoseris dichotoma*, both of which are abundant on rocky surfaces along shaded floors of *Macrocystis* forests between 5–15 m depths. *Polycoryne* is uncommon and is usually represented by either tetrasporophytes or male gametophytes. Only two female gametophytes were collected during the present survey.

NOTES: Depending upon the reproductive status of the plants, branches arising from basal cushions are short and acicular (tetrasporangial and spermatangial thalli) or longer and thicker (female thalli). In most instances, branches of mature plants lack uniseriate portions, become corticated and thickened immediately behind the apices, and often appear conical, much as Zinova (1963) originally described for *P. compacta*. Her illustrations (1963, fig. 6) of *P. compacta*, however, illustrate thalli that are more wart-like and compact than the Macquarie specimens of *P. radiata* (Zinova, 1963, fig. 10; pers. obs.).

Pseudophycodrys Skottsberg, 1923: 32

Type species: *P. phyllophora* (J. Agardh) Skottsberg, 1923: 32.

Members of this genus are erect and have fronds that are foliose, stipitate, and either undivided, irregularly lobed, or highly laciniate. Growth is from single transversely dividing apical cells and intercalary cell divisions that occur in primary cell rows. Young thalli have extensive monostromatic portions, whereas older blades are mostly polystromatic, with prominent midribs that produce oppositely paired or alternate, coarse lateral veins. Branching is from thallus margins. Microscopic veins are mostly lacking and major veins are composed of rhizoids intermixed with large rounded cells. Procarps are scattered across blade surfaces and usually develop adjacent to (but not directly on) veins. Tetrasporangia are aggregated in sori produced between veins and thallus margins.

Using a combination of vegetative and reproductive criteria, the Delesseriaceae can be divided into the two sub-families Delesserioideae and Nitophylloideae (Kylin, 1923, 1924; Wagner, 1954). *Pseudophycodrys* combines certain features that independently are considered to be characteristic for each sub-family. It was first aligned with the Delesserioideae (Skottsberg, 1923) but is now considered to be a member of the Nitophylloideae (Kylin, 1924, 1956; Wynne 1983). Skottsberg (1923) recognized some features in common with *Phycodrys quercifolia*, but considered *Pseudophycodrys* to be closer to *Delesseria* (in the Delesserioideae) because of similarities in apical construction and the presence of rhizoids in veins. Kylin (1924), however, considered *Pseudophycodrys* to be closer to *Phycodrys* (in the Nitophylloideae) because of similar branching patterns, presence of intercalary cell divisions in primary cell rows, and procarps scattered across blade surfaces. In the Nitophylloideae, *Pseudophycodrys* is probably most like the Argentine endemic genus *Pseudonitophylla* Mendoza (1975), a monotypic genus that produces procarps only on midribs and major veins of primary blades.

Pseudophycodrys is monotypic and restricted to the Subantarctic, although earlier accounts (Kylin, 1956) have credited the genus with two additional species. One of these, *P. rainosukei* Tokida (1932), first described from Japan, has been placed by Wynne (1970) in a new genus, *Tokidadendron*. *Pseudophycodrys pulcherrima* Baardseth (1941), previously believed to be endemic to Tristan da Cunha, is considered in the present survey to be synonymous with the type, *P. phyllophora*.

Pseudophycodrys phyllophora (J. Agardh) Skottsberg, 1923: 32

Fig. 118

BASIONYM: *Delesseria phyllophora* J. Agardh, 1872: 55.

SYNONYMS:
Delesseria davisii sensu Kylin & Skottsberg, 1919, 39.
Pseudophycodrys pulcherrima Baardseth, 1941: 102.

TYPE LOCALITY: Falkland Is.

LECTOTYPE: Herb. Agardh, LD, 31276 (female) (N.B. Skottsberg (1923) refers to the anatomy of this specimen as typical for the genus.)

DISTRIBUTION: Falkland Is (J. Agardh, 1872); Fuegia (Hariot, 1889); Îles Kerguelen (Kylin & Skottsberg, 1919); Macquarie I.; Tristan da Cunha (Baardseth, 1941).

MATERIAL EXAMINED:
Pseudophycodrys phyllophora
1. Falkland Is: Lectotype (Hooker, 1842): LD, 31276; additional specimens: LD 31272–31275, 31277.
2. Macquarie I.: collections MA-34, 227, 248, 271, 298, 302–304, 378; Herbarium specimens: MELU, R 0173–0175, 2437–2439, 2458, 3409, 3410, 3460–3462.
Pseudophycodrys pulcherrima
Tristan da Cunha: Holotype and isotype. Station 13, Julia Point, in rock pools (Baardseth, 21 December 1937): O, 80 (N.B. holotype illustrated by Baardseth (1941, fig. 54).)

DESCRIPTION: Thalli erect, to 15 cm tall, reddish-orange to reddish-brown to purplish-red, and with fronds foliose, stipitate, mostly monostromic and simple when young, becoming polystromatic, laciniate, and irregularly branched with age (Fig. 118a, b, e). Growth from transversely dividing apical cells (Fig. 118c) and intercalary cell divisions occurring in primary cell rows. Apices of mature blades generally worn away, with thallus margins becoming irregular and sometimes proliferous (Fig. 118d). Primary blades of mature plants usually highly eroded, with midribs remaining as central stipes from which secondary blades arise; lateral veins of primary blades serving as midribs for secondary lateral blades. Microscopic veins absent; macroscopic veins extending close to thallus margins and composed of internal rhizoids and larger isodiametric cells; percurrent midrib usually present, alternately or oppositely bearing simple or dichotomizing lateral veins that sometimes anastomose. Procarps scattered across blade surfaces; mature cystocarps ostiolate, embedded in thallus but protuberant, and located near or on veins. Cystocarp maturation progressive from immediately behind blade margins to lower thallus; gonimoblasts with basal fusion cells bearing outwardly radiating chains of carposporangia (Fig. 118h, i). Tetrasporangia tetrahedrally divided, embedded in cortex, and aggregated either in linear sori bordering veins or rounded sori scattered between veins and frond margins (Fig. 118f, g). Spermatangia unknown.

Plants are uncommon at Macquarie I., but have been collected from shaded subtidal habitats at both the northern and southern ends of the island. *Pseudophycodrys phyllophora* sometimes grows epiphytically on *Desmarestia chordalis*, but usually occurs on rocky surfaces between 5–18 m depths along the floors of *Macrocystis* forests. It is often associated with other delesseriaceous algae such as *Myriogramme livida*, and sometimes is partly covered by *Dasyptilon pellucidum*, *Pterothamnion simile*, and bryozoan colonies, the last generally covering portions of the lower thallus.

Tetrasporophytes are most common, with female gametophytes infrequent and male gametophytes unknown.

NOTES: Baardseth (1941) considered *P. pulcherrima* to be closely related to *P. phyllophora*, but he distinguished the former as having larger thalli, less prominent veins, and procarps scattered across blade surfaces instead of located close to veins (as occurs in *P. phyllophora*). Based on present observations, thalli of *P. phyllophora* from Macquarie I. are comparable in size (known maximum lengths of 15 cm vs. 18 cm) and morphology to *P. pulcherrima*. Vein development and branching patterns appear to be variable and dependent partly on thallus age and reproductive status of plants; female gametophytes generally have coarser and more highly branched veins

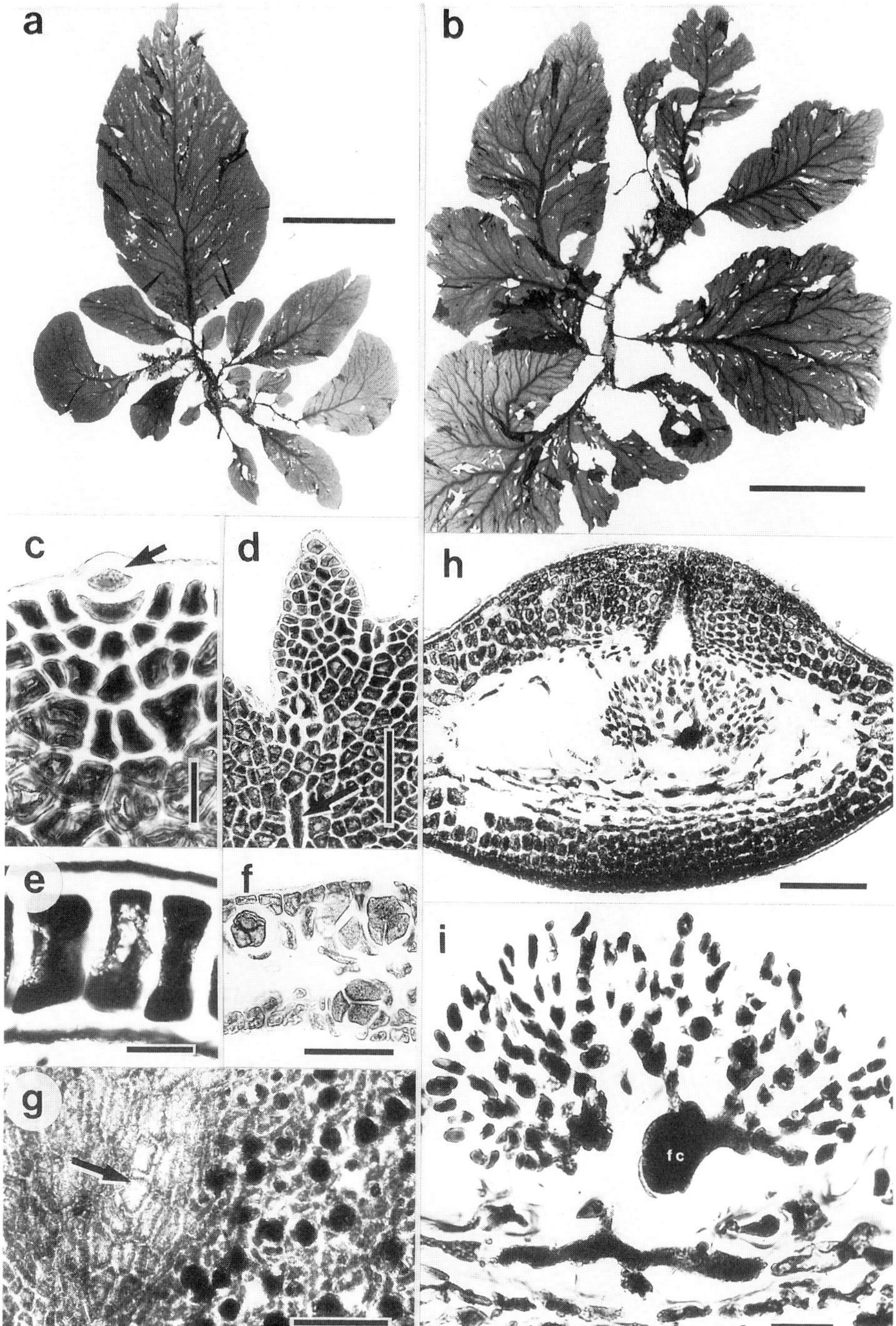
a
b
c
d
h
e
f
i
g
fc

than those of juveniles, sterile thalli and tetrasporophytes. Procarps on young female gametophytes are often scattered and not associated with veins, whereas cystocarps of mature fronds are usually positioned adjacent to or on macroscopic veins that apparently develop after procarps have been formed. This differs from *Pseudonitophylla* Mendoza (1975), in which procarps initially develop on veins.

The present observations do not support Baardseth's distinctions of two species, nor do they offer other criteria to separate *P. pulcherrima* from the type species. Consequently, only the latter is recognized.

Schizoseris Kylin, 1924: 67

Type species: *S. laciniata* Kylin, 1924: 67 (= *S. condensata* (Reinsch) R. W. Ricker comb. nov.).

Members of this genus are erect and foliose, with fronds of older individuals usually highly laciniate and appearing to be irregularly branched or extremely proliferous. Growth is from a marginal zone of meristemic cells and intercalary cell divisions that occur in primary cell rows. Blades are monostromatic above and thicker below and have numerous coarse veins. Microscopic veins are absent, and macroscopic veins, which lack internal rhizoids, extend throughout the blades and usually dichotomize repeatedly before disappearing just behind the frond margins. Branches arise by accelerated growth at certain points along blade margins or by laciniation and subsequent erosion of fronds back to major veins, with bladelets often proliferating along eroded veins. Procarps are scattered across blade surfaces. Mature carposporophytes have basal fusion cells with outwardly radiating lobes that are terminated by chains of carposporangia. Spermatangia and tetrasporangia are aggregated in sori that develop between and on veins.

Schizoseris and *Myriogramme* produce similar thalli, except that those of the latter generally lack dichotomizing veins and have fronds that supposedly do not become as deeply split (Kylin, 1956). Present observations, however, have revealed extensive laciniation of fronds in species of both genera, and consequently the two are separated in the present survey largely on the presence (*Schizoseris*) or absence (*Myriogramme*) of veins throughout the blades.

Schizoseris is predominantly a Subantarctic genus, although three species have been reported from the Northern Hemisphere (*Index Nominum Algarum: S. minima* Kaneko & Masaki—Japan, *S. subdichotoma* (Segawa) Yamada—Korea, *S. pygmaea* Dawson—Gulf of Mexico). Of the eight species recorded from the Southern Ocean*, only three are recognized at present because *S. davisii, S. kerguelensis, S. papenfussii, S. polydactyla* and *S. pseudodichotoma* are considered in this report to be synonymous with *S. dichotoma*. The genus is represented at Macquarie I. by this species and the type, *S. condensata*.

Schizoseris condensata (Reinsch) R. W. Ricker comb. nov.

Fig. 119

Replaced name: *Delesseria laciniata* Kützing, 1866: 8.

Synonyms:

Delesseria condensata Reinsch, 1888*b*: 150.
Nitophyllum affine Reinsch, 1888*b*: 153.

* Including *S. condensata* (Reinsch) R. W. Ricker, *S. davisii* (Hooker f. & Harvey) Kylin (1924), *S. dichotoma* (Hooker f. & Havey) Kylin (1924), *S. kerguelensis* A. Zinova (1963), *S. multifoliata* Baardseth (1941), *S. papenfussii* Baardseth (1941), *S. polydactyla* (Reinsch) Skottsberg (1941*a*) and *S. pseudodichotoma* A. Zinova (1963).

Fig. 118 *Pseudophycodrys phyllophora* (J. Agardh) Skottsberg: MELU, R 0173, 0175, 3409, 3410. **a.** Habit of tetrasporophyte. **b.** Habit of female gametophyte. **c.** Apex (arrow) of young blade. **d.** Proliferation along blade margin. (N.B. small vein (arrow) extending close to frond edge.) **e.** T.S. monostromatic blade. **f.** Section through tetrasporangial sorus. **g.** Blade surface showing tetrasporangial sorus adjacent to vien (arrow). **h.** Section through cystocarp. **i.** Detail of carposporophyte showing basal fusion cell (fc) and radiating chains of carposporangia. Scales: a, b = 5 cm; c, e = 20 μm; d, f = 100 μm; g, h = 200 μm; i = 50 μm.

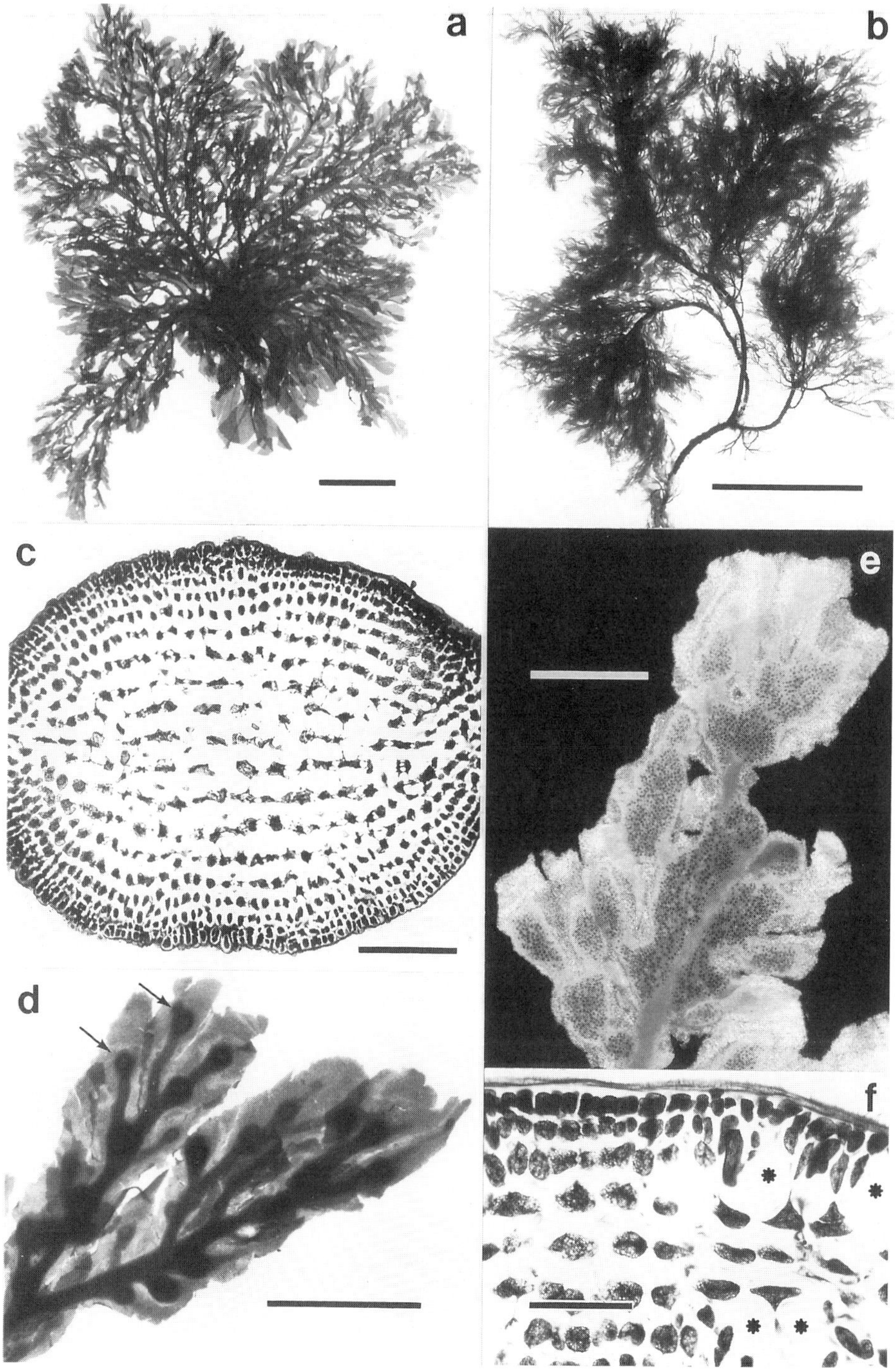
a
b
c
d
e
f

Schizoseris laciniata Kylin, 1924: 67.
Myriogramme macquariensis Zinova, 1963: 63.
[N.B. see Papenfuss (1964*a*) under *Schizoseris laciniata* for additional synonyms.]

TYPE LOCALITY: South Georgia.

HOLOTYPE: not designated; Syntype in M, unnumbered specimen.

DISTRIBUTION: Falkland Is (Skottsberg, 1923); Fuegia (Kützing, 1866; Levring, 1960); Heard and Macquarie Is; Îles Crozet and Kerguelen (Levring, 1944); New Zealand (Laing, 1901—as *Nitophyllum affine*); South Georgia (Reinsch, 1888*b*; Kylin & Skottsberg, 1919).

MATERIAL EXAMINED:
Delesseria condensata
South Georgia: Syntype. 'Nordstrand der Landzunge' (Will, February 1883): M, unnumbered specimen.
Nitophyllum affine
South Georgia: Syntype. 'Klippen nächst dem Strande an der Grenze des Niedrigwassers' (Will, 20 December 1882): M, unnumbered specimen.
Schizoseris condensata
1. Heard I.: Atlas Cove (Chittleborough, 11 September 1949): MELU, R 3725 (female), 3726 (female).
2. Macquarie I.: collections MA-11, 34, 46, 111, 134, 137, 215, 222, 238, 271, 364; Herbarium specimens: MELU, R 0792, 1218, 1731, 3508, 3509.

DESCRIPTION: Thalli erect, to 33 cm tall, reddish-orange to reddish-brown to purplish-black, and with fronds foliose, stipitate, mostly monostromatic, highly laciniate, becoming irregularly branched, usually proliferating along eroded veins, and often appearing feathery (Fig. 119a, b). Growth from numerous marginal initials. Microscopic veins absent; macroscopic veins dichotomously or trichotomously forked, extending from thallus base to just behind frond margins, lacking internal rhizoids (Fig. 119c). Procarps scattered across blade surfaces, with cystocarps eventually on or adjacent to veins (Fig. 119d); cystocarps protuberant, hemispherical, and each with one centrally positioned ostiole surrounded by raised collar of thickened pericarp tissue. Mature gonimoblasts with large candelabra-like fusion cells having central bulbous mass and outwardly radiating fusion processes (long arms) spreading across floor of cystocarp; carposporangia rounded to clavate and borne in chains from upper surfaces and ends of long arms. Tetrasporangia tetrahedrally divided, embedded in cortex, and aggregated in irregularly shaped sori extending over entire frond surfaces of the upper thallus (Fig. 119e). Anatomy of mature sori with cavities where tetrasporangia have been shed (Fig. 119f). Spermatangia not studied.

Plants abound in pools and surge channels from the mid- to lower-intertidal along most stretches of Macquarie's shoreline but also extend to 14 m depths in shaded habitats of *Macrocystis* forests. *Schizoseris condensata* occasionally is epiphytic on species such as *Phyllophora appendiculata* and *Desmarestia chordalis* but more often grows on rocky sufaces where *Acrosiphonia pacifica*, *Heterosiphonia berkeleyi*, *Delesseria lancifolia*, and *Myriogramme livida* also occur. In some collections, *S. condensata* was epiphytized by *Ulva*, *Ectocarpus constanciae*, *Petalonia fascia*, and various members of the Delesseriaceae.

Tetrasporophytes are most abundant, with female gametophytes common and male gametophytes seemingly rare.

NOTES: The intended basionym for *Schizoseris laciniata* Kylin (1924) was *Delesseria laciniata* Kützing (1866), which is a later homonym of *Delesseria laciniata* (Hudson) Greville (1824), the

Fig. 119 *Schizoseris condensata* (Reinsch) R. W. Ricker: MELU, R 1218, 3508, 3509. **a.** Habit of sterile thallus. **b.** Habit of tetrasporophyte. **c.** T.S. stipe. **d.** Habit of female gametophyte frond bearing cystocarps (arrows). **e.** Blade surface showing tetrasporangial sori. **f.** Section through tetrasporangial sorus showing cavities (*) where tetrasporangia have been shed. Scales: a, e = 2 cm; b = 4 cm; c = 200 μm; f = 100 μm.

basionym for *Callophyllis laciniata* (Hudson) Kützing. According to Article 72.1 of the *International Code of Botanical Nomenclature*, the name *Schizoseris laciniata* is to be treated as a *nomen novum*, which is validated by Kützing's (1866) description of *D. laciniata*. Any names of species synonymous with *S. laciniata* published before Kylin's (1924) erection of this species have priority over the epithet *laciniata*. Reinsch's (1888*b*) *Delesseria condensata* is identical to *S. laciniata* and appears to be the earliest (available) name for the species.

The present observations on *S. condensata* agree with Mendoza's (1969*b*, pls. I–III) description and illustrations of tetrasporophytes and female gametophytes from South America. Kylin (1924, fig. 53) first illustrated gonimoblast structure in this species, but his illustration was based on an immature cystocarp. More recent work by Mendoza (1969*b*, pl. III, fig. 7) shows the structure of a mature cystocarp, which has an extensive candelabra-like fusion cell bearing chains of carposporangia.

Schizoseris condensata and *S. multifoliata* Baardseth (1941) are extremely similar, if not conspecific and have the most dissected fronds of any species in the genus. Baardseth (1941) considered the two to be closely related, but he distinguished the former as having coarser and larger fronds with numerous proliferations developed from their stipes. Fronds of *S. dichotoma* also become extremely laciniate but generally retain some flabellate portions, do not appear feathery, and lack proliferations as produced by *S. condensata*. Additionally, tetrasporangial sori of *S. condensata* are larger, extending across veins to thallus margins, and more irregular than those of *S. dichotoma*.

Zinova (1963) erected *Myriogramme macquariensis* for one small (2 cm tall) sterile intertidal plant with dissected blades having prominent veins that are forked. Her description and illustration (1963, fig. 8) of this species matches the morphology of juvenile thalli from both *S. condensata* and *S. dichotoma*. Based on this resemblance and the presence of *M. macquariensis* in the intertidal, it is believed to be an immature plant of *S. condensata*.

Schizoseris dichotoma (Hooker f. & Harvey) Kylin, 1924: 68

Figs 120, 121, 122

BASIONYM: *Delesseria dichotoma* Hooker f. Harvey *in* Harvey & Hooker f. *1845: 184.*

SYNONYMS:
Delesseria davisii Hooker f. & Harvey, 1845: 252.
Nitophyllum multinerve Hooker f. & Harvey, 1845: 255.
Delesseria polydactyla Reinsch, 1888*b*: 150.
Schizoneura dichotoma (Hooker f. & Harvey) J. Agardh, 1898: 168.
Myriogramme multinervis (Hooker f. & Harvey) Kylin, 1924: 60.
Schizoseris davisii (Hooker f. & Harvey) Kylin, 1924: 68.
Schizoseris polydactyla (Reinsch) Skottsberg, 1941*a*: 89.
Schizoseris papenfussii Baardseth, 1941: 107.
Schizoseris kerguelensis A. Zinova, 1963: 59.
Schizoseris pseudodichotoma A. Zinova, 1963: 64.

TYPE LOCALITY: Not designated in the original account. Lectotype from Auckland I.

LECTOTYPE: K *in* BM; specimen 89 (basis for Hooker illustration in *Flora Antarctica* vol. 1(1), pl. 71, fig. 2) selected by M. H. Hommersand (November 1978) as provisional lectotype (Price, pers. comm.).

Fig. 120 *Schizoseris dichotoma* (Hooker f. & Harvey) Kylin: MELU, R 0849, 0857, 1264, 1401, 1704, 1908, 1930. **a–c.** Habits of sterile thalli. **a.** *'Schizoseris papenfussii'* form epiphytic on *Desmarestia chordalis*. **b.** *'Schizoseris polydactyla'* form. **c.** Eroded thallus. **d–h.** Venation patterns of blades. **d, e.** Primarily parallel venation. **f.** Dichotomously branched veins. **g.** Anastomosing veins. **h.** Percurrent main vein with laterally borne veins. Scales: a, b = 2 cm; c–h = 1 cm.

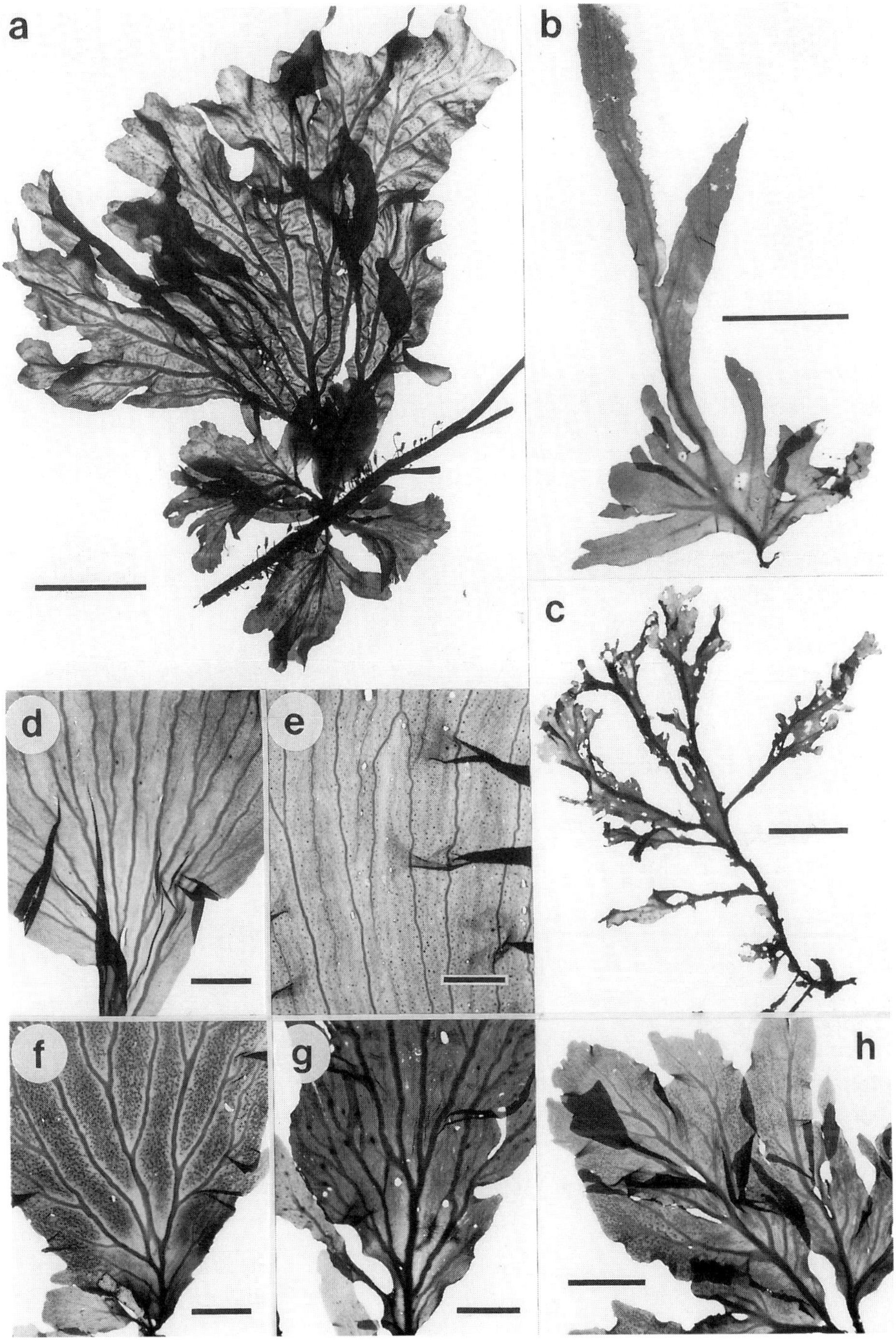
a
b
c
d
e
f
g
h

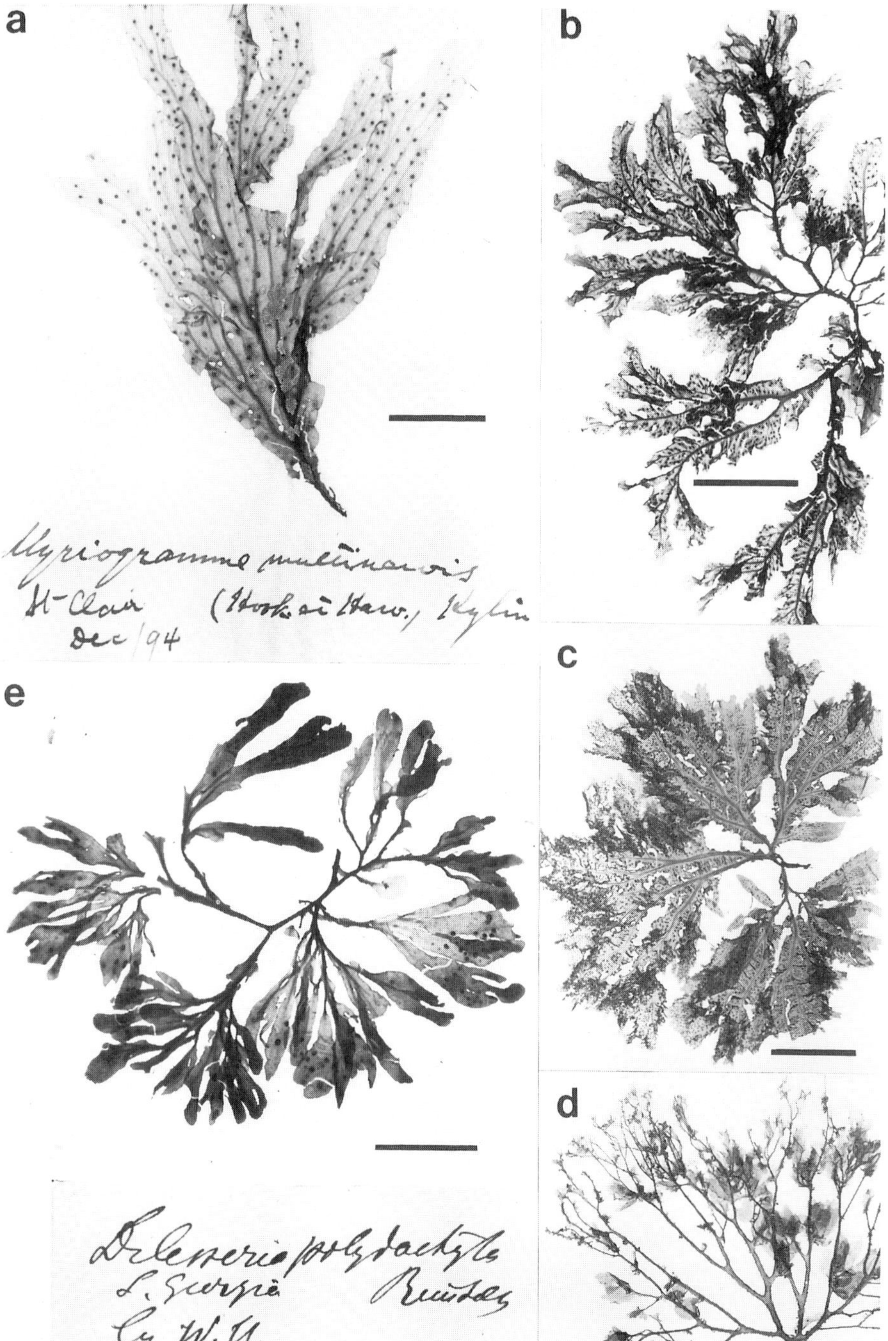
a
b
c
d
e

Distribution: Auckland and Campbell Is (Hooker f. & Harvey, 1845); Falkland Is (Kylin & Skottsberg, 1919; Skottsberg, 1923); Fuegia (Hariot, 1889); Îles Kerguelen (Reinbold, 1907; Zinova, 1963); Macquarie I. (Zinova, 1963); New Zealand (Laing, 1901—as *Schizoneura dichotoma* and *Nitophyllum multinerve*); South Georgia (Reinsch, 1888*b*, 1890; Kylin & Skottsberg, 1919); Tristan da Cunha (Baardseth, 1941).

Material examined:
Myriogramme multinervis
New Zealand: St. Clair (Laing ?, December 1894): LD, unnumbered specimen.
Schizoseris davisii
New Zealand: Lindauer (no collection data): L, 047, 192 . . . 175.
Schizoseris dichotoma
Macquarie I.: collections MA-11, 34, 39, 46, 140, 222, 237, 244, 248, 271, 283, 309, 364, 377, 378; Herbarium specimens: MELU, R 0159, 0847, 0849, 0857, 0862, 0863, 1264, 1401, 1704, 1908, 1930.
Schizoseris papenfussii
Tristan da Cunha: Holotype (female). Station 2, Quest Bay (Baardseth, 1937): O, no. 100 (N.B. holotype illustrated by Baardseth (1941, fig. 59) is filed with other (isotype?) specimens in one envelope. In contrast to collection data written on the holotype sheet, a common label on the envelope indicates that all plants were collected from Station 14, Julia Point, littoral zone, on *Halopteris funicularis* (Baardseth, 22 December 1937).)
Schizoseris polydactyla
South Georgia. Lectotype. 'Nordstrand der Landzunge' (Will, 22 March 1883): M, 67.83/41 (female); Syntypes (all female): M, 67.83/40, 67.83/42, 67.83/43 (two specimens), 67.83/44, 67.83/45, 67.83/46 (eight specimens).

Description: Thalli erect, to 37 cm tall, pink to reddish-purple to reddish-brown, and with blades foliose, stipitate, flattened or undulating, mostly monostromatic (Fig. 122c) but thicker near veins (Figs 120a, 122a, b), and highly laciniate (Fig. 120b) or completely eroded to veins (Fig. 120c). Growth from numerous marginal initials. Microscopic veins absent; macroscopic veins fine or coarse, branched, lacking internal rhizoids, and with central layer of large cells surrounded by layers of smaller cells (Fig. 122d); veins usually dichotomizing (Fig. 120f) but sometimes either anastomosing (Fig. 120g), predominantly parallel (Fig. 120d, e), or alternately to sub-oppositely branched (Fig. 120h). Spermatangia in irregularly shaped sori spread between veins in upper thallus. Procarps scattered across blade surfaces, with each supporting cell bearing one 4-celled carpogonial branch (Fig. 122e, f). Mature gonimoblasts with large bulbous fusion cells bearing chains of rounded carposporangia (Fig. 122g). Cystocarps protuberant, hemispherical, mostly scattered across blade surfaces but sometimes restricted to veins (Fig. 122a), and having ostioles generally flush with surfaces of pericarps, although sometimes surrounded by raised collars of thickened pericarp tissue. Tetrasporangia tetrahedrally divided, embedded in cortex, and aggregated in small, rounded sori developing on and between veins (Fig. 122h).

Plants sometimes grow epiphytically on the strap-like fronds of *Desmarestia chordalis* but are generally more abundant on shaded rocky surfaces between 3–15 m depths along the floors of *Macrocystis* forests around Macquarie I. In these subtidal habitats, *S. dichotoma* often grows next to or is epiphytized by algae such as *Ballia callitricha*, *Heterosiphonia berkeleyi*, *Delesseria lancifolia* and *Myriogramme livida*. *Polycoryne radiata* infrequently occurs on the stipes and apophyses of *S. dichotoma* as well as on the lower portions of *Myriogramme livida*.

Tetrasporophytes and female gametophytes are common, whereas male gametophytes are seemingly rare.

Fig. 121 *Schizoseris dichotoma* (Hooker f. & Harvey) Kylin: **a.** Habit of plant from New Zealand determined by Kylin as *Myriogramme multinervis* (LD, unnumbered specimen). **b.** Habit of plant from New Zealand determined by Lindauer as *Schizoseris davisii* (L, 047, 192 . . . 175). **c.** Habit of holotype of *Schizoseris papenfussii* (O, 100). **d.** Habit of isotype of *S. papenfussi* (O). **e.** Syntype of *Delesseria polydactyla* (M, 67.83/46). Scales: a, c–e = 2 cm; b = 4 cm.

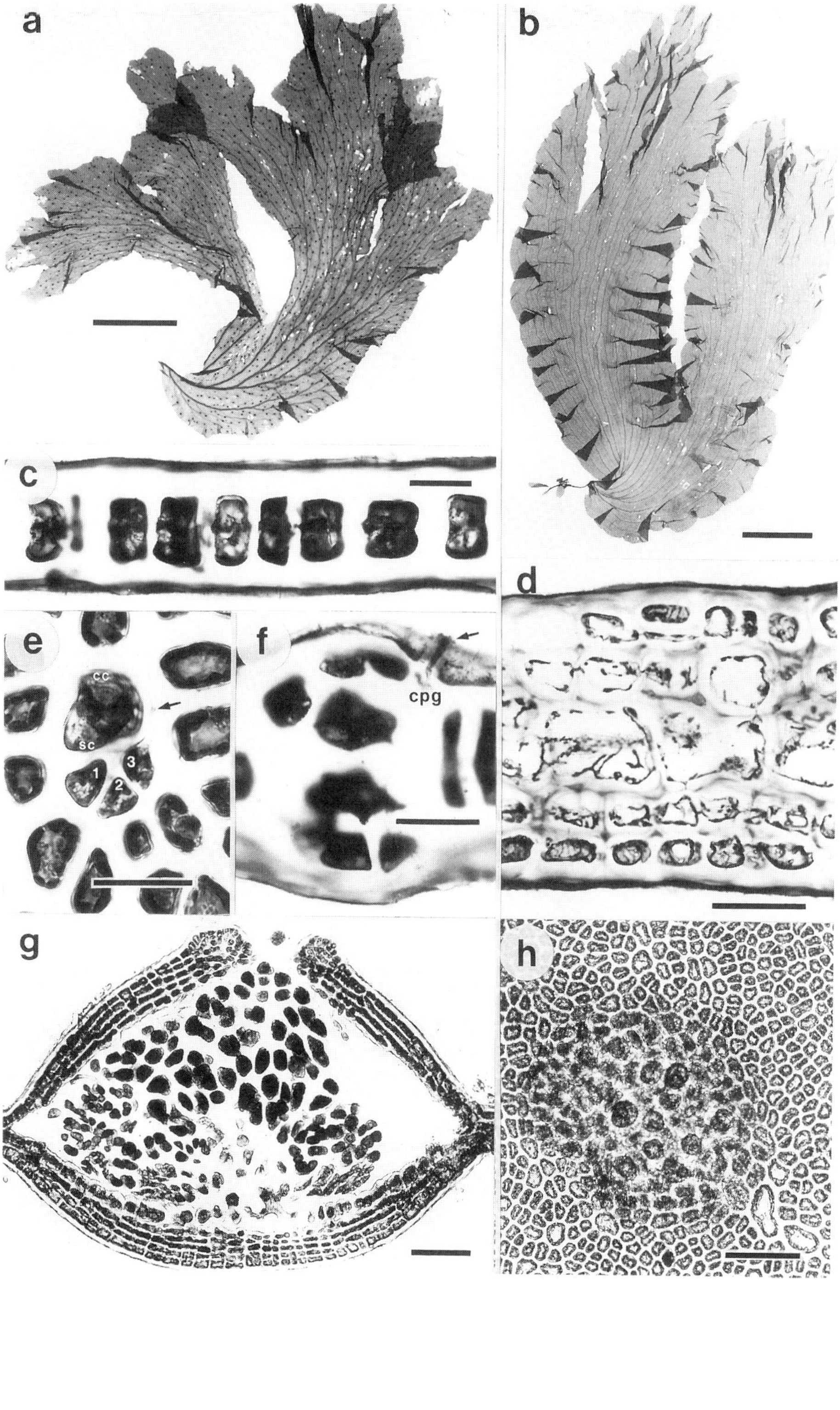
a
b
c
d
e
cc
sc
1
2
3
f
cpg
g
h

NOTES: Most *Schizoseris* species are separated by differences in thallus habit, although *S. pseudodichotoma* was circumscribed largely by the shape of its cystocarps. Zinova (1963) described this species from Macquarie I. as morphologically similar to, possibly even conspecific with, the widely distributed *S. dichotoma*. Based on only one female gametophyte with asymmetrical cystocarps and no lateral veins, however, she distinguished *S. pseudodichotoma* from *S. dichotoma*. My findings do not confirm this distinction. Most plants studied produced symmetrical cystocarps, although a few that are morphologically similar to *S. dichotoma* had asymmetrical cystocarps comparable to those illustrated by Zinova (1963, fig. 9a, b). Present observations on cystocarps of *S. dichotoma* reveal many having ostioles without collars and less elaborate fusion cells than those produced by *S. condensata*, although the illustrations of Baardseth (1941, fig. 57c, d—for *S. papenfussii*) indicate similar cystocarp construction as in *S. condensata*. Both cystocarp morphology and gonimoblast structure appear to be distinctive for the genus but not for species, which are generally better distinguished by various vegetative features.

During the present study of over 200 individuals, forms comparable to *S. davisii*, *S. dichotoma*, *S. kerguelensis*, *S. papenfussii* and *S. pseudodichotoma* have been observed; each 'species' representing a particular degree of blade laciniation and vein development. The boundaries between these taxa, however, are obscured by numerous transition forms, which have led to the present concept of a single, morphologically variable species, *S. dichotoma*. The basic thallus form, from which the various 'species' could be derived, would be an undivided foliose frond, similar to some blades produced by *S. papenfussii*. Most fronds of this species, however, are dissected to some degree (Fig. 121c) and are morphologically identical to those typical of *Myriogramme multinervis* (Fig. 121a). The two genera *Myriogramme* and *Schizoseris* have similar vegetative anatomy and reproductive features but supposedly differ in venation, species of the former lacking dichotomizing veins throughout their thalli (Kylin, 1956). *Myriogramme multinervis* has coarse, dichotomizing veins throughout its fronds, and is thus more like *Schizoseris* than *Myriogramme*.

Fronds of *S. papenfussii* (Fig. 121d) sometimes become highly eroded and strongly resemble those of *Schizoseris polydactyla* (Fig. 121e). Intermediate stages of frond erosion can result in thalli with morphologies similar to *Schizoseris davisii* (Fig. 121b) and *S. kerguelensis* (cf. Zinova, 1963, fig. 5).

Rhodomelaceae J. E. Areschoug, 1847: 260

The Rhodomelaceae is that largest family of red algae and contains approximately 125 genera, totalling some 600-650 species (Kraft, 1981), that have been variously grouped in tribes (e.g., Scagel, 1953; Kylin, 1956). The tribes (= Gruppen) recognized by Kylin (1956) have been revised somewhat and their constituent genera are listed by Hommersand (1963). These groupings are based mostly on the presence or absence of trichoblasts and differences in pericentral cell numbers, degree of dorsiventrality, and free-living vs. parasitic modes of life.

This family is well represented in Antarctic and Subantarctic waters, with at least three genera, *Antarctocolax*, *Picconiella* and *Sporoglossum*, endemic to the southern high latitudes. Papenfuss (1964*a*) credited 47 rhodomelaceous taxa from 10 tribes (as circumscribed by Hommersand (1963)) to the Southern Ocean. The Rhodomelaceae is represented at Macquarie I. by four genera, each with one species, representing three tribes (Bostrychieae—*Bostrychia*; Lophothalieae—*Sporoglossum*; Polysiphonieae—*Lophurella*, *Polysiphonia*).

Fig. 122 *Schizoseris dichotoma* (Hooker f. & Harvey) Kylin: MELU, R 0159, 0849, 0862, 0863. **a, b.** Habits of *'Myriogramme multinervis'* form. **a.** Female gametophyte with cystocarps. **b.** Tetrasporophyte. **c.** Section through upper thallus away from veins. **d.** Section through vein in lower thallus. **e.** Blade surface with procarp showing central cell (cc) bearing one pericentral cell (=supporting cell 'sc') in plane of focus; three cells of carpogonial branch visible, with terminal trichogyne (arrow) colourless. **f.** Section through procarp showing carpogonium (cpg) and pore (arrow) through which trichogyne usually protrudes. **g.** Section through mature cystocarp. **h.** Blade surface showing tetrasporangial sorus. Scales: a, b = 5 cm; c, e, f = 20 μm; d = 50 μm; g, h = 100 μm.

Bostrychia Montagne *in* Ramon de la Sagra, 1842*a*: 39 (*nom. cons.*)

Type species: *B. scorpioides* (Hudson) Montagne *in* Dict. Univ. Sci. Nat. II, 1842*c*: 661.

Members of this genus are dorsiventrally organized and have terete to flattened, corticate or ecorticate, branched axes that are prostrate and erect. Hommersand (1963) considered most features in this genus to be primitive with regard to others in the Rhodomelaceae and proposed the sub-family Bostrychioideae to accommodate *Bostrychia*. His descriptions of the sub-family and genus are similar and include the following characteristics: (1) branching of thalli is alternate-distichous, with uniaxial branches having pericentral cells that do not elongate as much as axial cells and eventually are comparable to the periaxial cells of *Ceramium* (Womersley, 1978); (2) the pericentral cells divide transversely to form tiers, with each tier cell capable of producing cortical filaments in corticated species; (3) trichoblasts are entirely absent; (4) Spermatangia are produced from superficial cells on ordinary branches; (5) procarps, each containing only one sterile group, develop along unspecialized branches from any pericentral cell except the first formed; auxiliary cells give rise to more than one gonimoblast initial; mature carposporphytes are sympodially branched, lack fusion cells, form anastomoses by numerous secondary pit connections, and bear terminal carposporangia; a pericarp, similar in appearance to cortical tissue, is formed after fertilization; (6) tetrasporangia are formed in whorls on unspecialized indeterminate branches (Hommersand, 1963) or stichidia (Kylin, 1956).

Falkenberg (1901) recognized 14 species, which he distinguished by branching patterns, number of pericentral cells, presence or absence and type of cortication, and number of tetrasporangia per whorl. These are still the main features used to separate species.

Bostrychia is principally a tropical genus that typically occurs in high intertidal habitats, with most representatives in brackish waters. Some species, however, grow in temperate to colder regions in either freshwater or undiluted seawater. Estimations on the number of species in the genus vary considerably. Farlow (1881) calculated a total of some 20 species, whereas Post (1936) recognized only 13. Kylin (1956) later placed the total around 25. Three species are listed from the Subantarctic and none from the Antarctic (Papenfuss, 1964*a*), with one reported by Zinova (1958) from Macquarie I.

Bostrychia vaga Hooker f. & Harvey, 1845: 270

Fig. 123

Synonyms: see Post, 1936.

Type locality: Betsy Cove, Christmas Harbour, Îles Kerguelen.

Type: BM, no. 647.

Distribution: Falkland Is (Skottsberg, 1923); Fuegia (Hariot, 1889); Heard I.; Îles Kerguelen (Hooker f. & Harvey, 1845; Dickie, 1879); Macquarie I. (Zinova, 1958); South Georgia (Kylin & Skottsberg, 1919).

Material examined:

1. Heard I.: Atlas Cove (Chittleborough, 30 August 1949): MELU, R 3727, 3728; Transect no. 1, north of Wharf Point, Atlas Cove, eulittoral (Smith, 6 February 1983): MELU, R 3733.
2. Îles Kerguelen: Type specimens. Betsy Cove, Christmas Harbour (J. D. Hooker, 1839–1843): BM, nos 646, 647.
3. Macquarie I.: collections MA-45, 98, 119, 130, 278; Herbarium specimens: MELU, R 0063, 1285, 1286, 1288, 1290–1300.

Description: Thalli comprising prostrate and erect axes, to 1–2 cm tall, greenish-purple to dark purple, and alternately to unilaterally branched, with upper branches strongly arcuate (Fig. 123a, b). Axes constructed of (4-)6(-7) pericentral cells (Fig. 123c), with each dividing

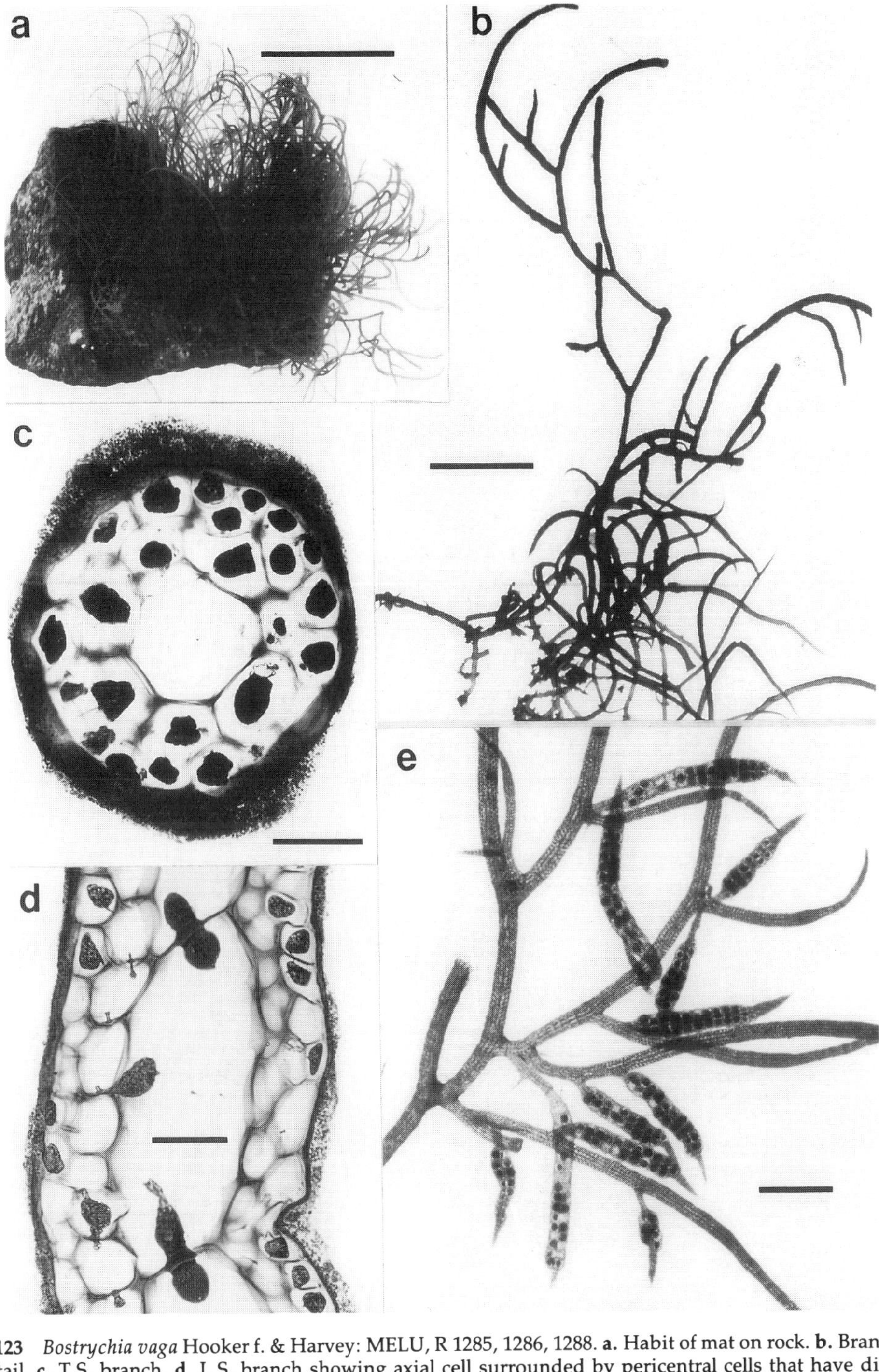

Fig. 123 *Bostrychia vaga* Hooker f. & Harvey: MELU, R 1285, 1286, 1288. **a.** Habit of mat on rock. **b.** Branching detail. **c.** T.S. branch. **d.** L.S. branch showing axial cell surrounded by pericentral cells that have divided into six tier cells, which subsequently have produced cortical cells. **e.** Tetrasporangial stichidia. Scales: a = 1 cm; b = 3 mm; c, d = 50 μm; e = 500 μm.

transversely into 5 or 6 tier cells that subsequently produce cortical cells in a single layer (Fig. 123d). Tetrasporangia tetrahedrally divided, two per fertile segment, borne in stipitate stichidia that taper gradually or abruptly toward acute apices (Fig. 123c). Sexual reproductive structures not seen.

Plants are extremely abundant along the shores of Macquarie I., where they form dense springy mats on rocky surfaces in the upper intertidal. In one instance, however, *B. vaga* was growing adjacent to a thick felt of *Acrochaetium daviesii* (Fig. 64a) on a piece of driftwood collected from the mid-intertidal. Filaments of *Rhizoclonium ambiguum* often are entangled with branches of *Bostrychia* and *Rhodochorton purpureum* frequently colonizes the lower parts of *Bostrychia* mats (Fig. 67a). An unidentified ascomycete also infects *B. vaga* and was common on most plants studied.

NOTES: *Bostrychia vaga* is distributed more widely throughout the Subantarctic than either of the other two species, *B. mixta* Hooker f. & Harvey and *B. scorpioides* (Hudson) Montagne, listed by Papenfuss (1964*a*). It can be distinguished from these taxa by its arcuate corticated branches that are constructed with 4–7 pericentral cells (Post, 1936).

Lophurella Schmitz *in* Schmitz & Falkenberg, 1897: 440

Type species: *L. periclados* (Sonder) Schmitz *in* Schmitz & Falkenberg, 1897: 441.

Lophurella and *Polysiphonia* are closely related, with species of both having radially organized thalli that branch along all sides of terete axes (Kylin, 1956). The two genera produce colourless deciduous trichoblasts and develop tetrasporangia singly per fertile segment, often in spiral series (Falkenberg, 1901). *Lophurella*, however, is distinguished by its firm cartilaginous branches (caused by heavy cortication around an unvarying core of an axial filament plus four pericentral cells), and *Polysiphonia* by its generally lax and lightly corticated thalli, which have 4–20 pericentral cells. Additionally, tetrasporangia in *Lophurella* are borne in either shortened lateral branches or specialized stichidia, whereas tetrasporangia in *Polysiphonia* are produced mostly in seemingly unspecialized branches.

The Southern Hemisphere genus *Lophurella* has had six species credited to it, most of which have been distinguished by differences in habit. Three, *L. comosa*, *L. hookeriana* and *L. patula*, are listed from various Subantarctic localities (Papenfuss, 1964*a*), none from the Antarctic, and two from the cold temperate coasts of southern Australia [*L. periclados* (Lucas & Perrin, 1947)] and New Zealand [*L. caespitosa* (Falkenberg, 1901)]. A sixth species (*L. gaimardii*), reported from the Falkland Is. (Cotton, 1915), is likely to prove conspecific with one of the other *Lophurella* species when critically studied (Papenfuss, 1964*a*). The genus is represented at Macquarie I. by *L. hookeriana*, which is the most widespread of these species.

Lophurella hookeriana (J. Agardh) Falkenberg, 1901: 58

Fig. 124

BASIONYM: *Rhodomela hookeriana* J. Agardh, 1863: 880.

SYNONYMS: see Papenfuss, 1964*a*.

TYPE LOCALITY: Not indicated. Original collections from Cape Horn and the Falkland Is.

TYPE: presumably in LD (Koster, 1969), but specimens were not seen.

Fig. 124 *Lophurella hookeriana* (J. Agardh) Falkenberg: MELU, R 0550, 1794. **a.** Habit of female gametophyte. **b.** Section through cortex. **c.** T.S. main axis showing axial cell (ax) and four pericentral cells. **d.** Lateral branch showing apical cell (arrow) and deciduous trichoblasts. **e.** Habits of two cystocarps. **f.** Upper branches of cystocarp-bearing female gametophyte. **g.** Tetrasporangial stichidia. Scales: a = 2 cm; b, c = 50 µm; d = 100 µm; e = 500 µm; f = 2 mm; g = 1 mm.

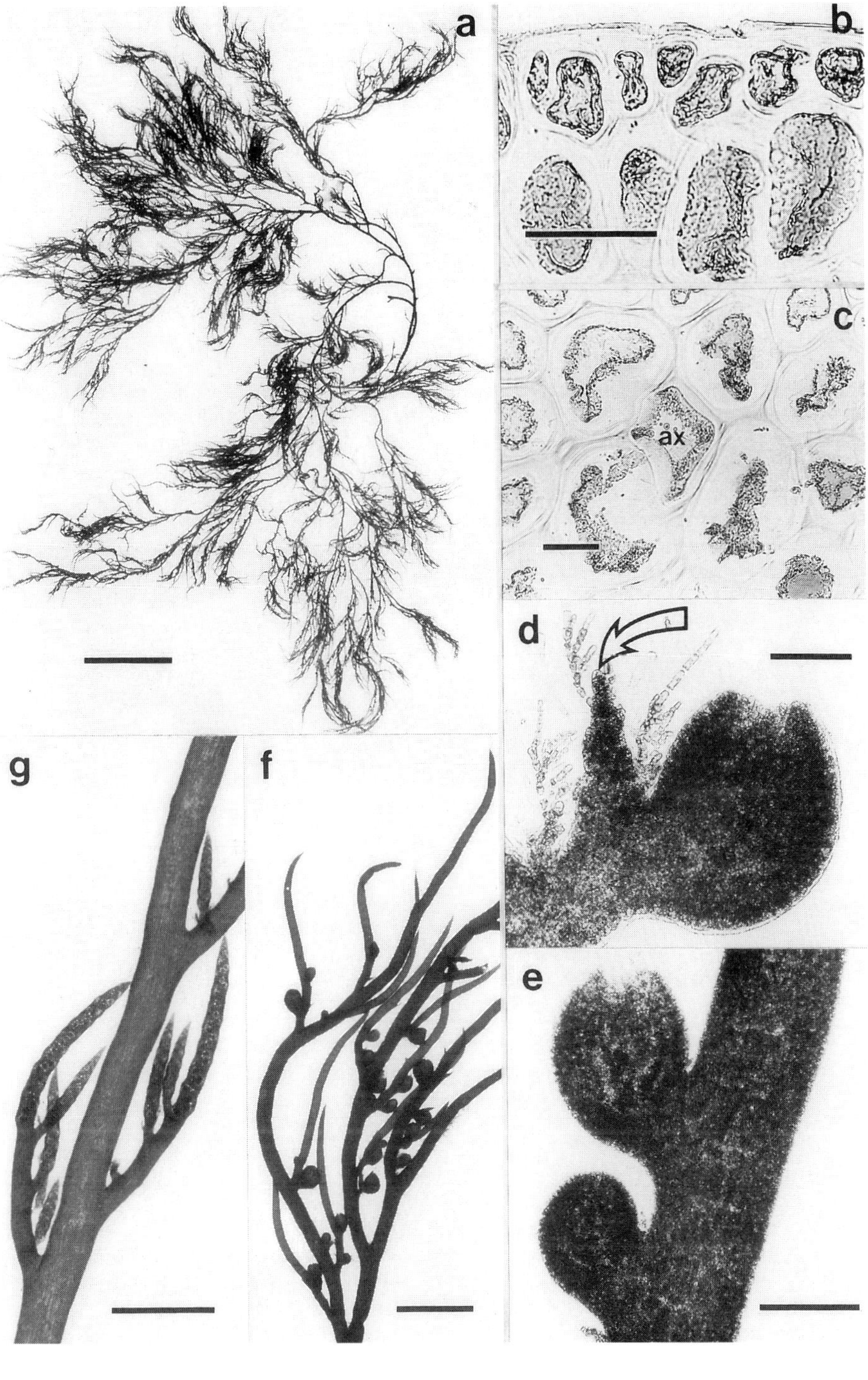
a
b
c
ax
d
e
f
g

DISTRIBUTION: Campbell I. (Levring, 1945); Falkland Is, Fuegia (Skottsberg, 1923); Heard and Macquarie Is; Îles Crozet and Kerguelen (Levring, 1944); Stewart I. (Adams *et al.*, 1974). (Presumably on the basis of Falkenberg's (1901) account of a single specimen sent to him by Baron Ferdinand von Mueller, Lucas & Perrin (1947) reported this species from Australia (Victoria). The presence of *L. hookeriana* in Victorian waters has not been confirmed by subsequent algal collections and it seems likely that Mueller's plant may have been a morphological variant of *L. periclados*, which is common in this region (pers. obs.).)

MATERIAL EXAMINED:
1. Heard I.: Atlas Cove (Chittleborough, 11 September 1949): MELU, R 3723, 3724.
2. Macquarie I.: collections MA-111, 213, 215, 222, 244, 295, 302, 309; Herbarium specimens: MELU, R 0093, 0550, 1794, 1971, 1972.

DESCRIPTION: Thalli erect, to 18 cm tall, purple-black, and with axes stiff, cartilaginous, terete, irregularly branched from all sides, and arising from discoid bases (Fig. 124a); branches uniaxial, with four pericentral cells surrounding axial cell (Fig. 124c); cortication dense, beginning immediately behind apices (Fig. 124d). Medulla extensive, colourless, composed of large (50–70 μm diameter) nearly isodiametric cells; cortex monostromatic, pigmented, composed of smaller diameter cells (Fig. 124b). Trichoblasts colourless, diciduous and, if present, only near branch apices (Fig. 124d). Cystocarps globuse to urceolate, with short stipes and apical ostioles (Fig. 124d–f). Tetrasporangia tetrahedrally divided, one per fertile segment, forming spiral series in upper parts of shortened lateral branches or along specialized stichidia often developing in clusters on adaxial surfaces of lateral branches (Fig. 124g). Spermatangia not seen.

Lophurella hookeriana is not abundant around Macquarie I., but is present along most shores. It grows on rocky surfaces in mid- to lower-intertidal pools and also 3–17 m depths. In the subtidal, plants sometimes occur adjacent to mats of *Codium subantarcticum*, which grow on small boulders generally partially shaded by overhead canopies of *Macrocystis* fronds or an overstory of *Desmarestia willii*. *Lophurella* also grows alongside various red algae such as *Plocamium* and members of the Delesseriaceae and sometimes is epiphytized by *Rhodophyllis acanthocarpa*. Most plants, however, lack epiphytes. The red algal 'parasite' *Sporoglossum lophurellae* occasionally infects host plants and can be abundant on branches in mid-thallus.

NOTES: Branching habit and thallus consistency of *L. hookeriana* distinguish it from all other red algae at Macquarie I. The presence of *Sporoglossum* (in this vicinity at least) also is diagnostic for *Lophurella*. Kylin & Skottsberg (1919) reported specimens of *L. hookeriana* from the Falkland Is that were infected by two seemingly specific red algal 'parasites', *Colacopsis lophurellae* and *Sporoglossum lophurellae*, although the latter has been observed more recently to grow also on *Aphanocladia* Mendoza (1970*a*).

Polysiphonia Greville, 1824: lxvii, 308 (***nom. cons.***)

Type species: *P. urceolata* (Dillwyn) Greville, 1824: 309.

Members of this genus have terete, radially organized branched thalli that are mostly erect (although often with some prostrate portions), constructed with pericentral cells, and sometimes laterally bearing colourless trichoblasts. Indeterminate branches generally arise either from oblique divisions of apical or sub-apical cells prior to pericentral cell formation (exogenous branching—Falkenberg, 1901) or from basal cells of trichoblasts along the lower thallus (cicatrigenous branching—Hollenberg, 1942). Less commonly, however, branches develop from axial cells after pericentral cells have formed (endogenous branching—Falkenberg, 1901). Tetrasporangia are produced one per segment and form series in the upper parts of mostly unmodified vegetative branches.

Reproductive structures generally are uniform throughout the genus, although some species are distinguished by the presence or absence of sterile cells near the apices of spermatangial stichidia, the disposition of tetrasporangia, and the shape of tetrasporangial branches. Vegetative criteria are usually used to separate species. These include differences in thallus habit, branch origins, dimensions and proportions of thalli and branch segments, numbers of pericentral cells, presence or absence of and degree of cortication, and presence or absence of rhizoids produced from pericentral cells. Womersley (1979) considered other features such as pericarp structure too uniform and trichoblast presence or absence too variable for taxonomic purposes.

Polysiphonia, with some 200 species, is one of the largest red algal genera and is represented along coasts of nearly all seas (Womersley, 1979). Papenfuss (1964*a*) listed 15 species from the Subantarctic, with only one of these, *P. abscissa* Hooker f. & Harvey, extending into the Antarctic (Skottsberg, 1953). Womersley (1979), however, reduced the number of species in the Subantarctic to 14 by synonymizing *P. cancellata* Harvey with *P. decipiens* Montagne. Species numbers are likely to be reduced still further by monographic study of all Subantarctic representatives. The genus is represented at Macquarie I. by *Polysiphonia anisogona* Hooker f. & Harvey.

Polysiphonia anisogona Hooker f. & Harvey, 1845: 265

Fig. 125

SYNONYM: see Papenfuss, 1964*a*.

TYPE LOCALITY: Not designated in the original account. Lectotype from the Falkland Is.

LECTOTYPE: K *in* BM; specimen selected by J. H. Price (1 March 1985); Herbarium sheet with J. D. Hooker annotation: 'Berkeley Sound, East Falkland I. September 1842. D. L.[yall].' (Price, pers. comm.).

DISTRIBUTION: Falkland Is, Fuegia (Hooker f. & Harvey, 1845); Îles Crozet (Levring, 1944); Îles Kerguelen (Dickie, 1879); Macquarie I.; South Georgia (Reinsch, 1890; Kylin & Skottsberg, 1919).

MATERIAL EXAMINED: Macquarie I.: collections MA 14, 46, 51, 83, 94, 113–116, 137, 140, 215, 222, 227, 236–238, 244, 248, 264, 271, 283, 295, 298, 302, 303, 309, 364, 366, 378; Herbarium specimens: MELU, R 0584, 0585, 0901–0906, 1123–1133.

DESCRIPTION: Thalli delicate, tuft to skein-forming (with skeins to 40 cm long), reddish-brown to purplish, and with axes filamentous, lax, terete, radially branched, uncorticated, and usually lacking distinct points of attachment (Fig. 125a). Filaments polysiphonous, with 8–12 pericentral cells (Fig. 125e) *c.* three times as long as broad (to 80 µm long) in upper branches but becoming extremely elongate (3–4 mm long) in lower branches (Fig. 125g). Indeterminate axes 100–125 µm diameter in mid-thallus, to 200 µm diameter in lower thallus, bearing short, hook-like, determinate laterals that interlock adjacent indeterminate axes (Fig. 125f). Trichoblasts absent or rare in sterile and tetrasporangial plants, present but deciduous and simple or 1–2 times forked in gametophytes (especially male thalli). Spermatangial stichidia terete, gently tapering, to 300 µm long, with monosiphonous bases, lacking sterile apical cells, replacing one branch of a trichoblast (Fig. 125d). Carposporophytes with large basal fusion cells bearing terminal, pyriform to clavate carposporangia inside well developed, subspherical to urceolate (*c.* 550 µm broad × 600 µm long), ostiolate pericarps on polysiphonous stalks (Fig. 125c). Tetrasporangia cruciately divided, one per segment, developing in laxly spiraled series in upper branches (Fig. 125b).

Plants are abundant around Macquarie I., where they grow on rocks or entangled with other algae in rock pools in the mid- to lower-intertidal down to 20 m depths. In the subtidal, *P. anisogona* often forms part of a red algal understory that is shaded by fronds of *Desmarestia* or *Macrocystis*. Other algae included in this understory community include various members of the Delesseriaceae and *Iridaea cordata*.

NOTES: Tetrasporangia produced by species of *Polysiphonia* generally are tetrahedrally divided. Those produced by *P. anisogona* are unusual in their cruciate, sometimes decussate, division patterns.

Levring (1944, figs 18c–e, 19) described and illustrated *Polysiphonia urbanioides* from the Crozet and Kerguelen Is. This species appears to be closely related to, if not identical with, *P. anisogona*,

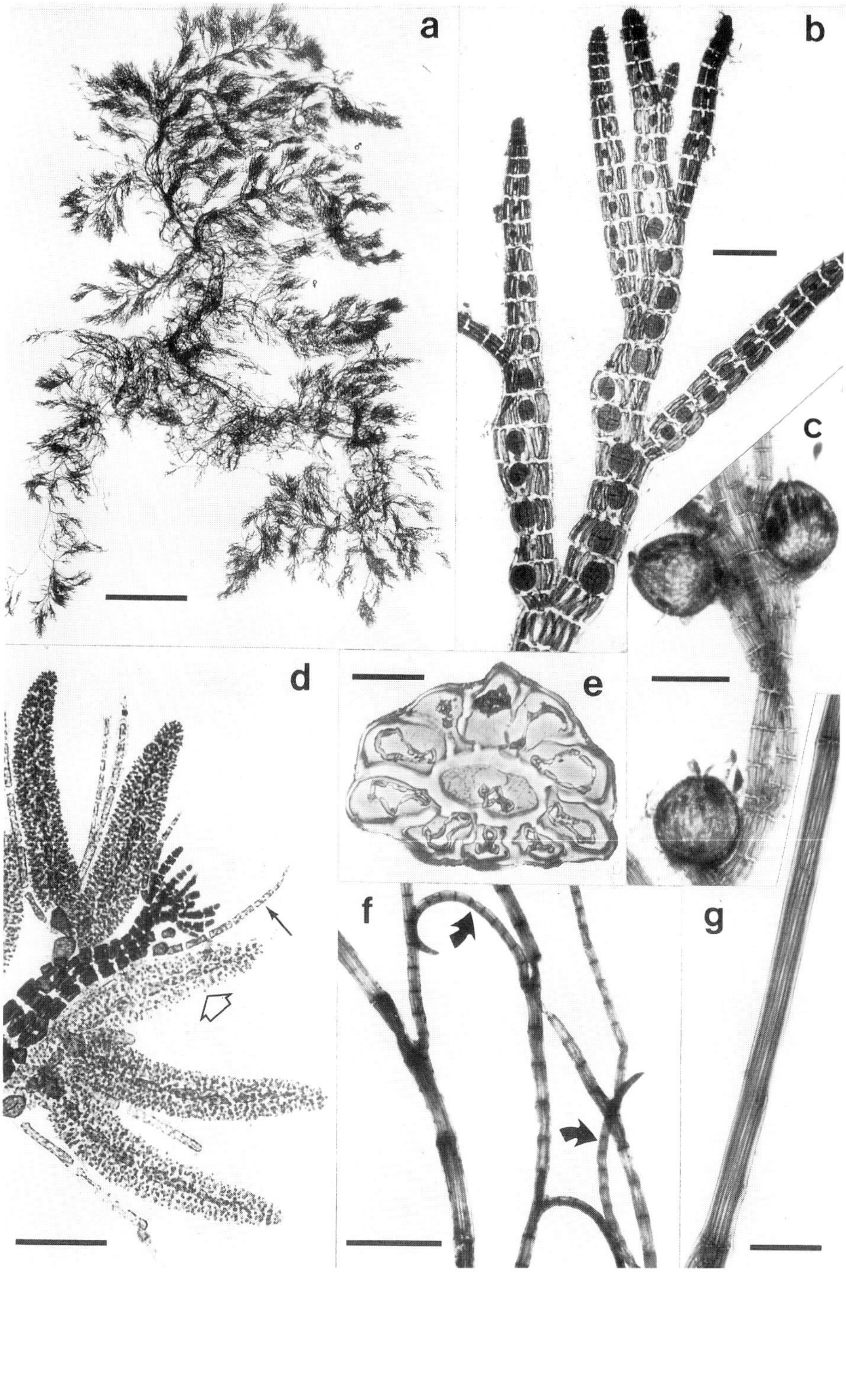
a
♂
♀
b
c
d
e
f
g

which also occurs around these islands as well as throughout the Subantarctic. Levring suggested that the two species might be closely related, as both have uncorticated axes composed of a similar number of pericentral cells and supposedly develop lateral branches from the basal cells of trichoblasts, but he considered the former species to be more robust and morphologically distinct. His habit photograph of *P. urbanioides* (Fig. 19), however, strongly resembles some Macquarie specimens of *P. anisogona*. The development of lateral branches from the basal cells of trichoblasts, as described by Levring (1944), has not been confirmed in the present studies, which have revealed a general absence of trichoblasts from most thalli. Tetrasporangium division pattern can not be compared between the two species because tetrasporophytes of *P. urbanioides* are not known. Clarification of the relationship of *P. anisogona* to *P. urbanioides* awaits critical study of type material of the latter, which was not available during this survey.

Sporoglossum Kylin *in* Kylin & Skottsberg, 1919: 57

Type species: *S. lophurellae* Kylin *in* Kylin & Skottsberg, 1919: 57.

Sporoglossum is a monotypic genus that Kylin (*in* Kylin & Skottsberg, 1919) described as 'parasitic' on *Lophurella hookeriana*. He distinguished plants by their flattened, tongue-like tetrasporangial stichidia (for which the genus was named). Each fertile segment produces two tetrahedrally divided tetrasporangia which are oppositely paired so that the sporangia form identical rows on both faces of the flattened sporophylls. Each stichidium is subtended by a polysiphonous stipe with four pericentral cells. Various other features such as a distichous branching pattern and a production of spermatangial stichidia with monosiphonous bases suggest a close taxonomic relationship with the genus *Picconiella* De Toni (Kylin & Skottsberg, 1919; Kylin, 1956).

Along with *Sporoglossum*, Kylin (*in* Kylin & Skottsberg, 1919) described *Colacopsis lophurellae*, another red algal 'parasite' that lives on *Lophurella hookeriana*. The two can be distinguished easily, as the latter has a different habit, terete tetrasporangial stichidia with monosiphonous bases, and globular spermatangial stichidia with polysiphonous bases. Of the two, however, only *Sporoglossum* occurs at Macquarie I.

Sporoglossum lophurellae Kylin *in* Kylin & Skottsberg, 1919: 57

Fig. 126

Type locality: Falkland Is.

Type: not located in LD; location unknown.

Distribution: Argentina (Mendoza, 1970*a*); Campbell I. (Levring, 1945); Falkland Is (Kylin & Skottsberg, 1919); Fuegia (Skottsberg, 1923); Macquarie I.; Stewart I. (Adams *et al.*, 1974).

Material examined: Macquarie I.: collections MA-213, 244; Herbarium specimen: MELU, R 0549.

Description: Thalli 'parasitic' on *Lophurella hookeriana* (Fig. 126a), forming dense clusters (to 1–2 mm diameter) of terete to flattened, whitish to light purple-brown coloured fronds arising from basal cushions composed of compacted filaments that anchor in host medullary tissues with penetrating rhizoids (Fig. 126b, c). Tetrasporangia tetrahedrally divided, oval, and borne in flattened stichidia subtended by polysiphonous stipes with four pericentral cells (Fig. 126d). Each

Fig. 125 *Polysiphonia anisogona* Hooker f. & Harvey: MELU, R 0901–0903. **a.** Habits of entangled male and female gametophytes. **b.** Upper branches of tetrasporophyte showing single tetrasporangium per tier of pericentral cells. **c.** Habits of three cystocarps. **d.** Male gametophyte bearing trichoblasts (solid arrow) and spermatangiophores (hollow arrow). **e.** T.S. upper branch. **f.** Main axis bearing hook-like determinate laterals (arrows). **g.** Lower axis with elongate pericentral cells. Scales: a = 3 cm; b = 250 µm; c, g = 500 µm; d = 100 µm; e = 30 µm; f = 1 mm.

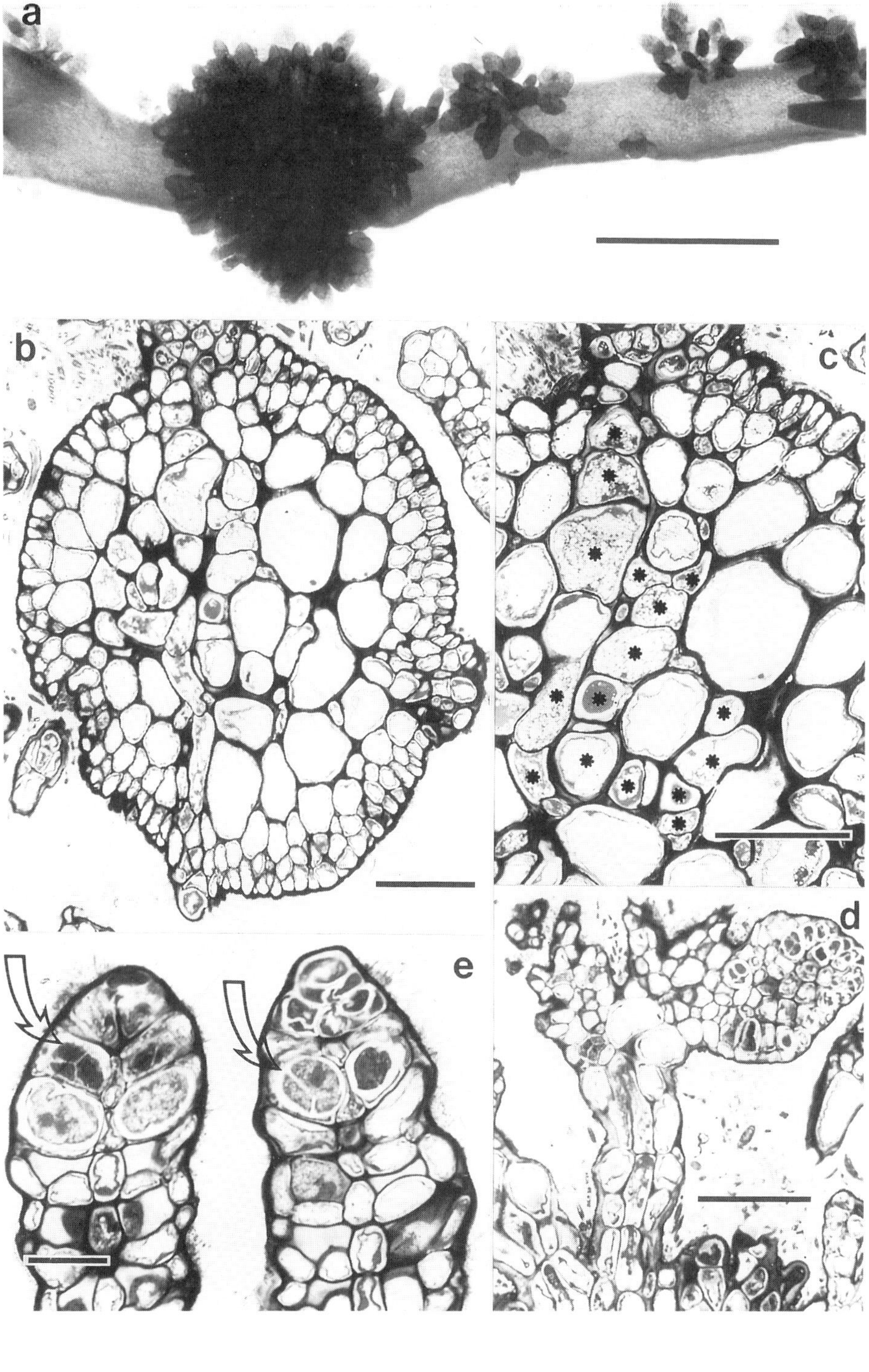
a
b
c
d
e

fertile segment with two tetrasporangia oppositely paired (Fig. 126e). Sexual reproductive structures not seen.

Sporoglossum was collected only twice—one collection coming from rock walls lining mid- to lower-littoral pool on the west coast and the second from partially shaded boulders on the floor of a *Macrocystis* forest in 15 m depth on the east coast. Although the host species *L. hookeriana* is not uncommon and is distributed around the entire island, *S. lophurellae* seems to be a rare and localized component of Macquarie's flora.

NOTES: Mendoza (1970*a*, pls 3, 4) described and illustrated male and female gametophytes and tetrasporophytes, all of which were abundant on the host alga *Aphanocladia robusta* Pujals growing in the mid- to lower-intertidal.

Fig. 126 *Sporoglossum lophurellae* Kylin: MELU, R 0549, Slide R/M 0572. **a.** Habit of tetrasporophytes on *Lophurella hookeriana*. **b.** T.S. host branch at site of infection. **c.** Detail of parasite cells (*) penetrating host tissue. **d.** Optical section of flattened tetrasporophyll. **e.** L.S. tetrasporophyll showing oppositely paired tetrasporangia (arrows) borne along monosiphonous filament. Scales: a = 1 mm; b–d = 100 μm; e = 30 μm.

3: Characterization and analysis of flora

There is further an inherent tendency in every one occupied with specialities to exaggerate the value of his materials and labours, whence it happens, that botanists engaged exclusively upon local floras are at issue with those of more extended experience, the former considering as species what the latter call varieties, and what the latter suspect to be an introduced plant the former are prone to consider a native.

J. D. Hooker *Flora Novae-Zelandiae*, 1853: xiii

3.1 Abundance, luxuriance and seasonality

Sealers provided the first indications of a rich growth of seaweeds around Macquarie I., although their accounts referred almost entirely to the offshore beds of *Macrocystis* and the lush covering of *Durvillaea* on intertidal boulders. Kenny & Haysom (1962) and Simpson (1976) described zonation of plants and animals in the intertidal at several localities around the island, but their studies did not include any quantitative data for seaweeds. The present report is the first floristic survey undertaken in the Subantarctic that is based on extensive collections using SCUBA, and consequently has revealed a luxuriant growth of seaweeds in the subtidal that hitherto was undescribed. Most descriptions of Macquarie's constituent green, brown, and red algae given here are also limited to qualitative comments about the relative abundances of species in this vicinity; quantitative studies based on either biomass or actual plant numbers per meter square have yet to be done.

Most rocky surfaces in the intertidal around Macquarie I. are carpeted with extensive, luxuriant growths of macroalgae. In some regions, however, the rocks are barren. The scouring action of sand and gravel often appears to be responsible for the lack of macrophytes, but in other instances, the passage of elephant seals over plant-covered intertidal reefs seems to result in small patches denuded of algae. In addition to these irregularly produced barren patches, a rocky band generally lacking seaweeds or on which the algae have been closely cropped is present along most stretches of intertidal reef. When reporting on intertidal zonation at Macquarie I., Kenny & Haysom (1962) described this band as a 'bare' zone separating an upper horizontal band dominated by *Porphyra* from a lower band dominated by *'Rhodymenia'* (= *Palmaria*—considered to be typical of Kenny's & Haysom's 'upper red zone'). Limpets are the most abundant organism in this zone and are probably the cause for the paucity of macrophytes. Comparable zones along shores in lower latitudes would usually be dominated by sessile organisms such as barnacles and mussels, which with few exceptions are lacking at Macquarie I. (Simpson, 1976). Kenny & Haysom (1962) attributed the absence of such sessile animals partly to disturbance caused by *Durvillaea* fronds which sometimes extend into the 'bare' zone. Simpson (1976) disagreed with this explanation, and considered a lack of dispersal to this remote location as the cause for their absence.

Many algal species at Macquarie I. are present in both the littoral and sublittoral zones, although numerous members of the Rhodophyta occur only in shaded habitats of the latter (see Chapter 2.2, Systematic Overview). In most subtidal localities surveyed, rocky surfaces generally were covered by plants and seaweed densities appeared to be controlled largely by the availability of rocky surfaces for thallus attachment.

Seasonality is mostly unknown for Macquarie's marine flora. The present survey has relied largely upon summer collections, and consequently lacks detail on seasonal changes in the flora. Ecological studies by Kenny & Haysom (1962) and Simpson (1976), however, were based on collections made over 12 or more months and include some data pertaining to seasonal changes in certain biological and physical phenomena. Simpson indicated that the composition of conspicuous species in his quadrats and the relative positions of these organisms in the intertidal did not change throughout the year, although seasonal differences in the abundances of various seaweeds were noted. Populations of *Adenocystis utricularis, Chaetangium fastigiatum* and *Porphyra columbina*, for example, became denser during Spring (September to December). However,

Simpson's findings, however, largely suggested maintenance of the status quo, with dominant organisms present all year round. Some of these dominant plants, judging by their large sizes and coarse thalli (e.g., *Desmarestia chordalis, Durvillaea antarctica, Macrocystis pyrifera*), appear to be perennial species, while others with more delicate thalli are likely to be annual species. The latter include some filamentous (e.g., various members of the Ceramiales) and foliose (e.g., *Monostroma sp., Porphyra columbina, Porphyropsis coccinea, Schizoseris dichotoma*) algae. In the case of *Porphyra columbia*, which is considered to be an annual species with a foliose stage present throughout the year, plants presumably are partly aseasonal. That is, reproduction of some individuals is not synchronized to particular seasons and thalli mature at various times throughout the year to maintain a continuous presence of foliose fronds.

The uniformity in water and air temperatures around Macquarie I. probably provides a reasonably stable environment which might allow a particular composition of species to be maintained throughout the year. Light levels, however, fluctuate markedly between summer and winter seasons and are likely to control the growth of various annual species as well as effect changes in the reproductive status of both annual and perennial species, although no data from this locality exist for such speculations.

3.2 Species diversity

The diversity of species in a marine flora is determined largely by the length of coastline to be colonized, the number of habitats available, the range of water temperatures in the region, and the rates of immigration and emigration of individual plants. Immigration rates usually reflect the effectiveness of dispersal mechanisms in the region and the distance to surrounding floras that might serve as sources for immigrant species.

The degree of diversity in a flora, in spite of being based on an absolute number of species, is a relative value derived from the comparison of two or more floras. The southern coastline of Australia, for example, has some 1100 species of green, brown, and red algae, and is the richest known flora for any comparable length of coastline exposed to a similarly narrow (±5°C) temperature range (Womersley, 1981*b*). Japan's marine flora is richer still, but extends across a broader range of temperatures, i.e., from subarctic conditions in the north to subtropical conditions in the south (Womersley, 1981*a*). Considering Macquarie's small size and remote location, in addition to the uniform water temperatures surrounding the island, its benthic algal flora is seemingly diverse, with 43 families totalling 103 species from 81 genera. The majority (60) of these species belong to the Rhodophyta, although the Phaeophyta, with 28 representatives, contains the most prominent algae in both intertidal (*Durvillaea antarctica*) and subtidal (*Macrocystis pyrifera, Desmarestia chordalis* and *D. willii*) environments. The Chlorophyta includes 15 species from 12 genera, of which *Acrosiphonia pacifica* and species of *Enteromorpha* and *Ulva* dominant in certain intertidal localities; *Codium subantarcticum* is often prominent on shallow, rocky surfaces in the subtidal. An additional ten species, all encrusting members of the Corallinaceae, have also been reported but not confirmed by the present survey. Such overall diversity far exceeds that reported for any other Subantarctic island of comparable latitude (such as Bouvet, Crozet, Heard and Marion Is). These differences in species diversity, however, most likely reflect collecting intensity rather than an unusually diverse flora at Macquarie I.

3.3 Analysis of floristic components

Species of marine algae occurring at Macquarie I. can be grouped according to their presence in only this vicinity or throughout one or more broad geographic regions. Such regions are defined largely by discontinuities in environmental conditions including light and temperature regimes, salinity levels, pH, and dissolved nutrients. The arrangement of groups as below, however, does not necessarily represent the tolerance limits of species to certain physical conditions. The following phytogeographic groups are based on both confirmed and unconfirmed distribution records, but the latter may be in error because of incorrect plant identifications. Additionally, concepts of phytogeographic groups reflect the intensity of plant collecting in particular areas, where floras are often poorly studied or entirely unknown. Consequently, the present composition of these groups is likely to be modified upon re-examination of some old collections and integration of new data from future floristic surveys.

A total of 113 species are listed below for the seaweed flora of Macquarie I. Ten are encrusting members of the Corallinaceae reported by Zaneveld & Sanford (1980) but not found during the present survey. Each is designated by an asterisk (*). Species with distribution ranges throughout both Northern and Southern Hemispheres, although not necessarily extending from tropical to polar waters, are considered to be cosmopolitan and comprise 19 species or 17% of the flora. Five species, or 4% of Macquarie's flora, are distributed throughout southern high latitudes and occur in Cold Temperate, Subantarctic and Antarctic waters. Ten species, or 9% of the flora, grow only in the Southern Cold Temperate and Subantarctic Zones. Species and infraspecific taxa occurring in both Subantarctic and Antarctic Zones form the second largest group, 17 species or 15% of the flora, of benthic macrophytes at Macquarie I. The largest group, 46 species or 41% of the flora, includes plants with ranges restricted to the Subantarctic Zone. Twelve species and infraspecific taxa, constituting 11% of the flora, apparently are endemic to Macquarie I., and an additional four species, or slightly less than 4% of the flora, are inadequately represented to permit specific identification.

Cosmopolitan

CHLOROPHYTA:

Blidingia minima
Enteromorpha intestinalis
Ulva angusta
Ulva rigida
Urospora penicilliformis

PHAEOPHYTA:

Ectocarpus siliculosus
Macrocystis pyrifera
Petalonia fascia
Pilayella littoralis
Scytosiphon lomentaria

RHODOPHYTA:

Acrochaetium daviesii
Ahnfeltia plicata
Ceramium strictum
*Leptophytum foecundum**
Melobesia membranacea
Porphyropsis coccinea
Pterothamnion plumula
Rhodochorton purpureum
Rhodochorton variabile

Southern Cold Temperate/Subantarctic/Antarctic

RHODOPHYTA:

Ballia callitricha
Delisea pulchra
Mesophyllum patena
Pachymenia sp. cf. *stipitata*
Rhodochorton concrescens

Southern Cold Temperate/Subantarctic

PHAEOPHYTA:

Asteronema ferruginea
Chordariopsis capensis
Microzonia velutina
Scytothamnus fasciculatus

RHODOPHYTA:

Antithamnionella ternifolia
Dasyptilon pellucidum
Hymenena decumbens
Porphyra woolhousiae
Ptilonia willana
Pugetia delicatissima

Subantarctic/Antarctic

CHLOROPHYTA:

Acrosiphonia pacifica
Endophyton atroviridis
Enteromorpha bulbosa
Prasiola crispa subsp. *antarctica*

PHAEOPHYTA:

Adenocystis utricularis
Alethocladus corymbosus
Desmarestia willii
Elachista antarctica
Myrionema incommodum
Ralfsia australis

RHODOPHYTA:

Delesseria lancifolia
Iridaea cordata
*Leptophytum coulmanicum**
Palmaria decipiens
Palmaria georgica
Phyllophora appendiculata (?= *Gymnogongrus turquetii*)
Pterothamnion simile

Subantarctic

CHLOROPHYTA:

Codium subantarcticum
Entocladia maculans
Rhizoclonium ambiguum

PHAEOPHYTA:

Cladothele striarioides
Desmarestia chordalis
Durvillaea antarctica
Ectocarpus constanciae
Geminocarpus geminatus
Gononema pectinatum
Gononema ramosum
Papenfussiella tristanensis
Sphacelaria bornetii

RHODOPHYTA:

*Antarcticophyllum aequabile**
Bostrychia vaga
Callithamnion gracile
Callocolax neglectus
Callophyllis variegata
Cenacrum subsutum
Chaetangium fastigiatum
Cladodonta lyallii
Colacodasya inconspicua

RHODOPHYTA (*cont.*):

Heterosiphonia berkeleyi
Hildenbrandia lecannellieri
Hymenena laciniata
*Leptophytum asperulum**
*Lithophyllum atalyense**
*Lithothamnion neglectum**
Lophurella hookeriana
Medeiothamnion flaccidum
Mesophyllum schmitzii
Myriogramme livida
Plocamium hookeri
Plocamium secundatum
Plumariopsis eatoni
Polycoryne radiata
Polysiphonia anisogona
Porphyra columbia
*Pseudolithophyllum almanense**
Pseudolithophyllum consociatum
Pseudolithophyllum discoideum
Pseudophycodrys phyllophora
Rhodophyllis acanthocarpa
Rhodymenia subantarctica
Schizoseris condensata
Schizoseris dichotoma
Sporoglossum lophurellae

Endemic Species

CHLOROPHYTA:

Derbesia furcata

PHAEOPHYTA:

Dilophus decumbens

RHODOPHYTA:

Antithamnionella alternans
Iridaea remuliformis
*Lithothamnion macquariensis**
*Lithothamnion zaneveldii**

Endemic Species

PHAEOPHYTA (*cont.*):
Haplogloia moniliformis
Lithoderma piliferum var. *debile*
Streblonema codiatile

RHODOPHYTA (*cont.*):
Phymatolithon lenormandii f. *macquariensis**
Porphyra ionae
Porphyra plocamiestris

Insufficiently Known Species

CHLOROPHYTA:

Monostroma sp. 'a'
Monostroma sp. 'b'

RHODOPHYTA:
Porphyra sp.
Dasyptilon sp.

3.4 Biogeography in the Southern Ocean

Biogeography is concerned largely with analysis of distribution patterns of organisms throughout various regions of the world. Breaks in plant and animal distributions that coincide with environmental discontinuities generally are used to formulate biogeographic regions (Knox, 1979). Delineation of sharp geographic boundaries between such regions, however, is usually not possible because of overlapping distributions of some species that are used to typify separate regions (Michanek, 1979). Interruptions in species' distributions can usually be explained by either the tolerance limits of organisms to environmental conditions, various dispersal factors, or degrees of collecting intensity in particular areas. All three explanations are applicable to the distribution of marine algae in the Southern Ocean.

Following Skottsberg's (1960, 1964) biogeographic studies of the Southern Ocean, most subsequent accounts of this region also recognize the Subantarctic as a biologically and physically distinct zone separating Cold Temperate waters to the north from Antarctic waters to the south. The present observation of 46 strictly Subantarctic species of benthic marine algae at Macquarie I. further supports this viewpoint. Three species of brown algae that are prominent in their respective zones also exemplify phytogeographic differences in the Southern Ocean. *Macrocystis pyrifera* is abundant in Cold Temperate and Subantarctic waters but absent from the Antarctic Zone (Nicholson, 1979; pers. obs.), whereas *Durvillaea antarctica* occurs only in the Subantarctic Zone (Hay, 1979*a*; pers. obs.) and *Himantothallus grandifolius* only in the Antarctic Zone (Moe & Silva, 1981; pers. obs.). An additional biological distinction for coastlines in southern high latitudes is the general absence of articulated members of the Corallinaceae (Rhodophyta), which typically occur along most shores in other regions of the world.

If one accepts the Subantarctic as a biologically distinct zone, then questions arise regarding the distributions of organisms within this circumpolar belt. Are plants and animals uniformly distributed, or do they form discrete clusters that would warrant subdivisions of this zone into smaller sections or biological provinces? Previous schemes such as those of Powell (1955, 1960), Knox (1960) and Skottsberg (1960) recognized from 3–4 biological provinces. The present studies on Macquarie's seaweed flora, however, support recognition of one circumpolar phytogeographic province for most Subantarctic islands excepting those immediately south of New Zealand, which appear to have marine floras composed of a large percentage of New Zealand taxa.

Interpretation of relationships between floras is best done by comparing the number of shared species. Endemic taxa, although highly distinctive for particular areas, reveal little, if anything, about floristic relationships between separate localities. Cosmopolitan taxa, for other reasons, also do not clarify floristic relationships. This is because their distributions extend across and blur many biogeographic boundaries. Of the remaining phytogeographic groups that constitute Macquarie's flora, the largest contains species that occur only in the Subantarctic. Consequently, this group is considered to be most appropriate for assessing floristic relationships within the Subantarctic Zone. The distributions of its 46 species have been analyzed to arrive at the present concept of a single, circumpolar phytogeographic province for the entire Subantarctic Zone.

Nearly three quarters of these species also occur in the vicinity of Fuegia and the Falkland Is., and presumably will be found in most intermediate localities following intensive floristic surveys at the various Subantarctic islands situated to the east of South America and west of Macquarie I. Considering the highly homogeneous distribution of seaweed taxa throughout the Subantarctic, there is little support for subdividing this zone into smaller biogeographic regions. The circumpolar distributions of seaweeds in the Southern Ocean are explained most simply by attributing plant dispersal to water currents of the West Wind Drift. Because algal diversity is greater along the shores of the southern tip of South America than anywhere else in the Subantarctic Zone (Santelices & Abbott, 1978; Santelices, 1980), South America is believed to be the main source from whence seaweeds have dispersed to other Subantarctic localities.

The distributions of five species, *Callithamnion gracile*, *Cenacrum subsutum*, *Ptilonia willana*, *Pugetia delicatissima* and *Rhodophyllis acanthocarpa*, suggest some similarity between the marine floras of Macquarie I. and the New Zealand Subantarctic islands. By over-emphasizing the importance of these shared species, one might place all of these islands in a single biogeographic province, a conclusion similar to that reached earlier by Skottsberg (1960). Such an interpretation, however, is contrary to the data (representing distributions of nearly ten times as many taxa) presented above and the opinion held by South (1979*a*), both of which support the hypothesis that Macquarie's flora is controlled largely by West Wind Drift dispersal. Additionally, the findings of Lowry & Fenwick (1983) and Lowry & Stoddart (1983) on amphipod species, which often inhabit benthic algae, show little relationship between the faunas of Macquarie I. and various islands in the New Zealand Subantarctic.

3.5 Conclusions

> . . . for a good observer is one thing, and the knowledge and experience required to make use of facts for purposes of generalization, another: minute differences however, when long dwelt upon, become magnified and assume undue value, and the general botanist must always receive with distrust the conclusions deduced from a few species of a large genus, or from a few specimens of a widely distributed plant.
>
> J. D. Hooker *Flora Novae-Zelandiae*, 1853: xiii

Division of the Southern Ocean into three major zones, Cold Temperate, Subantarctic and Antarctic, is supported by discontinuities in both biological and physical features. Based on seaweed distributions, the various biological subdivisions or provinces of the Subantarctic Zone that were recognized previously do not seem relevant. A single circumpolar phytogeographic region is proposed presently for the Subantarctic Zone because of the largely homogeneous distribution of species throughout this area. Macquarie's marine flora has been divided into six groups based on the breadth of its species' distributions, of which the largest group contains 46 species with ranges extending throughout but not beyond the Subantarctic Zone. A relatively high percentage of endemism (11%) has been observed, but is not unexpected for a remote locality such as Macquarie I. This number of seemingly endemic species, however, will probably be reduced as a result of further studies of other Subantarctic floras. At present, ignorance of plant distributions in many Southern Ocean localities is still the largest obstacle to phytogeographic studies in this region.

References

Abbott, I. A. 1979. Some tropical species related to *Antithamnion* (Rhodophyta, Ceramiaceae). *Phycologia* **18**: 213–227.

Abbott, I. A. & Hollenberg, G. J. 1976. *Marine Algae of California*. vii–xii + 827 pp. Stanford Univ. Press, Stanford, California.

Abbott, I. A. & Norris, R. E. 1966. Studies on *Callophyllis* (Rhodophyceae) from the Pacific Coast of North America. *Nova Hedwigia* **10**: 67–84, 14 pls.

Acleto, C. O. 1973. Las algas marinas del Peru. *Bol. Soc. Peruana Bot.* **6**: 1–164.

Adams, N. M., Conway, E. & Norris, R. E. 1974. The marine algae of Stewart Island—a list of species. *Rec. Dom. Mus.* **8**: 185–245.

Adey, W. H. 1970. A revision of the Foslie crustose coralline herbarium. *K. Norske Vid. Selsk. Skr.* **1970**(1): 46 pp.

Adey, W. H. & Johansen, H. W. 1972. Morphology and taxonomy of Corallinaceae with special reference to *Clathromorphum*, *Mesophyllum*, and *Neopolyporolithon* gen. nov. (Rhodophyceae, Cryptonemiales). *Phycologia* **11**: 159–180.

Agardh, C. A. 1811. Dispositio algarum sueciae, quam publico examini subjiciunt Carl Adolph Agardh Ph. Doct. Bot. Pt. II, pp. 17–26. Lund.

Agardh, C. A. 1820–1823. *Species algarum* . . . 1(1): i–iv + (1)–168 (Lund), **1**(2): v–vi + 169–398 (Lund), vii–viii + (399)–531.

Agardh, C. A. 1824. *Systema algarum*. xxxviii + 312 pp. Lund.

Agardh, C. A. 1828. *Species algarum* . . . **2**: lxxvi + 1–189.

Agardh, J. G. 1841. In historiam algarum symbolae. *Linnaea* **15**: 1–50, 443–457.

Agardh, J. G. 1846. *Anadema*, ett nytt slägte bland Algerne. Öfvers *K. [svenska] Vetensk-Akad. Förh.* **3**: 103–104.

Agardh, J. G. 1851. *Species genera et ordines algarum*. Vol. 2(1) i–xii + 1–351. Gleerup, Lund.

Agardh, J. G. 1852. *Species genera et ordines algarum*. Vol. 2(3:1), 337–786. Gleerup, Lund.

Adardh, J. G. 1863. *Species genera et ordines algarum*. Vol. 2(3:2), 787–1291. Gleerup, Lund.

Agardh, J. G. 1872. Bidrag till Florideernas systematik. Lunds Univ. Arsskr. Afd. 2 **8**(6). 60 pp.

Agardh, J. G. 1876. *Species genera et ordines algarum*. Vol. 3(1), vii + 724 pp. Epicrisis systematis floridearum. Leipzig.

Agardh, J. G. 1878. De Algis Novae Zelandiae marinis. *Lunds Univ. Arsskr. Afd. 2.* **14**(4): 32 pp.

Agardh, J. G. 1882. Till algernes systematik. *Lunds Univ. Arsskr. Afd. 2.* **17**(4): 136 pp., 3 pls.

Agardh, J. G. 1890. Till algernes systematik: xi. Florideae. *Lunds Univ. Arsskr. Afd. 2.* **26**(3): 125 pp., 3 pls.

Agardh, J. G. 1892. Analecta algologica. *Lunds Univ. Arsskr.* Afd. 2. **28**(6): 182 pp., 3 pls.

Agardh, J. G. 1894. Analecta algologica, cont. II. *Lund Univ. Arsskr.* Afd. 2. **30**(7): 99 pp., 1 pl.

Agardh, J. G. 1896. Analecta algologica, cont. III. *Lunds Univ. Arsskr.* Afd. 2 **32**(2): 140 pp., 1 pl.

Agardh, J. G. 1898. *Species genera et ordines algarum*. **3**(3): 239 pp. De Dispositione Delesseriearum mantissa algologica, Lund.

Allender, B. M. 1980. *Dictyotopsis propagulifera* (Phaeophyta)—an algal enigma. *Phycologia* **19**: 234–236.

Allender, B. M. & Kraft, G. T. 1983. The marine algae of Lord Howe Island (New South Wales): The Dictyotales and Cutleriales (Phaeophyta). *Brunonia* **6**: 73–130.

Andrews, J. 1976. Pathology of marine algae. *Biol. Rev.* **51**: 211–253.

Archer, A. & Burrows, E. M. 1960. Heteromorphic life-histories as a family criterion in the Cladophorales. *Br. phycol. Bull.* **2**: 31–33.

Areschoug, J. E. 1843. Algarum (Phycearum) minus rite cognitarum pugillus secundus. *Linnaea* **17**(3): 257–269, pl. 9.

Areschoug, J. E. 1850. Phycearum, quae in maribus Scandinaviae crescunt, enumeratio. Sectio posterior Ulvaceas continens. *Nova Acta Soc. Sci. Upsal.* Ser. 2. **14**: 385–454, 3 pls.

Areschoug, J. E. 1866. Observationes phycologicae. Particula prima. De Confervaceis nonnullis. *Nova Acta Soc. Sci. Upsal.* Ser. 3. **6**: 26 pp., 4 pls.

Areschoug, J. E. 1874. Observationes phycologicae. Particula secunda. De Urospora mirabili Aresch. et de Chlorozoosporarum coplatione. *Nova Acta Soc. Sci. Upsal.* Ser. 3. **9**(1): 12 pp.

Areschoug, J. E. 1875. Observationes phycologicae. Particula tertia. De algis nonnullis scandinavicis et de conjuctione Phaeozoosporarum Dictyosiphonis hippuroidis. *Nova Acta Soc. Sci. Upsal.* Ser. 3 **10**(1): 36 pp., 3 pls.

Arnaud, P. & Delépine, R. 1964. Une nouvelle rhodophycée marine de Terre Adelie (Antarctique) *Rhodoglossum Schotteri* Delépine (Gigartinale). *CNFRA-Biologie* **1**(3): 79–91.

Asensi, A. 1966*a*. Guia para reconocer los géneros de algas pardas de la Argentina. *Contr. Inst. Antar. Arg.* **103**: 51 pp.

Asensi, A. 1966*b*. La presencia de Pylaiella littoralis (L.) Kjellman en la Antartida. *Contr. Inst. Antar. Arg.* **101**: 14 pp.

Asensi, A. 1971. Un orden de algas pardas nuevo para la Argentina (Cutleriales). *Darwiniana* **16**: 435–442.

Asensi, A. 1975. La estructura, la distribucion y el cultivo de *Scytothamnus fasciculatus* (Hook. et Harv.) Cotton (Phaeophyta). *Physis* (A) **34**: 269–282.

Asensi, A. 1976. Observaciones sobre *Cladothele decaisnei* Hook. et Harv. (Phaeophyta). *Physis* (A) **35**: 1–14.

Ashton, D. H. 1965. Regeneration pattern of *Poa foliosa* Hook. f. on Macquarie Island. *Proc. R. Soc. Vic.* **79**: 215–234.

Ashton, D. H. & Gill, A. M. 1965. Pattern and process in a Macquarie Is. feldmark. *Proc. R. Soc. Vic.* **79**: 235–247.

Askenasy, E. 1888. Algen mit Unterstützung der Herren E. Bornet, E. Grunow, P. Hariot, M. Moebius, O. Nordstedt bearbeitet. *In* A. Engler (Ed.), *Forschungsreise S.M.S. 'Gazelle'* Theil **4**: Bot. 58 pp., 12 pls. Berlin.

Baardseth, E. 1941. *The Marine Algae of Tristan da Cunha* (Results of the Norwegian Scientific Expedition to Tristan da Cunha 1937–1938, No. 9). 173 pp. Oslo.

Bailey, J. W. & Harvey, W. H. 1874. Algae. In *United States Exploring Expedition . . . 1838–1842*. Vol 17, 153–192, 9 pls. Sherman & Co. Philadelphia.

Barber, H. N., Dadswell, H. E. & Ingle, H. D. 1959. Biology: transport of driftwood from South America to Tasmania and Macquarie Is. *Nature* **184**: 203–204.

Batters, E. A. L. 1895. On some new British marine algae. *Ann. Bot., Lond.* **9**: 307–321, pl. 11.

Bennett, I. 1971. *Shores of Macquarie Island.* Rigby Limited, Adelaide.

Black, D. 1931. Paleogeography and polar shift. A study of hypothetical projections. *Bull. Geol. Soc. China* **10**: 105–157.

Blackman, F. F. & Tansley, A. G. 1902. A revision of the classification of the green algae. *New Phytol.* **1**: 17–24, 47–48, 67–72, 89–96, 114–120, 133–144, 163–168, 189–192, 213–220, 238–244.

Bliding, C. 1928. Studien über die Florideenordnung Rhodymeniales. *Lunds Univ. Arsskr. N.F.* Avd. 2. **24**(3): 74 pp.

Bliding, C. 1963. A critical survey of European taxa in Ulvales. I. *Capsosiphon, Percursaria, Blidingia, Enteromorpha. Opera Bot.* **8**(3): 160 pp.

Bliding, C. 1968. A critical survey of European taxa in Ulvales. II. *Ulva, Ulvaria, Monostroma, Kornmannia. Bot. Notiser* **121**: 535–629.

Bold, H. C. & Wynne, M. J. 1978. *Introduction to the Algae.* 706 pp. Prentice-Hall Inc., Englewood Cliffs, N. J.

Bonin, D. R. 1981. *Systematics and life histories of New Zealand Bonnemaisoniaceae (Rhodophyta).* i–xviii + 126 pp. MSc Thesis, Univ. Auckland.

Boraso, A. L. 1975. Los géneros *Enteromorpha, Blidingia,* y *Percursaria* (Algae—Chlorophyta) en las costas atlanticas argentinas. *Darwiniana* **19**: 285–311.

Boraso, A. L. 1977. El género *Ulva* (Algae, Chlorophyta) en Argentina. I. *Ulva* en Puerto Deseado (Provincia de Santa Cruz). *Darwiniana* **21**: 162–171.

Bory de Saint-Vincent, J. B. 1823. *Dictionnaire Classique d'Histoire Naturelle* **4**: 393. Paris.

Bory de Saint-Vincent, J. B. 1826*a*. *Durvillea ('Durvillaea'). Dictionnaire Classique d'Histoire Naturelle* **9**: 192–193. Paris.

Bory de Saint-Vincent, J. B. 1826*b*. Cryptogamia. *In* J. Dumont-D'Urville (Ed.), Flore des Malouines. *Mem. Soc. Linn. Paris,* **4**: 573–621.

Bory de Saint-Vincent, J. B. 1826*c*. *Iridea. Dictionnaire Classique d'Histoire Naturelle* **9**: 15–16. Paris.

Bory de Saint-Vincent, J. B. 1827–1829. Cryptogamie. *In* L. I. Duperrey, *Voyage autour du monde . . . 'La Coquille'*, pp. 1–96 (1827), 97–200 (1928), 201–301 (1829), atlas of 38 pls. Paris.

Brauner, J. F. 1979. Developmental morphology and taxonomy of *Medeiothamnion flaccidum* (Hooker & Harvey) comb. nov. (Ceramiaceae, Rhodophyta) from South America. *Phycologia* **18**: 338–346.

Bunt, J. S. 1965. Observations on the fungi of Macquarie Is. *ANARE Sci. Rep. Ser. B,* Vol. 2. Bot. No. **78**: 22 pp.

Cameron, I. 1974. *Antarctica. The last continent.* 256 pp. Little, Brown & Co., Boston.

Cantacuzène, A. 1930. *Contribution à l'étude des tumeurs bacteriennes chez les Algues marines.* Thèse presentée à la Faculté des Sciences de l'Université de Paris.

Castagne, L. 1851. Supplément au catalogue des plantes qui croissent naturellement aux environs de Marseille. Aix, Imprimerie de Nicot et Pardigon, sur le Cours, **55**: 125 pp., pls. 8–11.

Cerro, M. del, Cogen, J. & Cerro, C. del. 1980. Stevenel's Blue, an excellent stain for optical microscopical study of plastic embedded tissues. *Microscopica Acta* **83**: 117–121.

Chadefaud, M. 1960. *Traite de botanique systematique.* Tome I. Les vegetaux non vasculaires (Cryptogamie). 1016 pp. Paris.

Chamberlain, Y. M. 1963. The identity of *Monostroma endiviifolium* A. and E. S. Gepp. *Nova Hedwigia* **5**: 151–155.

Chapman, A. R. O. 1972. Morphological variation and its taxonomic implications in the ligulate members of the genus *Desmarestia* occurring on the west coast of North America. *Syesis* **5**: 1–20.

Chapman, A. R. O. & Burrows, E. M. 1971. Field and culture studies of *Desmarestia aculeata* (L.) Lamour. *Phycologia* **10**: 63–76.

Chapman, V. J. 1956. *The marine algae of New Zealand.* Pt. I. Myxophyceae and Chlorophyceae. *J. Linn. Soc., Bot.* **55**: 333–500.

Chapman, V. J. 1969. *The marine algae of New Zealand.* Pt. II. Rhodophyceae. Issue 1: Bangiophyceae and Florideophycidae (Nemalionales, Bonnemaisoniales, Gelidiales). pp. 1–113. Cramer, Lehre.

Chapman, V. J. 1979. *The marine algae of New Zealand.* Pt. III. Rhodophyceae. Issue 4: Gigartinales. pp. 279–506. Cramer, Lehre.

Chapman, V. J. & Chapman, D. J. 1973. *The Algae.* 2nd edn, 497 pp. Macmillan Press Ltd.

Chapman, V. J. & Dromgoole, F. I. 1970. *The marine algae of New Zealand.* Pt. III. Rhodophyceae. Issue 2: Florideophycidae: Rhodymeniales. pp. 115–154. Cramer, Lehre.

Chapman, V. J. & Parkinson, P. G. 1974. *The marine algae of New Zealand.* Pt. III. Rhodophyceae. Issue 3: Cryptonemiales. pp. 155–278, pls 51–94. Cramer, Lehre.

Cheeseman, T. F. 1919. The vascular flora of Macquarie Island. Australasian Antarctic Expedition 1911–14 . . . *Sci. Rep. Ser. C: Zoo. and Bot.* **7**(3): 1–63.

Chemin, E. 1937*a*. Le développement des spores chez les Rhodophycées. *Rev. gén. Bot.* **49**: 300–327.

Chemin, E. 1937*b*. Rôle des bacteries dans la formation des galles chez les Floridées. *Annls Sci. nat.* ser. 10, Bot. **19**: 61–71.

Chen, L. C.-M. 1977. The sporophyte of *Ahnfeltia plicata* (Huds.) Fries (Rhodophyceae, Gigartinales) in culture. *Phycologia* **16**: 163–168.

Chiang, Y.-M. 1970. Morphological studies of red algae of the family Cryptonemiaceae. *Univ. Calif. Publs Bot.* **58**: 1–95.

Chihara, M. & Yoshizaki, M. 1971. Morphological study of the thallus and reproductive organs of *Delisea fimbriata* (Florideophycidae, Rhodophyta). *Phycologia* **10**: 149–154.

Chihara, M. & Yoshizaki, M. 1972. Bonnemaisoniaceae: their gonimoblast development, life history, and systematics. *In* I. A. Abbott & M. Kurogi (Eds), *Contributions to the Systematics of Benthic Marine Algae of the North Pacific.* pp. 243–251. Jap. Soc. Phycol., Kobe.

Chihara, M. & Yoshizaki, M. 1978. Anatomical and reproductive features of *Ptilonia okadai* (Rhodophyta, Bonnemaisoniaceae). *Phycologia* **17**: 382–387.

Chilton, C. 1909. The biological relations of the sub-Antarctic islands of New Zealand: summary of results. *In* C. Chilton (Ed.), *The Sub-Antarctic Islands of New Zealand.* Vol. 2, pp. 793–807. Wellington.

Chittleborough, R. G. 1956. The settlement of marine organisms on submerged surfaces at Heard Island, 1949. *ANARE Interim Reports* **15**: 67–81.

Choris, L. 1822. *Voyage pittoresque autour du monde . . .* [Part I]. vi + 17pp., 12pls. Paris.

Clayton, M. N. 1974. Studies on the development, life history, and taxonomy of the Ectocarpales (Phaeophyta) in southern Australia. *Aust. J. Bot.* **22**: 743–813.

Clayton, M. N. 1981. Phaeophyta. *In* M. N. Clayton & R. J. King (Eds), *Marine Botany An Australian Perspective.* pp. 104–137. Longman Cheshire, Melbourne.

Clifford, H. T. 1953. The mosses of Macquarie Is. and Heard Is. *ANARE Rep. Ser. B,* Vol. 2, Bot.: 12 pp., 1 pl.

Codomier, L. 1978. Recherches sur les *Kallymenia* (Cryptonemiales) III. Problèmes de la délimitation du genre *Kallymenia. Rev. algol., N.S.* **13**: 241–249.

Colhoun, E. A. & Goede, A. 1973. Fossil penguin bones, ^{14}C dates and the raised marine terrace of Macquarie Island: some comments. *Search* **4**: 499–501.

Conway, E. & Knaggs, F. W. 1966. Contributions to our knowledge of the genus *Rhodochorton:* I. *R. purpureum. In* H. Barnes (Ed.), *Some Contemporary Studies in Marine Science.* pp. 195–203. London.

Conway, E., Mumford, T. F. & Scagel, R. F. 1975. The genus *Porphyra* in British Columbia and Washington. *Syesis* **8**: 185–244.

Cotton, A. D. 1908. xv.—The New Zealand species of *Rhodophyllis. Kew, R. Bot. Gardens, Bull. Misc. Info.* **3**: 97–102.

Cotton, A. D. 1915. Cryptogams from the Falkland Islands collected by Mrs. Vallentin. *J. Linn. Soc., Bot.* **43**: 137–231, pls 4–10.

Cribb, A. B. 1954. The algal vegetation of Port Arthur, Tasmania. *Pap. & Proc. R. Soc. Tas.* **88**: 1–44, 10 pls.

Cumpston, J. S. 1968. *Macquarie Island.* xii + 380 pp. Govt. Printing Office, Canberra.

Currie, R. I. (Ed.) 1968. *Symposium on Antarctic Oceanography.* [SCAR/SCOR/IAPO/IUBS] Santiago, Chile, 13–16 September 1966. xi + 268 pp. Scott Polar Res. Inst. Cambridge.

Dawson, E. Y. 1941. Field observations on the algae of the Gulf of California. A review of the genus *Rhodymenia* with descriptions of new species. *Allan Hancock Pac. Exped.* **3**: 115–181, 13 pls.

Dawson, E. Y. 1944. Some new and unreported sublittoral algae from Cerros Island, Mexico. *Bull. S. Calif. Acad. Sci.* **43**: 102–112.

Decaisne J. 1842. Essais sur une classificatione des Algues et des Polypiers calcifères de Lamouroux *Ann. Sci. Nat. Bot.* Ser. 2 **17**: 297–380.

Debenham, F. 1945. *The Voyage of Captain Bellingshausen to the Antarctic Seas 1819–1821.* (English translation). Vols 1 & 2, vii–xxx + 474 pp. Hakluyt Soc., London.

DeCew, T. C. & West, J. A. 1977. Culture studies on the marine red algae *Hildenbrandia occidentalis* and *H.*

prototypus (Cryptonemiales, Hildenbrandiaceae). *Bull. Jap. Soc. Phycol.* **25**(Suppl. Mem. Iss. Yamada): 31–41.

DeCew, T. C. & West, J. A. 1982. A sexual life history in *Rhodophysema* (Rhodophyceae), a reinterpretation. *Phycologia* **21**: 67–74.

Delépine, R. 1966. La végétation marine dans l'Antarctique de l'Ouest comparée à celle des Îles australes françaises. Conséquences biogéographiques. *C. R. Soc. Biogeogr.* **374**: 52–68.

Delépine, R. 1967. Sur un nouveau genre de Chlorophycées antarctiques, *Lambia. C. R. Acad. Sci. Paris.* **264**(D): 1410–1413.

Delépine, R. 1968. Sur la morphologie et la mode de croissance de deux Algues marines antarctiques. *Mem. Soc. Bot. France.* **115**: 120–129.

Delépine, R. & Asensi, A. 1975. *Asteronema* nov. gen. nouveau genre de Phéophycée australe. *Bull. Soc. Bot. France.* **122**: 295–304.

Delépine, R. & Asensi, A. 1978. Réactions écophysiologiques et variations morphogénétiques chez *Adenocystis* et *Utriculidium* (Phéophycées). *Rev. algol., N. S.* **13**: 43–45.

Delépine, R., Asensi, A. & Guglielmi, G. 1976. Nouveau types d'ultrastructure plastidiale chez les Phéophycées. *Phycologia* **15**: 425–434.

Delépine, R., Delesalle, B. & Lambert, C. 1979. [Algologie.] Sur l'existence d'un tétrasporophyte dans le cycle de la Rhodophycée *Chaetangium fastigiatum* (Bory) J. Ag. aux îles Kerguelen. *C. R. Acad. Sc. Paris* **289**: 595–598.

Delépine, R., Lamb, I. M. & Zimmermann, M. H. 1966. Preliminary report on the marine vegetation of the Antarctic Peninsula. *Proc. 5th Intl. Seaweed Symp.* **5**: 107–116.

Delépine, R., Lamb, I. M. & Zimmermann, M. H. 1970. Sur les algues marines antarctiques rapportées au genre *Monostroma* Thuret. *C. R. Acad. Sci. Paris* **270**: 1973–1976.

Delépine, R. & Lambert, C. 1981. Botanique.-L'appareil femelle de la Rhodophycée *Chaetangium fastigiatum* et l'existence d'un gonophore chez cette espèce. *C. R. Acad. Sci. Paris.* **292**: 127–130.

Dell, R. K. 1972. Antarctic benthos. *Advances in Marine Biology* **10**: 1–216.

Denizot, M. 1968. *Les Algues Floridées Encroutantes.* 310 pp. Lab. de Cryptogamie, Mus. Nat. Hist., Paris.

Derbès, [A. A.], & Solier, [A. J. J.]. 1850. Sur les organes reproducteurs des Algues. *Ann. Sci. Nat. Bot.* III, **14**(5): 261–282, pls 32 37.

Desikachary, T. V. 1963. Status of the order Chaetangiales (Rhodophyta). *J. Indian Bot. Soc.* **42A**: 16–26.

De Toni, G. B. 1891. Systematische Übersicht der bisher bekannten Gattungen der echten Fucoideen. *Flora* **74**: 171–182.

De Toni, G. B. 1895. *Sylloge algarum.* Vol. 3 (Fucoideae). xvi + 638 pp. Padua.

De Toni, G. B. 1897. *Sylloge algarum.* Vol. 4(1). (Florideae). xx + lxi + 386 pp.

De Toni, G. B. 1900. *Sylloge algarum.* Vol. 4(2). (Florideae). pp. 387–774.

De Toni, G. B. 1903. *Sylloge algarum.* Vol. 4(3). (Florideae). pp. 775–1521.

De Toni, G. B. 1905. *Sylloge algarum.* Vol. 4(4). (Florideae). pp. 1523–1973.

De Toni, G. B. 1924. *Sylloge algarum.* Vol. 6(5). (Florideae). pp. 1–767.

De Toni, G. B. & Forti, A. 1922. Alghe di Australia, Tasmania e Nuova Zelanda. *Mem. R. Ist. Veneto Sc. Lett. ed Arti* **29**(3): 183 pp., 10 pls.

Dickie, G. 1876*a*. Notice of some marine algae from Kerguelen Island. *J. Bot.* **14**: 50–51.

Dickie, G. 1876*b*. Algae collected by Mr. Moseley at Marion Island, in 40 fathoms. *J. Linn. Soc., Bot.* **15**: 42–43.

Dickie, G. 1876*c*. Marine algae collected by Mr. Moseley at the island of Kerguelen. *J. Linn. Soc., Bot.* **15**: 43–47.

Dickie, G. 1876*d*. Algae collected by Mr. Moseley at Heard Island, 250 miles s. of Kerguelen. *J. Linn. Soc., Bot.* **15**: 47–48.

Dickie, G. 1876*e*. Notes on algae found at Kerguelen Land by the Rev. A. E. Eaton. *J. Linn. Soc., Bot.* **15**: 198–204.

Dickie, G. 1877. Supplemental notes on algae collected by N. H. Moseley, M.A., of H.M.S. Challenger, from various localities. (4) Algae from Port Louis, Falkland Islands, and Magellan Straits. *J. Linn. Soc., Bot.* **15**: 487.

Dickie, G. 1879. Marine algae (exclusive of the Diatomaceae). *In* An account of the petrological, botanical, and zoological collections made in Kerguelen's Land and Rodriguez during the Transit of Venus Expeditions . . . 1874–75. The collections from Kerguelen Island. *Phil. Trans. R. Soc.* **168**(extra vol.): 53–54.

Dillwyn, L. W. 1802–1809. *British Confervae.* 87 pp., 115 pls. London.

Dixon, P. S. 1960. Notes on important algal herbaria, II. The herbarium of John Ellis (? 1710–1776). *Br. Phycol. Bull.* **2**(1): 28–31.

Dixon, P. S. 1964. Taxonomic and nomenclatural notes on the Florideae iv. *Bot. Notiser* **117**: 56–78.

Dixon, P. S. 1966. Notes on important algal herbaria, iv. The herbarium of Lewis Weston Dillwyn (1778–1885). *Br. Phycol. Bull.* **3**: 19–22.

Dixon, P. S. 1967. The typification of *Fucus cartilagineus* L. and *F. corneus* Huds. *Blumea* **15**: 55–62.

Dixon, P. S. 1973. *Biology of the Rhodophyta.* xiii + 251 pp. Hafner Press, New York.

Dixon, P. S. 1982. Rhodophycota. *In* S. P. Parker (Ed.), *Synopsis and Classification of Living Organisms. pp. 61–79. McGraw-Hill, New York.*

Dixon, P. S. & Irvine, L. M. 1977. *Seaweeds of the British Isles.* Vol. 1 Rhodophyta. Pt. 1 Introduction,

Nemaliales, Gigartinales. xi + 252 pp. Br. Mus. (Nat. Hist.). London.
Dixon, P. S. & Price, J. H. 1981. The genus *Callithamnion* (Rhodophyta: Ceramiaceae) in the British Isles. *Bull. Br. Mus. nat. Hist.* (Bot.) **9**(2): 99–141.
Dodge, C. W. & Rudolph, E. D. 1955. Lichenological notes on the flora of the Antarctic Continent and the subantarctic islands. i–iv. *Ann. Missouri Bot. Gard.* **42**: 131–149.
Doty, M. S. 1947. The marine algae of Oregon. I. Chlorophyta and Phaeophyta. *Farlowia* **3**: 1–65, 10 pls.
Drew, K. 1928. A revision of the genera *Chantransia, Rhodochorton,* and *Acrochaetium. Univ. Calif. Publs Bot.* **14**: 139–224.
Duby, J. E. 1830. Bot. Gall. Pars secunda plantes Cellulares continens. pp. (vi) + 545–1068 + lviii. Paris.
Dumortier, B.-C. 1822. Commentationes botanicae. Observations botaniques, dédiées à la Société d'Horticulture de Tournay. 116[117] pp. Tournay.
Earle, S. A. 1969. Phaeophyta of the eastern Gulf of Mexico. *Phycologia* **7**: 71–254.
Edelstein, T., Chen, L. C.-M. & McLachlan, J. 1970. The life cycle of *Ralfsia clavata* and *R. borneti. Can. J. Bot.* **48**: 527–531.
Ellis, J. 1768. De sex Confervarum speciebus. *Phil Trans.* **57**: 426 pp.
Erskine, D. 1955. Reproduction and affinities of *Dasyptilon* (Ceramiaceae,Rhodophyceae). *Pac. Sci.* **9**: 292–297.
Esper, E. J. C. 1788–1830. *Die Pflanzenthiere in Abbildungen nach der Natur mit Farben erleuchtet nebst Beschreibungen.* Vol. 1, 320 pp. + 149 pls, Vol. 2, 305 pp. + 130 pls; Vol. 3, 407 pp. + 19 pls.
Etcheverry, H. D. 1960. Algas marinas de las islas oceanicas Chilenas (Juan Fernandez, San Felix, San Ambrosio, Pascua). *Rev. Biol. Mar.* **10**: 83–132, 6 pls.
Falkenberg, P. 1901. Die Rhodomelaceen des Golfes von Neapel und der angrenzenden Meeresabschnitte. *Fauna Flora Golf. Neapel* **26**: xvi + 754 pp., pls 1–24.
Fan, K.-C. 1961. Morphological studies of the Gelidiales. *Univ. Calif. Publs Bot.* **32**: 315–368.
Farlow, W. G. 1881. The marine algae of New England. *Rep. U.S. Fish. Commiss.* **1879**. 210 pp.
Farr, E. R., Leussink, J. A. & Stafleu, F. A. (Eds). 1979. Index Nominum Genericorum (Plantarum). *Reg. Veg.* **100–102**: 1–1896.
Feldmann, G. 1950. Sur quelques Céramiacées de Nouvelle-Zelande (suite). *Bull. Mus. Hist. Nat. (Paris)* Ser. 2. **22**: 307–314.
Feldmann, J. 1937. Les algues marines de la côte des Albères. I–III. Cyanophycées, Chlorophycées, Pheophycées. *Rev. algol.* **9**: 141–335, pls. 8–17.
Feldmann, J. 1952. Les cycles de reproduction des Algues et leurs rapports avec la phylogenie. *Rev. Cytol. et Biol. Veget.* **13**(1/2): 4–44.
Feldmann, J. 1953. L'évolution des organes femelles chez les Floridées. *Proc. First Intl. Symp. Edinburgh, 1952.* pp. 11–12. Inst. Seaw. Res. Inveresk, Scotland.
Feldmann, J. & Feldmann, G. 1942. Recherches sur les Bonnemaisoniacées et leur alternance de générations. *Ann. Sci. Nat. Bot.* Ser. 11, **3**: 75–175.
Feldmann, J. & Feldmann, G. 1951. Un nouveau genre de Rhodophycée parasite d'une Delesseriacée. *C. R. Acad. Sci. Paris* **233**: 1137–1139.
Feldmann, J. & Feldmann, G. 1958. Recherches sur quelques Floridées parasites. *Rev. gén. Bot.* **65**: 49–124, 2 pls.
Feldmann-Mazoyer, G. 1940. *Recherches sur les Céramiacées de la Mediterranée occidentale.* 510 pp., 4 pls. Algiers.
Fletcher, R. L. 1978. Studies on the family Ralfsiaceae (Phaeophyta) around the British Isles. *In* D. E. G. Irvine & J. H. Price (Eds), *Modern Approaches to the Taxonomy of Red and Brown Algae.* Systematics Association Special Vol. 10. pp. 371–398. Academic Press, London.
Foslie, M. 1900. Calcareous algae from Fuegia. *In* O. Nordenskjöld (Ed.), *Wissenschaftliche Ergebnisse der schwedischen Expedition nach den Magellansländern 1895–97.* Vol. 3. pp. 65–75. Stockholm.
Foslie, M. 1905. Den botaniske samling. *K. Norske Vid. Selsk. Skr.* **1904**(6): 15–18.
Foslie, M. 1906. Algologiske notiser. II. *K. Norske Vid. Selsk. Skr.* **1906**(2): 1–28.
Foslie, M. 1909. Algologiske notiser. VI. *K. Norske Vid. Selsk. Skr.* **1909**(2): 1–63.
Foslie, M. 1912. Calcareous algae. *In* W. S. Bruce (Ed.), *Report of the Scientific Results of the Voyage S.Y. 'Scotia' During the Years 1902, 1903, and 1904.* Vol. 3: Botany. pp. 89–91. Edinburgh.
Frakes, L. A. & Kemp, E. M. 1972. Influence of continental positions on early tertiary climates. *Nature* **240**: 97–100.
Friedmann, I. & Manton, I. 1960. Gametes, fertilization and zygote development in *Prasiola stipitata* Suhr. I. Light microscopy. *Nova Hedwigia* **1**: 443–462.
Fries, E. 1835. *Flora Scanica. Corpus florarum provincialium Sueciae.* xiv + 394 pp. Upsaliae.
Fritsch, F. E. 1935. *The Structure and Reproduction of the Algae.* Vol. 1, 791 pp. Cambridge Univ. Press.
Fritsch, F. E. 1945. *The Structure and Reproduction of the Algae.* Vol. 2, xii + 939 pp. Cambridge Univ. Press.
Gain, L. 1911*a*. Une nouvelle espèce de *Monostroma* prouvenant de la région antarctique sud-américaine. *C. R. Acad. Sci. Paris.* **152**: 724–726.
Gain, L. 1911*b*. Note sur trois espèces nouvelles d'algues marines prouvenant de la région antarctique sud-américaine. *Bull. Mus. Hist. Nat. (Paris)* **17**: 482–484.

Gain, L. 1912. La flore algologique des régions antarctiques et subantarctiques. *In* J. Charcot (Ed.), *Deuxième Expédition Antarctique Française (1908–1910)*. 218 pp., 5 pls. Sciences Naturelles, Paris.
Garbary, D. J. 1978. On the phylogenetic relationships of the Acrochaetiaceae (Rhodophyta). *Br. phycol. J.* **13**: 247–254.
Garbary, D. J. 1979. Numerical taxonomy and generic circumscription in the Acrochaetiaceae (Rhodophyta). *Bot. Mar.* **22**: 477–492.
Garbary, D. J., Hansen, G. I. & Scagel, R. F. 1980*a*. A revised classification of the Bangiophyceae (Rhodophyta). *Nova Hedwigia* **33**: 145–166.
Garbary, D. J., Hansen, G. I. & Scagel, R. F. 1980*b*. The marine algae of British Columbia and northern Washington: Division Rhodophyta (red algae), Class Bangiophyceae. *Syesis* **13**: 137–195.
Garbary, D. J., Hansen, G. I. & Scagel, R. F. 1982. The marine algae of British Columbia and northern Washington: Division Rhodophyta (red algae), Class Florideophyceae, Orders Acrochaetiales and Nemaliales. *Syesis* **15**(Suppl. 1): 102 pp.
Gardner, N. L. 1909. New Chlorophyceae from California. *Univ. Calif. Publs Bot.* **3**: 371–375.
Gaudichaud, C. 1825. (see Mirbel).
Gaudichaud, C. 1826. Botanique. *In* L. Freycinet (Ed.), *Voyage autour du monde, entrepons par ordre du Roi, . . . exécuté sur les corvettes de S. M. l''Uranie' et la 'Physicienne', pendant les années 1817, 1818, 1819 et 1820*. vii + 522 pp., atlas of 96 pp. Paris.
Gepp, A. & Gepp, E. S. 1905*a*. Antarctic algae. *J. Bot.* **43**: 105–109, pl. 470.
Gepp, A. & Gepp, E. S. 1905*b*. *Leptosarca*: a correction. *J. Bot.* **43**: 162.
Gepp, A. & Gepp, E. S. 1905*c*. More Antarctic algae. *J. Bot.* **43**: 193–196, pl. 426.
Gepp, A. & Gepp, E. S. 1907. *Marine Algae*. I. Phaeophyceae and Florideae. [British] National Antarctic Expedition, Natural History, Vol. 3 (Art. 16) 15 pp., 4 pls. London.
Gepp, A. & Gepp, E. S. 1912. Marine algae of the Scottish National Antarctic Expedition. *In* W. S. Bruce (Ed.), *Report of the Scientific Results of the Voyage of S. Y. 'Scotia' During the Years 1902, 1903, and 1904* . . . Vol 3: Botany. pp. 73–83, 2 pls. Edinburgh.
Godley, E. J. 1965. Botany of the southern zone. Exploration to 1843. *Tuatara* **13**: 140–181.
Goff, L. J. 1982. The biology of parasitic red algae. *In* F. E. Round & D. J. Chapman (Eds), *Progress in Phycological Research*, Vol. 1, pp. 289–369. Elsevier Biomedical Press B. V.
Gordon, E. M. 1972. Comparative morphology and taxonomy of the Wrangelieae, Spondylothamnieae, and Spermothamnieae (Ceramiaceae, Rhodophyta). *Aust. J. Bot.*, Suppl. Ser. 4. 180 pp.
Gordon-Mills, E. [M.] 1977. Two new species of marine algae from Stewart Island, New Zealand, *Medeiotham nion norrisii* and *Ptilothamnion rupicolum* (Ceramiaceae, Rhodophyta). *Phycologia* **16**: 79–85.
Gordon-Mills, E. M. & Kraft, G. T. 1981. The morphology of *Radiathamnion speleotis* gen. et sp. nov., representing a new tribe in the Ceramiaceae (Rhodophyta) from southern Australia. *Phycologia* **20**: 122–130.
Greville, R. K. 1824. *Flora edinensis* lxxxi + 478 pp. Edinburgh.
Greville, R. K. 1827. *Scottish cryptogamic flora*. Vol. V. t.p. + pls 241–300. Edinburgh.
Greville, R. K. 1830. *Algae Britannicae*. lxxxviii + 218 pp., 19 pls. MacLachlan & Stewart, Edinburgh.
Guiry, M. D. 1974. A preliminary consideration of the taxonomic position of *Palmaria palmata* (Linnaeus) Stackhouse = *Rhodymenia palmata* (Linnaeus) Greville. *J. mar. biol. Ass. U.K.* **54**: 509–528.
Guiry, M. D. 1975. An assessment of *Palmaria palmata* forma *mollis* (S&G) comb. nov. (=*Rhodymenia palmata* forma *mollis* S. et G.) in the eastern North Pacific. *Syesis* **8**: 245–261.
Guiry, M. D. 1977. Studies on marine algae of the British Isles. 10. The genus *Rhodymenia*. *Br. phycol. J.* **12**: 385–425.
Guiry, M. D. 1978*a*. Notes on some family names of Florideophyceae (Rhodophyta). *Taxon* **27**: 191–195.
Guiry, M. D. 1978*b*. The importance of sporangia in the classification of the Florideophycideae. *In* D. E.G. Irvine & J. H. Price (Eds), *Modern Approaches to the Taxonomy of Red and Brown Algae*. Systematics Association Special Vol. 10, pp. 111–144. Academic Press, London.
Guiry, M. D. 1982. *Devaleraea*, a new genus of the Palmariaceae (Rhodophyta) in the North Atlantic and North Pacific. *J. mar. biol. Ass. U.K.* **62**: 1–13.
Hamel, G. 1928. Sur quelques *Porphyra* des mers australes. *Ann. Cryptog. Exot.* **1**: 51–57.
Hamel, G. 1931–1939. *Phéophycées de France*. xlvii + 432 pp., 10 pls. Paris
Hamel, G. & Lemoine, P. 1953. Corallinacées de France et d'Afrique du Nord. *Archs Mus. nat. Hist. Paris*, Ser. 7, **1**: 15–136
Hamilton, A. 1894. Notes on a visit to Macquarie Island. *Trans N. Z. Inst.* **27**: 560–579.
Hamilton, H. 1926. Ecological notes and illustrations of the flora of Macquarie Island. Government Printer, NSW, A. A. E. 1911–1914. *Sci. Rep. Ser. C, Zool. & Bot.* **7**(5): 1–10, 19 pls.
Hansen, G. I. 1980. A morphological study of *Fimbrifolium*, a new genus in the Cystocloniaceae (Gigartinales, Rhodophyta). *J. Phycol.* **16**: 207–217.
Hariot, P. 1887*a*. Algues Magellaniques nouvelles. *J. Bot.* **1**: 55–59, 72–74.
Hariot, P. 1887*b*. Note sur le genre *Mastodia*. *J. Bot.* **1**: 231–234.
Hariot, P. 1889. *Algues. Mission scientifique de Cap Horn, 1882–1883*. Vol. 5: Botanique. pp. 3–109, pls 1–9. Paris.
Hariot, P. 1892. Complément à la flore algologique de la Terre de Feu. *Notarisia* **7**: 1427–1435.

Hariot, P. 1895. Nouvelle contribution à l'étude des algues de la région Magellanique. *Journ. de Bot.* **9**: 95–99.

Hariot, P. 1907. Algues. *In* J. Charcot (Ed.), *Expedition Antarctique Française (1903–1905).* Sciences Naturelles . . . Botanique. Fasc. 1, Pt. 2. 9 pp. Paris.

Hartog, C. den. 1959. *The epilithic algal communities occurring along the coast of the Netherlands.* pp xi + 241. Academisch proefschrift. Amsterdam, North-Holland.

Harvey, W. H. 1840. Description of *Ballia,* a new genus of algae. *J. Bot.* **2**: 190–193, pl. 60.

Harvey, W. H. 1847. *Nereis Australis or Algae of the Southern Ocean.* vii + 124 pp. Reeve, London.

Harvey, W. H. 1849–1851. *Phycologia Britannica.* Vol. 2, pp. (i)–vi + pls 121–240 & Vol. 3, pp. (i)–xlv + pls 241–360.

Harvey, W. H. 1849*a*. *A Manual of the British Marine Algae: Containing Generic and Specific Descriptions of All the Known British Species of Sea-weeds.* lii + 252 pp. London.

Harvey, W. H. 1855. Algae. *In* J. D. Hooker (Ed.), *The Botany of the Antarctic Voyage of H. M. Discovery Ships 'Erebus' and 'Terror' in the Years 1839–1843, under the command of Captain Sir James Clark Ross . . . Vol. II. Flora Novae-Zelandiae. Pt. II Flowerless plants.* pp. 211–266, pls 107–121. London.

Harvey, W. H. 1858. *Phycologica Australica.* Vol. 1, xi, viii pp., pls i–lx. Lovell Reeve & Co., London.

Harvey W. H. 1860. Algae. *In* J. D. Hooker (Ed.), *The Botany of the Antarctic Voyage of H. M. Discovery Ships 'Erebus' and 'Terror' in the Years of 1839–1843, under the command of Captain Sir James Clark Ross . . . Vol. III. Flora Tasmaniae. Pt. II Monocotyledons and Acotyledons* pp. 282–343, pls 185–196. London.

Harvey, W. H. 1863. *Phycologia Australica.* Vol. 5, pls. 241–300, synop. pp. 1–73.

Harvey, W. H. & Hooker, J. D. 1845. Algae. *In* J. D. Hooker (Ed.), *The Botany of the Antarctic Voyage of H. M. Discovery Ships 'Erebus' and 'Terror' in the Years 1839–1843, under the command of Captain Sir James Clark Ross . . . Vol. 1. Flora Antarctica. Pt. I. Botany of Lord Auckland's Group and Campbell's Island.* (1844–1845). pp. 175–193, pls 69–78. London.

Hauck, F. 1883–1885. Die Meeresalgen Deutschlands und Oesterreichs. *In* Dr L. Rabenhorst's *Kryptogamen-Flora von Deutschland, Oesterreich und der Schweiz.* 2nd Edn. Vol. 2, pp. 1–320 (1883), 321–513 (1884), 513–575 + i–xxiv (1885).

Hawkes, M. W. (1978). Sexual reproduction in *Porphyra gardneri* (Smith et Hollenberg) Hawkes (Bangiales, Rhodophyta). *Phycologia* **17**: 329–353.

Hawkes, M. W. 1983. Anatomy of *Apophlaea sinclairii* – an enigmatic red alga endemic to New Zealand. *Jap. J. Phycol. (Sôrui)* **31**: 55–64.

Hay, C. H. 1979*a*. A phytogeographical account of the southern bull kelp seaweeds *Durvillaea* spp. Bory 1826 (Durvilleales Petrov 1965). *Proc. Intl. Symp. Mar. Biogeogr. & Evol. in the S. Hemis., Auckland, N. Z., 17–20 July 1978.* **2**: 443–454.

Hay, C. H. 1979*b*. Nomenclature and taxonomy within the genus *Durvillaea* Bory (Phaeophyceae: Durvilleales Petrov). *Phycologia* **18**: 191–202.

Hay, C. H. 1979*c*. Some factors affecting the upper limit of the southern bull kelp *Durvillaea antarctica* (Chamisso) Hariot on two New Zealand shores. *J. R. Soc. N. Z.* **9**: 279–289.

Hedgpeth, J. W. 1969. Introduction to Antarctic Zoogeography. *Amer. Geogr. Soc., Antarc. Map Folio Ser.* **11**: 1–9.

Hedgpeth, J. W. 1970. Marine biogeography of the Antarctic regions. *In* M. Holdgate (Ed.), *Antarctic Ecology.* pp. 97–103. Academic Press, London & New York.

Hemsley, W. B. 1884. Report on the botany of the Bermudas and various other islands of the Atlantic and Southern oceans. *In Report on the scientific results of the voyage of H. M. S. 'Challenger' during . . . 1873–76 . . . Botany.* Vol. 1. part 2. 299 pp., pls 14–53. London.

Henry, E. C. 1977. (Notes on technique) A method for obtaining ribbons of serial sections of plastic embedded specimens. *Stain Tech.* **52**: 59–60.

Henry, E. C. 1984. Syringodermatales ord. nov. and *Syringoderma floridana* sp. nov. (Phaeophyceae). *Phycologia* **23**: 419–426.

Henry, E. C. & Müller, D. G. 1983. Studies on the life history of *Syringoderma phinneyi* sp. nov. (Phaeophyceae). *Phycologia* **22**: 387–393.

Herriott, E. M. 1923. Some morphological notes on the New Zealand giant kelp *Durvillea antarctica* (Chamisso) Hariot. *Trans. Proc. N. Z. Inst.* **54**: 549–564.

Heydrich, F. 1901. Die Lithothamnien des Muséum d'Histoire naturelle in Paris. *Engl. Bot. Jahrb.* **28**: 529–545.

Heydrich, F. 1907. Über *Sphaeranthera lichenoides* (Ell. et Sol.) Heydr. mscr. *Beih. Bot. Centralbl.* **22**: 222–230, pls. x & xi.

Hoek, C. van den. 1981. Chlorophyta: morphology and classification. *In* C. Lobban & M. J. Wynne (Eds), *The Biology of Seaweeds.* Bot. Monogr. Vol 17, pp. 86–132. Univ. Calif. Press.

Hoek, C. van den & Flinterman, A. 1968. The life-history of *Sphacelaria fucigera* Kütz. (Phaeophyceae). *Blumea* **16**: 193–242.

Hohenacker, R. F. 1852–1862. Algae marinae siccatae. Lief. vi, nos 251–300 (1857).

Holdgate, M. W. (Ed.) 1970. *Antarctic Ecology.* Vol. 1, xx + (1)–604 pp.; Vol 2, x + (605)–998 pp. Academic Press, London & New York.

Hollenberg, G. J. 1942. An account of the species of *Polysiphonia* on the Pacific Coast of North America. I. *Oligosiphonia. Amer. J. Bot.* **29**: 772–785.

Hollenberg, G. J. 1970. Phycological notes. IV. Including new marine algae and new records for California. *Phycologia* **9**: 61–72.

Holmes, E. M. 1905. Some South Orkney algae. *J. Bot.* **43**: 196–198.

Holmes, E. M. 1912. Some South Orkney algae. *In* W. S. Bruce (Ed.), *Report of the Scientific Results of the Voyage of S. Y. 'Scotia' During the Years 1902, 1903, and 1904.* Vol. 3: Botany. pp. 87–88. Edinburgh.

Holmgren, P. K. & Keuken, W. 1974. *Index Herbariorum. Pt. I. The Herbaria of the World.* 6th edn, vii + 397 pp. Utrecht.

Hommersand, M. H. 1963. The morphology and classification of some Ceramiaceae and Rhodomelaceae. *Univ. Calif. Publs Bot.* **35**: 165–366.

Hooker, J. D. 1844–1845. *The Botany of the Antarctic Voyage of H. M. Discovery Ships 'Erebus' and 'Terror' in the Years 1839–1843 under the command of Captain Sir James Clark Ross . . . Vol. I. Flora Antarctica. Pt. I, Botany of Lord Auckland's Group and Campbell's Island* pp. [1]–208, pls 1–80. London.

Hooker, J. D. 1846–1847. *The Botany of the Antarctic Voyage of H. M. Discovery Ships 'Erebus' and 'Terror' in the Years 1839–1843 under the command of Captain Sir James Clark Ross . . . Vol. I. Flora Antarctica. Pt. II, Botany of Fuegia, the Falklands, Kerguelen's Land,* etc. (1845–1847). Algae. pp. 454–502, pls 165–194. London.

Hooker, J. D. 1853. *The Botany of the Antarctic Voyage of H. M. Discovery Ships 'Erebus' and 'Terror' in the Years 1839–1843 under the command of Captain Sir James Clark Ross . . . Vol II. Flora Novae-Zelandiae. Pt. I, Flowering Plants.* p. xiii. London.

Hooker, J. D. 1867. *Handbook of the New Zealand Flora.* [Pt. II.] *Order IX. Algae.* pp. 638–721. London.

Hooker, J. D. & Harvey, W. H. 1845. Algae antarcticae, being characters and descriptions of the hitherto unpublished species of algae, discovered in Lord Auckland's Group, . . . *Lond. J. Bot.* **4**: 249–276, 293–298.

Hooper, R. & South, G. R. 1974. A taxonomic appraisal of *Callophyllis* and *Euthora* (Rhodophyta). *Br. phycol. J.* **9**: 423–428.

Hooper, R. & South, G. R. 1977. Distribution and ecology of *Papenfussiella callitricha* (Rosenv.) Kylin (Phaeophyceae, Chordariaceae). *Phycologia* **16**: 153–157.

Hornemann, J. V. (Ed.) 1818. *Flora danica.* Vol. 9, fase 27.

Horning, D. S. 1979. Exploring Macquarie Island, Pt. I: Southern wildlife outpost. *Aust. Nat. Hist.* **19**: 236–243.

Howe, M. A. 1914. The marine algae of Peru. *Mem. Torrey Bot. Club.* **15**: 1–185, 66 pls.

Hudson, P. R. & Waaland, J. R. 1974. Ultrastructure of mitosis and cytokinesis in the multinucleate green alga *Acrosiphonia. J. Cell Biol.* **62**: 274–294.

Hudson, W. 1762. *Flora Anglica.* 1st edn, 506 pp. London.

Huizing, H. J. & Rietema, H. 1975. Xylan and mannan as cell wall constituents of different stages in the life-histories of some siphoneous green algae. *Br. phycol. J.* **10**: 13–16.

Hylmö, D. E. 1919. Zur Kenntnis der subantarktischen und antarktischen Meeresalgen. III. Chlorophyceen. *In* O. Nordenskjöld (Ed.), *Wissenschaftliche Ergebnisse der schwedischen Südpolar-Expedition 1901–1903 . . .* Vol. 4:2(16), 20 pp. Stockholm.

Hylmö, D. E. 1938. Botanische Ergebnisse der schwedischen Expedition nach Patagonien und dem Feuerlande 1907–1909. XI. Meeresalgen 3. Chlorophyceae. *K. svenska Vetensk-Akad. Handl.* Ser. 3. **17**(1): 23 pp.

Inagaki, K.-I. 1958. A systematic study of the order Chordariales from Japan and its vicinity. *Sci. Pap. Inst. algol. Res. Hokkaido Univ.* **4**: 87–197, 11 pls.

Irvine, L. M. 1983. *Seaweeds of the British Isles.* Vol 1. Rhodophyta. Pt. 2A Cryptonemiales (*sensu stricto*), Palmariales, Rhodymeniales. xii + 115 pp. Br. Mus. (Nat. Hist.). London.

Itono, H. 1977. Studies on the ceramiaceous algae (Rhodophyta) from southern parts of Japan. *Biblthca phycol.* **35**: 599 pp.

Jenkin, J. F. 1972. *Studies on plant growth in a subantarctic environment.* Ph.D. Thesis, Univ. of Melbourne.

Jenkin, J. F. 1975. Macquarie Island, Subantarctic. Structure and function of tundra ecosystems. *Ecol. Bull. (Stockholm)* **20**: 375–397.

Jenneborg, L. H. 1977. *Eurychasma*-infection of marine algae. Changes in algal morphology and taxonomical consequences. *Bot. Mar.* **20**: 499–507.

Johansen, H. W. 1976. Phycological Reviews 4: Current status of generic concepts in coralline algae (Rhodophyta). *Phycologia* **15**: 221–244.

Johansen, H. W. 1981. *Coralline Algae, A First Synthesis.* 239 pp. CRC Press, Boca Raton, Florida.

Jönsson, B. 1890. Beiträge zur Kenntnis des Dickenzuwachses der Rhodophyceen. *Lunds Univ. Arsskr. N F.* Avd. 2. **27**(4): 41 pp.

Jönsson, B. 1901. Zur Kenntnis des Baues und der Entwicklung des Thallus bei den Desmarestieen. *Lunds. Univ. Arsskr. N. F.* Avd. 2. **37**(6): 42 pp., 3 pls.

Jónsson, S. 1959. Le cycle de dévelopment du *Spongomorpha lanosa* (Roth) Kütz. et la nouvelle familie des Acrosiphoniacées. *C. R. Acad. Sci. Paris.* **248**: 1565–1567.

Kennett, J. P. 1977. Cenozoic evolution of Antartica glaciation, the circum-antarctic Ocean, and their impact on global paleoceanography. *J. Geophys. Res.* **82**: 3843–3860.

Kennett, J. P. 1978. The development of planktonic biogeography in the Southern Ocean during the Cenozoic. *Mar. Micropaleontology* **3**: 301–345.

Kenny, R. & Haysom, N. 1962. Ecology of rocky shore organisms at Macquarie Island. *Pac. Sci.* **16**: 245–263.

Kim, D. H. 1976. A study of the development of cystocarps and tetrasporangial sori in Gigartinaceae (Rhodophyta, Gigartinales). *Nova Hedwigia* **27**: 1–146.

Kirk, T. 1891. The flora of Macquarie Island. *Report of the third meeting of the Australasian Association for the Advancement of Science.* pp. 226–227.

Kjellman, F. R. 1872. *Bidrag till kännedomen om Skandinaviens Ectocarpeer och Tilopterider.* 112 pp., 2 pls. Stockholm.

Kjellman, F. R. 1880. Rhodospermeae O. Fucoideae. *Points-förteckning öfrer Skandinoviens växter* **4**: 2–9.

Kjellman, F. R. 1883. The algae of the Arctic Sea. *K. svenska Vetensk-Akad. Handl.* **20**(5): 351 pp., 31 pls.

Kjellman, F. R. 1890. *Handbok i Skandinaviens hafsalgtlora I. Fucoidea* (iv) + 103 pp. Stockholm.

Knaggs, F. W. 1967. Review of the world distribution and ecology of *Rhodochorton purpureum* (Lightf.) Rosenvinge. *Nova Hedwigia* **14**: 549–570.

Knebel, G. 1935. Monographie der Algenreihe der Prasiolales, insbesondere von *Prasiola crispa. Hedwigia* **75**: 1–120.

Knox, G. A. 1960. Marine biology: Littoral ecology and biogeography of the southern oceans. *Proc. R. Soc.*, Ser. B. **152**: 577–624.

Knox, G. A. 1979. Distribution patterns of Southern Hemisphere marine biotas: some comments on their origins and evolution. *Proc. Intl. Symp. Mar. Biogeogr. & Evol. in the S. Hemis., Auckland, N. Z., 17–20 July 1978.* **1**: 43–81.

Koeman, R. P. T. & Hoek, C. van den. 1981. The taxonomy of *Ulva* (Chlorophyceae) in the Netherlands. *Br. phycol. J.* **16**: 9–35.

Kohlmeyer, J. 1973. Spathulosporales, a new order and possible missing link between Laboulbeniales and Pyrenomycetes. *Mycologia* **65**: 614–647.

Kohlmeyer, J. & Kohlmeyer, E. 1975. Biology and geographical distribution of *Spathulospora* species. *Mycologia.* **67**: 629–637.

Kohlmeyer, J. & Kohlmeyer, E. 1979. *Marine mycology. The higher fungi.* xiv + 690 pp. Academic Press.

Kornmann, P. 1962*a*. Der Lebenzyklus von *Desmarestia viridis. Helgoländer Wiss. Meeresuntersuch.* **8**: 287–292.

Kornmann, P. 1962*b*. Eine Revision der Gattung *Acrosiphonia. Helgoländer Wiss. Meeresuntersuch.* **8**: 219–242.

Kornmann, P. 1963. Die Ulotrichales, neu geordnet auf der Grundlage entwicklungsgeschichtlicher Befunde. *Phycologia* **3**: 60–68.

Kornmann, P. 1965. Ontogenie und Lebenszyklus der Ulotrichales in phylogenetischer Sicht. *Phycologia* **4**: 163–172.

Kornmann, P. 1970*a*. Der Lebenszyklus von *Acrosiphonia grandis* (Acrosiphoniales; Chlorophyta). *Mar. Biol.* **7**: 324–331.

Kornmann, P. 1970*b*. Phylogenetische Beziehungen in den Grünalgengattungen *Acrosiphonia. Helgoländer Wiss. Meeresuntersuch.* **21**: 292–304.

Kornmann, P. 1973. Codiolophyceae, a new class of Chlorophyta. *Helgoländer Wiss. Meeresuntersuch.* **25**: 1–13.

Kornmann, P. & Sahling, P.-H. 1962. Zur Taxonomie und Entwicklung der *Monostroma*-Arten von Helgoland. *Helgoländer Wiss. Meeresuntersuch.* **8**: 302–320.

Kornmann, P. & Sahling, P.-H. 1974. Prasiolales (Chlorophyta) von Helgoland. *Helgoländer Wiss. Meeresuntersuch.* **26**: 99–133.

Kornmann, P. & Sahling, P.-H. 1977. Meeresalgen von Helgoland. Benthische Grün-, Braun- und Rotalgen. *Helgoländer Wiss. Meeresuntersuch.* **29**: 1–289.

Koster, J. 1955. The genus *Rhizoclonium* Kütz. in the Netherlands. *Pubbl. Staz. Zool. Napoli* **27**: 335–357.

Koster, J. 1969. Type collections of algae. *Taxon* **18**: 549–559.

Kraft, G. T. 1977. The morphology of *Grateloupia intestinalis* from New Zealand with some thoughts on generic criteria within the family Cryptonemiaceae. *Phycologia* **16**: 43–51.

Kraft, G. T. 1981. Rhodophyta: morphology and classification. *In* C. Lobban & M. J. Wynne (Eds), *The Biology of Seaweeds.* Bot. Monogr. Vol. 17. pp. 6–51. Univ. Calif. Press.

Krauss, F. 1846. Pflanzen des Cap- und Natal-Landes, gesammelt und zusammengestellt von Dr. Ferdinand Krauss. (Schluss.) *Flora* **29**(14): 209–219.

Kuckuck, P. 1929. Fragmente einer Monographie der Phaeosporeen. Nach dem Tode des Verfassers herausgegeben von Wilhelm Nienburg. *Wiss. Meeresuntersuch. (Helgoland)* **17**(4): 93 pp.

Kumano, S. 1978. Notes on freshwater red algae from West Malaysia. *Bot. Mag. Tokyo* **91**: 97–107.

Kunieda, H. 1934. On the life-history of *Monostroma. Proc. Imp. Acad. Japan.* **10**: 103–106.

Kuntze, O. 1891–1898. *Revisio generum plantarum.* Pars 1 & 2. clvi + 1011 pp. Pars 3. clvii–ccccxxii + i–iv + 576 pp. Würzburg.

Kurogi, M. 1972. Systematics of *Porphyra* in Japan. *In* I. A. Abbott & M. Kurogi (Eds), *Contributions to the Systematics of Benthic Marine Algae of the North Pacific.* pp. 167–191, 4 pls. Jap. Phycol. Soc., Kobe.

Kühneman, O. 1970. Vegetación marina de la Ría de Puerto Deseado. *Op. Lilloana* **17**: 1–169, 132 pls.

Kützing, F. T. 1841. Ueber *Ceramium* Ag. *Linnaea* **15**: 227–246.
Kützing, F. T. 1843. *Phycologia generalis*. xvi + 1–144 + xvii–xxxii + 145–458 + [1] pp., 80 pls. Leipzig.
Kützing, F. T. 1847. Diagnosen und Bermerkungen zu neuen oder kritischen Algen. *Bot. Ztg.* **5**: 22–25.
Kützing, F. T. 1849. *Species algarum*. vi–922 pp. Brockhaus, Leipzig.
Kützing, F. T. 1854. *Tabulae phycologicae*. Vol. 5. 23 pp., 100 pls. Nordhausen.
Kützing, F. T. 1860. *Tabulae phycologicae*. Vol. 10. 39 pp., 100 pls. Nordhausen.
Kützing, F. T. 1866. *Tabulae phycologicae*. Vol. 16. 35 pp., 100 pls. Nordhausen.
Kylin, H. 1923. Studien über die Entwicklungsgeschichte der Florideen. *K. svenska Vetensk-Akad. Handl.* **63**(11): 139 pp.
Kylin, H. 1924. Studien über die Delesseriaceen. *Lunds Univ. Arsskr. N. F.* Avd. 2. **20**(6): 111 pp.
Kylin, H. 1925. The marine red algae in the vicinity of the biological station at Friday Harbor, Wash. *Lunds Univ. Arsskr. N. F.* Avd. 2. **21**(9): 87 pp.
Kylin, H. 1928. Entwicklungsgeschichtliche Florideenstudien. *Lunds Univ. Arsskr. N. F.* Avd. 2. **84**(4): 127 pp.
Kylin, H. 1929. Die Delesseriaceen Neu-Seelands. *Lunds Univ. Arsskr. N. F.* Avd. 2. **25**(2): 15 pp., 12 pls.
Kylin, H. 1930. Ueber die Entwicklungsgeschichte der Florideen. *Lunds Univ. Arsskr. N. F.* Avd. 2. **26**(6): 104 pp.
Kylin, H. 1931. Die Florideenordnung Rhodymeniales. *Lunds Univ. Arsskr. N. F.* Avd. 2. **27**(11): 48pp., 20 pls.
Kylin, H. 1932. Die Florideenordnung Gigartinales. *Lunds Univ. Arsskr. N. F.* Avd. 2. **28**(8): 88 pp, 28pls.
Kylin, H. 1933. Ueber die Entwicklungsgeschichte der Phaeophyceen. *Lunds Univ. Arsskr. N. F.* Avd. 2. **39**(7): 102 pp. + 2 pls.
Kylin, H. 1940. Die Phaeophyceenordung Chordariales. *Lunds Univ. Arsskr N. F.* Avd. 2. **36**(9): 67 pp., 8 pls.
Kylin, H. 1941. Californische Rhodophyceen. *Lunds Univ. Arsskr. N. F.* Avd. 2. **37**(2): 51 pp., 13 pls.
Kylin, H. 1947*a*. Die Phaeophyceen der schwedischen Westküste. *Lunds Univ. Arsskr. N. F.* Avd. 2. **43**(4): 100 pp, 18 pls.
Kylin, H. 1947*b*. Ueber die Fortpflanzungsverhältnisse in der Ordnung Ulvales. *K. fysiogr. Sällsk. Lund Förh.* **17**(17): 9 pp.
Kylin, H. 1949. Die Chlorophyceen der schwedischen Westküste. *Lunds Univ. Arsskr. N. F.* Avd. 2. **45**(4): 79 pp.
Kylin, H. 1956. *Die Gattungen der Rhodophyceen*. xv + 673 pp. Lund.
Kylin, H. & Skottsberg, C. 1919. Zur Kenntnis der subantarktischen und antarktischen Meeresalgen. II. Rhodophyceen. *In* O. Nordenskjöld (Ed.), *Wissenschaftliche Ergebnisse der schwedischen Süd-polar Expedition 1901–1903*. Vol. 4: **2**(15). 88 pp., 1 pl. Stockholm.
Laing, R. M. 1896. Notes on several species of *Delesseria*, one being new. *Trans. & Proc. N. Z. Inst.* **29**: 446–450, pls 27, 28.
Laing, R. M. 1901. Revised list of New Zealand seaweeds, Pt. II. *Trans. & Proc. N. Z. Inst.* **34**: 327–359.
Laing, R. M. 1905. On the New Zealand species of Ceramiaceae. *Trans. & Proc. N. Z. Inst.* **37**: 384–408, 8 pls.
Laing, R. M. 1909. The marine algae of the sub-Antarctic islands of New Zealand. *In* C. Chilton (Ed.), *The sub-Antarctic islands of New Zealand*. Vol. 2, pp. 493–527, pls 21–24. Wellington.
Lamb, I. M. & Zimmermann, M. H. 1977. Benthic marine algae of the Antarctic Peninsula. *Antarctic Res. Ser.* Washington **23**: 129–229.
Lamouroux, J. V. F. 1812. Extrait d'un mémoire sur la classification des Polypiers coralligènes, non entièrement pierreux. *Nouveau Bull. Sci. Soc. Philomatique, Paris* **3**: 181–188.
Lamouroux, J. V. F. 1813. Essai sur les genres de la famille des Thalassiophytes non articulées. *Annls Mus. Hist. nat. Paris* **20**: 21–47.
Lamouroux, J. V. F. 1819. *Delisea*. *In* S. Leman (Ed.), *Dictionnaire des Sciences Naturelles*. **13**: 41–42. Levrault, Strasbourg.
Law, P. G. & Burstall, T. 1956. Macquarie Island. *ANARE Interim rep.* 14: pp. i–iii + 1–26.
Lebednik, P. A. 1977. Postfertilization development in *Clathromorphum*, *Melobesia*, and *Mesophyllum* with comments on the evolution of the Corallinaceae and the Cryptonemiales (Rhodophyta). *Phycologia* **16**: 379–406.
Lebednik, P. A. 1978. Development of male conceptacles in *Mesophyllum* Lemoine and other genera of the Corallinaceae (Rhodophyta) *Phycologia* **17**: 388–395.
Lee, I. K. 1978. Studies on Rhodymeniales from Hokkaido. *J. Fac. Sci., Hokkaido Univ.*, Ser. V (Bot.). **11**: 1–203.
Lee, I. K. & Kurogi, M. 1977. On the taxonomic position of *Rhodymenia cuneifolia* Okamura (Rhodophyta.). *Bull. Jap. Soc. Phycol.* **25**(Suppl. Mem. Iss. Yamada): 113–118.
Lee, I. K. & Kurogi, M. 1978. *Neohalosacciocolax aleutica* gen. et sp. nov. (Rhodophyta), parasitic on *Halosaccion minjaii* I. K. Lee from the north Pacific. *Br. phycol. J.* **13**: 131–139.
Leister, G. L. 1977. *Taxonomy and reproductive morphology of Iridaea cordata (Turner) Bory and Iridaea crispata Bory (Gigartinaceae, Rhodophyta) from southern South America*. Ph.D. Thesis, Duke Univ. 186 pp.
Le Jolis, A. 1863. Liste des algues marines de Cherbourg. *Mém. Imp. Soc. Sci. Nat. Cherbourg* **10**: 1–168, 6 pls.
Lemoine, P. 1913. Melobesiées, révision des Melobesiées Antarctiques. *In* J. Charcot (Ed.), *Deuxième Expédition Antarctique Française (1908–1910)*. 67 pp., 2 pls. Sciences Naturelles. Paris.
Lemoine, P. 1915. Melobesieae. *In* A. D. Cotton, Cryptogams from the Falkland Islands collected by Mrs. Vallentin. *J. Linn. Soc., Bot.* **43**: 193–200, pls 9, 10.

Lemoine, P. 1928. Un nouveau genre de Melobesiées: *Mesophyllum*. *Bull. Soc. bot. France*. **75**: 251–254.

Levring. T. 1939. Ueber die Phaeophyceengattungen *Myriogloia* Kuck. und *Haplogloia* nov. gen. *Bot. Notiser* **1939**: 40–52.

Levring, T. 1940. Die Phaeophyceengattungen *Chlanidophora, Distromium*, und *Syringoderma*. *K. Fysiogr. Salls. I. Lund Forhandl*. **10**(20): 11 pp.

Levring, T. 1941. Die Meeresalgen der Juan Fernandez Inseln. *In* C. Skottsberg (Ed.), *Natural History of Juan Fernandez and Easter Island*. Vol. 2. pp. 601–670, pls 49–53. Uppsala.

Levring, T. 1944. Meeresalgen von den Crozet-Inseln und Kerguelen. *Ark. Bot*. **31a**(8): 1–31.

Levring, T. 1945. Marine algae from some Antarctic and Subantarctic islands. *Lunds Univ. Arsskr. N. F.* Avd. 2 **41**(7): 36 pp., 1 pl.

Levring, T. 1953. The marine algae of Australia. I. Rhodophyta: Goniotrichales, Bangiales, and Nemalionales. *Ark. Bot*. **2**: 457–530.

Levring, T. 1955. Contributions to the marine algae of New Zealand. I. *Ark. Bot*. ser. 2, **3**: 407–432.

Levring, T. 1960. Contributions to the marine algal flora of Chile. *Lunds Univ. Arsskr. N. F.* Avd. 2. **56**(10): 85 pp.

Lightfoot, J. 1977. *Flora Scotica*. Vol. 2. London.

Lindauer, V. W. 1947. *[Erratum to] algae Nova Zelandicae Exsiccatae*. fasc. ix, no. 218. (N.B. specimen originally labelled as *Ptilonia magellanica*.)

Lindauer, V. W., Chapman, V. J. & Aiken, M. 1961. The marine algae of New Zealand. II. Phaeophyceae. *Nova Hedwigia* **3**: 129–350, pls 57–97.

Link, H. F. 1820. Epistola . . . de algis aquaticis in genera disponendis. *In* C. G. Nees ab Esenbeck (Ed.), *Horae physicae berolinenses*. pp. 1–8, 1 pl. Bonn.

Link, H. F. 1833. *Handbuch zur Erkennung der nutzbarsten und am häufigsten vorkommenden Gewächse*. Pt. 3, xviii + 536 pp. Berlin.

Linnaeus, C. 1753. Species plantarum. 1st edn. Vol. 2., pp. 561–1200. Stockholm.

Linnaeus, C. 1771. *Mantissa plantarum*. Vol. 2, p. 311.

Littler, M. M. 1972. The crustose Corallinaceae. *Oceanogr. Mar. Biol. Ann. Rev*. **10**: 311–347.

Lokhorst, G. M. 1978. Taxonomic studies on the marine and brackish water species of *Ulothrix* (Ulotrichales, Chlorophyceae) in Western Europe. *Blumea* **24**: 191–299.

Lokhorst, G. M. & Trask, B. J. 1981. Taxonomic studies on *Urospora* (Acrosiphonales, Chorophyceae) in Western Europe. *Acta Bot. Neerl*. **30**: 353–431.

Lokhorst, G. M. & Vroman 1974*a*. Taxonomic studies on the genus *Ulothrix* (Ulotrichales, Chlorophyceae). II. *Acta Bot. Neerl*. **23**: 361–398.

Lokhorst, G. M. & Vroman 1974*b*. Taxonomic studies on the genus. *Ulothrix* (Ulotrichales, Chlorophyceae). III. *Acta Bot. Neerl*. **23**: 561–602.

Lowry, J. K. 1979. Exploring Macquarie Island, Part 2: Subantarctic refuge. *Aust. Nat. Hist*. **19**: 242–245.

Lowry, J. K., and Fenwick, G. D. 1983. The shallow-water gammaridean Amphipoda of the subantarctic islands of New Zealand and Australia: Melitidae, Hadziidae. *J. R. Soc. N. Z*. **13**: 201–260.

Lowry, J. K., Horning, D. S., Poore, G. C. B. & Ricker, R. W. 1978. *The Australian Museum Macquarie Island expedition, summer 1977–1978*. 152 pp. Austr. Mus. Trust.

Lowry, J. K. & Stoddart, H. E. 1983. The shallow-water gammaridean Amphipoda of the subantarctic islands of New Zealand and Australia: Lysianassoidea. *J. R. Soc. N. Z*. **13**: 279–294.

Lucus, A. H. S. 1919. The algae of Commonwealth Bay. Australasian Antarctic Expedition 1911–1914. *Sci. Rep. Ser. C.: Zool. & Bot*. **7**(2): 1–18, pls 1–9.

Lucas, A. H. S. 1936. *The Seaweeds of South Australia. Part I. Introduction and the Green and Brown Seaweeds*. pp. 1–106. Govt. Printer., Adelaide.

Lucas, A. H. S. & Perrin, F. 1947. *The Seaweeds of South Australia. Part. II. The Red Seaweeds*. pp. 112–458. Govt. Printer, Adelaide.

Lund, S. 1959. The marine algae of East Greenland. I. Taxonomic Part. *Meddel. Grønland* **156**(1): 247 pp.

Lyle, L. 1922. *Antithamnionella*, a new genus of algae. *J. Bot*. **60**: 346–350.

Lyngbye, H. C. 1819. *Tentamen Hydrophytologiae Danicae*. 248 pp., 70 pls. Copenhagen.

Martin, M. T. 1939. The structure and reproduction of *Chaetangium saccatum* (Lam.) J. Ag. — I. Vegetative structure and male plants. *Proc. Linn. Soc. London*, Session 148, Pt. 3: pp. 165–181, pl. 10.

McFadden, M. E. 1911. On a *Colacodasya* from Southern California. *Univ. Calif. Publs Bot*. **4**: 143–150, 1 pl.

Meer, J. P. van der & Todd, E. R. 1980. The life history of *Palmaria palmata* in culture. A new type for the Rhodophyta. *Can. J. Bot*. **58**: 1250–1256.

Mendoza, M. L. 1969*a*. Estudio sistematico y ecologico de las Ceramiaceae (Algae-Rhodophyta) de Puerto Deseado Provincia de Santa Cruz (Argentina). *Darwiniana* **15**: 287–362, 10 pls.

Mendoza, M. L. 1969*b*. Las Delesseriaceae (Rhodophyta) de Puerto Deseado, Provincia de Santa Cruz, Argentina. I. Estudio sistematico y ecologico de los géneros *Schizoseris* Kylin, *Cladodonta* Skotts. e *Hymenena* Greville. *Physis* **28**: 419–441.

Mendoza, M. L. 1969*c*. Las Delesseriaceae (Rhodophyta) de Puerto Deseado, Provincia de Santa Cruz, Argentina. II. Estudio sistematico y ecologico del género *Myriogramme* Kylin. *Physis* **29**: 245–260.
Mendoza, M. L. 1970*a*. Dos especies de algas rojas hemiparasitas de Puerto Deseado, Provincia de Santa Cruz, Argentina. *Physis* **29**: 363–371.
Mendoza, M. L. 1970*b*. Las Delesseriaceae (Rhodophyta) de Puerto Deseado, Provincia de Santa Cruz, Argentina. III. Estudio sistematico del género *Gonimophyllum* Batters. *Physis* **29**: 372–378.
Mendoza, M. L. 1973. El estado sexual masculino de *Pseudolaingia larsenii* (Skotts.) Levring (Rhodophyta) y su presencia en Tierra del Fuego e Isla de los Estados. *Physis* **32**: 211–217.
Mendoza, M. L. 1974. El estado sexual masculino de *Delesseria fuegiensis* Skottsberg y la presencia de los géneros *Delesseria* Lamouroux y *Schizoseris* Kylin (Rhodophyta) en Tierra del Fuego e Isla de los Estados. *Physis* **33**: 483–504.
Mendoza, M. L. 1975. *Pseudonitophylla*, neuvo género de la familia Delesseriaceae (Rhodophyta). *Physis* **34**: 59–65.
Mendoza, M. L. 1976*a*. *Antarcticophyllum*, nuevo género para las Corallinaceae. *Bol. Soc. Argent. Bot.* **17**: 252–261.
Mendoza, M. L. 1976*b*. *Odontolaingia*, nuevo género de la familia Delesseriaceae (Rhodophyta). *Bol. Soc. Argent. Bot.* **17**: 190–198.
Mendoza, M. L. 1976*c*. Presencia del género *Pseudolithophyllum* (Corallinaceae) en Argentina. *Contr. Cien. Centro Invest. Biol. Mar.* No. 140: 1–9.
Mendoza, M. L. 1977. Las Corallinaceae (Rhodophyta) de Puerto Deseado, Provincia de Santa Cruz, Argentina. I. Géneros *Dermatolithon* y *Mesophyllum*. *Physis* **36**: 21–29.
Meneghini, G. 1838. Cenni sulla organographia e fisiologia delle Alghe. *Nuovi Saggi Imp. Regia Accad. Sci. Padova* **4**: 325–388.
Michanek, G. 1979. Phytogeographic provinces and seaweed distribution. *Bot. Mar.* **22**: 375–391.
Mikami, H. 1965. A systematic study of the Phyllophoraceae and Gigartinaceae from Japan and its vicinity. *Mem. Fac. Fish. Hokkaido Univ.* **5**: 181–285.
Min-Thein, U. & Womersley, H. B. S. 1976. Studies on southern Australian taxa of Solieriaceae, Rhabdoniaceae, and Rhodophyllidaceae (Rhodophyta). *Aust. J. Bot.* **24**: 1–166.
Mirbel, C. F. Brisseau de. 1825. Rapport sur la flore des îles Malouines; par M. Gaudichaud. *Annls Sci. nat.* **5**: 89–110, pls 2, 3.
Moe, R. L. & DeLaca, T. E. 1976. Occurrence of macroscopic algae along the Antarctic Peninsula. *Antarct. J. U. S.* **11**: 20–24.
Moe, R. L. & Henry, E. C. 1982. Reproduction and early development of *Ascoseira mirabilis* Skottsberg (Phaeophyta), with notes on Ascoseirales Petrov. *Phycologia* **21**: 55–66.
Moe, R. L. & Silva, P. C. 1977*a*. Antarctic marine flora: Uniquely devoid of kelps. *Science* **196**: 1206–1208.
Moe, R. L. & Silva, P. C. 1977*b*. Sporangia in the brown algal genus *Desmarestia* with special reference to Antarctic *D. ligulata*. *Bull. Jap. Soc. Phycol.* **25**(Suppl. Mem. Iss. Yamada): 159–167.
Moe, R. L. & Silva, P. C. 1979. Morphological and taxonomic studies on Antarctic Ceramiaceae (Rhodophyceae). I. *Antarcticothamnion polysporum* gen. et sp. nov. *Br. phycol. J.* **14**: 385–405.
Moe, R. L. & Silva, P. C. 1980. Morphological and taxonomic studies on Antarctic Ceramiaceae (Rhodophyceae). II. *Pterothamnion antarcticum* (Kylin) comb. nov. (*Antithamnion antarcticum* Kylin). *Br. phycol J.* **15**: 1–17.
Moe, R. L. & Silva, P. C. 1981. Morphology and taxonomy of *Himantothallus* (including *Phaeoglossum* and *Phyllogigas*), an Antarctic member of the Desmarestiales (Phaeophyceae). *J. Phycol.* **17**: 15–29.
Moe, R. L. & Silva, P. C. 1983. Morphological and taxonomic studies on Antarctic Ceramiaceae (Rhodophyceae). III. *Georgiella* and *Plumariopsis* (Tribe Ptiloteae). *Br. phycol. J.* **18**: 275–298.
Montagne, [J. F.] C. 1839. Cryptogames de la Patagonie. *In* A. D'Orbigny (Ed.), *Voyage dans l'Amérique Méridionale (le Brésil, la republique . . .) exécuté pendant les années 1826, 1827, 1828, 1829, 1830, 1831, 1832 et 1833. Vol. 7: Botanique.* Part I. Sertum Patagonium. 19 pp, 4 pls. Paris.
Montagne, [J. F.] C. 1842*a*. Algae. *In* Ramon de la Sagra, *Histoire physique, politique et naturelle de l'Ile de Cuba. Botanique-plantes cellulaires.* pp. 1–104, pls 1–5. Paris.
Montagne, [J. F.] C. 1842*b*. *Prodromus generum specierumque phycearum novarum, in itinere ad polum antarcticum.* 16 pp. Paris.
Montagne, [J. F.] C. 1842*c*. *Bostrychia. Dict. Univ. Nat. His.* [Orbigny] **2**: 660–661.
Montagne, [J. F.] C. 1844. Quelques observations touchant la structure et la fructification des genres *Ctenodus*, *Delisea* et *Lenormandia* de la famille des Floridées. *Ann. Sci. Nat.* Ser. 3, Bot. **1**: 151–161.
Montagne, [J. F.] C. 1845. Plantes cellulaires. *Voyage au Pole Sud et dans l'Océanie sur les corvettes l'Astrolabe et la Zélée sous le commandment de J. Dumont D'Urville* . . . pendant . . . 1837, 1838, 1839, 1840 . . . Botanique. Vol. 1. xiv + 349 pp. + atlas of 20 pls. Paris.
Montagne, [J. F.] C. 1846. Phyceae. *In* C. Gaudichaud (Ed.), *Voyage autour du monde exécuté pendant les années 1836 et 1837 sur la corvette 'La Bonite' commandée par M. Vaillant. Botanique. Vol. 1. Cryptogames cellulaires et vasculaires.* pp. 1–112. Paris.

Murray, G. 1981. On *Cladothele* Hook. f. et Harv. (*Stictyosiphon* Kütz.). *J. Bot.* **29**: 193–196, pl. 306.
Müller, D. G. 1984. Culture studies on the life history of *Adenocystis utricularis* (Phaeophyceae, Dictyosiphonales). *Phycologia* **23**: 87–94.
Müller, D. G. & Meel, H. 1982. Culture studies on the life history of *Arthrocladia villosa* (Desmarestiales, Phaeophyceae). *Br. phycol. J.* **17**: 419–425.
Müller, O. F. 1778. *Flora danica* Vol. 5, fasc 13: 7, pl. 768.
Nakamura, Y. 1972. A proposal on the classification of the Phaeophyta. *In* I. A. Abbott & M. Kurogi (Eds), *Contributions to the Systematics of Benthic Marine Algae of the North Pacific.* pp. 147–156. Jap. Soc. Phycol., Kobe.
Nakamura, Y. & Tatewaki, M. 1975. The life history of some species of the Scytosiphonales. *Sci. Pap. Inst. algol. Res. Hokkaido Univ.* **6**: 57–93, 5 pls.
Nardo, G. D. 1834. De nuovo genere Algarum cui nomen est *Hildenbrandtia prototypus. Isis* (editied by L. Chen) **1834**: 675–676.
Naylor, M. N. 1949. Observations on the anatomy of *Durvillea antarctica* (Chamisso) Hariot. *Ann. Bot. N. S.* **13**: 286–308.
Naylor, M. N. 1955. The life history of *Adenocystis utricularis* (Bory) H. et H. *Trans. R. Soc. N. Z.* **83**: 295–301, pls 11–13.
Nägeli, C. 1847. Die neueren Algensysteme und Versuch zur Begründung eines eigenen Systems der Algen und Florideen. *Nene Deukschr. Allg. Schweiz. Ges. Gesammten Naturwiss.* **9**(2): 275 pp.
Nägeli, C. 1861. Beiträge zur Morphologie und Systematik der Ceramiaceae. *Sber. bayer. Akad. Wiss.* **2**: 297–415.
Nägeli, C. & Cramer, C. 1855–1858. Pflanzenphysiologische Untersuchungen. Fredrich Schulthess, Zürich.
Nees, ab Esenbeck, C. G. 1820. *Horae physicae Berolinenses collectae ex symbolis vivorum doctorum.* Adolphi Marcus, Bonnae. 123 pp, [index pp. 1–4], 27 pls.
Nelson, W. A. 1982. A critical review of the Ralfsiales, Ralfsiaceae and the taxonomic position of *Analipus japonicus* (Harv.) Wynne (Phaeophyta). *Br. phycol. J.* **17**: 311–320.
Nelson, W. A. & Adams, N. M. 1983. A taxonomic revision of the families Chordariaceae and Chordariopsidaceae (Phaeophyta) in New Zealand. *N. Z. J. Bot.* **21**: 77–92.
Neushul, M. 1960. Diving in Antarctic waters. *Polar Record* **10**: 353–358.
Neushul, M. 1968. Benthic marine algae. *Amer. Geogr. Soc., Antarc. Map Folio Ser.* **10**: 9–10.
Newton, L. 1931. *A Handbook of the British Seaweeds.* xiii + 478 pp. Br. Mus. (Nat. Hist.), London.
Nicholson, N. L. 1979. Evolution within *Macrocystis*: Northern and southern hemisphere taxa. *Proc. Intl. Symp. Mar. Biogeogr. & Evol. in the S. Hemis, Auckland, N. Z. 17–20 July 1978.* **2**: 433–442.
Nielsen, R. 1979. Culture studies on the type species of *Acrochaete, Bolbocoleon* and *Entocladia* (Chaetophoraceae, Chlorophyceae). *Bot. Notiser* **132**: 441–449.
Nielsen, R. 1983. Culture studies of *Acrochaete leptochaete* comb. nov. and *A. wittrockii* comb. nov. (Chaetophoraceae, Chlorophyceae). *Nord. J. Bot.* **3**: 689–694.
Nizamuddin, M. 1968. Observations on the order Durvilleales J. Petrov. 1965. *Bot. Mar.* **11**: 115–117.
Norris, R. E. 1957. Morphological studies on the Kallymeniaceae. *Univ. Calif. Publs Bot.* **28**: 251–333, pls 28–40.
Norris, R. E. 1964. The morphology and taxonomy of South African Kallymeniaceae. *Bot. Mar.* **7**: 90–129.
O'Donnell, E. H. J. 1954. The structure and taxonomic position of the brown alga *Microzonia. Amer. J. Bot.* **41**: 380–384.
O'Kelly, C. J. 1980. *The taxonomy and systematics of certain marine algae referable to the Chaetophoraceae and Chaetosiphonaceae (Chlorophyta).* i–x + 284 pp., 28 pls. Ph.D. Thesis, Univ. Wash.
O'Kelly C. J. & Yarish, C. 1980. Observations on marine Chaetophoraceae (Chlorophyta). I. Sporangial ontogeny in the type species of *Entocladia* and *Phaeophila. J. Phycol.* **16**: 549–558.
O'Kelly, C. J. & Yarish, C. 1981. Observations on marine Chaetophoraceae (Chlorophyta). II. On the circumscription of the genus *Entocladia* Reinke. *Phycologia* **20**: 32–45.
Ogle, W. 1912. De partibus animalium. *In* J. A. Smith and W. D. Ross (Eds), *The Works of Aristotle Translated into English.* Oxford Univ. Press.
Okamura, K. 1907–1942. *Icones of Japanese Algae.* Vols 1–7, 345 pls. Tokyo.
Okamura, K. 1934. Notes on algae dredged from the Pacific coast of Tiba Prefecture. *Rec. Oceanogr. Works Japan* **6**(1): 13–18, pl. vii.
Oltmanns, F. 1889. Beiträge zur Kenntnis der Fucaceen. *Biblioth. Bot.* **3**(14): 94 pp.
Oltmanns, F. 1904. *Morphologie und Biologie der Algen.* Vol. I. Jena.
Oltmanns, F. 1922. *Morphologie und Biologie der Algen.* Vol. II. Jena.
Papenfuss, G. F. 1945. Review of the *Acrochaetium-Rhodochorton* complex of the red algae. *Univ. Calif. Publs Bot.* **18**: 299–334.
Papenfuss, G. F. 1947. Further contributions toward an understanding of the *Acrochaetium-Rhodochorton* complex of red algae. *Univ. Calif. Publs Bot.* **18**: 433–447.
Papenfuss, G. F. 1950. Review of the genera of algae described by Stackhouse. *Hydrobiol.* **2**: 181–208.
Papenfuss, G. F. 1951. Phaeophyta. *In* G. M. Smith (Ed.), *Manual of Phycology—An Introduction to the Algae and*

Their Biology ('A new series of plant science books', Vol. 27) pp. 119–158. Chonica Botanica Co., Waltham, Mass.

Papenfuss, G. F. 1958. Notes on algal nomenclature IV. Various genera and species of Chlorophyceae, Phaeophyceae, and Rhodophyceae. *Taxon* **7**: 104–109.

Papenfuss, G. F. 1960. On the genera of the Ulvales and the status of the order. *J. Linn. Soc., Bot.* **56**: 303–318.

Papenfuss, G. F. 1964*a*. Catalogue and bibliography of Antarctic and sub-Antarctic benthic marine algae. *Antarc. Res. Ser.* Washington, **1**: pp. 1–76.

Papenfuss, G. F. 1946*b*. Problems in the taxonomy and geographic distribution of Antarctic marine algae. *Biologie antarctique. Comptes-rendus, Premier symposium organise par le S.C.A.R.* , 1962, pp. 155–160.

Papenfuss, G. F. 1977. Review of the genera of Dictyotales (Phaeophycophyta). *Bull. Jap. Soc. Phycol.* **25**(Suppl. Mem. Iss. Yamada): 271–287.

Parke, M. & Dixon, P. S. 1964. A revised check-list of British marine algae. *J. mar. biol. Ass. U. K.* **44**: 499–542.

Parke, M. & Dixon, P. S. 1968. Check-list of British marine algae—second revision. *J. mar. biol. Ass. U. K.* **48**: 783–832.

Parke, M. & Dixon, P. S. 1976. Check-list of British marine algae—third revision. *J. mar. biol. Ass. U. K.* **56**: 527–594.

Parsons, M. J. 1975. Morphology and taxonomy of the Dasyaceae and the Lophothalieae (Rhodomelaceae) of the Rhodophyta. *Aust. J. Bot.* **23**: 549–713.

Pedersen, P. M. 1974. On the systematic position of *Delamarea attenuata* (Phaeophyceae). *Br. phycol. J.* **9**: 313–318.

Pedersen, P. M. 1976. Culture studies on marine algae from West Greenland II. *Coelocladia arctica* (Dictyosiphonales, Coelocladiaceae fam. nov.). *Norw. J. Bot.* **23**: 243–249.

Pedersen, P. M. 1978. Culture studies on marine algae from West Greenland III. The life histories and systematic positions of *Pogotrichum filiforme* and *Leptonematella fasciculata* (Phaeophyceae). *Phycologia* **17**: 61–68.

Pedersen, P. M. 1981. The life histories in culture of the brown algae *Gononema alariae* sp. nov. and *G. aecidioides* comb. nov. from Greenland. *Nord. J. Bot.* **1**: 263–270.

Pedersen, P. M. 1984. Studies on primitive brown algae (Fucophyceae). *Opera Bot.* **74**: 1–76.

Perestenko, L. P. 1973. *De specibus novis Rhodymeniae Grev. et Odonthaliae Grev. notula.* (In Russian.) *Nov. Sist. Nizsh. Rast.* **10**: 61–68.

Perestenko, L. P. 1980. *Vodorosli zaliva Petra Velikogo.* (In Russian.) 232 pp. Akad. Nauk SSSR, Dal'nevost. Nauchn. Tsentr, Inst. Biol. Morya.

Petrov. Ju. E. 1965. De positione familiae Durvilleacearum et systematica classis cyclosporophycearum (Phaeophyta). *Nov. Sist. Vysshikh. Rast.; novitates systematicae plantarum non vascularium.* pp. 70–72. Akad. Nauk. SSSR. Bot. Inst., Moscow.

Pfeiffer. L. 1874. *Nomenclator botanicus* Vol. 2, pt. I, p. 350.

Piccone, A. 1889. Nuove alghe del viaggio di circumnavigazione della Vettor Pisani. *Mem. R. Accad. Lincei, Cl. Sc. Fis. Mat. e Nat.*, Ser. 4, **6**: 10–63.

Post, E. 1936. Systematische und pflanzengeographische Notizen zur *Bostrychia-Caloglossa* Association. *Rev. algol.* **9**: 1–84.

Powell, A. W. B. 1951. Antarctic and subantarctic Mollusca. *Discovery Rep.* **26**: 47–196.

Powell, A. W. B. 1955. Mollusca of the southern islands of New Zealand. *Cape Exped. Ser.*, Bull. 15: pp. 1–152. DSIR, Wellington.

Powell, A. W. B. 1960. Antarctic and subantarctic Mollusca. *Rec. Auck. Inst. & Mus.* **5**: 117–193.

Prescott, G. W. 1979. A contribution to a bibliography of Antarctic and Subantarctic algae—together with a checklist of freshwater taxa reported to 1977. *Biblio. Phycol.* **45**: 1–312.

Price, I. R. 1969. The structure and classification of *Scytothamnus australis* (J. Ag.) Hooker and Harv. (Phaeophyta). *Phycologia* **8**: 37–41.

Prud'homme van Reine, W. F. 1974. Geographic distribution of European *Sphacelaria*-species in the world (Phaeophyceae, Sphacelariales). *Bull. Soc. phycol. fr.* **19**: 171–177.

Prud'homme van Reine, W. F. 1982. *A taxonomic revision of the European Sphacelariaceae (Sphacelariales, Phaeophyta).* Bot. Ser. Vol. 6, 293 pp. E. J. Brill/Leiden Univ. Press, Leiden.

Pueschel, C. M. & Cole, K. M. 1982. Rhodophycean pit plugs: an ultrastructural survey with taxonomic implications. *Amer. J. Bot.* **69**: 703–720.

Pujals, C. 1963. Catálogo de Rhodophyta citadas para la Argentina. *Rev. Mus. Argent. Cienc. Nat. Bernardino Rivadavia (Bot.)* **3**: 1–139.

Pujals, C. 1970. *Medeiothamnion* neuvo género de Ceramiaceae. *Rev. Mus. Argent. Cienc. Nat. Bernardino Revadavia (Bot.)* **3**: 287–299.

Pujals, C. 1977. Adiciones y correcciones al 'Catálogo de Rhodophyta citadas para la Argentina'. *Rev. Mus. Argent. Cienc. Nat. Bernardino Revadavia (Bot.)* **5**: 123–163.

Pujals, C. 1981. Identidad de *Ballia scoparia* (Hook. f. et Harv.) Harv. (Rhodophyta, Florideophycidae. *Rev. Mus. Argent. Cienc. Nat. Bernardino Revadavia (Bot.)* **6**: 1–22.

Quilty, P. G, Rubenach, M. & Wilcoxon, J. A. 1973. Miocene ooze from Macquarie Island. *Search* **4**: 163–164.

Rabenhorst, L. 1863. *Kryptogamen-Flora von Sachsen*. 1. Abth. xx + 653 pp. Leipzig.

Rabenhorst, L. 1878. Beitrag zur Meeresalgen-Flora der Auckland-Inseln. *Hedwigia* **17**: 65–77.

Ravanko, O. 1970. Morphological, developmental and taxonomic studies in the *Ectocarpus* complex (Phaeophyceae). *Nova Hedwigia* **20**: 179–252.

Reinbold, T. 1907. *Die Meeresalgen der deutschen Tiefsee-Expedition 1898–1899. Wissenschaftliche Ergebnisse der deutschen Tiefsee-Expedition auf dem Dampfer 'Valdivia' 1898–1899*. Vol. 2, pp. 551–586, pls 55–58. Jena.

Reinbold, T. 1908. Die Meeresalgen der deutschen Südpolar-Expedition 1901–1903. *In* E. von Drygalski (Ed.), *Deutsche Südpolar Expedition 1901–1903* . . . Vol. 8: Botanik. pp. 177–202. Berlin.

Reinke, J. 1879. Zwei parasitische Algen. *Bot. Ztg*. **37**: 475–478.

Reinke, J. 1890. Uebersicht der bisher bekannten Sphacelariaceen. *Ber. Deutsch. Bot. Ges*. **8**: 201–215.

Reinsch, P. F. 1874–1875. *Contributiones ad algologiam et fungologiam*. xii + 104 pp., 130 pls. Nürnberg.

Reinsch, P. F. 1888*a*. Ueber einige neue Desmarestien. *Flora* **71**: 188–192.

Reinsch, P. F. 1888*b*. Species et genera nova algarum ex insula Georgia Australi. *Ber. Deutsch. Bot. Ges*. **6**: 144–156.

Reinsch, P. F. 1890. Zur Meeresalgenflora von Süd-Georgien. *In* G. Neumayer (Ed.), *Internationale Polarforschung, 1882–1883. Die deutschen Expeditionen und ihre Ergebnisse* Vol. 2. Beschreibende Naturwissenschaften. pp. 366–449, pls 1–19. Berlin.

Ricker, R. W. & Kraft, G. T. 1979. Morphology of the subantarctic red alga *Cenacrum subsutum* gen. et sp. nov. (Rhodymeniales) from Macquarie Island. *J. Phycol*. **15**: 434–444.

Rosenberg, T. 1933. Studien über Rhodomelaceen und Dasyaceen. Akad. Abhandl. Lund. 87 pp.

Rosenvinge, L. K. 1900. Note sur une Floridée aérienne (*Rhodochorton islandicum* nov. sp.). *Bot. Tidsskr*. **23**(1): 61–81.

Rosenvinge, L. K. 1909–1931. The marine algae of Denmark. Rhodophyceae. *K. Danske Vidensk. Selsk. Skr., 7* Raekke, Afd. **7**: 1–637, 8 pls.

Rosenvinge, L. K. & Lund, S. 1941. The marine algae of Denmark. Vol II. Phaeophyceae. I. Ectocarpaceae and Acinetosporaceae. *K. Danske Vidensk. Selsk. Skr., 7* Raekke, Afd. **1**(4): 1–79.

Roth, A. G. 1797–1806. *Catalecta Botanica*. vii + 244 + [10] pp., 8 pls. Leipzig.

Round, F. E. 1981. *The Ecology of Algae*. vii + 653 pp. Cambridge Univ. Press.

Rueness, J. 1978. Hybridization in red algae. *In* D. E. G. Irvine & J. H. Price (Eds), *Modern Approaches to the Taxonomy of Red and Brown Algae*. Systematics Association Special Vol. 10. pp. 247–262. Academic Press, London.

Rueness, J. & Rueness, M. 1975. Genetic control of morphogenesis in two varieties of *Antithamnion plumula* (Rhodophyceae, Ceramiales). *Phycologia* **14**: 81–85.

Ruprecht, F. J. 1850. *Algae Ochotenses. Die ersten sicheren Nachrichten über die Tange des Ochotskischen* St. Petersburg. (Preprint of 1851 article: Tange des Ochotskischen Meeres.) 243 pp., 10 pls.

Russell, G. & Fletcher, R. L. 1975. A numerical taxonomic study of the British Phaeophyta. *J. Mar. Biol. Ass. U. K*. **55**: 763–783.

Santelices, B. 1980. Phytogeographic characterization of the temperate coast of Pacific South America. *Phycologia* **19**: 1–12.

Santelices, B. & Abbott, I. A. 1978. New records of marine algae from Chile and their effect on phytogeography. *Phycologia* **17**: 213–222.

Sauvageau, C. 1900–1914. Remarques sur les Sphacélariacées. Publ. 1900, *J. Bot*. Paris. **14**; 1901, *J. Bot*. Paris **15**; 1902, *J. Bot*. Paris **16**; 1903, *J. Bot*. Paris **17**; 1904, *J. Bot*. Paris **18**: & publ. separately; 1914; publ. separately.

Sauvageau, C. 1929. Sur le développment de quelques phaeosporées. *Bull. Stn. biol. Arcachon* **26**: 253–418.

Scagel, R. F. 1953. A morphological study of some dorsiventral Rhodomelaceae. *Univ. Calif. Publs Bot*. **27**: 1–108.

Schmitz, [C. J.] F. 1889. Systematische Uebersicht der bisher bekannten Gattungen der Florideen. *Flora* **72**: 435–456, pl. 21.

Schmitz, [C. J.] F. 1892. [6. Klasse Rhodophyceae] 2. Unterklasse Florideae. *In* A. Engler (Ed.), *Syllabus der Vorlesungen über specielle und medicinische–pharmaceutische Botanik. Grosse Ausgabe*. pp. 16–23. Berlin.

Schmitz, [C. J.] F. 1896. *Kleinere Beiträge zur Kenntnis der Florideen*. Vol. VI. Nuova Notarisia, Padova.

Schmitz, [C. J.] F. & Falkenberg, P. 1897. Rhodomelaceae. *In* A. Engler & K. Prantl (Eds), *Die natürlichen Pflanzenfamilien*. Vol. 1: part 2, pp. 421–480. Leipzig.

Schreiber, E. 1932. Ueber die Entwicklungsgeschichte und die systematische Stellung der Desmarestiaceen. *Zeitschr. Bot*. **25**: 561–582.

Scott, J. H. 1883. Macquarie Island. *Trans. N. Z. Inst*. **15**: 484–493, pl. 39.

Searles, R. B. 1980. Observations on the morphology and systematics of *Chordaria linearis* (J. D. Hooker et Harvey) Cotton (Phaeophyta, Chordariales) from Chile. *Phycologia* **19**: 194–201.

Searles, R. B. & Leister, G. L. 1980. North Carolina marine algae. IX. *Onslowia endophytica* gen. et sp. nov. (Phaeophyta, Spacelariales) and notes on other new records for North Carolina. *J. Phycol*. **16**: 35–40.

Semba, R. 1979. Contribution to semithin sectioning on a conventional rotary microtome. *Stain Tech.* **54**: 251–255.
Seppelt, R. D. 1977. Studies on the Bryoflora of Macquarie Island. I. Introduction and checklist of species. *The Bryologist* **80**: 167–170.
Setchell, W. A. 1923. Parasitic Florideae, II. *Univ. Calif. Publs Bot.* **10**: 393–396.
Setchell, W. A. 1937. The *Codiums* of the Juan Fernandez Islands. *In* C. Skottsberg (Ed.), *The natural history of Juan Fernandez and Easter Island.* Vol 2. pp. 587–600, pls 34–48. Uppsala.
Setchell, W. A. & Gardner, N. L. 1920. Phycologica contributions. I. *Univ. Calif. Publs Bot.* **7**: 279–324.
Setchell, W. A. & Gardner, N. L. 1922. Phycological contributions. VI. New Species of *Ectocarpus. Univ. Calif. Publs Bot.* **7**: 403–426.
Setchell, W. A. & Gardner, N. L. 1924*a*. New marine algae from the Gulf of California. *Proc. Calif. Acad. Sci.*, Ser. 4, **12**: 695–949, 77 pls.
Setchell, W. A. & Gardner, N. L. 1924*b*. Phycological contributions, vii. *Univ. Calif. Publs Bot.* **13**: 1–13.
Setchell, W. A. & Gardner, N. L. 1925. The marine algae of the Pacific Coast of North America. Pt. III. Melanophyceae. *Univ. Calif. Publs Bot.* **8**: 383–898.
Setchell, W. A. & Gardner, N. L. 1936. *Iridophycus* gen. nov. and its representation in South America. *Nat. Acad. Sci. Proc.* **22**: 469–473.
Siddique, M. & Faridi, M. A. F. 1977. The life history of *Cladophora crispata* (Roth) Ag. *Pak. J. Bot.* **9**: 159–162.
Silva, P. C. (unpublished) *Index Nominum Algarum.* Bot. Dept. Univ. Calif. Berkeley, California.
Silva, P. C. 1952. A review of nomenclatural conservation in the algae from the point of view of the type method. *Univ. Calif. Publs Bot.* **25**: 241–324.
Silva, P. C. 1957. Remarks on algal nomenclature. *Taxon* **6**: 141–145.
Silva, P. C. 1979. The benthic algal flora of central San Francisco Bay. *In* T. J. Conomos (Ed.), *San Francisco Bay: the Urbanized Estuary.* pp. 287–345. Pac. Div., Amer. Ass. for the Advancement of Sci., San Francisco, Calif.
Silva, P. C. 1980*a*. Names of classes and families of living algae. *Reg. Veg.* **103**: 1–156.
Silva, P. C. 1980*b*. Remarks on algal nomenclature vi. *Taxon* **29**: 121–145.
Silva, P. C. 1982. Chlorophyta. *In* S. P. Parker (Ed.), *Synopsis and Classification of Living Organisms.* pp. 133–161, McGraw-Hill, New York.
Silva, P. C. & Womersley, H. B. S. 1956. The genus *Codium* (Chlorophyta) in southern Australia. *Aust. J. Bot.* **4**: 261–289.
Simpson, R. D. 1976. The shore environment of Macquarie Island. *ANARE Rep.* Ser. B(1) Zool. No. 125: pp. 1–41.
Skinner, S. & Womersley, H. B. S. 1983. New records (possibly introductions) of *Striaria, Stictyosiphon* and *Arthrocladia* (Phaeophyta) for southern Australia. *Trans. R. Soc. S. Aust.* **107**: 59–68.
Skottsberg, C. 1906. Observations on the vegetation of the Antarctic Sea. **In** *Botaniska Studier tillägnade F. R. Kjellman den 4 November 1906.* pp. 245–264, pls vii-ix, 1 map. Uppsala.
Skottsberg, C. 1907. Zur Kenntnis der subantarktischen und antarktischen Meeresalgen. I. Phaeophyceen. *In* O. Nordenskjöld (Ed.), *Wissenschaftliche Ergebnisse der schwedischen Südpolar-Expedition 1901–1903.* Vol. 4:1(6), 172 pp., 10 pls, 1 map. Stockholm.
Skottsberg, C. 1921. Botanische Ergebnisse der schwedischen Expedition nach Patagonien und dem Feuerlande 1907–1909. VIII. Marine algae. I. Phaeophyceae. *K. svenska Vetensk-Akad. Handl.* **61**(11): 56 pp.
Skottsberg, C. 1923. Botanische Ergebnisse der schwedischen Expedition nach Patagonien und dem Feuerlande 1907–1909. IX. Marine algae. 2. Rhodophyceae. *K. svenska Vetensk-Akad. Handl.* **63**(8): 70 pp.
Skottsberg, C. 1941*a*. Communities of marine algae in sub-Antarctic and Antarctic waters. *K. svenska Vetensk-Akad. Handl.* ser. 3. [i.e., ser. 5] **19**(4): 92 pp., 3 pls.
Skottsberg, C. 1941*b*. Marine algal communities of the Juan Fernandez Islands, with remarks on the composition of the flora. *In* C. Skottsberg (Ed.), *The Natural History of Juan Fernandez and Easter Island.* Vol. 2, pp. 671–696, 1pl. Uppsala.
Skottsberg, C. 1953. On two collections of Antarctic marine algae. *Ark. Bot.* Ser. 2. **2**: 531–566, 1 pl.
Skottsberg, C. 1960. Remarks on the plant geography of the southern cold temperate zone. *Proc. R. Soc.* Ser. B. **152**: 447–457.
Skottsberg, C. 1964. Antarctic phycology. *In* R. Carrick, M. Holdgate & J. Prevost (Eds), *Biologie antarctique.* pp. 147–154. Hermann & Cie.
Skuja, H. 1939. Versuch einer systematischen Einteilung der Bangioideen oder Protoflorideen. *Acta Horti Bot. Univ. Latv.* **11/12**: 23–40.
Smith, G. M. 1942. Notes on some brown algae from the Monterey Peninsula, California. *Amer. J. Bot.* **29**: 645–653.
Smith, G. M. 1955. *Cryptogamic Botany*, Vol. 1, ix + 546 pp. McGraw-Hill, New York.
Smith, J. E. & Sowerby, J. 1843. *English botany suppl.* Vol. 3; t 2866. Longman, London.
Solier, A. 1846. Sur deux Algues zoosporées formant le nouveau genre *Derbesia. Rev. Bot. [Duchartre]* **1**(10): 452–454.

South, G. R. 1974. *Herpodiscus* gen. nov. and *Herpodiscus durvilleae* (Lindauer) comb. nov., a parasite of *Durvillea antarctica* (Chamisso) Hariot endemic to New Zealand. *J. R. Soc. N. Z.* **4**: 455–461.

South, G. R. 1979*a*. Biogeography of benthic marine algae of the southern oceans. *Proc. Intl. Symp. Mar. Biogeogr. & Evol. in the S. Hemis., Auckland, N.Z., 17–20 July 1978.* **1**: 85–108.

South, G. R. 1979*b*. The genus *Plocamium* Lamouroux (Rhodophyta, Gigartinales) in New Zealand. *Proc. Intl. Symp. Mar. Biogeogr. & Evol. in the S. Hemis, Auckland, N.Z. 17–20 July 1978.* **2**: 423–432.

South, G. R. & Adam, N. M. 1979. A revision of the genus *Plocamium* Lamouroux (Rhodophyta, Gigartinales) in New Zealand. *Phycologia* **18**: 120–132.

Sparling, S. R. 1957. The structure and reproduction of some members of the Rhodymeniaceae. *Univ. Calif. Publs Bot.* **29**: 319–396.

Stackhouse, J. 1797. *Nereis britannica*. Fasc. 2. ix–xxiv. Bath.

Stackhouse, J. 1801. *Nereis britannica* Fasc. 3. xi + pp. 71–112. Bath.

Stegenga, H. 1978. Life histories of *Rhodochorton purpureum* and *R. floridulum* (Rhodophyta, Nemaliales) in culture. *Br. phycol. J.* **13**: 279–289.

Stegenga, H. 1979. *Life Histories and Systematics of the Acrochaetiaceae.* iii–34 pp. Total Photo/Total Print, Amsterdam, Holland.

Stegenga, H. & Vroman, M. 1977. The significance of life history characters for the sytematics of the Acrochaetiaceae. *J. Phycol.* **13**(suppl.): 64.

Stewart, K. D. & Mattox, K. R. 1975. Comparative cytology, evolution and classification of the green algae with some consideration of the origin of other organisms with chlorophylls a and b. *Bot. Rev.* **41**: 104–135.

Suhr, J. N. 1839. Beiträge zur Algenkunde. Bot. Ztg. **22**(5): 65–75, pls i–iv.

Suneson, S. 1937. Studien über die Entwicklungsgeschichte der Corallinaceen. *Lunds Univ. Arsskr. N.F.* Avd. 2 **33**(2): 102 pp., 4 pls.

Suneson, S. 1943. The structure, life history and taxonomy of the Swedish Corallinaceae. *Lunds Univ. Arsskr. N.F.* Avd. 2 **39**(9): 66 pp., 9 pls.

Tanner, C. E. 1980. *Chloropelta* gen. nov., an ulvaceous green alga with a different type of development. *J. Phycol.* **16**: 128–137.

Tanner, C. E. 1981. Chlorophyta: life histories. *In* C. Lobban & M. J. Wynne (Eds), *The Biology of Seaweeds.* Bot. Monogr. Vol. 17, pp. 218–247. Univ. Calif. Press.

Tatewaki, M. 1972. Life history and systematics in *Monostroma. In* I. A. Abbott & M. Kurogi (Eds), *Contributions to the Systematics of Benthic Marine Algae of the North Pacific.* pp. 1–15. Jap. Soc. Phycol., Kobe.

Tatewaki, M. & Iima, M. 1984. Life histories of *Blidingia minima* (Chlorophyceae), especially sexual reproduction. *J. Phycol.* **20**: 368–376.

Taylor, B. W. 1955. The flora, vegetation and soils of Macquarie Island. *ANARE Rep.* Ser. B. Vol. 2. Bot. 192 pp., 2 maps.

Taylor, W. R. 1937. *Marine Algae of the Northeastern Coast of North America.* ix + 427 pp., 60 pls. Ann Arbor, Michigan.

Taylor, W. R. 1939. Algae collected by the "Hassler", "Albatross" and Schmitt Expeditions II. Marine algae from Uruguay, Argentina, the Falkland Is., and the Strait of Magellan. *Pap. Mich. Acad. Sc. Arts. Lett.* **24**: 127–164, pls 1–7.

Taylor, W. R. 1957. *Marine Algae of the Northeastern Coast of North America.* Rev. edn. ix + 509 pp. Ann Arbor, Michigan.

Thuret, G. 1854. Note sur la synonymie des *Ulva lactuca* et *latissima* L. suivie de quelques rémarques sur la tribu des Ulvacées. *Mém. Imp. Soc. Sci. Nat. Cherbourg* **2**: 17–32.

Tokida, J. 1932. The marine algae from Robben Island. (Kaihyô-tô), Saghalien. *Bull. School Fish. Hokkaido Univ.* **2**: 1–34.

Tokida, J. 1934. Phycological observations I. *Trans. Sapporo Nat. Hist. Soc.* **13**: 196–202.

Townsend, R. A. 1979. *Synarthrophyton*, a new genus of Corallinaceae (Cryptonemiales, Rhodophyta) from the southern hemisphere. *J. Phycol.* **15**: 251–259.

Tupa, D. D. 1974. An investigation of certain chaetophoralean algae. *Beih. Nova Hedw.* No. 46: 155 pp.

Turner, D. 1809. *Fuci sive plantarum fucorum generi a botanicis ascriptarum icones descriptiones et historia.* 2. 162 pp., 63 pls. London.

Varne, R., Gee, R. D. & Quilty, P. G. J. 1969. Macquarie Island and the cause of oceanic linear magnetic anomalies. *Science* **166**: 230–233.

Waern, M. 1952. Rocky-shore algae in the Öregrund Archipelago. *Acta Phytogeogr. Suecica* **30**: 136–148.

Wagner, F. S. 1954. Contributions to the morphology of the Delesseriaceae. *Univ. Calif. Publs Bot.* **27**: 279–345.

Walker, D. C. & Henry, E. C. 1978. Unusual reproductive structures in *Syringoderma abyssicola* (S. & G.) Levring. *New Phytol.* **80**: 193–197.

Warming, E. 1984. Haandbog i den systematiske botanik. Amden gjennemste indgave. [iv] + 434 pp. Kjobenhavn.

West, J. A. 1970. The life history of *Rhodochorton concrescens* in culture. *Br. phycol. J.* **5**: 179–186.

West, J. A. 1972. Environmental regulation of reproduction in *Rhodochorton purpureum. In* I. A. Abbott & M.

Kurogi (Eds), *Contributions to the Systematics of Benthic Marine Algae of the North Pacific.* pp. 213–230. Jap. Soc. Phycol., Kobe.

West, J. A. & Hommersand, M. H. 1981. Rhodophyta: life histories. *In* C. Lobban & M. J. Wynne (Eds), *The Biology of Seaweeds.* Bot. Monogr. Vol. 17, pp. 133–193. Univ. Calif. Press.

Wittrock, V. B. 1866. Försök till en Monographi ofver Algslägtet *Monostroma.* 66 pp., 4 pls. Stockholm.

Woelkerling, W. J. 1971. Morphology and taxonomy of the *Audouinella* complex (Rhodophyta) in southern Australia. *Aust. J. Bot.*, Suppl. Ser. 1. 91 pp.

Woelkerling, W. J. 1973. The morphology and systematics of the *Audouinella* complex (Acrochaetiaceae, Rhodophyta) in North-eastern United States. *Rhodora* **75**: 529–621.

Woelkerling, W. J. 1983. Phycological Reviews 8: The *Audouinella (Acrochaetium-Rhodochorton)* complex (Rhodophyta): present perspectives. *Phycologia* **22**: 59–92.

Wollaston, E. M. 1968. Morphology and taxonomy of southern Australian genera of Crouanieae Schmitz (Ceramiaceae, Rhodophyta). *Aust. J. Bot.* **16**: 217–417.

Wollaston, E. M. 1971*a*. *Antithamnion* and related genera occurring on the Pacific Coast of North America. *Syesis* **4**: 73–92.

Wollaston, E. M. 1971*b*. Generic features of *Antithamnion* (Ceramiaceae, Rhodophyta) in the Pacific Region. *Proc. 7th Intl. Seaweed Symp.* pp. 142–145. Univ. of Tokyo Press.

Wollaston, E. M. 1974. Sexual reproduction in *Ballia mariana* Harvey and *Ballia ballioides* (Sonder) Wollaston (Ceramiaceae, Rhodophyta). *Phycologia* **13**: 21–26.

Wollaston, E. M. 1979. Recognition of *Pterothamnion* Nägeli with taxonomic notes on *P. simile* (Hook. et Harv.) Nägeli and *Platythamnion nodiferum* (J. Ag.) Wollaston (Rhodophyta, Ceramiaceae). *Phycologia* **103**: 191–196.

Womersley, H. B. S. 1954. The species of *Macrocystis* with special reference to those on southern Australian coasts. *Univ. Calif. Publs Bot.* **27**: 109–132.

Womersley, H. B. S. 1956. A critical survey of the marine algae of southern Australia. I. Chlorophyta. *Aust. J. Mar. & Freshw. Res.* **7**: 344–383.

Womersley, H. B. S. 1967. A critical survey of the marine algae of southern Australia. II. Phaeophyta. *Aust. J. Bot.* **15**: 189–270.

Womersley, H. B. S. 1971*a*. New records and taxa of marine Chlorophyta in southern Australia. *Trans. R. Soc. S. Aust.* **95**: 113–120.

Womersley, H. B. S. 1971*b*. *The genus Plocamium* (Rhodophyta) in southern Australia. *Trans. R. Soc. S. Aust.* **95**: 9–27.

Womersley, H. B. S. 1978. Southern Australian species of *Ceramium* Roth (Rhodophyta). *Aust. J. Mar. & Freshw. Res.* **29**: 205–257.

Womersley, H. B. S. 1979. Southern Australian species of *Polysiphonia* Greville (Rhodophyta). *Aust. J. Bot.* **27**: 459–528.

Womersley, H. B. S. 1981*a*. Aspects of the distribution and biology of Australian marine macro-algae. *In* J. S. Pate & A. J. McComb (Eds), *The Biology of Australian Plants.* pp. 294–306. Univ. of W. Aust. Press.

Womersley, H. B. S. 1981*b*. Biogeography of Australasian marine macroalgae. *In* M. N. Clayton & R. J. King (Eds), *Marine Botany An Australasian perspective.* pp. 292–307. Longman Cheshire, Melbourne.

Womersley, H. B. S. 1984. *The Marine Benthic Flora of Southern Australia.* Pt. I. 329 pp. Govt. Printer, South Australia.

Womersley, H. B. S. & Conway, E. 1975. *Porphyra* and *Porphyropsis* (Rhodophyta) in southern Australia. *Trans. R. Soc. S. Aust.* **99**: 59–70.

Womersley, H. B. S. & Edmonds, S. J. 1957. General account of the intertidal ecology of South Australian coasts. *Aust. J. Mar. & Freshw. Res.* **9**: 217–257.

Womersley, H. B. S. & Norris, R. E. 1971. The morphology and taxonomy of Australian Kallymeniaceae (Rhodophyta). *Aust. J. Bot.*, Suppl. Ser. 2., 62 pp.

Womersley, H. B. S. & Shepley, E. A. 1982. Southern Australian species of *Hypoglossum* (Delesseriaceae, Rhodophyta). *Aust. J. Bot.* **30**: 321–346.

Wynne, M. J. 1969. Life history and systematic studies of some Pacific North American Phaeophyceae (brown algae). *Univ. Calif. Publs Bot.* **50**: 1–88.

Wynne, M. J. 1970. Marine algae of Amchitka Island (Aleutian Islands). I. Delesseriaceae. *Syesis* **3**: 95–144.

Wynne, M. J. 1972. Culture studies of Pacific coast Phaeophyceae. *Mém. Soc. Bot. France.*, **1972**: 129–144.

Wynne, M. J. 1973. A key to the unbranched, cylindrical brown algae. *Contr. Mar. Sci.* **17**: 133–152.

Wynne, M. J. 1982*a*. Observations on four species of Delesseriaceae (Rhodophyta) from the South Sandwich Islands, the Antarctic. *Contr. Univ. Mich. Herb.* **15**: 325–337.

Wynne, M. J. 1982*b*. Phaeophyceae. *In* S. P. Parker (Ed.), *Synopsis and Classification of Living Organisms.* pp. 115–125. McGraw-Hill, New York.

Wynne, M. J. 1983. The current status of genera in the Delesseriaceae (Rhodophyta). *Bot. Mar.* **26**: 437–450.

Wynne, M. J. & Kraft, G. T. 1981. [Appendix] Classification summary. *In* C. Lobban & M. J. Wynne (Eds), *The Biology of Seaweeds.* Bot. Monogr. Vol. 17, pp. 743–750. Univ. Calif. Press.

Wynne, M. J. & Loiseaux, S. 1976. Phycological Reviews 5: Recent advances in life history studies of the

Phaeophyta. *Phycologia* **15**: 435–452.
Yamanouchi, S. 1912. The life history of *Cutleria*. *Bot. Gaz.* **54**: 441–502.
Yamanouchi, S. 1913. The life history of *Zanardinia*. *Bot. Gaz.* **56**: 1–35.
Yarish, C. 1975. A cultural assessment of the taxonomic criteria of selected marine Chaetophoraceae (Chlorophyta). *Nova Hedwigia* **26**: 385–430.
Yoshida, T. 1974. Sur un genre nouveau, *Tokidaea* (Ceramiacées, Rhodophytes) du nord du Japon. *Bull. Mus. natn. Hist. nat., Bot.* **9**: 61–70.
Zaneveld, J. S. 1964. The benthic algal vegetation of Antarctica. *Bull. U.S. Antarc. Proj. Officer* **5**(10): 66–69.
Zaneveld, J. S. 1969. *Iconography of Antarctic and Sub-Antarctic benthic marine algae. Pt. I. Chlorophycophyta and Chrysophycophyta.* pp. i–xii, 117 pls. Cramer.
Zaneveld, J. S. & Sanford, R. B. 1980. Crustose corallinaceae algae (Rhodophyta) of the New Zealand and United States Scientific Expedition to the Ross Sea, Balleny Islands, and Macquarie Ridge, 1965. *Blumea* **26**: 205–231.
Zinova, A. D. 1953. Opredelitel' burykh vodoroslej severnykh morej SSSR. 226 pp. Moskva & Leningrad: Akad. Nauk SSSR.
Zinova, A. D. 1954. Novye smejstvo, rod i vid u burykh vodoroslej. *Trudy Bot. Inst. Akad. Nauk SSSR*, Ser. 2, Sporov. Rast. **9**: 223–244.
Zinova, A. D. 1958. [Composition and character of the algal flora near the shores of the Antarctic continent and in the vicinity of the Kerguelen and Macquarie Islands.] (In Russian.) *Inform. Byull. Soviet Antarktic Ekspeds.* **3**: 47–49. (N.B. English translation published in 1964.)
Zinova, A. D. 1963. Delesseriaceae apud Insulas Kerguelen et Macquarie. (In Russian and Latin.) *Trudy Bot. Inst. Akad. Nauk. SSSR*, Ser. 2 Sporov. Rast. **16**: 52–67.
Zinova, A. D. 1964. The composition and character of algal flora at the Antarctic coast and in the vicinity of Kerguelen and Macquarie Islands. *Sov. Antarc. Exped. Inf. Bull.* **1**: 123–125.
Zinova, A. D. 1966. New data on marine algae, Subantarctic and Antarctic. (In Russian) *Nov. Sist. Nizsh. Rast.* **1966**: 103–109.
Zinova, A. D. 1972. Representatives of the family Delesseriaceae (Rhodophyta) in the northern part of the Pacific Ocean. (In Russian) *Nov. Sist. Nizsh. Rast.* **9**: 65–82.
Zinova, A. D. 1973. Marine algae from Hopeful Bay, Kerguelen Island. (In Russian.) *Nov. Sist. Nizsh. Rast.* **10**: 44–49.
Zinova, A. D. 1976. Some species of the family Delesseriaceae (Rhodophyta) in the northern part of the Pacific Ocean. (In Russian.) *Nov. Sist. Nizsh. Rast.* **13**: 7–10.
Zinova, A. D. 1981. On the systematic position of the red alga *Nitophyllum (Myriogramme) yezoense* (Yamada et Tokida) Mikami (Delesseriaceae). (In Russian.) *Nov. Sist. Nizsh. Rast.* **18**: 10–15.

Appendix 1
Collection data

The following is an abbreviated version of the collection records previously published by the Australian Museum Trust as a preliminary report on the Museum's expedition to Macquarie I. (Lowry *et al.*, 1978). Only those collections that have been referred to in the present survey are listed below.

MA–10 Garden Bay, Macquarie I., 20 November 1977; col. D. S. Horning & J. K. Lowry (1–3 m depths). Assortment of red, brown and green algae.

MA–11 Garden Bay, Macquarie I., 22 November 1977; col. R. W. Ricker (1–10 m depths). Assortment of red, brown and green algae.

MA–14 Near Hayter Rock, Garden Bay, Macquarie I., 23 November 1977; col. J. K. Lowry (10 m depth). Red algae on rocky bottom in *Macrocystis pyrifera* forest.

MA–15 Near Hayter Rock, Garden Bay, Macquarie I., 23 November 1977; col. J. K. Lowry (10 m depth). Brown algae, mostly *Desmarestia*, on rocky bottom in *Macrocystis pyrifera* forest.

MA–34 Small bay south of Tern Rock, Macquarie I., 29 November 1977; col. R. W. Ricker (1–10 m depths). Spot collections of algae from rocky substrate.

MA–35 Small bay south of Tern Rock, Macquarie I., 29 November 1977; col. G. C. B. Poore & R. W. Ricker (7 m depth). *Macrocystis pyrifera* holdfasts and individuals of *Papenfussiella*.

MA–39 South end of Goat Bay, Tottan Head, Macquarie I., 30 November 1977; col. R. W. Ricker (14 m depth). *Codium subantarcticum* from intertidal rock reef, and various red algae from subtidal on vertical face of offshore rock.

MA–44 Catch-Me-Cave, Macquarie I., 2 December 1977; col. R. W. Ricker. *Rhodochorton* from shaded rocks in high intertidal.

MA–45 Between Aerial Cove and Catch-Me-Cave, Macquarie I., 2 December 1977; col. R. W. Ricker. 1. Rocks covered with moss and short green and golden algal turfs taken from undersides of large boulders in upper intertidal. 2. *Iridaea remuliformis* from unshaded rock surface.

MA–46 Aerial Cove, Macquarie I., 2 December 1977; col R. W. Ricker (0–5 m depths). Algae from *Codium*-dominated bottom.

MA–53 Isthmus of Buckles Bay, Macquarie I., 4 December 1977; col. R. W Ricker. *Durvillaea* from intertidal rocks.

MA–54 Aerial Cove, Macquarie I., 4 December 1977; col. R. W. Ricker (3 m depth). Algae from near vertical wall shaded by *Durvillaea*.

MA–55 Aerial Cove, Macquarie I., 4 December 1977; col. R. W. Ricker & D. S. Horning (4.1 m depth). Unshaded rocky surfaces dominated by *Pseudolithophyllum consociatum*.

MA–56 Aerial Cove, Macquarie I., 4 December 1977; col. R. W. Ricker (6 m depth). Horizontal rock surface with full light exposure on edge of sand plain. Encrusting coralline algae dominant.

MA–83 Aerial Cove, Macquarie I., 6 December 1977; col. R. W. Ricker (6 m depth). Encrusting coralline algae, *Codium*, and foliose red algae.

MA–84 Aerial Cove, Macquarie I., 6 December 1977; col. R. W. Ricker. *Ulva, Acrosiphonia* and *Scytosiphon* from rock substrate in *Durvillaea* zone.

MA–94 Anchor Rock, Macquarie I., 8 December 1977; col. R. W. Ricker (20 m depth). Foliose and filamentous (ceramiaceous) red algae from south-facing, vertical rock wall.

MA–95 Anchor Rock, Macquarie I., 8 December 1977; col. R. W. Ricker (13 m depth). Algae from lower end of *Durvillaea* and *Desmarestia* zone, on 70° slope facing south.

MA–97 Aerial Cove, Macquarie I., 6 December 1977; col. R. W. Ricker. Drift *Durvillaea* on north side of cove.

MA–98 Aerial Cove, Macquarie I., 9 December 1977; col. R. W. Ricker (5 m depth). Plants growing under large overhang shading base of rock wall. Many rocks covered with mat of filamentous red algae.

MA–99 Aerial Cove, Macquarie I., 9 December 1977; col. J. K. Lowry (5 m depth). Tunicates growing abundantly on underside of rock overhang.

MA–111 Aerial Cove, Macquarie I., 12 December 1977; col. R. W. Ricker. Algae from intertidal rocks in *Durvillaea* zone.

MA–112 Catch-Me-Cave, Macquarie I., 13 December 1977; col. R. W. Ricker. *Durvillaea* (plants reproductive) on rocks in sheltered intertidal zone.

MA–113 Garden Bay, Macquarie I., 14 December 1977; col. R. W. Ricker & J. K. Lowry (7 m depth). Transect from Hayter Rock directly east for about 45 m, from an intertidal *Durvillaea* belt into a dense *Macrocystis* forest. Quadrat 1, on rock bottom, exposed to slight surge and broken lighting by overhead *Macrocystis* and *Desmarestia*.

MA–114 Garden Bay, Macquarie I., 14 December 1977; col. R. W. Ricker & J. K. Lowry (5 m depth). Transect from Hayter Rock directly east for about 45 m, from an intertidal *Durvillaea* belt into a dense *Macrocystis* forest. Quadrat 2, on large rock boulder, exposed to slight surge and broken lighting by large individual of *Desmarestia*. (Only one half of quadrat completed, see MA-116 for second half.)

MA–115 Garden Bay, Macquarie I., 14 December 1977; col. R. W. Ricker. *Desmarestia* collection from rocky floor of *Macrocystis* forest; slight surge.

MA–116 Garden Bay, Macquarie I., 15 December 1977; col. R. W. Ricker & J. K. Lowry (5 m depth). Second half of quadrat 2 (see MA-114) atop large boulder; moderate surge (greater than on 14 December 1977) and broken sunlight by overhead *Macrocystis* canopy.

MA–117 Garden Bay, Macquarie I., 15 December 1977; col. R. W. Ricker & J. K. Lowry (4 m depth). Transect from Hayter Rock directly east for about 45 m, from an intertidal *Durvillaea* belt into a dense *Macrocystis* forest. Quadrat 3, from a *Desmarestia*/*Plocamium* association on an 80° slope facing east. Rock face covered with foliaceous red algae.

MA–118 Garden Bay, Macquarie I., 15 December 1977; col. R. W. Ricker. Two foliaceous red algae from drift.

MA–119 Hasselborough Bay, Macquarie I., 16 December 1977; col. R. W. Ricker,. 1. Intertidal collection 1 m above upper limit of *Durvillaea*. 2. *Durvillaea* blade with bacterial gall from lower intertidal.

MA–120 Hasselborough Bay, Macquarie I., 16 December 1977; col. R. W. Ricker. Algae from mid- to upper-intertidal, *c.* 2–4 m above upper limit of *Durvillaea*.

MA–121 Hasselborough Bay, Macquarie I., 16 December 1977; col. R. W. Ricker. *Rhodochorton* from intertidal, *c.* 0.3 m above upper limit of *Durvillaea*, amongst boulders protected from sun, wind and surf.

MA–129 Hasselborough Bay, Macquarie I., 17 December 1977; col. R. W. Ricker (2–4 m depths). Algae from rocks and boulders bordering sand plain.

MA–130 Catch-Me-Cave Point, Macquarie I., 18 December 1977; col. R. W. Ricker. Algae from 1–2 m above upper limit of *Durvillaea*. *Bostrychia vaga* dominant, forming mat on sheltered, shaded rocks in upper littoral/supralittoral zone. Other plants (e.g., *Iridaea*, *Porphyra*, *Acrosiphonia*) from sun-exposed rocks.

MA–131 Between Catch-Me-Cave and Aerial Cove, Macquarie I., 18 December 1977; col. R. W. Ricker. *Pilayella littoralis* and *Urospora penicilliformis* growing in nearly pure stands on separate rocks in mid-intertidal.

MA–133 North Head, Macquarie I., 18 December 1977; col. R. W. Ricker. *Enteromorpha*, *Ulva*, *Porphyra* and *Adenocystis* from tide pools exposed to low salinities, direct sunlight and minimal surf.

MA–134 North of Garden Bay, Macquarie I., 20 December, 1977 col. R. W. Ricker. *Hildenbrandia*, *Durvillaea* and *Haplogloia moniliforme* (superficially similar to *Papenfussiella* from MA-35) growing in tide pools protected from surge.

MA–137 North of Garden Bay, Macquarie I., 21 December 1977; col. R. W. Ricker. *Durvillaea* from lower intertidal, red algae and *Papenfussiella* from 0.6 m above upper margin of *Durvillaea* zone.

MA–140 Buckles Bay, Macquarie I., 22 December 1977; col. D. S. Horning (15 m depth). Red and brown algae from rocks on sandy bottom.

MA–143 Southeast corner of Gorilla Head Rock, Macquarie I., 23 December 1977; col D. S. Horning (12 m depth). Red algae (at least 2 species) on vertical rock wall; exposed to strong currents.

MA–213 Bauer Bay, Macquarie I., 27 December 1977; col. R. W. Ricker. Algae from various depths in tide pools in mid- to lower-intertidal exposed to full sun but partially protected from surf.

MA–215 Mawson Point (shallow inlet), Macquarie I., 28 December 1977; col. R. W. Ricker (1–3 m depths). Rock/sand bottom with coarse-grained sand separating rocks. *Codium* on rock walls at entrance to inlet and along sides of boulders inside inlet; *Hildenbrandia* abundant and often growing singly on rock surfaces.

MA–217 North of Aurora Cave, Macquarie I., 29 December 1977; col. R. W. Ricker. *Enteromorpha* (2 forms) and one filamentous green alga from high intertidal rock pool above *Porphyra* zone.

MA–222 Sandy Bay (near offshore rocks directly in front of hut), Macquarie I., 30 December 1977; col. R. W. Ricker (9 m depth). Red algae, including *Delesseria, Iridaea, Polysiphonia anisogona* and *Schizoseris dichotoma*, grew on boulders surrounded by coarse gravel. Brown algae largely represented by *Desmarestia* and *Macrocystis*, less so by *Papenfussiella*. Green algae were absent.

MA–227 Sandy Bay (off rock platform 200 m south of hut), Macquarie I., 31 December 1977; col. R. W. Ricker (9 m depth). Red algae from shaded, vertical wall of surge channel between two subtidal rocky reefs. Reef crests covered with *Desmarestia chordalis.*

MA–236 Sandy Bay (east of rock platform 200 m south of hut), Macquarie I., 1 January 1978; col. D. S. Horning (12 m depth). Red algae covering rocks on gravel-sand bottom.

MA–237 Sandy Bay (east of rock platform 200 m south of hut), Macquarie I., 1 January 1978; col. D. S. Horning (15 m depth). Red algae covering rocks on gravel-sand bottom.

MA–238 Sandy Bay, Macquarie I., 3 January 1978; col. R. W. Ricker. Intertidal algae from tide pools at *Porphyra* zone about 0.3 m above upper limit of *Durvillaea*. 1. *Codium* from upper margins of ≥2 m deep cauldron-shaped pools. 2. Encrusting corallines from ≤10 cm deep pools. All areas exposed to sun and surf at high tides.

MA–244 Green Gorge (in front of hut), Macquarie I., 7 January 1978; col. R. W. Ricker (14 m depth). Sand bottom with submerged rocky islands. Sparsely-populated *Macrocystis* forest comprising minimal canopy with mostly subsurface fronds. 1. Red algae from tops and sides of rocks 0.3–0.6 m diameter. 2. *Codium* from rock shaded by *Desmarestia chordalis.*

MA–248 Green Gorge (off rock platform 500 m south of hut), Macquarie I., 8 January 1978; col. R. W. Ricker (8 m depth). Sand bottom with submerged rocky islands. Sparsely-populated *Macrocystis* forest comprising minimal canopy with subsurface fronds. Abundance of filamentous and foliaceous red algae. *Desmarestia willii* shading most rocks. *Dilophus* growing prostrate mainly on rocks under *Desmarestia* or epiphytic on *Macrocystis* holdfasts.

MA–255 Green Gorge (rock platform 500 m south of hut), Macquarie I., 8 January 1978; col. D. S. Horning. Red algae on rock wall 0.7 m below water surface in intertidal pool.

MA–264 Green Gorge (rock platform 500 m south of hut), Macquarie I., 9 January 1978; col. R. W. Ricker. Algae from rims of tide pools at upper limit of *Durvillaea* zone. Most wave force broken by outer edge of reef; semi-protected habitat with full sun exposure.

MA–271 Green Gorge (reef in front of hut), Macquarie I., 10 January 1978; col. R. W. Ricker (5–10 m depths). Algae from rocky substrates exposed to various degrees of lighting and surge.

MA–278 Green Gorge, Macquarie I., 11 January 1978; col. R. W. Ricker. Algae from intertidal above *Durvillaea* zone on rock outcrop near mouth of stream.

MA–283 Green Gorge (off rock platform 500 m south of hut), Macquarie I., 12 January 1978; col. R. W. Ricker & D. S. Horning (9 m depth). Transect quadrat 1, submerged rock islands on sand bottom at base of vertical rock wall. Detached *Desmarestia chordalis* loose in quadrat. Few plants present in quadrat, but *Macrocystis*, *Desmarestia*, *Codium*, and foliose red algae abundant on surrounding rocks.

MA–284 Green Gorge (off rock platform 500 m south of hut), Macquarie I., 12 January 1978; col. R. W. Ricker & D. S. Horning (10 m depth). Transect quadrat 2, rock face of vertical wall under overhang. 100% *Codium* cover with epiphytic filamentous red algae. *Macrocystis* and *Desmarestia willii* abundant. *Codium* mats extensive on vertical wall; no overstory plants directly shading quadrat.

MA–295 Green Gorge (*Macrocystis* forests east of hut), Macquarie I., 13 January 1978; col R. W. Ricker (17 m depth). Sand plain with submerged rocky islands. 1. Brown algae (e.g., *Dilophus decumbens*) on 0.3–1 m diameter rocks shaded by *Desmarestia willii* and several large individuals plus numerous juveniles of *Macrocystis*. 2. Assortment of red algae from rocky habitats.

MA–298 Caroline Cove (off large rock on southwest side of cove), Macquarie I., 15 January 1978; col. J. K. Lowry (8 m depth). *Macrocystis* holdfasts, *Microzonia*, *Plocamium* and a red tunicate from boulder-covered floor of dense *Macrocystis* forests.

MA–300 Caroline Cove (off large rock on southwest side of cove), Macquarie I., 15 January 1978; col. J. K. Lowry (8 m depth). *Codium*, *Plocamium*, *Ophiocantha* and a scallop collected by R. W. Ricker from boulders on floor of dense *Macrocystis* forests.

MA–302 Caroline Cove (northwest of large rock on southwest side of cove), Macquarie I., 15 January 1978; col. R. W. Ricker (8 m depth). Habitat mostly shaded, rocky and exposed to waves. Dense *Macrocystis* forest comprising fertile plants *c.* 6 m long whose fronds do not form a surface canopy. Bottom nearly covered by *Codium* mats, but with some barren gravel patches. *Microzonia velutina* densely crowded in imbricate layers on upper surface of large boulder adjacent to gravel patch. *Mesophyllum schmitzii* also epilithic and forming layers of calcified fronds alongside *Microzonia*. *Delisea pulchra* and *Plocamium* growing on rocks in dense (shaded) part of forest. *Codium* covered with red epiphytes.

MA–303 Caroline Cove (three rocky outcrops at mouth of cove), Macquarie I., 16 January 1978; col. R. W. Ricker (18 m depth). *Codium subantarcticum*, *Microzonia velutina*, *Hymenena decumbens*, and various filamentous and foliose red algae from rocks on floor of dense *Macrocystis* forest. *Macrocystis* thalli 3–4.5 m tall.

MA–304 Caroline Cove, Macquarie I., 16 January 1978; col. R. W. Ricker (1.5–3 m depths). Epiphytes on *Desmarestia chordalis* shaded by overhead *Durvillaea* canopy.

MA–306 Caroline Cove (three rocky outcrops at mouth cove), Macquarie I., 16 January 1978; col. J. K. Lowry (13 m depth). *Macrocystis pyrifera* holdfasts, seastars, tunicates from dense *Macrocystis* forest.

MA–309 Caroline Cove (inner part of cove north of hut), Macquarie I., 17 January 1978; col. R. W. Ricker (14 m depth). Large expanse of *Codium* adjacent to *Macrocystis* forest. Filamentous red algae (mostly *Ceramium strictum*) epiphytic on *Codium* mat. *Microzonia* growing on exposed ridge (at 12 m depth) without overhead canopy. Many foliose red algae from 14 m depth.

MA–312 Caroline Cove (near Southwest Point), Macquarie I., 19 January 1978; col. R. W. Ricker. *Rhodochorton* from cave walls *c.* 3 m above upper limit of *Durvillaea* and 10 m landward from water's edge.

MA–313 Lusitania Bay (north of hut), Macquarie I., 20 January 1978; col. R. W. Ricker. *Rhodochorton* near opening of cave on gravel beach.

MA–352 North of Garden Bay, Macquarie I., 26 January 1978; col. R. W. Ricker. *Rhodochorton* and various green algae growing under rock about 0.6 m above upper limit of *Durvillaea*.

MA–364 South end of Goat Bay, Tottan Head, Macquarie I., 30 January 1978; col. R. W. Ricker (12–14 m depths). Collections from rocky bottom of *Macrocystis* forest and on vertical face of offshore rock. Red algae embedded in sponge; *Codium* with delicate, epiphytic, red film; *Microzonia* and foliose red seaweeds abundant.

MA–365 South end of Goat Bay, Tottan Head, Macquarie I., 30 January 1978; col. R. W. Ricker (2.4 m depth). *Ectocarpus* epiphytic on *Desmarestia chordalis*. Not shaded from overhead sunlight.

MA–366 South end of Goat Bay, Tottan Head, Macquarie I., 30 January 1978; col. D. S. Horning (12 m depth). Large sponge on vertical wall of offshore rock.

MA–367 South end of Goat Bay, Tottan Head, Macquarie I., 30 January 1978; col. D. S. Horning (10 m depth). Yellow sponges on exposed vertical wall of offshore rock.

MA–377 East side of offshore rocks near tip of Garden Bay Peninsula, Macquarie I., 1 February 1978; col R. W. Ricker (9 m depth). Algae from vertical wall with slight overhang. *Durvillaea* canopy overhead; *Macrocystis/Desmarestia* association on wall; *Codium* and foliose red algae abundant.

MA–378 Offshore rocks near tip of Garden Bay Peninsula, Macquarie I., 1 February 1978; col. R. W. Ricker (11–14 m depths). Sponges, foliose red algae, *Microzonia*, ceramiaceous algae in sponge matrix and on *Codium* removed from rocky surfaces.

MA–381 Aerial Cove, Macquarie I., 2 February 1978; col. D. S. Horning & J. K. Lowry (3 m depth). *Ulva* and encrusting coralline algae from rock substrate.

MA–397 Garden Bay, Macquarie I., 4 February 1978; col. R. W. Ricker. Entire *Macrocystis* plant and numerous specimens of *Desmarestia chordalis* from 2–4 m depth. *Ahnfeltia plicata* growing in $\leqslant$1 m depth, forming dense clumps trapping sediment.

Appendix 2
Artificial key to species documented in the present survey of Macquarie I. Seaweeds

1 Plant colour grass green .... I Chlorophyta (p.24)
1 Plant colour yellowish-brown to olive-green, sometimes nearly black
II Phaeophyta (p. 63)
1 Plant colour pink, red, reddish-orange, reddish-purple, blackish-purple, sometimes greenish .... III Rhodophyta (p. 139)

I Chlorophyta

1 Plant habit wholly prostrate or endophytic 2
1 Plant habit with erect portions 4
2 Plants mat-forming, epilithic .... *Codium subantarcticum* (p. 61)
2 Plants endophytic or epiphytic, filamentous 3
3 Plants endophytic in fleshy red algae .... *Endophyton atroviridis* (p. 25)
3 Plants endophytic in or epiphytic on membranaous red algae (members of the Delesseriaceae) .... *Entocladia maculans* (p. 27)
4 Thalli foliaceous, flattened and/or tubular 5
4 Thalli filamentous 11
5 Thalli tubular 6
5 Thalli blade-like, sometimes saccate 8
6 Tubes without basal rhizoids .... *Blidingia minima* (p. 30)
6 Tubes with attachment rhizoids at base 7
7 Tubes branched near plant base, to 3 cm long .... *Enteromorpha bulbosa* (p. 36)
7 Tubes unbranched, to 15 cm long .... *Enteromorpha intestinalis* (p. 39)
8 Blades one cell thick 9
8 Blades two cells thick 10
9 Plants mostly growing in spray zone, often in areas with high levels of guano; blades flat (uninfected) or saccate (fungal-infected), with surface cells clustered in packets
Prasiola crispa subsp. *antarctica* (p. 46)
9 Plants growing in either intertidal or subtidal; cells of blade surfaces not in packets
Monostroma spp (pp. 32, 34)
10 Fronds linear; mid- to upper-thallus sections with cells nearly equal height to width
Ulva angusta (p. 41)
10 Fronds broad and expansive; mid- to upper-thallus sections with cells 2.5–5 times height to width .... *Ulva rigida* (p. 41)
11 Thalli unicellular (siphonous) .... *Derbesia furcata* (p. 58)
11 Thalli multicellular 12
12 Filaments simple .... *Urospora penicilliformis* (p. 52)
12 Filaments branched (at least basally) 13
13 Axes simple, except at bases; thalli dimorphic, with some fronds prostrate and wiry, others erect, lax and skein-forming .... *Rhizoclonium ambiguum* (p. 55)
13 Axes branched throughout; determinate branches short, hook-like interlocking adjacent filaments into matted skeins .... *Acrosiphonia pacifica* (p. 49)

II. Phaeophyta

1 Plants entirely filamentous; filaments mostly uniseriate, sometimes partly bi- or tri-seriate 2
1 Plants with parenchymatous or pseudoparenchymatous tissue 8
2 Thalli mostly endophytic 3
2 Thalli epilithic, epiphytic, or epizoic; not endophytic 4

3 Filaments forming lattice around utricles of host (*Codium*) thallus *Streblonema codiatile* (p. 76)
3 Filaments of mature thallus growing among cortical files of (*Adenocystis*) host thallus *Myrionema incommodum* (p. 83)
4 Reproductive structures intercalary .... *Pilayella littoralis* (p. 74)
4 Reproductive structures lateral or terminal 5
5 Filaments in part pluriseriate; reproductive structures paired and borne oppositely *Geminocarpus geminatus* (p. 72)
5 Filaments wholly uniseriate 6
6 Plastids clustered in stellate arrangement .... *Asteronema ferruginea* (p. 64)
6 Plastids band-shaped or discoid 7
7 Plants to 1.5 cm long; with hook-like determinate laterals .... *Ectocarpus constanciae* (p. 67)
7 Plants to 10 cm long; lacking hook-like laterals .... *Ectocarpus siliculosus* (p. 70)
8 Thalli wholly or predominantly crustose 9
8 Thalli largely erect or lacking crustose portions 11
9 Young plants with endophytic (in *Adenocystis*) basal disc bearing erect filaments; mature plants lacking basal disc .... *Myrionema incommodum* (p. 83)
9 Plants encrusting and epilithic 10
10 Crusts with monostromatic basal layer; rhizoids lacking *Lithoderma piliferum* var. *debile* (p. 79)
10 Crusts with basal layer of horizontally-running filaments giving rise to assurgent filaments above; rhizoids present but infrequent .... *Ralfsia australis* (p. 81)
11 Thalli minute, to 2 mm diameter, with free filamentous upper portions and basal, compacted, pseudoparenchymatous cushions 12
11 Thalli larger, erect, without basal cushions 14
12 Assimilatory filaments simple, dimorphic, either long (to 5 mm) and lax or short and slightly curved; phaeophycean hairs lacking .... *Elachista antarctica* (p. 86)
12 Assimilatory filaments branched; phaeophycean hairs present 13
13 Plants with pectinately-branched assimilators; epilithic .... *Gononema pectinatum* (p. 88)
13 Plants without pectinate branching; growing inside cryptostomata of *Adenocystis* *Gononema ramosum* (p. 90)
14 Thalli with terete to flattened branches 15
14 Thalli tubular or saccate; unbranched 23
14 Thalli with foliose portions 24
15 Branching strictly opposite to subopposite 16
15 Branchilng irregular, radial, or alternate (sometimes partly opposite) 17
16 Mature thalli with coarse, flattened, cord-like fronds .... *Desmarestia chordalis* (p. 120)
16 Plants remaining delicate when mature; thalli with feathery to wiry, terete fronds *Desmarestia willii* (p. 124)
17 Branches with microscopically-conspicuous, darkly-pigmented apical cells; cells below apex arranged in successive tiers 18
17 Branches lacking prominent, darkly-pigmented apical cells; thallus without tiered construction 19
18 Lateral branches derived directly from apical cell .... *Alethocladus corymbosus* (p. 107)
18 Lateral branches derived from subapical cells .... *Sphacelaria bornetii* (p. 104)
19 Plants with hairy surface and gelatinous consistency 20
19 Plants with smooth surface and firm consistency 21
20 Plants to 10 cm tall, with uniaxial construction; cells of assimilatory filaments strongly constricted at cross-walls, appearing barrel-shaped and forming bead-like chains to 200 μm long .... *Haplogloia moniliformis* (p. 93)
20 Plants to 50 cm tall, with multiaxial construction; most cells of assimilatory filaments weakly constricted at cross-walls; assimilatory filaments hair-like to 1.5 mm long *Papenfussiella tristanensis* (p. 96)
21 Cortex composed of outwardly-swollen and inflated cells; branches with monosiphonous upper portions .... *Cladothele striarioides* (p. 110)
21 Cortex without large, inflated cells; branches without monosiphonous portions 22

22 Plants with wiry branches; fertile cortex with 1–2-celled paraphyses and spindle-shaped unilocular sporangia .... *Chordariopsis capensis* (p. 101)
22 Plants bushy; branches consticted at bases; fertile cortex with pyriform to oval unilocular sporangia; paraphyses absent .... *Scytothamnus fasciculatus* (p. 98)
23 Plants tubular, to 30 cm long by 7 mm diameter, without solid stipe *Scytosiphon lomentaria* (p. 117)
23 Plants saccate, to 9 cm tall, with solid stipe ≤5 mm long .... *Adenocystis utricularis* (p. 112)
24 Stipes terete, longer than foliose portions; pneumatocysts present *Macrocystis pyrifera* (p. 133)
24 Blades longer than stipes; pneumatocysts lacking 25
25 Fronds ≥30 cm long, coarse, rubbery to leathery, usually with internal air-chambers *Durvillaea antarctica* (p. 136)
25 Fronds <10 cm tall, not rubbery or leathery, lacking internal air-chambers 26
26 Growth from microscopically-conspicuous apical cells 27
26 Growth diffuse or intercalary, without distinct apical cells; medulla with 2–4 layers of large, colourless cells .... *Petalonia fascia* (p. 115)
27 Blades fan-shaped, with marginal row of apical cells .... *Microzonia velutina* (p. 127)
27 Blades dichotomously branched, with single apical cells .... *Dilophus decumbens* (p. 130)

III Rhodophyta

1 Cell walls impregnated with calcium carbonate; thalli encrusting and stony 2
1 Cell walls not impregnated with calcium carbonate; thalli either encrusting, erect, prostrate, or pulvinate 6
2 Crusts with vegetative portions 2–3 cells thick .... *Melobesia membranacea* (p. 170)
2 Crusts with vegetative portions >10 cells thick 3
3 Tetrasporangial conceptacles multiporate 4
3 Tetrasporangial conceptacles uniporate 5
4 Hypothallium coaxial; plants usually epiphytic on *Ballia callitricha* *Mesophyllum patena* (p. 173)
4 Hypothallium non-coaxial; plants mostly epilithic, forming leafy, overlapping fronds *Mesophyllum schmitzii* (p. 175)
5 Crust surfaces usually convoluted and knobby; perithallum extensive, to several mm thick, with cells rectangular, aligned in vertical rows, laterally connected to cells of adjacent files by numerous cytoplasmic fusions; hypothallium indistinct if present *Pseudolithophyllum consociatum* (p. 178)
5 Crust surfaces usually smooth, often forming ridges at confluence of two crusts; perithallium less developed, with cells rounded to barrel shaped, aligned in vertical rows with files less compacted than those of *P. consociatum*; hypothallium distinct as single layer of oblong cells .... *Pseudolithophyllum discoideum* (p. 180)
6 Plants parasitic, forming irregularly shaped cushions or roughly spherical thalli with spikey branches 7
6 Plants not parasitic; habits of free-living thalli not pulvinate or ball-like 9
7 Thalli irregularly lobed; medulla with small rhizoidal cells; parasitic upon *Callophyllis* *Callocolax neglectus* (p. 188)
7 Thalli hemispherical to nearly spherical, with spikey surfaces or branches arising from cushion; parasitic on members of the Ceramiales 8
8 Basal cushions reduced; thalli composed mostly of terete branches similar to those of host *Heterosiphonia berkeleyi* .... *Colacodasya inconspicua* (p. 256)
8 Basal cushions reduced, with branches terete below and flattened above; construction dissimilar to that of host *Lophurella hookeriana* .... *Sporoglossum lophurellae* (p. 299)
8 Basal cushions prominent, whitish, with spikey surfaces, 'parasitic' upon various members of the Delesseriaceae .... *Polycoryne radiata* (p. 277)
9 Thalli encrusting, composed of vertical files of small, cuboidal cells *Hildenbrandia lecannellieri* (p. 185)
9 Thalli either erect or prostrate, not encrusting 10

10 Plants filamentous, delicate, with uniseriate axes either naked or corticated 11
10 Plants thicker, either terete, compressed, or foliaceous 24
11 Axes uncorticated 12
11 Axes at least in part corticated 20
12 Axial cells of major branches with whorls of 2–4 lateral branches 13
12 Axial cells of major branches bearing laterals alternately or irregularly, not in whorls 16
13 Laterals paired; axial cells of main branches to 3 mm long; gland cells absent; tetrasporangia tetrahedrally divided, sessile, borne on adaxial surfaces of laterals *Medeiothamnion flaccidum* (p. 244)
13 Laterals paired; axial cells of main branches to 35 µm long; gland cells solitary, covering single cells on adaxial surfaces of laterals .... *Antithamnionella alternans* (p. 225)
13 Laterals borne in whorls of (2–)3–4 14
14 Axes prostrate and erect; whorl branches simple, 3–4; tetrasporangia tetrahedrally divided, sessile, solitary, adaxial on whorl branches .... *Antithamnionella ternifolia* (p. 227)
14 Axes erect; whorl branches branched, 2–4; tetrasporangia cruciately divided, stipitate and sometimes clustered 15
15 Plants bushy; whorl branches usually 4; gland cells solitary, lenticular *Pterothamnion plumula* (p. 249)
15 Plants flattened in one plane; whorl branches 2–3(–4), with 2 long and 1 (2) short; gland cells in series of 2–4, nearly spherical .... *Pterothamnion simile* (p. 252)
16 Branching alternate distichous from each axial cell of main axes ... *Callithamnion gracile* (p. 232)
16 Branching sparse and not regularly alternate 17
17 Plants with basal crusts; epiphytic on *Ballia callitricha* .... *Rhodochorton concrescens* (p. 154)
17 Plants erect and lacking basal crusts 18
18 Branching mostly pectinate; plants soft, tuft-forming, often epiphytic on large brown algae .... *Rhodochorton variabile* (p. 159)
18 Branching irregular, sometimes partly alternate 19
19 Plants forming stiff, wiry clumps; cells with single plate-like, parietal plastid; tetrasporangia cruciately divided, generally borne in small groups of 2–4 *Acrochaetium daviesii* (p. 152)
19 Plants mostly epilithic, forming red felts on rocky surfaces; cells with numerous discoid plastids; tetrasporangia cruciately divided, borne in large clusters of *c.* 15–30 *Rhodochorton purpureum* (p. 156)
20 Cortication of main and lateral branches restricted to nodes .... *Ceramium strictum* (p. 235)
20 Cortication completely covering main axes 21
21 Apical cells of indeterminate axes with transverse divisions perpendicular to branch axis; upper portions of branches linear with pyriform axial cells .... *Ballia callitricha* (p. 230)
21 Apical cells of indeterminate axes with transverse divisions oblique to branch axis, with orientation of division plane alternating from side to side; upper portions of branches having zig-zag appearance and without pyriform axial cells 22
22 Carpogonial branches wrapping around and aligned perpendicular to fertile branch axes (bearing supporting cells); tetrasporangia cruciately divided .... *Dasyptilon sp.* (p. 240)
22 Carpogonial branches wrapping around supporting cells and aligned parallel to fertile branches; tetrasporangia cruciately divided 23
23 Plants to 10 cm tall; heavily corticated, with corticating rhizoids initially developing in chevron pattern; spermatangia borne in adaxial and abaxial clusters on lateral branches *Plumariopsis eatoni* (p. 247)
23 Plants to 3 cm tall; lightly corticated, with corticating rhizoids (near branch apices) initially developing parallel to branch axis; spermatangial mother cells catenate, adaxial on lateral branches, bearing spermatangia in whorls of 4 .... *Dasyptilon pellucidum* (p. 238)
24 Thalli terete, with cells arranged in tiers from apices to bases (polysiphonous construction) 25
24 Thalli terete, compressed, or foliaceous; if terete, then lacking polysiphonous construction 28
25 Axes densely entangling to form skeins; tetrasporangia cruciately divided, borne in vegetative branches with one sporangium per tier .... *Polysiphonia anisogona* (p. 297)

25 Axes sometimes loosely entangling but not forming skeins; tetrasporangia tetrahedrally divided, borne in stichidia 26
26 Axes mostly uncorticated except in lower parts; branching sympodial, with all branches lying in one plane .... *Heterosiphonia berkeleyi* (p. 258)
26 Axes entirely corticated; branching monopodial and radial 27
27 Plants forming dense mats in upper intertidal; branches strongly arcuate; trichoblasts not produced .... *Bostrychia vaga* (p. 292)
27 Plants forming loose, erect clumps in subtidal; branches lax, not arcuate; trichoblasts deciduous .... *Lophurella hookeriana* (p. 294)
28 Thalli terete to compressed, with cross-sections circular to elliptical in outline 29
28 Thalli flat and branched or broad and foliose with simple to dissected fronds; cross-sections linear in outline 34
29 Plants uniaxial, with central axis visible at least near apices and sometimes forming prominent, percurrent rib 30
29 Plants multiaxial, with medulla consisting of loose or compact mass of filaments 32
30 Frond margins strongly dentate; teeth produced alternately along thallus edges *Delisea pulchra* (p. 161)
30 Fronds without marginal teeth; sometimes bearing irregularly shaped lateral proliferations 31
31 Main axes distinct, linear; branching irregular, with marginal proliferations common; tetrasporophytes presumably crustose .... *Ptilonia willana* (p. 164)
31 Main axes distinct, with zig-zag appearance; branching with series of 2–4 laterals produced alternately along frond margins; tetrasporophytes erect, common, with tetrasporangia produced in stichidia 32
32 Determinate laterals broad and leaf-like; indeterminate laterals in upper thallus pectinately bearing series of 2–3(–4) branchlets; tetrasporangial stichidia knobby to irregularly branched .... *Plocamium hookeri* (p. 205)
32 Determinate laterals narrow, not leafy; indeterminate laterals in upper thallus pectinately bearing series of 4–6 branchlets; tetrasporangial stichidia pinnately branched .... *Plocamium secundatum* (p. 207)
33 Thalli terete, stiff, wiry, solid, with compact construction .... *Ahnfeltia plicata* (p. 200)
33 Thalli terete to compressed, cartilaginous, hollow to solid, with loose, internal, filamentous construction .... *Chaetangium fastigiatum* (p. 166)
34 Fronds delicate, membranous, narrow to broadly foliaceous, simple or branched 35
34 Fronds coarse, usually cartilaginous, narrow to broadly foliaceous, simple to branched 48
35 Blades entirely monostromatic 36
35 Blades >1 cell thick, at least toward bases 40
36 Attachment cushion lacking rhizoids; fronds extremely delicate, <10 mm tall, with cells widely separated by thick gelatinous walls .... *Porphyropsis coccinea* (p. 149)
36 Attachment by rhizoidal bases; constituent cells of fronds more closely spaced, lacking thick gelatinous walls 37
37 Reproductive structures borne over entire thallus surfaces 38
37 Reproductive structures restricted to frond margins 39
38 Fronds rose-pink in colour, to 40 cm tall, 70–80 μm thick, composed of cells taller (25–36 μm) than broad (8–18 μm) .... *Porphyra ionae* (p. 142)
38 Fronds generally epiphytic on *Plocamium*, rose in colour, to 14 cm long, to 30 μm thick, composed of rounded to box-like cells 8–14 μm diameter .... *Porphyra plocamiestris* (p. 144)
39 Fronds usually silvery-purple, common on rocky surfaces in the upper intertidal; cells generally in tetrads arranged in rows perpendicular to one another *Porphyra columbina* (p. 140)
39 Fronds pink to rose to burgundy in colour, on rocky surfaces in the subtidal; cells generally paired in upper thallus, solitary in mid-thallus .... *Porphyra woolhousiae* (p. 146)
40 Cortical cells arranged in rosettes; cystocarps with spiney projections *Rhodophyllis acanthocarpa* (p. 209)
40 Cortical cells not in rosettes 41

41 Blades with single apical cell 42
41 Blades with marginal zones of apical cells 44
42 Procarps borne on midrib of fertile blade; primary cell rows lacking intercalary cell divisions .... *Delesseria lancifolia* (p. 264)
42 Procarps scattered over fertile blade surfaces; primary cell rows with intercalary cell divisions 43
43 Fronds prostrate, often thickened along margins; main veins without descending rhizoids .... *Cladodonta lyallii* (p. 261)
43 Fronds erect, without thickened margins; main veins with descending rhizoids *Pseudophycodrys phyllophora* (p. 280)
44 Thalli erect or prostrate; microscopic veins present 45
44 Thalli erect, microscopic veins absent 46
45 Thalli prostrate, to 2 cm long; blade margins smooth .... *Hymenena decumbens* (p. 268)
45 Thalli erect, to 25 cm tall; blade margins often ruffled or bearing proliferations *Hymenena laciniata* (p. 270)
46 Venation restricted to lower thallus, with veins simple or once forked *Myriogramme livida* (p. 272)
46 Venation persistent throughout entire thallus, with veins repeatedly dichotomous to irregularly branched 47
47 Fronds highly laciniate, appearing somewhat feathery, not broad and foliaceous; veins not anastomosing .... *Schizoseris condensata* (p. 283)
47 Fronds usually less dissected, often broad and foliaceous; veins sometimes anastomosing *Schizoseris dichotoma* (p. 286)
48 Medulla entirely filamentous 49
48 Medulla partially or entirely filled with large, rounded cells 51
49 Cortical cells with protuberances, often fusing laterally to adjacent cells; tetrasporangia produced diffusely over blade surfaces; plants non-procarpic, with auxiliary cells and carpogonial branches borne is clusters of specialized branches (ampullae) *Pachymenia 'stipitata'* (p. 183)
49 Cortical cells without protuberances, not fusing laterally; tetrasporangia in sori; plants procarpic, with supporting cells (= auxiliary cells) borne on unspecialized filaments in the vegetative thallus 50
50 Fronds strongly iridescent, to 35 cm tall, without thickened margins ..... *Iridaea cordata* (p. 195)
50 Fronds not or only slightly iridescent, to 3 cm tall, with thickened margins *Iridaea remuliformis* (p. 198)
51 Medulla with mixture of filaments and large, rounded cells 52
51 Medulla composed solely of large, rounded cells 54
52 Fronds broad, foliaceous, cartilaginous; medulla with cavities (near apices), filaments, and rounded cells; tetrasporangia in sori; cystocarps bulging above frond surfaces *Cenacrum subsutum* (p. 212)
52 Fronds usually narrow, dissected to various degrees, if broad, mostly delicate and not cartilaginous; medulla lacking cavities, mostly filled with large, rounded cells 53
53 Thalli cartilaginous and highly dissected with strap-like irregularly branched fronds *Callophyllis variegata* (p. 189)
53 Thalli somewhat delicate, not cartilaginous, less dissected, often broad and foliose *Pugetia delicatissima* (p. 192)
54 Plants narrow to broadly foliaceous, softening after fixation in 4% formalin-seawater; tissues rehydrating poorly after drying; tetrasporangia with stalk cells 55
54 Plants narrow and strap-like, not foliaceous, remaining firm and cartilaginous after fixation in 4% formalin-seawater; tissues rehydrating well after drying; tetrasporangia presumably without stalk cells 56
55 Thalli to 60 cm tall, usually broadly foliaceous, growing on exposed rocky surfaces in the lower intertidal .... *Palmaria decipiens* (p. 218)
55 Thalli to 6 cm tall, highly dissected with dichotomous to irregular branching *Palmaria georgica* (p. 222)
56 Lower thallus often terete and wiry; upper thallus flattened; branching dichotomous,

in one plane; cortical cells near blade apices irregularly sized and loosely arranged over subcortical cells .... *Rhodymenia subantarctica* (p. 215)

56 Lower and upper thallus flattened; branching irregular from blade faces and margins; cortical cells near blade apices somewhat uniformly sized and closely packed, blocking view of subcortical cells .... *Phyllophora appendiculata* (p. 202)

Taxonomic index

Numbers in bold indicate Figure numbers.